Number Theory Informed by Computation

IAS/PARK CITY
MATHEMATICS SERIES
Volume 29

Number Theory Informed by Computation

Jennifer S. Balakrishnan
Bjorn Poonen
Akshay Venkatesh
Editors

American Mathematical Society
Institute for Advanced Study

Rafe Mazzeo, Series Editor
Jennifer S. Balakrishnan, Bjorn Poonen, and Akshay Venkatesh, Volume Editors.

IAS/Park City Mathematics Institute runs mathematics education programs that bring together high school mathematics teachers, researchers in mathematics and mathematics education, undergraduate mathematics faculty, graduate students, and undergraduates to participate in distinct but overlapping programs of research and education. This volume contains the lecture notes from the Graduate Summer School program.

2020 *Mathematics Subject Classification.* Primary 11-06, 11Yxx.

Library of Congress Cataloging-in-Publication Data

Cataloging-in-Publication Data has been applied for by the AMS.
See http://www.loc.gov/publish/cip/.

Library of Congress Cataloging-in-Publication Data

Cataloging-in-Publication Data has been applied for by the AMS.
See http://www.loc.gov/publish/cip/.

♾ The paper used in this book is acid-free and falls within the guidelines
established to ensure permanence and durability.
Visit the AMS home page at `https://www.ams.org/`

10 9 8 7 6 5 4 3 2 1 30 29 28 27 26 25

Contents

- Volume 13: *Geometric Combinatorics* (2004)
- Volume 14: *Mathematical Biology* (2005)
- Volume 15: *Low Dimensional Topology* (2006)
- Volume 16: *Statistical Mechanics* (2007)
- Volume 17: *Analytical and Algebraic Geometry* (2008)
- Volume 18: *Arithmetic of L-functions* (2009)
- Volume 19: *Mathematics in Image Processing* (2010)
- Volume 20: *Moduli Spaces of Riemann Surfaces* (2011)
- Volume 21: *Geometric Group Theory* (2012)
- Volume 22: *Geometric Analysis* (2013)
- Volume 23: *Mathematics and Materials* (2014)
- Volume 24: *Geometry of Moduli Spaces and Representation Theory* (2015)
- Volume 25: *The Mathematics of Data* (2016)
- Volume 26: *Random Matrices* (2017)
- Volume 27: *Harmonic Analysis* (2018)
- Volume 28: *Quantum Field Theory and Manifold Invariants* (2021)

The American Mathematical Society publishes occasional material from the Undergraduate Summer School in their Student Mathematical Library; material from the Teacher Leadership Program led to a number of volumes in the IAS/PCMI—the Teacher Program series.

After more than 30 years, PCMI retains its intellectual vitality and continues to draw remarkable groups of participants each year from across the entire spectrum of mathematics, from Fields Medalists to undergraduates newly embarking on their mathematical studies.

Rafe Mazzeo
PCMI Director
May 2026

Preface

The IAS/Park City Mathematics Institute (PCMI) was founded in 1991 as part of the Regional Geometry Institute initiative of the National Science Foundation. In mid-1993, the program found an institutional home at the Institute for Advanced Study (IAS) in Princeton, New Jersey, where it remains to this day. It is now funded by the Simons Foundation and the National Science Foundation.

The IAS/Park City Mathematics Institute encourages research, training and education in mathematics and fosters interaction between these different emphases. The three-week summer institute offers programs for researchers and postdoctoral scholars, graduate students, undergraduate students and faculty at undergraduate facing institutions, and has also included programs for K-12 teachers, international teachers and education researchers.

One of the main goals of PCMI is to increase the awareness amongst all participants of the full range of activities that occur in research, education and the training of mathematicians. To that end, daily events are scheduled during the summer institute devoted to talks, demonstrations and activities aimed at the broader audience of all participants; these are meant to encourage interaction between all participants. Topics range from broader themes in recent research to mathematical art, lessons in immersive filmmaking and presentations by the undergraduate participants.

Each year the Research Program and Graduate Summer School focuses on a different mathematical area, chosen to represent some major thread of current mathematical interest. Activities in the Undergraduate Summer School and Undergraduate Faculty Program are also linked to this topic, the better to encourage interaction between participants at all levels. Lectures notes from the Summer School are published each year. The prior volumes are:

- Volume 1: *Geometry and Quantum Field Theory* (1991)
- Volume 2: *Nonlinear Partial Differential Equations in Differential Geometry* (1992)
- Volume 3: *Complex Algebraic Geometry* (1993)
- Volume 4: *Gauge Theory and the Topology of Four-Manifolds* (1994)
- Volume 5: *Hyperbolic Equations and Frequency Interactions* (1995)
- Volume 6: *Probability Theory and Applications* (1996)
- Volume 7: *Symplectic Geometry and Topology* (1997)
- Volume 8: *Representation Theory of Lie Groups* (1998)
- Volume 9: *Arithmetic Algebraic Geometry* (1999)
- Volume 10: *Computational Complexity Theory* (2000)
- Volume 11: *Quantum Field Theory, Supersymmetry, and Enumerative Geometry* (2001)
- Volume 12: *Automorphic Forms and their Applications* (2002)

IAS/Park City Mathematics Series
Volume **29**, 2025
https://doi.org/10.1090/pcms/029/00888

Introduction

Jennifer Balakrishnan, Bjorn Poonen, and Akshay Venkatesh

In 2022, the three of us organized the IAS/Park City Mathematics Institute program "Number Theory Informed by Computation". The motivation for the program was to bring together people working at the interface between number theory and computation, both terms interpreted broadly. Indeed, number theory has always drawn inspiration from computational experiments, and the difficulty of carrying out such computations has itself been technologically important.

Our program was originally scheduled for 2020, but was grounded by the pandemic. We were able to organize a small virtual workshop during the summer of 2021, while the original in-person program was moved to 2022. Although the pandemic remained very much present (the lecture hall was filled with a sea of masked faces), the delight among many younger participants to *finally* be able to share mathematics face-to-face was very palpable. We are very grateful to PCMI director Rafe Mazzeo and also Dena Vigil and the rest of the IAS/PCMI staff for their hard work in making our program a success.

The number-theoretic parts of the program included a research program, a graduate summer school, and an undergraduate summer school (the other parts were a teacher leadership program, an undergraduate faculty program, and a workshop on rehumanizing mathematics). Many of the graduate summer school lecturers graciously wrote detailed notes of their lecture series. These, ordered by author's last name, and preceded by a write-up of a public lecture given by Hendrik Lenstra, comprise the current volume. The volume is aimed at students with a background in graduate-level number theory, though undergraduate abstract algebra may be sufficient for some of the expositions, and algebraic geometry is needed for others.

Hendrik Lenstra's public lecture, aimed at a general audience, demonstrates how modern algebra, despite often being called "abstract algebra", has the power to explain concrete properties of integers. He illustrates this by showing how it yields efficient and elegant methods for testing when numbers of the form $2^n \pm 1$ are prime.

Hendrik Lenstra's lecture series, written up by Daniël M. H. van Gent, contains a careful, systematic development of a suite of polynomial-time algorithms for solving many problems in algebraic number theory. One such algorithm is coprime basis factorization, which expresses the positive integers in a given finite list as products of powers of a set of coprime positive integers; this can be used to check whether two products of powers of arbitrary integers are equal in polynomial time, without having to factor the integers, and without evaluating the products, which may be too large to write down. Finitely generated abelian groups are presented as the cokernel of an integer matrix, and algorithms for all the operations one would expect possible in such an abelian category are given. Next, algorithms for finite commutative rings are given, modulo the difficult problem of computing the radical of a positive integer, which is equivalent to the problem of finding the ring of integers of a given number field. Since one might not know how to compute the ring of integers of a number field, it is worthwhile to work with orders that are not necessarily maximal, so algorithms for computing with their ideals are given. Finally, the algorithms for finitely generated abelian groups are applied to the computation of Legendre symbols, Jacobi symbols, and their generalizations.

David Harvey gives an account of a breakthrough in computational number theory and arithmetic geometry that he has been involved with, namely, a method for counting points on the mod p reduction of a variety for many primes p that is faster than computing these counts one p at a time. Along the way, he surveys recent breakthroughs in fast multiplication of integers and polynomials, points out open problems, and invites the reader to flesh out the details of some previously unpublished algorithms.

Melanie Matchett Wood gives an introduction to two major problems in arithmetic statistics. One is the problem of asymptotically counting number fields of given degree and bounded discriminant, possibly with constraints on the Galois group, a setting in which there are predictions given by Malle's conjecture. The other problem concerns the distribution of class groups when a number field varies in a family — this is the subject of the Cohen–Lenstra heuristics and its generalizations, which include predictions for nonabelian unramified extensions of number fields. These problems are studied using a variety of tools, including parametrizations, geometry of numbers, and class field theory.

Joseph H. Silverman presents the theory of lattices and explains how hard problems about lattices can be used to develop cryptography protocols. In particular, he discusses the shortest vector problem and the closest vector problem, and explains how LLL lattice reduction can find reasonably good bases for lattices in polynomial time. He then explains the history of cryptography, especially public key cryptography based on the supposedly hard problems of number theory that were discovered to be no longer hard if one has a full-scale quantum computer. Finally, he discusses various schemes relying instead on hard lattice problems that might remain secure even if quantum computers become viable.

Bianca Viray gives a guide to the Brauer–Manin obstruction, which provides a way to prove that certain varieties over a number field k have no k-point even when they have points over every completion over k. In addition to presenting the theory, she illustrates through examples how the obstruction can often be computed, using Hilbert symbols. She also mentions some open questions in the area.

Jan Vonk's contribution presents the theory of rigid cocycles for $\mathbb{Z}[1/p]$. This theory attempts to provide p-adic analytic constructions of abelian extensions of real quadratic fields, analogous to the classical analytic constructions of abelian extensions of imaginary quadratic fields given by singular moduli (j-invariants of elliptic curves with complex multiplication). He discusses how early machine computation played a role in the understanding of singular moduli in $\mathbb{C}$, and discusses

p-adic calculations that lead to conjectural analogues over real quadratic fields. The *geometry* that might explain such conjectures remains elusive, however.

Olivier Wittenberg discusses classical and modern approaches to the inverse Galois problem, a problem dating from the 19th century, asking whether every finite group G is the Galois group of a finite extension of $\mathbb{Q}$. First, there is the connection to Noether's problem asking when the quotient of an affine space by a linear action of G is a *rational* variety. Next there is the approach of finding G-covers of a rational curve via the Riemann existence theorem and rigidity, which has succeeded in realizing many finite groups, including the monster group. Finally, he discusses how the property of weak approximation of rational points for certain varieties would imply a positive answer to the inverse Galois problem, and he presents cases in which this approach has had success, such as all finite supersolvable groups.

We thank Ian Morrison for his assistance with the technical aspects of putting this volume together. We also thank all the referees who put in time to help improve the expositions. The three of us hope that you find the volume to be of value as an introduction to many topics in this beautiful and rich area of mathematics!

IAS/Park City Mathematics Series
Volume **29**, 2025
https://doi.org/10.1090/pcms/029/00889

Algebra in the Real World

Hendrik Lenstra

Notes by Samuel Tiersma

Abstract. Since the earliest days of arithmetic, Fermat numbers and Mersenne numbers have held a special fascination for the numerologically minded. One of the intriguing properties of these numbers is that there are easy rules to test them for primality. Such rules were discovered no earlier than the late nineteenth century, and they are more easily stated than explained. Providing a perfectly satisfactory understanding of why they work is one of the merits of modern algebra.

1. Introduction

In the early days, my business friends would ask me, "How does it feel to be in the real world?" I would say, and I still feel, that mathematics seems much more real to me than business—in the sense that, well, what's the reality in a McDonald's stand? It's here today and gone tomorrow. Now, the integers—that's reality. When you prove a theorem, you've really done something that has a substance to it, to which no business venture can compare for reality. (Jim Simons, The Emissary, June 1998)

The title of my lecture is intended to convey that what Americans call *abstract algebra* is in fact very concrete. It tells us not only that the integers—and that is reality!—have certain properties that we care about, but also *why* they have those properties. I was, as a student, first vividly struck by this power of algebra when I learnt of their application to Fermat numbers and Mersenne numbers: invoking basic results about groups, rings and fields one after the other, one obtains astonishingly efficient and elegant methods for deciding whether a number of the form $2^m \pm 1$ is a prime number. The details are in my lecture. Ever since, I have been using algebra as a means towards understanding the world I live in: the real world of mathematics that is here today and that will not be gone tomorrow.

2020 *Mathematics Subject Classification.* Primary 11A51; Secondary 97H40, 11A07.

Key words and phrases. Park City Mathematics Institute, motivating abstract algebra by elementary number theory, primality tests, Fermat primes, Pépin test, Mersenne primes, Lucas–Lehmer test, ring extensions.

2. Fermat primes

A *Fermat prime*, named after Pierre de Fermat [1607–1665], is a prime number of the shape $2^m + 1$ with $m \in \mathbb{Z}_{>0}$. Examples: 3, 5, 17, 257, 65537. Historically, these have been of interest in cyclotomy.

THEOREM 2.1 (Carl Friedrich Gauss [1777–1855]; Pierre Wantzel [1814–1848]). *A regular n-gon is constructible with compass and straightedge if and only if n equals a 2-power (including $2^0 = 1$) times a product of finitely many pairwise distinct Fermat primes (including the empty product 1).* □

It is not difficult to see that $2^m + 1$, with $m > 0$, can be prime only if $m = 2^k$ for some $k \in \mathbb{Z}_{\geq 0}$, so we will restrict our attention to exponents m that are a power of 2. We define for $k \in \mathbb{Z}_{\geq 0}$ the kth *Fermat number* to be $F_k = 2^{2^k} + 1$. It is conjectured that the only Fermat numbers F_k that are prime are the five mentioned above.

CONJECTURE 2.2. *F_k is prime $\iff$ $k \in \{0, 1, 2, 3, 4\}$.*

It is known that F_k is *not* prime for $5 \leq k \leq 32$, and for a handful of larger k, such as $k = 18\,233954$; for all but two of all these values of k, a *factor* of F_k is known, such as $641 \mid F_5$ (Euler; see Exercise 2.8). The two exceptions are $k = 20$ and $k = 24$. So how can we nonetheless be sure that $F_{20} = 2^{2^{20}} + 1$ and $F_{24} = 2^{2^{24}} + 1$ are not prime? After all, performing trial division up to $\sqrt{F_{20}} = 2.59\cdots \times 10^{157826}$ is out of the question.

THEOREM 2.3 (Théophile Pépin [1826–1904], 1877). *Let $m \in \mathbb{Z}_{\geq 2}$ and $n = 2^m + 1$. Define $r_i \in \mathbb{Z}/n\mathbb{Z}$ for $i \geq 0$ by $r_0 = (3 \bmod n)$, $r_{i+1} = r_i^2$ (again $\bmod\ n$). Then*

$$n \text{ is prime} \iff r_{m-1} = -1\,(= 2^m).$$

This involves only m ($\approx \frac{\log n}{\log 2}$) arithmetic operations, far fewer than $\sqrt{n}$.

Example. Let $m = 8$, so $n = 257$. Then we compute $r_0 = 3$, $r_1 = 9$, $r_2 = 81$, $r_3 = 6561 = 1421 = -121$, $r_4 = 14641 = 1791 = -8$, $r_5 = 64 = 2^6$, $r_6 = 2^{12} = -2^4$, $r_7 = 2^8 = -1$, so n is prime.

PROOF. We start by noting that $r_i = (3^{2^i} \bmod n)$ for all $i \geq 0$.

Case m is odd. Then $3 \mid n$ and $n > 3$, so n is not prime. On the other hand, $3 \mid r_i$ for all i, so $r_{m-1} \neq (-1 \bmod n)$. Thus both assertions are false and the equivalence holds.

Henceforth assume m is even. Then $2^m \equiv 1 \bmod 3$, so $n \equiv -1 \bmod 3$.

[$\Leftarrow$:] Suppose $r_{m-1} = -1$. Let $d > 1$ be a divisor of n; showing that $d = n$ will prove the primality of n. Now $3^{2^{m-1}}$ is congruent to -1 modulo n, hence also modulo d. So $3^{2^{m-1}} \not\equiv 1 \bmod d$ but $3^{2^m} \equiv 1 \bmod d$. Hence the order of $(3 \bmod d)$ in the unit group $(\mathbb{Z}/d\mathbb{Z})^*$ divides 2^m but not 2^{m-1}, so equals 2^m. Since the order of an element of a group is at most the order of the group, we have $2^m \leq \#(\mathbb{Z}/d\mathbb{Z})^* \leq d - 1$, whence $d \geq 2^m + 1 = n$. We conclude that $d = n$, as desired.

The converse is slightly more work, some of which is left to the reader in the form of the following exercise.

EXERCISE 2.4. Suppose A is a finite abelian group with precisely one element of order 2, say ϵ. Then $\#A$ is even, and for each $\alpha \in A$ it holds that

$$\alpha^{(\#A)/2} \neq \epsilon \iff \alpha^{(\#A)/2} = 1 \iff \exists \beta \in A : \alpha = \beta^2 .$$

[$\Rightarrow$:] Assume n is prime. Then $\mathbb{Z}/n\mathbb{Z}$ is a field with n elements, and we denote it by $\mathbb{F}_n$. In the exercise we take $A = \mathbb{F}_n^*$, of order $n - 1 = 2^m$. If $\epsilon \in A$ has order 2, then $(\epsilon - 1)(\epsilon + 1) = \epsilon^2 - 1 = 0$. Since $\epsilon - 1 \neq 0$ has an inverse in the field $\mathbb{F}_n$, it follows that $\epsilon = -1$, whence A satisfies the hypothesis of the exercise. Taking $\alpha = 3$, we find that $3^{2^{m-1}} \equiv -1 \bmod n$ if and only if 3 is not a square in $\mathbb{F}_n$.

Suppose there is a $\sqrt{3}$ (i.e. an element whose square is 3) in $\mathbb{F}_n$. There is also a $\sqrt{-1}$, namely $2^{m/2}$, so a $\sqrt{-3}$ exists as well, namely $\sqrt{3}\cdot\sqrt{-1}$. We contend that $\zeta = \frac{-1+\sqrt{-3}}{2}$ has order 3 in $\mathbb{F}_n^*$. Indeed, ζ satisfies $(2\zeta+1)^2 = -3$, so $4(\zeta^2+\zeta+1) = 0$. As 2 has an inverse modulo n, this gives $\zeta^2+\zeta+1 = 0$; thus $\zeta^3 = 1$ while $\zeta \neq 1$, proving our claim. Now Lagrange's Theorem—one of the first theorems one encounters in a course on group theory—states: the order of an element of a finite group divides the order of the group. This gives that $3 = \text{order}(\zeta) \mid \#\mathbb{F}_n^* = n - 1 = 2^m$, a contradiction which completes the proof. □

EXERCISE 2.5. Suppose $p > 2$ is prime and $\alpha \in \mathbb{F}_p^*$. Prove: $\mathbb{F}_p$ has a $\sqrt{\alpha} \iff \alpha^{(p-1)/2} = 1$.

EXERCISE 2.6. Suppose $p > 3$ is prime. Then $\mathbb{F}_p$ has a $\sqrt{-3} \iff 3 \mid p - 1$. Also, $\mathbb{F}_p$ has a $\sqrt{-1} \iff 4 \mid p - 1$. Conclude: $\mathbb{F}_p$ has a $\sqrt{3} \iff p \equiv \pm 1 \bmod 12$.

EXERCISE 2.7. Suppose p is a prime number dividing $F_k = 2^{2^k} + 1$.

(a) If $k \geq 0$, prove that $p \equiv 1 \bmod 2^{k+1}$. [Hint: $2^{2^k} \equiv -1 \bmod F_k$.]

(b) If $k \geq 2$, prove that $p \equiv 1 \bmod 2^{k+2}$. [Hint: $F_{k-1}^{2^{k+1}} \equiv -1 \bmod F_k$.]

EXERCISE 2.8. **(a)** Using the equalities $641 = 2^7 \cdot 5 + 1 = 5^4 + 2^4$, prove that $2^{32} \equiv -1 \bmod 641$. Deduce that 641 is a prime divisor of $F_5 = 2^{2^5} + 1$.

(b) Use the equalities $431 = 2^4 \cdot 3^3 - 1 = 2^9 - 3^4$ to prove in a similar way that 431 is a prime divisor of $2^{43} - 1$.

EXERCISE 2.9. **(a)** Replace in the first 32 rows of the Pascal triangle each even integer by a 0 and each odd integer by a 1. Show this yields the binary expansions of the first 32 odd n for which a regular n-gon is constructible (including $n = 1$).

(b) Show that the only n for which regular n-, $(n + 1)$- and $(n + 2)$-gons are all constructible with ruler and compass are $n = 1, 2, 3, 4, 15, 255, 65535$.

3. Mersenne primes

A *Mersenne prime* (named after Marin Mersenne [1588–1648]) is a prime number of the shape $2^m - 1$ with $m \in \mathbb{Z}_{\geq 2}$. For example $2^2 - 1 = 3$, $2^3 - 1 = 7$, $2^5 - 1 = 31$, $2^7 - 1 = 127$, and

$$2^{2^{2^{2^2-1}-1}-1} - 1 = 170\,141183\,460469\,231731\,687303\,715884\,105727.$$

Historically these have attracted interest due to their connection with perfect numbers. A *perfect number* is a positive integer n that equals the sum of its divisors $< n$ (e.g. $6 = 1+2+3$, $28 = 1+2+4+7+14$ and $496 = 1+2+4+8+16+31+62+124+248$).

THEOREM 3.1 (Euclid, $\Leftarrow$, 3rd century BC; Leonhard Euler [1707–1783], $\Rightarrow$, 1841). *Let $n \in \mathbb{Z}_{>0}$ be even. Then, n is perfect if and only if there exists $m \in \mathbb{Z}_{\geq 2}$ such that $2^m - 1$ is prime and $n = 2^{m-1}(2^m - 1)$.* □

Conjecturally, *odd* perfect numbers do not exist.

We leave as an exercise to show: if $2^m - 1$ is prime, then m is prime. The converse does not hold, e.g. $2^{11} - 1 = 2047 = 23 \cdot 89$.

It is conjectured that there are infinitely many Mersenne primes. Usually the largest known prime is a Mersenne prime, owing to the Lucas–Lehmer test to be discussed momentarily. As of September 29th, 2025 we know 52 Mersenne primes, and the largest known prime number is $2^{136\,279841} - 1$.

THEOREM 3.2 (Lucas–Lehmer test; François Édouard Anatole Lucas [1842–1891], 1878; Derrick Henry Lehmer [1905–1991], 1930). *Let $m \in \mathbb{Z}_{>2}$ and put $n = 2^m - 1$. Define $s_i \in \mathbb{Z}/n\mathbb{Z}$ $(i \geq 1)$ by $s_1 = (4 \bmod n)$ and $s_{i+1} = s_i^2 - 2$ for $i \geq 1$. Then*

$$n \text{ is prime} \iff s_{m-1} = 0.$$

EXERCISE 3.3. Let s_i $(i \geq 1)$ be defined as in the Lucas–Lehmer test, but with $\mathbb{R}$ replacing $\mathbb{Z}/n\mathbb{Z}$. Then for all $i \geq 1$,

$$s_i = (2 + \sqrt{3})^{2^{i-1}} + (2 - \sqrt{3})^{2^{i-1}}.$$

PROOF OF THEOREM 3.2. In the present proof, the case that m is even is the easy one, and we leave it to the reader. Therefore we assume m is odd. Before beginning the proof proper, we will establish a multiplicative interpretation for $s_{m-1} = 0$. We use a variant of the formula just given, but replacing $2 + \sqrt{3}$ by $(\sqrt{2} \cdot \frac{1+\sqrt{3}}{2})^2$. Although this complicates the bases, it simplifies both exponents to be just 2^i. Further, since we want to perform arithmetic not inside $\mathbb{R}$ but modulo n and its divisors, we work with a suitable ring R having an adequate supply of special elements.

Let R be a commutative ring $\neq 0$ with a $\sqrt{2}$, a $\sqrt{3}$ and a $\frac{1}{2}$. Define $\alpha, \beta \in R$ by $\alpha = \frac{\sqrt{2}(1+\sqrt{3})}{2}$ and $\beta = \frac{\sqrt{2}(1-\sqrt{3})}{2}$, so that $\alpha^2 = 2 + \sqrt{3}$ and $\beta^2 = 2 - \sqrt{3}$. Define $s_i \in R$ $(i \geq 0)$ by $s_i = \alpha^{2^i} + \beta^{2^i}$ (as before, $s_1 = 4$ and $s_{i+1} = s_i^2 - 2$ for $i \geq 1$). Because $\alpha\beta = -1$ and 2^{m-1} is even, we arrive at the promised multiplicative interpretation. Since

$$s_{m-1} = 0 \iff \alpha^{2^{m-1}} = -\beta^{2^{m-1}} = -\alpha^{-2^{m-1}} \iff \alpha^{2^m} = -1$$

we conclude that $s_{m-1} = 0$ implies that α has multiplicative order 2^{m+1}. (Note that $-1 \neq 1$ in R, since otherwise $0 = 2 \in R^*$, forcing $R = 0$, which we excluded).

The next step is to produce a suitable ring R. To do this, take a divisor $d > 1$ of $n = 2^m - 1$. The ring $\mathbb{Z}/d\mathbb{Z}$ already has a $\frac{1}{2}$, since n is odd. It also has a $\sqrt{2} = 2^{(m+1)/2}$, with indeed square $(\sqrt{2})^2 = 2^{m+1} = 2 \cdot 2^m = 2$; here we used that $d \mid 2^m - 1$, equivalently $2^m \equiv 1 \bmod d$. However, $\mathbb{Z}/d\mathbb{Z}$ may not have a $\sqrt{3}$ (see Exercise 3.5). Therefore we *adjoin* to $\mathbb{Z}/d\mathbb{Z}$ a $\sqrt{3}$, yielding an *extension ring*

$$R = (\mathbb{Z}/d\mathbb{Z}) \oplus (\mathbb{Z}/d\mathbb{Z}) \cdot \sqrt{3} = \{a + b\sqrt{3} : a, b \in \mathbb{Z}/d\mathbb{Z}\}$$

with componentwise addition, and multiplication for which $\sqrt{3} \cdot \sqrt{3} = 3$. Then R is a commutative ring $\neq 0$ having all required special elements, so we can start the actual proof of the asserted equivalence.

[$\Leftarrow$:] Assume that $s_{m-1} = 0$ in $\mathbb{Z}/n\mathbb{Z}$, then also in $\mathbb{Z}/d\mathbb{Z}$, hence in R. Therefore our multiplicative interpretation shows

$$n < 2(n+1) = 2^{m+1} = \text{order}(\alpha \in R^*) \leq \#R = d^2,$$

so $d > \sqrt{n}$. But if each divisor of n greater than 1 is greater than $\sqrt{n}$, then n is prime.

[$\Rightarrow$:] Suppose n is prime, then necessarily $d = n$. Now $\mathbb{Z}/n\mathbb{Z}$ is a field we denoted $\mathbb{F}_n$, and we contend that $R = \mathbb{F}_n \oplus \mathbb{F}_n\sqrt{3}$ is also a field. Namely, let $a + b\sqrt{3} \in R$ be nonzero, so $a \neq 0$ or $b \neq 0$. Note that $(a+b\sqrt{3})(a-b\sqrt{3}) = a^2 - 3b^2$. If $a^2 - 3b^2 = 0$ then $b \neq 0$, so $\frac{a}{b} = \sqrt{3}$. However, since $n = 2^m - 1 \equiv 1 \bmod 3$ and $n \equiv -1 \bmod 4$, Exercise 2.6 tells us that $\mathbb{F}_n$ does not have a $\sqrt{3}$. So $a^2 - 3b^2 \neq 0$ and, $\mathbb{F}_n$ being a field, it follows that $a + b\sqrt{3} \in R^*$, as contended.

Thus R is a finite field with n^2 elements, denoted $\mathbb{F}_{n^2}$, a *quadratic* extension of $\mathbb{F}_n$. We can now finish off the proof by invoking an elegant theorem from the theory of finite fields.

THEOREM 3.4. *Let p be a prime, and $k \in \mathbb{Z}_{\geq 1}$. Let $\mathbb{F}_{p^k}$ be a finite field with p^k elements. Then the automorphism group of $\mathbb{F}_{p^k}$ is cyclic of order k, generated by the* Frobenius automorphism Frob *given by* $\mathrm{Frob}(r) = r^p$ *for all $r \in \mathbb{F}_{p^k}$.*

On the other hand, we know that $R = \mathbb{F}_{n^2}$ has a *conjugation* automorphism $a + b\sqrt{3} \mapsto a - b\sqrt{3}$ of order 2, interchanging α and β. By the theorem this automorphism coincides with Frob, i.e. it sends every $r \in R$ to r^n. In particular, $\alpha^n = \mathrm{Frob}(\alpha) = \beta = -\alpha^{-1}$, whence $\alpha^{2^m} = \alpha^{n+1} = -1$. Thus, our multiplicative interpretation implies $s_{m-1} = 0$. □

This proof illustrates that problems about "ordinary" integers are often most easily solved by working in "extensions". Above we encountered the extensions $\mathbb{F}_n \subset \mathbb{F}_{n^2}$ and $\mathbb{Z}/d\mathbb{Z} \subset R$; for many other problems one uses extensions with base ring $\mathbb{Z}$ or $\mathbb{Q}$, which form the domain of *algebraic number theory*.

EXERCISE 3.5. Let $m \geq 3$ and $n = 2^m - 1$. Show that n is divisible by a prime $p \equiv \pm 5 \bmod 12$ and that for at least half of all divisors d of n, the ring $\mathbb{Z}/d\mathbb{Z}$ does *not* have a $\sqrt{3}$.

EXERCISE 3.6. Let p be a prime number and R a commutative ring in which $p \cdot 1 = 0$. Then the map Frob: $R \to R$ given by $\mathrm{Frob}(r) = r^p$ for all $r \in R$ is a ring homomorphism. Deduce this from the equality $X^p - 1 = (X-1)^p$ in $\mathbb{F}_p[X]$, which admits several neat proofs:

(a) Use Lagrange's theorem for $\mathbb{F}_p^*$ to show that $X^p - X = \prod_{a \in \mathbb{F}_p}(X - a)$. Show this implies $X^p - X = (X-1)^p - (X-1)$ and conclude.

(b) Show $1 \in \mathbb{F}_p$ has multiplicity greater than 1 as a zero of $f = X^p - 1$, and argue this excludes the existence of another zero $\zeta \neq 1$ in a splitting field of f.

(c) Let k be a splitting field of $f = X^p - 1$ over $\mathbb{F}_p$. Show $\gcd(p, \#k^*) = 1$ and deduce that 1 is the only zero of f in k.

EXERCISE 3.7. **(a)** Write $\sqrt{2}$ as a sum of two complex roots of unity.

(b) Suppose $p > 2$ is prime. Prove: $\mathbb{F}_p$ has a $\sqrt{2} \iff p \equiv \pm 1 \bmod 8$.

EXERCISE 3.8. Let p be an odd prime number and q a prime divisor of $2^p - 1$. Establish the congruences $q \equiv 1 \bmod p$ and $q \equiv \pm 1 \bmod 8$.

EXERCISE 3.9. **(a)** Use the equalities $223 = 15^2 - 2 = 7 \cdot 2^5 - 1 = 7^3 - 2^3 \cdot 15$ to prove that 223 is a prime divisor of $2^{37} - 1$.

(b) Which similar conclusion can you draw from the equalities $19^2 - 2^7 = 5^3 + 2^2 \cdot 3^3 = 5 + 19 \cdot 2^2 \cdot 3$?

EXERCISE 3.10. Let n be a positive integer. Prove that the following are equivalent:

(i) there exists a in $\mathbb{Z}/n\mathbb{Z}$ for which the ring $R = (\mathbb{Z}/n\mathbb{Z}) \oplus (\mathbb{Z}/n\mathbb{Z}) \cdot \sqrt{a}$ of cardinality n^2 has the property that the index of $(\mathbb{Z}/n\mathbb{Z})^*$ in R^* is a power of 2;

(ii) n equals a 2-power (including $2^0 = 1$) times a product of finitely many pairwise distinct Mersenne or Fermat primes (including the empty product 1).

UNIVERSITEIT LEIDEN, P.O. BOX 9500, 2300 RA LEIDEN, THE NETHERLANDS
Email address: hwl@math.leidenuniv.nl

IAS/Park City Mathematics Series
Volume **29**, 2025
https://doi.org/10.1090/pcms/029/00890

Polynomial-time Algorithms in Algebraic Number Theory: Based on lectures by Hendrik Lenstra

Daniël M. H. van Gent

ABSTRACT. There are many problems in algebraic number theory which one would like to solve algorithmically, for example computation of the maximal order $\mathcal{O}$ of a number field, and the many problems that are most often stated only for $\mathcal{O}$, such as inverting ideals and unit computations. However, there is no known fast, i.e. polynomial-time, algorithm to compute $\mathcal{O}$, which we motivate by a reduction to elementary number theory. We will instead restrict to polynomial-time algorithms, and work around this inaccessibility of $\mathcal{O}$. These notes are based on lectures by Clay Senior Scholar Hendrik Lenstra at the 2022 PCMI summer session.

CONTENTS

2020 *Mathematics Subject Classification.* Primary 11Y16; Secondary 13-XX, 11-01.
Key words and phrases. Park City Mathematics Institute.

1. Introduction

These notes are based on lectures by Clay Senior Scholar Hendrik Lenstra at the 2022 PCMI summer session.

In these notes we study methods for solving algebraic computational problems, mainly those involving number rings, which are fast in a mathematically precise sense. Our motivation is to prove the following theorem. We will however prove something stronger.

THEOREM 1.1. *There exists a polynomial-time algorithm that, given a number field K, non-zero elements $\alpha_1, \ldots, \alpha_m \in K$ and $n_1, \ldots, n_m \in \mathbb{Z}$, decides whether $\prod_i \alpha_i^{n_i} = 1$.*

As we will see in Section 2, even for $K = \mathbb{Q}$ this problem is not entirely trivial.

In Section 3 we work up from basic operations such as addition and multiplication of integers to computation in finitely generated abelian groups. Here we encounter lattices and the LLL-algorithm. Then, in Section 4 we give a polynomial-time algorithm to compute the blowup of an ideal. This allows us to generalize the proofs in Section 2 to decide whether $\prod_i \alpha_i^{n_i}$ is a unit in the maximal order of K. The final step in the proof of Theorem 1.1 brings us back to a lattice algorithm.

In Section 5 we exhibit some algorithmic number-theoretic problems which are deemed to be hard. We support this assertion by a reduction to the prime factorization problem. In Section 6 we give algorithms to compute Jacobi symbols, which are of independent interest.

Thanks are due to the Clay Mathematics Institute for funding the lectures of Hendrik Lenstra on which this article is based.

1.1. Algorithms. To be able to talk about polynomial-time algorithms, we should first define what an algorithm is. We equip the set of natural numbers $\mathbb{N} = \mathbb{Z}_{\geq 0}$ with a length function $l : \mathbb{N} \to \mathbb{N}$ that sends n to the number of digits of n in base 2, with $l(0) = 1$.

DEFINITION 1.2. A *problem* is a function $f : I \to \mathbb{N}$ for some set of inputs $I \subseteq \mathbb{N}$ and we call f a *decision problem* if $f(I) \subseteq \{0, 1\}$. An *algorithm* for a problem $f : I \to \mathbb{N}$ is a 'non-probabilistic method' to compute $f(x)$ for all $x \in I$. An algorithm for f is said to run in *polynomial time* if there exist $c_0, c_1, c_2 \in \mathbb{R}_{>0}$ such that for all $x \in I$ the time required to compute $f(x)$, called the *run-time*, is at most $(c_0 + c_1 l(x))^{c_2}$. We say a problem f is *computable* if there exists an algorithm for f.

This definition is rather empty: we have not specified what a 'non-probabilistic method' is, nor have we explained how to measure run-time. We will briefly treat this more formally. The reader for whom the above definition is sufficient can freely skip the following paragraph. The main conclusion is that we will not heavily rely on the formal definition of run-time in these notes.

In these notes, the word algorithm will be synonymous with the word *Turing machine*. For an extensive treatment of Turing machines, see Section 8.2 in [**14**]. A Turing machine is a model of computation described by Alan Turing in 1936 that defines an abstract machine which we these days think of as a computer. The main differences between a Turing machine and a modern day computer is that the memory of a Turing machine is a tape as opposed to random-access memory, and that a Turing machine has infinite memory. The run-time of a Turing machine is then measured as the number of elementary tape operations: reading a symbol on the tape, writing a symbol on the tape and moving the tape one unit forward or backward. It is then immediately clear that it is expensive for a Turing machine to move the tape around much to look up data, as opposed to the random-access memory model where the cost of a memory lookup is constant, regardless of where the data are stored in memory. This also poses a problem for our formal treatment of run-time, as it may depend on our model of computation. However, both models of computation are able to emulate each other in such a way that it preserves the property of computability in polynomial time, even though the constants c_0, c_1 and c_2 as in Definition 1.2 may increase drastically. We use this as an excuse to be informal in these notes about determining the run-time of an algorithm.

1.2. Basic computations. In these notes we build up our algorithms from basic building blocks. First and foremost, we remark that the basic operations in $\mathbb{Z}$ and $\mathbb{Q}$ are fast. Addition, subtraction, multiplication and division (with remainder in the case of $\mathbb{Z}$) can be done in polynomial time, as well as checking the sign of a number and whether numbers are equal. We assume here that we represent a rational number by a pair of integers, a numerator and a denominator. We may even assume the numerator and denominator are coprime: Given $a, b \in \mathbb{Z}$ we can compute their greatest common divisor $\gcd(a, b)$ and solve the Bézout equation $ax + by = \gcd(a, b)$ for some $x, y \in \mathbb{Z}$ using the extended Euclidean algorithm (see Exercise 2.17) in polynomial time. Applying these techniques in bulk we can also do addition, subtraction and multiplication of integer and rational matrices in polynomial time. Least trivially of our building blocks, using the theory of lattices we can compute bases for the kernel and the image of an integer matrix in polynomial time, which is the topic of Section 3.

1.3. Commutative algebra and number theory. We will go through some basic definitions from commutative algebra and number theory which we assume the reader to be familiar with in the coming sections. We will use [**2**] and [**17**] as a reference.

DEFINITION 1.3. All rings by definition have a multiplicative identity element denoted 1. Let R be a commutative ring.

(1) A *unit* of R is an $x \in R$ for which there exists an *inverse* $y \in R$ such that $xy = 1$, and we write R^* for the group of units of R. A *field* is a non-zero commutative ring of which every element but 0 is a unit.

(2) A *regular element* of R is an $x \in R$ for which multiplication by x is injective on R. A *domain* is a non-zero commutative ring of which every element but 0 is regular.
(3) A *nilpotent* of R is an element $x \in R$ for which there exists an $n \in \mathbb{Z}_{>0}$ such that $x^n = 0$, and we write $\mathrm{nil}(R)$ for the ideal of nilpotents of R. We say R is *reduced* if its only nilpotent is 0.

DEFINITION 1.4. Let R be a commutative ring. We say ideals $\mathfrak{a}$ and $\mathfrak{b}$ of R are *coprime* if $\mathfrak{a} + \mathfrak{b} = R$. We say an ideal $\mathfrak{p}$ of R is *prime* if $R/\mathfrak{p}$ is a domain and say an ideal $\mathfrak{m}$ of R is *maximal* if $R/\mathfrak{m}$ is a field. We say R is *local* if it has precisely one maximal ideal, and say R is *semi-local* if it has only finitely many maximal ideals. For a prime ideal $\mathfrak{p}$ of R we write $R_\mathfrak{p}$ for the *localization of R at $\mathfrak{p}$* (see p. 38 of [**2**]). Note that if R is a domain, then the zero ideal (0) is prime, the ring $R_{(0)}$ is a field and the natural map $R \to R_{(0)}$ is injective. In this case we refer to $R_{(0)}$ as the *field of fractions of R*.

DEFINITION 1.5. A *number field* is a field K containing the field of rational numbers $\mathbb{Q}$ such that the dimension of K over $\mathbb{Q}$ as a vector space is finite. A *number ring* is a ring isomorphic to a subring of a number field. An *order* is a domain whose additive group is isomorphic to $\mathbb{Z}^n$ for some $n \in \mathbb{Z}_{\geq 0}$. Note that some authors do not require an order to be a domain, but simply a commutative ring. Every order R is a number ring in the number field $R_{(0)} \cong \mathbb{Q} \otimes_\mathbb{Z} R$. Conversely, every number field K has a *maximal order* denoted $\mathcal{O}_K$ (Theorem I.1 in [**17**]).

EXERCISE 1.6. Let R be a commutative ring. Show that $\mathrm{nil}(R)$ is equal to the intersection of all prime ideals of R. Moreover, show that if $\mathrm{nil}(R)$ is finitely generated, then $\mathrm{nil}(R)$ is *nilpotent*, i.e. $\mathrm{nil}(R)^n = 0$ for some $n > 0$.

EXERCISE 1.7. Show that any finite commutative domain is a field. Conclude that in a general commutative ring prime ideals of finite index are maximal.

EXERCISE 1.8. Let R be a commutative ring and let $I_1, \ldots, I_m, J_1, \ldots, J_n \subseteq R$ be ideals such that for all i, j we have $I_i + J_j = R$. Show that $I_1 \cdots I_m + J_1 \cdots J_n = R$. Conclude that for any two distinct maximal ideals $\mathfrak{m}, \mathfrak{n} \subseteq R$ and any $m, n \in \mathbb{Z}_{\geq 0}$ we have $\mathfrak{m}^m + \mathfrak{n}^n = R$. Here $\mathfrak{m}^0 = \mathfrak{n}^0 = R$.

EXERCISE 1.9 (Chinese remainder theorem for ideals). Let R be a commutative ring and let $I_1, \ldots, I_n \subseteq R$ be pairwise coprime ideals. Show that $\bigcap_{i=1}^n I_i = \prod_{i=1}^n I_i$ and prove that the natural homomorphism

$$R/\Big(\bigcap_{i=1}^n I_i\Big) \to \prod_{i=1}^n (R/I_i)$$

is an isomorphism.

2. Coprime basis factorization

In this section we treat the following problem, which will be the motivation for the *coprime basis algorithm*.

THEOREM 2.1. *There is a polynomial-time algorithm that on input $t \in \mathbb{N}$, $q_1, \ldots, q_t \in \mathbb{Q}^*$ and $n_1, \ldots, n_t \in \mathbb{Z}$ decides whether*

$$\prod_{i=1}^t q_i^{n_i} = 1. \tag{2.1}$$

It is clear we can determine whether such a product has the correct sign: Simply take the sum of all n_i for which $q_i < 0$ and check whether the result is even. It is then sufficient to prove the following theorem instead.

THEOREM 2.2. *There is a polynomial-time algorithm that on input* $s, t \in \mathbb{N}$, $a_1, \ldots, a_s$, $b_1, \ldots, b_t \in \mathbb{Z}_{>0}$ *and* $m_1, \ldots, m_s, n_1, \ldots, n_t \in \mathbb{Z}_{\geq 0}$ *decides whether*

$$\prod_{i=1}^{s} a_i^{m_i} = \prod_{i=1}^{t} b_i^{n_i}. \tag{2.2}$$

In this form, the problem looks deceptively easy. Consider for example the most straightforward method to decide (2.2).

ALGORITHM 2.3. Compute $\prod_{i=1}^{s} a_i^{m_i}$ and $\prod_{i=1}^{t} b_i^{n_i}$ explicitly and compare the results.

This method is certainly correct in that it is able to decide (2.2). However, it fails to run in polynomial time even when $s = t = 1$. For $n \in \mathbb{Z}_{>0}$ the length $l(2^n)$ of 2^n as a string encoded in binary equals $n+1$. Hence $l(2^n)$ is not bounded by any polynomial in $l(n) \approx \log_2 n$. We wouldn't even have enough time to write down the number regardless of our proficiency in multiplication because the number is too long.

Another method uses the fundamental theorem of arithmetic, also known as unique prime factorization in $\mathbb{Z}$.

ALGORITHM 2.4. Factor $a_1, \ldots, a_s, b_1, \ldots, b_t$ into primes and for each prime that occurs compute the number of times it occurs in the products $\prod_{i=1}^{s} a_i^{m_i}$ and $\prod_{i=1}^{t} b_i^{n_i}$ and compare the results.

It is true that once we have factored all integers into primes only a polynomial number of steps remains. If we write x_{ip} for the exponent of the prime p in a_i, then we may compute $\sum_{i=1}^{t} m_i x_{ip}$, the exponent of p in $\prod_{i=1}^{t} a_i^{m_i}$, in polynomial time. Moreover, the number of prime factors of $n \in \mathbb{Z}_{>0}$ is at most $l(n)$, so the number of primes occurring is at most $\sum_{i=1}^{s} l(a_i) + \sum_{i=1}^{t} l(b_i)$, which is less than the length of the combined input. The problem lies in the fact that we have not specified how to factor integers into primes. As of January 23, 2026, nobody has been able to show that we can factor integers in polynomial time. Until this great open problem is solved, Algorithm 2.4 is out the window.

An interesting observation is that the main obstruction in Algorithm 2.3 lies in the exponents being large, while for Algorithm 2.4 the obstruction is in the bases. Our proof for Theorem 2.2 will be to slightly tweak Algorithm 2.4. Namely, observe that we do not need to factor into prime elements but that it suffices to factor into pairwise coprime elements. The following lemma follows readily from unique prime factorization.

LEMMA 2.5 (Unique coprime factorization). *Let* $k \in \mathbb{N}$ *and let* $c_1, \ldots, c_k \in \mathbb{Z}_{>1}$ *be pairwise coprime. If for* $m_1, \ldots, m_k, n_1, \ldots, n_k \in \mathbb{Z}$ *we have*

$$\prod_{i=1}^{k} c_i^{m_i} = \prod_{i=1}^{k} c_i^{n_i}, \tag{2.3}$$

then $m_i = n_i$ *for all* i. □

EXERCISE 2.6. Suppose $c_1, \ldots, c_k \in \mathbb{Z}_{>1}$ are pairwise coprime and let $m_1, \ldots, m_k \in \mathbb{Z}$. Then $\prod_i c_i^{m_i} \in \mathbb{Z}_{>0}$ if and only if $m_1, \ldots, m_k \geq 0$.

DEFINITION 2.7. For a subset $A \subseteq \mathbb{Z}_{\geq 1}$, a *coprime basis* is a subset $C \subseteq \mathbb{Z}_{>1}$ such that the elements of C are pairwise coprime and such that $\langle C \rangle \subseteq \mathbb{Q}^*$, the group generated by C, contains A. The set of coprime bases for A is partially ordered (Exercise 2.10) by the relation where $C \leq D$ if $\langle C \rangle \subseteq \langle D \rangle$.

EXAMPLE 2.8. For $A = \{30, 42\}$ a coprime basis is $C = \{5, 6, 7\}$. For $a, m, n \in \mathbb{Z}_{>1}$ the set $\{a^m, a^n\}$ has a coprime basis $C_d = \{a^d\}$ for any d dividing $\gcd(m, n)$, and $C_e \leq C_d$ if and only if $d \mid e$. For arbitrary $A \subseteq \mathbb{Z}_{\geq 1}$ the set of all primes is the unique maximal coprime basis.

We now propose the following algorithm for deciding (2.2).

METHOD 2.9. Factor $a_1, \ldots, a_s, b_1, \ldots, b_t$ into pairwise coprime $c_1, \ldots, c_k \in \mathbb{Z}_{>1}$. For each c_i compute the number of times it occurs in $\prod_{i=1}^s a_i^{m_i}$ and $\prod_{i=1}^t b_i^{n_i}$ and compare the results.

Now to prove Theorem 2.2 and in turn Theorem 2.1 it suffices to give a polynomial-time algorithm for computing coprime bases.

EXERCISE 2.10. Verify that the relation $\leq$ on the set of coprime bases is a partial order. In particular, show that $\langle C \rangle = \langle D \rangle$ implies $C = D$ for all coprime bases C and D.

LEMMA 2.11. *Let $A \subseteq \mathbb{Z}_{\geq 1}$. We write $\overline{A}$ for the closure of $A \cup \{1\}$ under multiplication, integer division (i.e. a division with integral result), and taking gcd's. Then*

(1) *$C \subseteq \mathbb{Z}_{>1}$ is a coprime basis for A if and only if it is a coprime basis for $\overline{A}$;*
(2) *the set $\operatorname{atom}(\overline{A})$ of minimal elements of $\overline{A} \setminus \{1\}$ with respect to the divisibility relation is the unique minimal (wrt. $\leq$) coprime basis for $\overline{A}$.*
(3) *if $C \subseteq \mathbb{Z}_{>1}$ satisfies $C \subseteq \overline{A} \subseteq \langle C \rangle$ and the elements of C are pairwise coprime, then $C = \operatorname{atom}(\overline{A})$.*

PROOF. (1) Suppose C is a coprime basis for A. For $a = \prod_{c \in C} c^{a_c}$ and $b = \prod_{c \in C} c^{b_c}$ in $\langle C \rangle \cap \mathbb{Z}$ we have

$$ab = \prod_{c \in C} c^{a_c + b_c}, \quad a/b = \prod_{c \in C} c^{a_c - b_c} \quad \text{and} \quad \gcd(a, b) = \prod_{c \in C} c^{\min\{a_c, b_c\}}.$$

Hence $ab, a/b, \gcd(a, b), \in \langle C \rangle$. As $A \subseteq \langle C \rangle$, we have $\overline{A} \subseteq \overline{\langle C \rangle} = \langle C \rangle$, so C is a coprime basis for $\overline{A}$.

Suppose C is a coprime basis for $\overline{A}$. Then $A \subseteq \overline{A} \subseteq \langle C \rangle$. If $C' \subseteq C$ is a coprime basis for A, then it is a coprime basis for $\overline{A}$ by the previous, so $C' = C$ by minimality. Hence C is a coprime basis for A.

(2) Firstly note that the elements of $C = \operatorname{atom}(\overline{A})$ are pairwise coprime: For $c, d \in C$ we have $\gcd(c, d) \in \overline{A}$, so by minimality of c and d we must have $\gcd(c, d) = 1$ or $c = d$.

Secondly, suppose $a \in \overline{A}$ is minimal such that $a \notin \langle C \rangle$. Clearly $1 \in \langle C \rangle$, so $a \neq 1$. Then some $c \in C$ satisfies $c \mid a$ by definition of C, so $a/c \in \overline{A}$. By minimality of a we have $a/c \in \langle C \rangle$, so $a = (a/c) \cdot c \in \langle C \rangle$. Hence $\overline{A} \subseteq \langle C \rangle$.

Thus C is a coprime basis for $\overline{A}$. Suppose $D \subseteq \mathbb{Z}_{>1}$ is a coprime basis for $\overline{A}$. Then $C \subseteq \overline{A} \subseteq \langle D \rangle$, so $\langle C \rangle \subseteq \langle D \rangle$. Hence C is the unique minimal coprime basis.

(3) Clearly C is a coprime basis, so by (2) it suffices to show that C is minimal. As $C \subseteq \overline{A}$ we have $\langle C \rangle \subseteq \langle \overline{A} \rangle \subseteq \langle C \rangle$, so $\langle C \rangle = \langle \overline{A} \rangle$. Any coprime basis D should satisfy $\langle C \rangle = \langle \overline{A} \rangle \subseteq \langle D \rangle$, so C is minimal. □

THEOREM 2.12 (Coprime basis factorization). *There is a polynomial-time algorithm that on input $s \in \mathbb{N}$ and $a_1, \ldots, a_s \in \mathbb{Z}_{>0}$ computes $k \in \mathbb{N}$ and the minimal coprime basis $\{c_1, \ldots, c_k\}$ of $\{a_1, \ldots, a_s\}$, as well as $(n_{ij}) \in \mathbb{Z}_{\geq 0}^{s \times k}$ such that $a_i = \prod_{j=1}^{k} c_j^{n_{ij}}$ for all i.*

We state the algorithm first and prove the theorem later. Note that we do not require the a_i in the theorem to be distinct.

ALGORITHM 2.13 (cf. Algorithm R in [**3**]). Remove all a_i for which $a_i = 1$. Construct a complete simple graph G on s vertices and label the vertices with $a_1, \ldots, a_s$. We call it a labeling because the map sending a vertex to its label need not be injective. While there are edges in G, iterate the following 5 steps:

(1) Choose an edge $\{U, V\}$ of G, delete it from the graph, and let u and v be the labels of U and V respectively.
(2) Compute $w = \gcd(u, v)$ using the Euclidean algorithm.
(3) Add a vertex W labeled w to G and connect it to U, V and those vertices which are neighbors of both U and V.
(4) Update the labels of U and V to u/w and v/w respectively.
(5) For each $S \in \{U, V, W\}$, if the label of S is 1, then delete S and its incident edges from G.

Once G has no more edges, the labels of the vertices in the graph are the required pairwise coprime elements. The remaining output can be computed in polynomial time.

In the graph that we construct and update, the edges represent the pairs of numbers of which we do not yet know whether they are coprime, while a missing edge denotes that we know the pair to be coprime.

EXAMPLE 2.14. We apply Algorithm 2.13 to $(a_1, a_2) = (4500, 5400)$. Since there are only two vertices, our graphs will fit on a single line. We denote the edge we choose in each iteration with a bullet and edges we have to erase are dotted. On the right we show how to keep track of the factorization of 4500 with minimal bookkeeping by writing it as a product of vertices in the graph.

Iteration 1:	4500							5400	4500
Iteration 2:	5				900			6	$5 \cdot 900$
Iteration 3:	1	5			180			6	$5^2 \cdot 180$
Iteration 4:		1	5		36			6	$5^3 \cdot 36$
Iteration 5:			5	1	36			6	$5^3 \cdot 36$
Iteration 6:			5		6		6	1	$5^3 \cdot 6 \cdot 6$
Iteration 7:			5		1	6	1		$5^3 \cdot 6^2$
Iteration 8:			5			6			$5^3 \cdot 6^2$

We obtain $(c_1, c_2) = (5, 6)$ and $4500 = 5^3 \cdot 6^2$. By trial division we obtain $5400 = 5^2 \cdot 6^3$.

EXAMPLE 2.15. We apply Algorithm 2.13 to $(a_1, a_2, a_3) = (15, 21, 35)$.

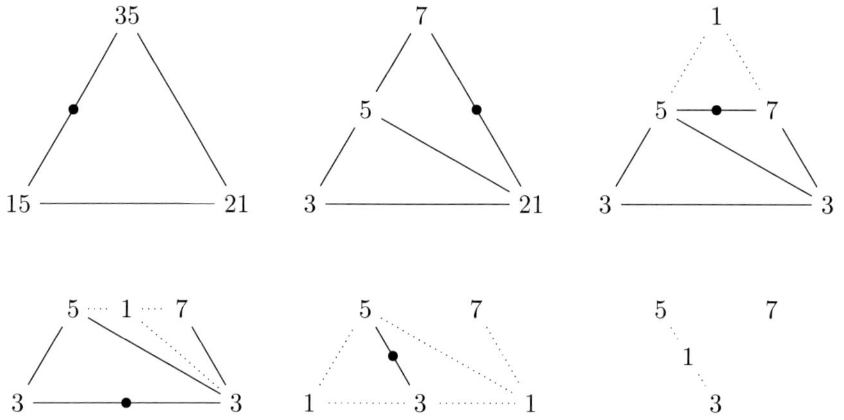

The resulting coprime basis is $(c_1, c_2, c_3) = (3, 5, 7)$. In the fifth graph something interesting happens. Vertex 7 suddenly becomes disconnected from the graph because we know it is coprime to one of the 3's.

PROOF OF THEOREM 2.12. We claim Algorithm 2.13 is correct and runs in polynomial time. One can show inductively that throughout the algorithm two vertices in the graph G have coprime labels when there is no edge between them. When the algorithm terminates because there are no edges in the graph, we may conclude that $c_1, \ldots, c_s$ are coprime. Additionally, one shows inductively that the numbers $a_1, \ldots, a_t$ can be written as some product of labels. Hence $c_1, \ldots, c_s$ is the minimal coprime basis of $a_1, \ldots, a_t$ by Lemma 2.11.3, as $c_1, \ldots, c_s$ are in the closure of $\{a_1, \ldots, a_t\}$. Thus Algorithm 2.13 is correct. It remains to show that it is fast.

Write P_n for the product of all labels of the vertices in the graph after iteration n. Note that $P_0 = a_1 \cdots a_s$ and that $P_{n+1} \mid P_n$ for all $n \in \mathbb{N}$. Since P_0 has at most $B := \log_2 P_0$ prime factors counting multiplicities, there are at most B iterations n for which $P_{n+1} < P_n$, and at most B vertices in G after every iteration in the algorithm. The iterations for which $P_n = P_{n+1}$ are those where the edge we chose is between coprime integers, meaning no vertices or edges are added to the graph and one edge is deleted. As the number of edges is at most B^2, then so is the number of consecutive iterations for which $P_n = P_{n+1}$. Hence the total number of iterations is at most B^3, which is polynomial in the length of the input. Lastly, note that each iteration takes only polynomial time because the labels of the vertices are bounded from above by 2^B and the Euclidean algorithm runs in polynomial time. Hence Algorithm 2.13 runs in polynomial time. □

The speed of this algorithm heavily depends on how fast the product of all vertices decreases. In each step we want to choose our edge $\{u, v\}$ such that $\gcd(u, v) \gg 1$. A heuristic for this could be to choose edges between large numbers. For those interested in the efficiency of coprime basis factorization we refer to [**4**] for a provably faster algorithm.

EXERCISE 2.16. Show that there exists a polynomial-time algorithm that, given $a, b, c, d \in \mathbb{Z}_{>0}$ such that $ab = cd$, computes $w, x, y, z \in \mathbb{Z}_{>0}$ such that $(a, b, c, d) = (wx, yz, wz, xy)$.

EXERCISE 2.17 (Modified Euclidean algorithm). Recall that $\gcd(0, 0) = 0$.

a. Show that for all $a, b \in \mathbb{Z}$ with $b \neq 0$ there exist $r, q \in \mathbb{Z}$ with $a = qb + r$ and $|r| \leq |b|/2$.
b. Show that for all $q, b, r \in \mathbb{Z}$ with $a = qb + r$ we have $\gcd(a, b) = \gcd(b, r)$ and $\gcd(a, 0) = |a|$.
c. Prove that there exists a polynomial-time algorithm that, given $a, b \in \mathbb{Z}$, computes $\gcd(a, b)$ as well as $x, y \in \mathbb{Z}$ such that $ax + by = \gcd(a, b)$.
d. Conclude that there exists a polynomial-time algorithm that, given $a, n \in \mathbb{Z}$ with $n \geq 1$, decides whether $(a \bmod n) \in (\mathbb{Z}/n\mathbb{Z})^*$ and if so computes some $a' \in \mathbb{Z}$ such that $aa' \equiv 1 \bmod n$.
e. Prove that there exists a polynomial-time algorithm that, given $a, b, m, n \in \mathbb{Z}$ with $n, m > 1$, decides whether there exists some $c \in \mathbb{Z}$ such that $c \equiv a \bmod m$ and $c \equiv b \bmod n$ and if so computes such a c.

EXERCISE 2.18. Show that there exists a polynomial-time algorithm that, given $k, n \in \mathbb{Z}_{>0}$ and $a_1, \ldots, a_k \in \mathbb{Z}$ satisfying $a_i^2 \equiv 1 \bmod n$ for all i, decides whether there exists some non-empty subset $I \subseteq \{1, \ldots, k\}$ such that $\prod_{i \in I} a_i \equiv 1 \bmod n$ and if so computes one such I.

Hint: Non-trivial roots of $X^2 - 1$ in $\mathbb{Z}/n\mathbb{Z}$ factor n.

EXERCISE 2.19. We equip $\mathbb{Q}^2 \setminus \{(0, 0)\}$ with an equivalence relation $\sim$ where $(x_1, y_1) \sim (x_2, y_2)$ if and only if there exists some $\lambda \in \mathbb{Q}^*$ such that $(\lambda x_1, \lambda y_1) = (x_2, y_2)$. Write $\mathbb{P}^1(\mathbb{Q}) = (\mathbb{Q}^2 \setminus \{(0, 0)\})/\sim$ for the *projective line* and write $(x : y)$ for the image of (x, y) in $\mathbb{P}^1(\mathbb{Q})$. Let $a, b \in \mathbb{Z}_{>0}$ and let $c_1, \ldots, c_n$ be the minimal coprime basis for a and b.

a. For $p \mid ab$ prime write $f(p) = (\mathrm{ord}_p(a) : \mathrm{ord}_p(b)) \in \mathbb{P}^1(\mathbb{Q})$. Show that every c_i naturally corresponds to a fiber of f and give the prime factorization of c_i in terms of the prime factorization of a and b.
b. Suppose $n = 7$. Show that $ab \geq 1485890406000$ and equality holds for exactly 8 pairs (a, b).
c. (difficult, Problem 2023-2/C in [**5**]) Give an asymptotic formula for the minimum of $\log(ab)$ in terms of n.

EXERCISE 2.20. Let $n \in \mathbb{Z}_{>0}$. We encode a matrix $\overline{M} = (\overline{m}_{ij})_{i,j}$ over $\mathbb{Z}/n\mathbb{Z}$ as a matrix $M = (m_{ij})_{i,j}$ over $\mathbb{Z}$ such that $0 \leq m_{ij} < n$ and $\overline{m}_{ij} \equiv m_{ij} \bmod n$ for all i, j. Show that there exist polynomial-time algorithms for the following problems:

a. given $n \in \mathbb{Z}_{\geq 0}$ and matrices M and N over $\mathbb{Z}/n\mathbb{Z}$, compute $M + N$ and $M \cdot N$ if well-defined;
b. given $n, k \in \mathbb{Z}_{>0}$ and a square matrix M over $\mathbb{Z}/n\mathbb{Z}$, compute M^k;
Note: An algorithm that takes k steps is not polynomial-time!
c. given $n \in \mathbb{Z}_{>0}$ and a matrix M over $\mathbb{Z}/n\mathbb{Z}$, compute a row-echelon form of M;
d. given $n \in \mathbb{Z}_{>0}$ and a square matrix M over $\mathbb{Z}/n\mathbb{Z}$, compute $\det(M)$ and $\mathrm{Tr}(M)$;
e. given $n \in \mathbb{Z}_{>0}$ and a matrix M over $\mathbb{Z}/n\mathbb{Z}$, decide whether M^{-1} exists and if so compute it.

You may use the following fact: For every $k, B \in \mathbb{Z}_{>0}$ and matrix $M = (m_{ij})_{i,j} \in \mathbb{Z}^{k \times k}$ with $|m_{ij}| \leq B$ for all i, j it holds that $|\det(M)| \leq B^k \cdot k^{k/2}$ (see Hadamard's inequality, Exercise 3.13).

f. Show that there exists a polynomial-time algorithm that, given a square integer matrix M, computes $\det(M)$ and $\mathrm{Tr}(M)$.

EXERCISE 2.21. Show that there exist polynomial-time algorithms for the following problems:

a. given $a, p, q \in \mathbb{Z}$ with p and q prime and $\gcd(a, p) = 1$, compute $e \in \mathbb{Z}_{\geq 0}$ such that the order of a in $(\mathbb{Z}/p\mathbb{Z})^*$ equals uq^e for some $u \in \mathbb{Z}_{>0}$ with $\gcd(u, q) = 1$;
b. given $a, p \in \mathbb{Z}$ with p prime, decide whether a is a square modulo p;
c. given $a, b, p \in \mathbb{Z}$ with p prime, a a square modulo p and b not a square modulo p, compute $c \in \mathbb{Z}$ such that $c^2 \equiv a \bmod p$;
d. given $a, b, p \in \mathbb{Z}$ with p prime, compute $c \in \mathbb{Z}$ such that c^2 equals a, b or ab modulo p.

EXERCISE 2.22. Show that there exists a polynomial-time algorithm that, given $a, b, k, n \in \mathbb{Z}$ with $k, n > 0$ and $a^k \equiv 1 \bmod n$ and $b^k \equiv -1 \bmod n$, computes some $c \in \mathbb{Z}$ such that $a \equiv c^2 \bmod n$.

Hint: First consider n odd and k a power of 2.

3. Finitely generated abelian groups

In this section we treat algorithms on finitely generated abelian groups. This will be used as a basis for algorithmic aspects of ring theory; for example, the additive group of orders R, as well as finitely generated modules over R, will be groups to which we can apply the methods of this section.

We begin by specifying a representation for our finitely generated groups. Recall that every finitely generated abelian group A fits in some exact sequence

$$\mathbb{Z}^m \xrightarrow{\alpha} \mathbb{Z}^n \xrightarrow{f} A \longrightarrow 0.$$

Namely, we obtain n and f by writing down some generators $a_1, \ldots, a_n \in A$ for A and let f map the i-th standard basis vector to a_i. For m and α we repeat the procedure with A replaced by $\ker(f)$. Note that α, being a morphism between free $\mathbb{Z}$-modules, has a natural representation as a matrix with integer coefficients. The isomorphism theorem gives $A \cong \mathbb{Z}^n/\ker(f) = \mathbb{Z}^n/\mathrm{im}(\alpha) = \mathrm{coker}(\alpha)$, so A is, up to isomorphism, completely defined by n and the image of α. Thus we choose to encode A as the matrix corresponding to α. We say representations $\alpha : \mathbb{Z}^k \to \mathbb{Z}^l$ and $\beta : \mathbb{Z}^m \to \mathbb{Z}^n$ are *equivalent* if $l = n$ and $\mathrm{im}(\alpha) = \mathrm{im}(\beta)$.

A morphism $f : A \to B$ of finitely generated abelian groups in terms of this representation gives a commutative diagram of exact sequences

$$\begin{array}{ccccccc} \mathbb{Z}^k & \xrightarrow{\alpha} & \mathbb{Z}^l & \longrightarrow & A & \longrightarrow & 0 \\ & & \downarrow{\scriptstyle \varphi} & & \downarrow{\scriptstyle f} & & \\ \mathbb{Z}^m & \xrightarrow{\beta} & \mathbb{Z}^n & \longrightarrow & B & \longrightarrow & 0. \end{array} \tag{3.1}$$

Here φ is any morphism that makes the diagram commute. We encode f by the matrix representing φ. Important to note is that not every φ defines a morphism $f : A \to B$. It defines a morphism precisely when $\mathrm{im}(\varphi \circ \alpha) \subseteq \mathrm{im}(\beta)$, however it is not immediately obvious how to test this. Computing the composition of morphisms and evaluating morphisms in this form is straightforward, as it is just matrix multiplication.

To work with abelian groups in our algorithms it takes more than just to specify an encoding. The following is a list, in no particular order, of operations we would like to be, and later will be, able to perform in polynomial time.

(1) decide whether a matrix encodes a morphism of given abelian groups;
(2) compute kernels, images and cokernels of homomorphisms;
(3) test if a homomorphism is injective/surjective;
(4) decide if two homomorphisms are equal;
(5) decide if there exists an element in the preimage of a group element under a homomorphism and if so compute such an element;
(6) compute direct sums, tensor products and homomorphism groups of pairs of abelian groups;
(7) compute the order of a given group element;
(8) compute the order/exponent of a finite abelian group;
(9) decide whether an exact sequence splits and if so compute a corresponding section and retraction;
(10) compute the torsion subgroup of an abelian group;
(11) write an abelian group as a direct sum of cyclic groups.

We will spend this section working towards the last entry of this list: An algorithmic version of the structure theorem of finitely generated abelian groups.

All algorithms for the above problems will be straight-forward. The only serious complication arises at the fundamentals, where we reduce to linear algebra over $\mathbb{Z}$. Here we encounter the LLL-algorithm, a lattice basis reduction algorithm.

As a rule of thumb, all problems on finitely generated abelian groups can be solved in polynomial time, unless the output of the problem is not polynomially bounded or one implicitly requires prime factorization (c.f. Algorithm 2.3 and Algorithm 2.4). For the first exception, consider the computation of the group A^k for an abelian group A and positive integer k. For the second, we should not expect to be able to factor a finite abelian group as a product of cyclic groups of prime-power order, since this would provide a factorization of n if we input $\mathbb{Z}/n\mathbb{Z}$.

Finally, we address an important subtlety that arises from our choice of encoding. As an exercise, one can prove the following lemma.

LEMMA 3.1. *Assuming the above problems have polynomial-time algorithms, we may solve the* discrete logarithm problem *in polynomial time. That is, given an abelian group A and elements $a, b \in A$, decide whether there exists some positive integer n such that $na = b$ and if so compute such n.*

It is well known that the discrete logarithm problem for $\mathbb{F}_q^*$ or elliptic curves over finite fields is difficult, i.e. not known to be solvable in polynomial time, even though both are finitely generated abelian groups (see Section 2 of [**6**]). The apparent paradox is solved by observing that the difficulty is in representing $\mathbb{F}_q^*$ and its elements in our encoding. For starters, we need to write down generators for $\mathbb{F}_q^*$ and subsequently write the input to our algorithms in terms of these generators. Doing so is almost equivalent to the discrete logarithm problem.

EXERCISE 3.2. Prove Lemma 3.1.

EXERCISE 3.3 (Product, Proposition 2.3.16 in [**9**]). Show that there exists a polynomial-time algorithm that, given finitely generated abelian groups A and B, computes the group $A \times B$ and the corresponding inclusions and projections.

3.1. Lattices and short bases. To understand general finitely generated abelian groups, we first need to understand the simplest instances, the free abelian groups. It will turn out to be fruitful to consider free abelian groups together with a positive-definite inner product. Any algorithm that keeps the vectors short with respect to the inner product, also keeps the vectors short with respect to length of the encoding, so that our algorithms have a chance to run in polynomial time. This will allow us later to compute images and kernels of linear maps.

DEFINITION 3.4. A *Euclidean (vector) space* is a finite-dimensional positive-definite real inner product space. For elements x and y we denote their inner product by $\langle x, y\rangle$, and will write $q(x) = \langle x, x\rangle$ for the *square norm* of x. Note that Euclidean spaces come with a natural metric, and in particular they are topological spaces. A *lattice* is a discrete subgroup of a Euclidean space, i.e. a subgroup Λ for which $\inf\{q(x) : x \in \Lambda \setminus \{0\}\} > 0$. An isomorphism of lattices is an inner product preserving isomorphism of groups.

EXAMPLE 3.5. **1.** For $n \in \mathbb{Z}_{\geq 0}$ the vector space $\mathbb{R}^n$ has an inner product given by

$$\langle (x_1, \ldots, x_n), (y_1, \ldots, y_n)\rangle = \sum_{i=1}^{n} x_i y_i.$$

Any basis (as $\mathbb{R}$-module) of $\mathbb{R}^n$ is a basis (as $\mathbb{Z}$-module) of a lattice in $\mathbb{R}^n$ (Proposition 3.3 in [**20**]). The standard basis generates the lattice $\mathbb{Z}^n$.

2. A Euclidean space one naturally encounters in number theory is $K \otimes_{\mathbb{Z}} \mathbb{R}$, for any number field K, which we equip with the inner product

$$\langle x, y\rangle = \frac{1}{[K:\mathbb{Q}]} \sum_{\sigma : K\otimes_{\mathbb{Z}}\mathbb{R}\to\mathbb{C}} \sigma(x)\cdot\overline{\sigma(y)},$$

where the sum ranges over all $\mathbb{R}$-algebra homomorphisms. In this Euclidean space every order of K is a lattice (Lemma 10.3 in [**27**]).

EXERCISE 3.6. Let V be a Euclidean vector space. Show that the inner product is uniquely determined by q, i.e. give a formula for $\langle x, y\rangle$ in terms of q.

DEFINITION 3.7. For a ring R and an R-module M we say M is free if $M \cong \bigoplus_{s\in S} R$ for some set S, i.e. it has a basis. If R is non-zero and commutative and M is free, then the cardinality of such S is unique, and we write $\operatorname{rk}_R M = \operatorname{rk} M$, the *rank* of M, for this cardinality.

PROPOSITION 3.8 (cf. Proposition 4.1 in [**20**]). *Any lattice Λ in a Euclidean space V is a free $\mathbb{Z}$-module with $\operatorname{rk}\Lambda \leq \dim V$ and the restriction of the inner product to Λ is $\mathbb{Z}$-bilinear, real-valued, symmetric and satisfies $\inf\{\langle x, x\rangle \mid x \in \Lambda \setminus \{0\}\} > 0$. Conversely, every free $\mathbb{Z}$-module Λ of finite rank equipped with a $\mathbb{Z}$-bilinear, real-valued, symmetric form φ for which $\inf\{\varphi(x, x) \mid x \in \Lambda \setminus \{0\}\} > 0$ can be embedded in a Euclidean vector space such that the inner product restricted to Λ equals φ.* □

We say a lattice Λ in a Euclidean space V is *full-rank* if $\operatorname{rk}\Lambda = \dim V$.

Proposition 3.8 shows that we have an equivalent definition of a lattice that does not require an ambient vector space. The bilinear form φ in the proposition is again naturally given by a matrix $(\varphi(b_i, b_j))_{1\leq i,j\leq n}$, the *Gram-matrix*, where $(b_1, \ldots, b_n)$ is the ($\mathbb{Z}$-)basis encoding Λ. For our computational purposes it is practical to restrict to Gram-matrices with rational entries. This will be our encoding for lattices. Unfortunately, the Gram-matrix of the inner product from Example 3.5.2 is generally not of this form.

An algorithmic problem we will encounter is computing a 'short basis' for a lattice $\Lambda \subseteq \mathbb{Z}^n$.

DEFINITION 3.9. Let Λ be a lattice. Let $(b_1, \ldots, b_n)$ be a basis for Λ and consider its Gram-matrix $M = (\langle b_i, b_j \rangle)_{1 \le i,j \le n}$. We define the *determinant* of Λ to be $\det(\Lambda) = |\det(M)|^{1/2}$.

EXERCISE 3.10. Show that the determinant of a lattice does not depend on the choice of basis and is thus well-defined.

Hint: Show that any two bases for Λ 'differ' by a matrix of unit determinant.

EXERCISE 3.11. Let $\Lambda \subseteq \mathbb{R}^n$ be a lattice of rank n and $S \subseteq \mathbb{R}^n$ a measurable bounded set, i.e. a set with a well-defined finite volume. Prove that if the addition map $\Lambda \times S \to \mathbb{R}^n$ is injective, then $\mathrm{vol}(S) \le \det(\Lambda)$ holds, and prove that if it is surjective, then $\mathrm{vol}(S) \ge \det(\Lambda)$ holds.

The determinant $\det(\Lambda)$ equals the volume of the parallelepiped spanned by a basis for Λ. Since the determinant is an invariant, finding a 'shorter' basis is equivalent to finding a 'more orthogonal' basis. In the lattice $\mathbb{Z}b_1 + \mathbb{Z}b_2$ below we can find a better basis.

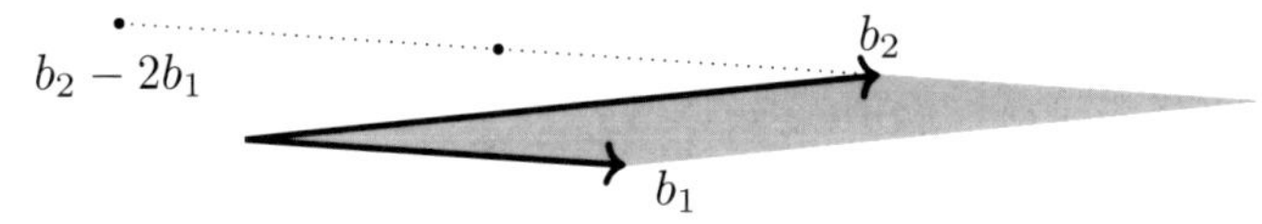

Taking $c_1 = b_1$ and $c_2 = b_2 - 2b_1$ produces the following basis:

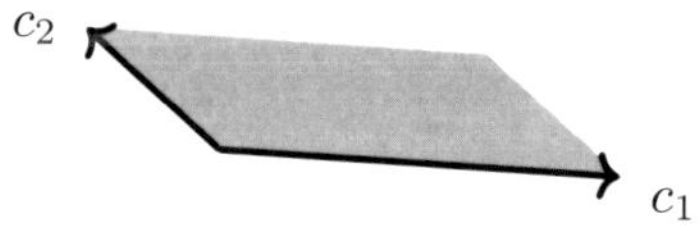

For a Euclidean space the Gram–Schmidt algorithm transforms a basis into an orthogonal one as follows.

DEFINITION 3.12. Let V be a Euclidean space with basis $B = (b_1, \ldots, b_n)$. We iteratively define

$$\mu_{ij} = \frac{\langle b_i, b_j^* \rangle}{\langle b_j^*, b_j^* \rangle} \quad \text{for } 1 \le j < i \text{ and } \quad b_i^* = b_i - \sum_{j<i} \mu_{ij} b_j^* \quad \text{for } 1 \le i \le n.$$

We call $B^* = (b_1^*, \ldots, b_n^*)$ and $(\mu_{ij})_{j<i}$ the *Gram–Schmidt basis* and *Gram–Schmidt coefficients*, respectively, corresponding to B.

When interpreting $M = (\mu_{ij})_{j<i}$ as an upper-triangular matrix, we note that that $(\mathrm{id} + M)B^* = B$. In particular $\det(B) = \det(B^*)$. Since $b_1^*, \ldots, b_n^*$ are indeed pairwise orthogonal, they form an orthogonal basis for V. Sadly the Gram–Schmidt coefficients will generally not be integers, meaning that if B is a basis for a lattice Λ, then generally B^* will not be. This is quite unsurprising, as not every lattice even has an orthogonal basis. A possible solution is to round the Gram-Schmidt coefficients to integers in every step, so that we are guaranteed to obtain a basis

for Λ. However, this does not yield the necessary bounds on our basis. In the next section we will state the existence of a better algorithm.

EXERCISE 3.13. **a.** Let $(b_1, \ldots, b_n)$ be a basis for a lattice Λ in a Euclidean space V. Write $\Lambda_k = \sum_{i \leq k} \mathbb{Z} b_i$ for all $0 \leq k \leq n$. Show that

$$q(b_i) \geq q(b_i^*) = \left(\frac{\det(\Lambda_i)}{\det(\Lambda_{i-1})}\right)^2.$$

b. Conclude *Hadamard's inequality* [**12**]: For an invertible matrix $B \in \mathbb{R}^{n \times n}$ with columns $b_1, \ldots, b_n$ we have

$$|\det(B)| \leq \prod_{i=1}^{n} \|b_i\|,$$

with equality if and only if the b_i are pairwise orthogonal.

EXERCISE 3.14. Let V be a Euclidean space. For a subspace $W \subseteq V$ we write $W^\perp = \{v \in V \mid \langle v, W \rangle = 0\}$.

a. Show that for all $W \subseteq V$ the natural map $W^\perp \to V/W$ is an isomorphism of vector spaces.

We equip V/W with the natural Euclidean vector space structure induced by $W^\perp$. Suppose $\Lambda \subseteq V$ is a full-rank lattice.

b. Show that if $W \subseteq V$ is a subspace and the image $\overline{\Lambda}$ of Λ in V/W is a lattice, then $\det(\Lambda) \geq \det(\overline{\Lambda}) \cdot \det(\Lambda \cap W)$.

Suppose that $\Lambda' \subseteq \Lambda$ is a subgroup such that Λ/Λ' is torsion free.

c. Show that the natural map $\Lambda/\Lambda' \to V/\mathbb{R}\Lambda'$ is injective and that its image is a lattice.

d. Show that $\det(\Lambda) = \det(\Lambda/\Lambda') \cdot \det(\Lambda')$.

EXERCISE 3.15. Let R be a commutative ring, $Z \subseteq R$ a subring, V a finitely generated free R-module, and $\varphi : V \times V \to R$ a symmetric R-bilinear map such that the natural map $V \to \mathrm{Hom}_R(V, R)$ given by $x \mapsto (y \mapsto \varphi(x, y))$ is an isomorphism. Let $b_1, \ldots, b_n$ be an R-basis for V and define $L = \sum_i Z \cdot b_i$. Consider $L^\dagger = \{x \in V \mid \varphi(x, L) \subseteq Z\}$. Show that $L^\dagger$ is generated as a Z-module by an R-basis $b_1^\dagger, \ldots, b_n^\dagger$ of V such that $\varphi(b_i, b_j^\dagger)$ equals 1 or 0 according as $i = j$ or not, and that $L^{\dagger\dagger} = L$.

EXERCISE 3.16. Suppose that Λ is a lattice. Show that $\Lambda^\dagger$, as in Exercise 3.15 with $Z = \mathbb{Z}$, $R = \mathbb{R}$ and φ the inner product on Λ, is a lattice with $\det(\Lambda)\det(\Lambda^\dagger) = 1$.

We call $\Lambda^\dagger$ the *dual lattice* or *polar lattice* of Λ.

EXERCISE 3.17. Let Λ be a lattice. Verify that $\mathrm{Hom}(\Lambda, \mathbb{Z})$ is a lattice when equipped with the square norm and inner product

$$q(f) = \sup_{x \in \Lambda \setminus \{0\}} \frac{f(x)^2}{q(x)} \quad \text{and} \quad \langle f, g \rangle = \frac{q(f+g) - q(f-g)}{4}$$

respectively. Show that $\mathrm{Hom}(\Lambda, \mathbb{Z})$ is isomorphic to $\Lambda^\dagger$, i.e. there exists an isomorphism of abelian groups that respects the inner products.

EXERCISE 3.18 (Section 5 in [**21**]). Let V_1 and V_2 be Euclidean vector spaces and let $f : V_1 \to V_2$ be a linear map.

a. Show that there exists a unique linear map $f^\dagger : V_2 \to V_1$, which we will call the *adjoint* of f, such that for all $x \in V_1$ and $y \in V_2$ we have $\langle f(x), y\rangle = \langle x, f^\dagger(y)\rangle$.
b. Suppose $\Lambda_i \subseteq V_i$ are full-rank lattices and $f\Lambda_1 \subseteq \Lambda_2$. Show that $f^\dagger \Lambda_2^\dagger \subseteq \Lambda_1^\dagger$.
Suppose that $g : \Lambda_1 \to \Lambda_2$ is a group homomorphism between lattices.
c. Conclude that g has a well-defined adjoint.
d. Show that

$$\det(\ker(g)) \cdot \det(\mathrm{im}(g)) \cdot \det(\Lambda_1^\dagger) = \det(\ker(g^\dagger)) \cdot \det(\mathrm{im}(g^\dagger)) \cdot \det(\Lambda_2).$$

3.2. The LLL-algorithm. In this section we state the existence the LLL-algorithm, which efficiently produces a 'small' basis for a given lattice. We will not prove its correctness, nor will we actually describe the algorithm. What we will do is define what we mean by 'small' bases in the context of the LLL-algorithm and derive some of their properties. In this section we will use [**21**] as our reference. Further references include [**23**] and [**25**].

DEFINITION 3.19 (Section 10 in [**21**]). Let Λ be a lattice with basis $B = (b_1, \dots, b_n)$ and let $(b_1^*, \dots, b_n^*)$ and $(\mu_{ij})_{j<i}$ be its corresponding Gram–Schmidt basis and coefficients as defined in Definition 3.12. Let $c \geq 1$. We say B is c-reduced if

(1) For all $1 \leq j < i \leq n$ we have $|\mu_{ij}| \leq \frac{1}{2}$.
(2) For all $1 \leq k < n$ we have $cq(b_{k+1}^*) \geq q(b_k^*)$.

Note that the first condition states that B is close to the Gram–Schmidt basis. The b_i^* can be interpreted as elements of the lattice Λ/Λ_{i-1}, with the notation as in Exercise 3.13.a and Exercise 3.14.c. The second condition, the *Lovász condition*, for $c = 1$ implies that b_i^* is in fact a shortest non-zero vector of Λ/Λ_{i-1}. For $c > 1$ this condition can be seen as an approximation. We may compute for $c > \frac{4}{3}$ a c-reduced basis, and in particular it always exists. That we may in fact compute it in polynomial time is non-trivial.

THEOREM 3.20 (LLL-algorithm [**18**]). *There exists an algorithm that, given $c > \frac{4}{3}$ and a lattice Λ, produces a c-reduced basis for Λ. For fixed c it runs in polynomial time.* □

We should warn the reader that the literature contains various definitions of a 'reduced basis', and even in the context of the LLL-algorithm there are at least two. The one we gave is sufficient for our purposes.

DEFINITION 3.21. Let Λ be a lattice of rank n. For $0 < i \leq n$ we define the *i-th successive minimum* to be the value

$$\lambda_i(\Lambda) = \min\{r \in \mathbb{R}_{\geq 0} \mid \mathrm{rk}\langle x \in \Lambda \mid q(x) \leq r\rangle \geq i\},$$

a quantity that indicates how large the generating vectors of a rank i sublattice need to be.

A basis of vectors attaining the successive minima is the gold standard of 'small' bases, although it does not always exist (Exercise 3.23). The following proposition gives bounds on how far away a reduced basis can be from these minima, which is the justification for calling reduced bases 'small'.

PROPOSITION 3.22. *Let $c \geq \frac{4}{3}$ and suppose $(b_1, \dots, b_n)$ is a c-reduced basis for a lattice Λ. Then for all $0 < i \leq n$ we have $c^{1-n} \cdot q(b_i) \leq \lambda_i(\Lambda) \leq c^{i-1} \cdot q(b_i)$.*

PROOF. See Exercise 3.24, or Section 10 in [**21**]. □

EXERCISE 3.23. Show that there exists a lattice Λ for which no basis $b_1, \ldots, b_n$ attains the successive minima, i.e. satisfies $q(b_i) = \lambda_i(\Lambda)$ for all i.

Hint: Consider $2\mathbb{Z}^n \subseteq \Lambda \subseteq \mathbb{Z}^n$.

EXERCISE 3.24. Let $c \geq \frac{4}{3}$ and $0 < i \leq n$ and suppose $(b_1, \ldots, b_n)$ is a c-reduced basis for Λ.

a. Show that $q(b_j^*) \leq c^{i-j} q(b_i^*)$ for all $j \leq i$.

b. Recall that $b_i = b_i^* + \sum_{j<i} \mu_{ij} b_j^*$. Show that $q(b_i) \leq c^{i-1} q(b_i^*)$.

c. Show that $q(b_j) \leq c^{i-1} q(b_i^*) \leq c^{i-1} q(b_i)$ for all $j \leq i$.

d. Conclude that $\lambda_i(\Lambda) \leq \max\{q(b_j) \mid j \leq i\} \leq c^{i-1} q(b_i)$.

Write $\Lambda_k = \sum_{j \leq k} \mathbb{Z} b_j$ for all $0 \leq k \leq n$.

e. Prove that for all $0 < k \leq n$ and $x \in \Lambda_k \setminus \Lambda_{k-1}$ we have $q(x) \geq q(b_k^*)$.

Write $S = \{x \in \Lambda \mid q(x) \leq \lambda_i(\Lambda)\}$ and let k be minimal such that $S \subseteq \Lambda_k$.

f. Show that $k \geq \operatorname{rk}\langle S \rangle \geq i$.

g. Conclude that $\lambda_i(\Lambda) \geq q(b_k^*) \geq c^{1-n} q(b_i)$.

EXERCISE 3.25. Let $c \geq \frac{4}{3}$ and suppose $(b_1, \ldots, b_n)$ is a c-reduced basis for a lattice Λ. Show that

$$\det(\Lambda)^2 \leq \prod_{i=1}^{n} q(b_i) \leq c^{\binom{n}{2}} \det(\Lambda)^2.$$

EXERCISE 3.26 (Proposition 1.12 in [**18**]). Let $c \geq \frac{4}{3}$, let $(b_1, \ldots, b_n)$ be a c-reduced basis of some lattice Λ and let $a_1, \ldots, a_m \in \Lambda$ be linearly independent.

a. For $1 \leq i \leq m$ write $a_i = \sum_j x_{ij} b_i$ and let $j(i)$ be maximal such that $x_{ij(i)} \neq 0$. Show that $q(b_{j(i)}^*) \leq q(a_i)$.

b. Suppose the a_i are ordered so that $j(1) \leq j(2) \leq \ldots \leq j(m)$. Show that $i \leq j(i)$ for all $1 \leq i \leq m$.

c. Show that for all $1 \leq i \leq m$ we have

$$q(b_i) \leq c^{n-1} \cdot \max\{q(a_1), q(a_2), \ldots, q(a_m)\}.$$

3.3. The kernel-image algorithm. The LLL-algorithm allows us to construct the kernel-image algorithm, which computes kernels and images of morphisms of finitely generated free abelian groups. Most of the algorithms for finitely generated abelian groups from the beginning of this section will follow from this. We first need a theorem from linear algebra.

THEOREM 3.27 (Cramer's rule, Theorem 4.1 in [**16**]). *Suppose $A \in \mathbb{R}^{n \times n}$ is an invertible matrix and let $b \in \mathbb{R}^n$. Write A_i for the matrix obtained from A by replacing the i-th column with b. Then there exists a unique $x \in \mathbb{R}^n$ such that $Ax = b$, and it is given by $x = \det(A)^{-1} \cdot (\det(A_1), \ldots, \det(A_n))$.* □

EXERCISE 3.28. Let $n \geq 0$. Suppose $N \subseteq M \subseteq \mathbb{Z}^n$ are subgroups such that $N \oplus P = \mathbb{Z}^n$ for some $P \subseteq \mathbb{Z}^n$. Show that $\operatorname{rk}(M) = \operatorname{rk}(N)$ if and only if $M = N$.

Hint: Split the exact sequence $0 \to N \to M \to \pi M \to 0$, for $\pi : \mathbb{Z}^n \to P$.

EXAMPLE 3.29. The polynomial ring $\mathbb{R}[\omega]$ is a totally ordered $\mathbb{R}$-algebra if by definition $f < g$ when the leading term of $f - g$ is negative. We think of ω as being infinity. We can define an *extended metric*, generalizing metrics, where the distance function can take values in $\mathbb{R}[\omega]$. For example the space $\mathbb{Z} \times \mathbb{R}$, the disjoint union of $\#\mathbb{Z}$ lines, admits an extended metric where $d((i,x),(j,y)) = |x-y| + |i-j|\cdot\omega$. If $\varphi : \mathbb{Z}^n \to \mathbb{Z}^m$ is a linear map, then $\langle x, x\rangle = \|x\|^2 + \|\varphi(x)\|^2 \cdot \omega$, where $\|-\|$ denotes the usual norm on $\mathbb{Z}^n$ and $\mathbb{Z}^m$, defines an 'extended inner product' on $\mathbb{Z}^n$, and $\ker(\varphi) = \{x \in \mathbb{Z}^n \mid q(x) < \omega\}$. Morally we use this concept in the following algorithm.

THEOREM 3.30 (Kernel-image algorithm, Section 14 of [**21**]). *There exists a polynomial-time algorithm that, given $m, n \in \mathbb{Z}_{\geq 0}$ and a linear map $\varphi : \mathbb{Z}^n \to \mathbb{Z}^m$, computes the rank r of φ and injective linear maps $\kappa : \mathbb{Z}^{n-r} \to \mathbb{Z}^n$ and $\iota : \mathbb{Z}^r \to \mathbb{Z}^n$ such that* $\operatorname{im}(\kappa) = \ker(\varphi)$ *and* $\operatorname{im}(\varphi \circ \iota) = \operatorname{im}(\varphi)$ *and the induced map $\mathbb{Z}^{n-r} \times \mathbb{Z}^r \to \mathbb{Z}^n$ is an isomorphism.*

PROOF. Write B for the largest absolute value of a coefficient of the matrix F defining φ. Compute

$$\omega = 2^{n-1} \cdot n^{n+1} \cdot B^{2n} + 1$$

and note that its length is polynomially bounded by the size of the input. Hence we can consider the lattice $\Lambda = \mathbb{Z}^n$ together with the bilinear form induced by $q(x) = \|x\|^2 + \|\varphi(x)\|^2 \cdot \omega$, where $\|-\|$ denotes the usual norm on $\mathbb{Z}^n$ and $\mathbb{Z}^m$. Using the LLL-algorithm, compute a 2-reduced basis $(b_1, \ldots, b_n)$ of Λ. We will show that this basis has the following properties:

(a) $q(b_i) < \omega$ for $0 < i \leq n - r$;
(b) $(b_1, \ldots, b_{n-r})$ is a basis for $\ker(\varphi)$;
(c) $q(b_i) \geq \omega$ for $n - r < i \leq n$;
(d) $(\varphi(b_{n-r+1}), \ldots, \varphi(b_n))$ is a basis for $\operatorname{im}(\varphi)$.

Then, we may compute r as the maximal $0 \leq r \leq n$ such that $q(b_{n+1-r}) \geq \omega$, and κ and ι as the natural maps with images $\langle b_1, \ldots, b_{n-r}\rangle$ and $\langle b_{n-r+1}, \ldots, b_n\rangle$ respectively.

CLAIM. For all $0 < i \leq n - r$ we have $\lambda_i(\Lambda) \leq n^{n+1}B^{2n}$.

PROOF. By Cramer's rule (Theorem 3.27), we may take linearly independent vectors $a_1, \ldots, a_{n-r} \in \ker(\varphi)$ for which the coefficients are determinants of $r \times r$ submatrices of F. Then by Hadamard's inequality (Exercise 3.13), each coefficient is bounded in absolute value by $n^{n/2}B^n$. Hence $q(a_i) = \|a_i\|^2 \leq n^{n+1}B^{2n}$ for all $0 < i \leq n - r$. The claim now follows from the independence of the a_i. ■

From Proposition 3.22 and the claim it follows that

$$q(b_i) \leq 2^{n-1} \cdot \lambda_i(\Lambda) \leq 2^{n-1} \cdot n^{n+1} \cdot B^{2n} < \omega$$

for all $0 < i \leq n - r$, proving (a). Clearly for all $x \in \Lambda$ such that $\omega > q(x) = \|x\|^2 + \omega\|\varphi(x)\|^2$ we have $\|\varphi(x)\|^2 = 0$ and thus $x \in \ker(\varphi)$. In particular, we have linearly independent $b_1, \ldots, b_{n-r} \in \ker(\varphi)$. From Exercise 3.28 we may conclude it is in fact a basis for $\ker(\varphi)$, proving (b). It follows from (b) that $b_i \notin \ker(\varphi)$ and thus $q(b_i) \geq \omega$ for all $n - r < i \leq n$, proving (c). Lastly, (d) follows from (b) and the homomorphism theorem. □

Note that in the proof of Theorem 3.30 we could have constructed a c-reduced basis for values of c other than 2. Moreover, the exact value of ω is not important, as long as it is sufficiently large while still being computable in polynomial time. Later we will encounter a version of the kernel-image algorithm for general finitely generated abelian groups (Exercise 3.41).

EXERCISE 3.31 (Order, cf. Proposition 2.6.1 in [**9**]). Let $n \in \mathbb{Z}_{\geq 0}$.

a. Let $M \in \mathbb{Z}^{n\times n}$ be an integer matrix. Show that there exists a $T \in \mathbb{Z}^{n\times n}$ such that $\det(T) = 1$ and MT is upper triangular.

b. Show that a linear map $\varphi : \mathbb{Z}^n \to \mathbb{Z}^n$ satisfies $\#\operatorname{coker}(\varphi) = |\det(\varphi)|$ if $\det(\varphi) \neq 0$.

c. Show that there exist a polynomial-time algorithm that, given a finitely generated abelian group A, decides whether A is finite and if so computes $\#A$.
Note: The linear map representing A need not be injective.

EXERCISE 3.32 (Cf. Theorem 2.1). Show that there exists a polynomial-time algorithm that, given $\alpha_1, \ldots, \alpha_m \in \mathbb{Q}^*$, computes the kernel of the map $\mathbb{Z}^m \to \mathbb{Q}^*$ that is given by $(k_1, \ldots, k_m) \mapsto \prod_i \alpha_i^{k_i}$.
Hint: Compute a coprime basis.

3.4. Applications of the kernel-image algorithm. In this subsection we provide a polynomial-time algorithm for most problems listed in the beginning of this section. The algorithms in the coming sections, and many more, can be found in Chapter 2 of [**9**]. An immediate consequence of the kernel-image algorithm is the following.

COROLLARY 3.33. *There exists a polynomial-time algorithm that, given a linear map $\varphi : A \to B$ of finitely generated free abelian groups, decides whether φ is injective/surjective.*

PROOF. For injectivity we check whether $\ker(\varphi) = 0$ and for surjectivity we may check using Exercise 3.31.c whether $\operatorname{coker}(\varphi) = 0$. ☐

PROPOSITION 3.34. *There exists a polynomial-time algorithm that, given linear maps $\varphi : A \to C$ and $\psi : B \to C$ of finitely generated free abelian groups, decides whether* $\operatorname{im}(\psi) \subseteq \operatorname{im}(\varphi)$.

PROOF. Using Theorem 3.30 compute the kernel $\kappa : K \to A \times B$ of $A \times B \to C$ as in the following diagram.

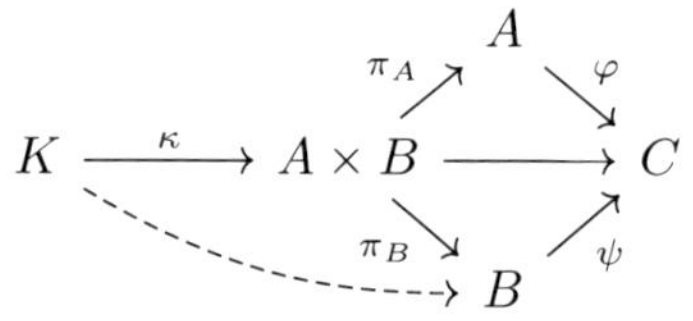

The image of $\pi_B \circ \kappa$ is precisely the set of elements $b \in B$ for which there exist $a \in A$ such that $\varphi(a) = \psi(b)$. Hence it suffices to decide using Corollary 3.33 whether $\pi_B \circ \kappa$ is surjective. ☐

COROLLARY 3.35. *There exists a polynomial-time algorithm that, given finitely generated abelian groups A and B, represented by linear maps $\alpha : A_0 \to A_1$ and $\beta : B_0 \to B_1$ respectively, and a linear map $\varphi : A_1 \to B_1$, decides whether φ represents a morphism $f : A \to B$.*

PROOF. Recall φ represents a morphism precisely when $\operatorname{im}(\varphi \circ \alpha) \subseteq \operatorname{im}(\beta)$. ☐

COROLLARY 3.36. *There exists a polynomial-time algorithm that, given finitely generated abelian groups A and B and morphisms $f, g : A \to B$, decides whether $f = g$.*

PROOF. Considering $h = f - g$ it suffices to be able to decide whether a morphism is zero. With η the matrix representing h and β the matrix representing B, we have $h = 0$ precisely when $\mathrm{im}(\eta) \subseteq \mathrm{im}(\beta)$. □

PROPOSITION 3.37 (Injectivity/surjectivity, Proposition 2.3.6 and Proposition 2.3.7. in [**9**]). *There exists a polynomial-time algorithm that, given a morphism $f : A \to B$ of finitely generated abelian groups, decides whether f is injective/surjective.*

PROOF. Write $\alpha : A_0 \to A_1$ and $\beta : B_0 \to B_1$ for the representatives of A and B respectively and φ for the representative of f. Note that f is surjective if and only if $B = \mathrm{im}(f) = \mathrm{im}(\varphi)/\mathrm{im}(\beta)$ if and only if $B_1 = \mathrm{im}(\varphi) + \mathrm{im}(\beta)$. It suffices to decide whether the map $A_1 \times B_0 \to B_1$ induced by φ and β is surjective, for which we have Corollary 3.33. Note that f is injective if and only if $\ker(\varphi) \subseteq \mathrm{im}(\alpha)$. Using Theorem 3.30 we compute a linear map $\kappa : K \to A_1$ with $\mathrm{im}(\kappa) = \ker(\varphi)$, and apply Proposition 3.34 to decide $\mathrm{im}(\kappa) \subseteq \mathrm{im}(\alpha)$. □

PROPOSITION 3.38. *There exists a polynomial-time algorithm that, given a linear map $\varphi : A \to B$ of free abelian groups A and B and $b \in B$, decides whether an element $a \in A$ exists such that $\varphi(a) = b$, and if so computes one.*

PROOF. Consider the linear map $\psi : A \times \mathbb{Z} \to B$ that sends (a, x) to $\varphi(a) + xb$.

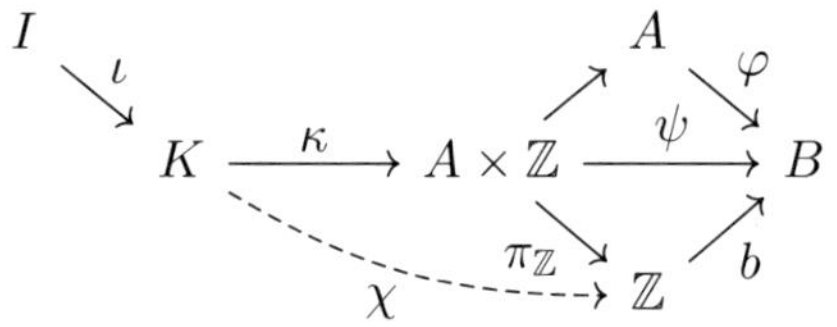

Compute using Theorem 3.30 the kernel $\kappa : K \to A \times \mathbb{Z}$ of ψ and in turn $\iota : I \to K$ a preimage of $\chi = \pi_{\mathbb{Z}} \circ \kappa$. Note that $\varphi(a) = b$ has a solution precisely when -1 is in the image of χ. Moreover, if we find $k \in K$ such that $\chi(k) = -1$, then its image under $K \to A \times \mathbb{Z} \to A$ gives an element $a \in A$ such that $\varphi(a) = b$. Note that $I \subseteq \mathbb{Z}$. If $I = 0$, then no solution to $\chi(k) = -1$ exists, and otherwise $k \in \iota(\{\pm 1\})$ gives a solution if it exists. □

COROLLARY 3.39 (Preimages). *There exists a polynomial-time algorithm that, given a homomorphism $f : A \to B$ of finitely generated abelian groups A and B and $b \in B$, decides whether an element $a \in A$ exists such that $f(a) = b$, and if so computes one.*

PROOF. Let $\varphi : A_1 \to B_1$ be the representative of f and $b_1 \in B_1$ of b. Then there exists some $a \in A$ such that $f(a) = b$ if and only if there exists some $a_1 \in A_1$ and $b_0 \in B_0$ such that $\varphi(a_1) = b_1 + \beta(b_0)$. If we consider again the induced map $A_1 \times B_0 \to B_1$, then it suffices to compute an inverse image of b_1 using Proposition 3.38. □

EXERCISE 3.40. Give a direct proof of Proposition 3.34 or Proposition 3.38 by giving an algorithm that applies the LLL-algorithm only once, similar to Theorem 3.30.

EXERCISE 3.41 (Kernels and images, cf. Proposition 2.3.15 in [**9**]).

a. Show that there exists a polynomial-time algorithm that, given morphisms $\delta : D_0 \to D_1$ and $\chi : C_1 \to D_1$ of finitely generated free abelian groups, computes a finitely generated free abelian group C_0 and an injection $\gamma : C_0 \to C_1$ such that $\mathrm{im}(\delta) = \mathrm{im}(\chi \circ \gamma)$.

b. Show that there exists a polynomial-time algorithm that, given a morphism $f : A \to B$ of finitely generated abelian groups, computes injective morphisms $k : K \to A$ and $i : I \to B$ such that $\mathrm{im}(k) = \ker(f)$ and $\mathrm{im}(i) = \mathrm{im}(f)$.
Hint: Take $(\delta_i, \chi_i) = (\beta, \varphi)$ and $(\delta_k, \chi_k) = (\alpha, \gamma_i)$.

3.5. Homomorphism groups of finitely generated abelian groups. Although we can algorithmically work with individual morphisms $f : A \to B$ of finitely generated abelian groups, we have yet to treat the group of homomorphisms $\mathrm{Hom}(A, B)$ as a whole. Recall that given representations $\alpha : A_0 \to \mathbb{Z}^a$ and $\beta : B_0 \to \mathbb{Z}^b$ of A and B respectively, where A_0 and B_0 are free, we represent a morphism $A \to B$ as a linear map $\mathbb{Z}^a \to \mathbb{Z}^b$ or, equivalently, as a matrix in $\mathbb{Z}^{b\times a}$.

DEFINITION 3.42. Let $\alpha : A_0 \to \mathbb{Z}^a$ and $\beta : B_0 \to \mathbb{Z}^b$ be representations of finitely generated abelian groups A and B respectively. We write

$$\mathcal{H}_1(A, B) = \{\varphi \in \mathbb{Z}^{b\times a} : \mathrm{im}(\varphi \circ \alpha) \subseteq \mathrm{im}(\beta)\} \quad \text{and}$$
$$\mathcal{H}_0(A, B) = \{\varphi \in \mathbb{Z}^{b\times a} : \mathrm{im}(\varphi) \subseteq \mathrm{im}(\beta)\}$$

for the group of matrices that represent a morphism $A \to B$ and the group of matrices that represent a zero-morphism $A \to B$ respectively, so that $\mathrm{Hom}(A, B) \cong \mathcal{H}_1(A, B)/\mathcal{H}_0(A, B)$. A *homomorphism group of A and B* is a pair (H, e) where H is a finitely generated abelian group, represented by say a map $H_0 \to H_1$, and $e : H_1 \to \mathbb{Z}^{b\times a}$ is a linear map, such that $\mathrm{im}(e) = \mathcal{H}_1(A, B)$ and such that e induces an isomorphism $H \to \mathcal{H}_1(A, B)/\mathcal{H}_0(A, B)$.

This definition is somewhat more formal than what we would like to work with. For example, A and B do not have a unique homomorphism group in the sense of the above definition, even though we would like to reason about it as being equal to the group $\mathrm{Hom}(A, B)$. Effectively, a homomorphism group of A and B is just a finitely generated abelian group H isomorphic to $\mathrm{Hom}(A, B)$ together with a means of evaluation e. We motivate being somewhat lax with the definition by Exercise 3.46.

THEOREM 3.43 (Hom, Proposition 2.4.1.2 in [**9**]). *There exists a polynomial-time algorithm that, given finitely generated abelian groups A and B, computes* $\mathrm{Hom}(A, B)$.

PROOF. Consider the case where A is a free abelian group and B is a finitely generated abelian group represented by $\beta : B_0 \to B_1$. We may compute the matrix $\beta_* : \mathrm{Hom}(A, B_0) \to \mathrm{Hom}(A, B_1)$ given by $f \mapsto \beta \circ f$. Since A is free we get an exact functor $\mathrm{Hom}(A, \_)$ such that

$$\left[\ B_0 \xrightarrow{\beta} B_1 \to B \to 0\ \right] \Rightarrow \left[\ \mathrm{Hom}(A, B_0) \xrightarrow{\beta_*} \mathrm{Hom}(A, B_1) \to \mathrm{Hom}(A, B) \to 0\ \right],$$

and thus $\mathrm{Hom}(A, B) \cong \mathrm{coker}(\beta_*)$. Note that $\mathrm{Hom}(A, B_0)$ and $\mathrm{Hom}(A, B_1)$ are free, so β_* represents the group $\mathrm{Hom}(A, B)$. Evaluating an element of $\mathrm{Hom}(A, B)$ in A reduces to evaluating an element of $\mathrm{Hom}(A, B_1)$ in A, which is just matrix multiplication. Hence we easily obtain the evaluation map.

Now consider the general case where A and B are general finitely generated abelian groups represented by $\alpha : A_0 \to A_1$ and β respectively. We note that the functor $\mathrm{Hom}(\_, B)$ is left-exact and contravariant. Applied to the exact sequence of A we get

$$\left[\, A_0 \xrightarrow{\alpha} A_1 \to A \to 0 \,\right] \Rightarrow \left[\, 0 \to \mathrm{Hom}(A, B) \to \mathrm{Hom}(A_1, B) \xrightarrow{\alpha^*} \mathrm{Hom}(A_0, B) \,\right].$$

Hence $\mathrm{Hom}(A, B) \cong \ker(\alpha^*)$. By the previous case we may compute $\mathrm{Hom}(A_1, B)$ and $\mathrm{Hom}(A_0, B)$ since A_1 and A_0 are free. It is not difficult to show we may then compute α^* and in turn its kernel using Exercise 3.41. Evaluation in A of elements in $\mathrm{Hom}(A, B)$ reduces to evaluation in A_1 of elements in $\mathrm{Hom}(A_1, B)$, which we may also do by the previous case. □

EXERCISE 3.44 (Exponent, Theorem 2.6.9 in [**9**]). Show that there exists a polynomial-time algorithm that, given a finitely generated abelian group A

a. and an element $a \in A$, decides whether a is torsion and if so computes $\mathrm{ord}(a)$;

b. decides whether A is finite and if so computes its exponent $\exp(A)$ and an $a \in A$ with $\mathrm{ord}(a) = \exp(A)$.

EXERCISE 3.45 (Exact sequences, cf. Proposition 2.5.1 in [**9**]). Show that there exists a polynomial-time algorithm that, given morphisms $f : A \to B$ and $g : B \to C$ of finitely generated abelian groups,

a. decides whether the sequence $0 \to A \xrightarrow{f} B \xrightarrow{g} C \to 0$ is exact;

b. if so, decides whether the sequence is split;

c. if so, produces a left-inverse of f and a right-inverse of g.

Hint: Consider the map $g_* : \mathrm{Hom}(C, B) \to \mathrm{Hom}(C, C)$.

Note that by taking $C = 0$ in Exercise 3.45 we may conclude that there exists a polynomial-time algorithm that, given a morphism of finitely generated abelian groups, decides whether it is an isomorphism and if so computes its inverse.

EXERCISE 3.46. Show that there exists a polynomial-time algorithm that, given finitely generated abelian groups A and B and two homomorphism groups (H, e) and (H', e') of A and B, computes an isomorphism $f : H \to H'$ that respects evaluation, i.e. so that for all $h \in H$ we have $e(h) \equiv e'(f(h)) \pmod{\mathcal{H}_0(A, B)}$.

EXERCISE 3.47 (Tensor, Proposition 2.4.1.1 in [**9**]). We naturally encode a *bilinear map* $A \times B \to C$ of finitely generated abelian groups as a map $A \to \mathrm{Hom}(B, C)$. Show that there exists a polynomial-time algorithm that, given finitely generated abelian groups A and B, computes a group T isomorphic to $A \otimes B$ as well as a bilinear map $A \times B \to T$ that induces the natural bilinear map $\otimes : A \times B \to A \otimes B$.

EXERCISE 3.48. Formulate and prove statements analogous to Exercise 3.47 for the bifunctors Tor_i and Ext_i.

3.6. Structure theorem for finitely generated abelian groups. The structure theorem for finitely generated abelian groups is the following.

THEOREM 3.49. *Suppose A is a finitely generated abelian group. Then there exists a unique sequence $(r, m, n_1, n_2, \ldots, n_m)$ of integers with $r, m \geq 0$ and $n_1, \ldots, n_m > 1$ such that $n_m \mid \cdots \mid n_2 \mid n_1$, for which*

$$A \cong \mathbb{Z}^r \times \prod_{k=1}^{m} (\mathbb{Z}/n_k\mathbb{Z}).$$

In this section we will prove its algorithmic counterpart, which also implicitly proves the structure theorem itself.

EXERCISE 3.50. Let $n > 0$ and $M \subseteq \mathbb{Q}^n$ a subgroup. We equip $\mathbb{Q}^n$ with the standard inner product. For subgroups $H \subseteq M$ write $H^\perp = \{x \in M \mid \langle x, H\rangle = 0\}$. Show that for all subgroups $N \subseteq M$ we have $(N^\perp)^\perp = (\mathbb{Q}N) \cap M$.

Hint: First consider the case where M and N are $\mathbb{Q}$-vector spaces.

LEMMA 3.51 (Torsion, cf. Theorem 2.6.2 in [**9**]). *There exists a polynomial-time algorithm that, given a finitely generated abelian group A, computes its torsion subgroup.*

PROOF. For a finitely generated abelian group H and map $h : H \to \mathbb{Z}^n$ we write $h^\perp : H^\perp \to \mathbb{Z}^n$ for (an encoding of) a kernel of the map $\mathbb{Z}^n \to \operatorname{Hom}(H, \mathbb{Z})$ given by $x \mapsto (y \mapsto \langle x, h(y)\rangle)$. Note that using Theorem 3.30 we may compute $h^\perp$ in polynomial time. In particular, we may compute $(\alpha^\perp)^\perp : T \to A_1$ for the representative $\alpha : A_0 \to A_1$ of A. It follows from Exercise 3.50 that $(\alpha^\perp)^\perp$ represents the torsion subgroup of A: Its image is precisely the set of those elements of A_1 for which a positive integer multiple is in $\alpha(A_0)$. □

THEOREM 3.52 (Structure theorem, Theorem 2.6.10 in [**9**]). *There exists a polynomial-time algorithm that, given a finitely generated abelian group A, computes integers r, m, $n_1, \ldots, n_m$ with $r, m \geq 0$ and $n_1, \ldots, n_m > 1$ such that $n_m \mid \cdots \mid n_1$ and computes for A an isomorphism*

$$A \cong \mathbb{Z}^r \times \prod_{k=1}^{m} (\mathbb{Z}/n_k\mathbb{Z}),$$

i.e. projections to and inclusions from the individual factors on the right hand side.

PROOF. We may compute using Lemma 3.51 the torsion subgroup T of A. Using the image algorithm of Exercise 3.41 we may compute an isomorphism $\mathbb{Z}^r \to A/T$. We have an exact sequence $0 \to T \to A \to A/T \to 0$ which splits, hence by Exercise 3.45 we may compute maps $A \to T$ and $A/T \to A$ inducing an isomorphism $A \cong T \times (A/T) \cong T \times \mathbb{Z}^r$. Replacing A by T we may now assume A is torsion. If $A = 0$ we are done. Using Exercise 3.44 we may compute an element $a \in A$ with order equal to the exponent e of A. Again we have an exact sequence $0 \to \mathbb{Z}a \to A \to A/(\mathbb{Z}a) \to 0$ which is split to which we apply Exercise 3.45. We proceed recursively with A replaced by $A/(\mathbb{Z}a)$. Note that the exponent of $A/(\mathbb{Z}a)$ is a divisor of the exponent of A, so indeed we will get $n_m \mid \cdots \mid n_1$. □

COROLLARY 3.53. *There exists a polynomial-time algorithm that, given a finitely generated abelian group A and a finite set S of positive integers, computes $r, m \in \mathbb{Z}_{\geq 0}$ and $c_1, \ldots, c_m \in \mathbb{Z}_{>1}$ and $t_1, \ldots, t_m \in \mathbb{Z}_{>0}$ such that any two c_i are either coprime or powers of the same integer, and every c_i either divides some power of an element of S or is coprime to all elements of S, and computes for A an isomorphism*

$$A \cong \mathbb{Z}^r \times \prod_{k=1}^{m} (\mathbb{Z}/c_k\mathbb{Z})^{t_k}.$$

PROOF. Apply Theorem 3.52 and compute a coprime basis from $S \cup \{n_1, \ldots, n_m\}$ using Theorem 2.12. This algorithm also expresses each n_i in terms of this basis, and we then proceed as in Theorem 3.52. □

EXERCISE 3.54. For a ring R and R-module M we say M is *projective* if there exists some R-module N such that $M \oplus N$ is a free R-module. Show for $n \in \mathbb{Z}_{>0}$ that n is square-free if and only if all $\mathbb{Z}/n\mathbb{Z}$-modules are projective.

EXERCISE 3.55. We define for finite sets of positive integers the function

$$\operatorname{rex}\Big\{\prod_p p^{s_{1,p}}, \dots, \prod_p p^{s_{m,p}}\Big\} = \prod_p p^{\gcd\{s_{1,p},\dots,s_{m,p}\}}.$$

a. Show that there exists a polynomial-time algorithm that computes rex.

Note: The inputs are just integers, and $\gcd\{\} = \gcd\{0\} = 0$.

For a finite abelian group $A \cong \prod_{k=1}^{m}(\mathbb{Z}/a_k\mathbb{Z})$ write $\operatorname{rex}(A) = \operatorname{rex}(a_1, \dots, a_m)$.

b. Show that $\operatorname{rex}(A)$ is well-defined (i.e. does not depend on the choice of isomorphism or a_i) and can be computed in polynomial time.

c. Show for $r = \operatorname{rex}(A)$ that $r^iA/r^{i+1}A$ is a projective $\mathbb{Z}/r\mathbb{Z}$-module for all $i \geq 0$, and when $A \neq 0$ that r is the unique maximal integer with this property.

3.7. Approximating lattices. This section is adapted from [**11**]. Let $\Lambda \subseteq \mathbb{R}^n$ be a non-zero lattice with generators $v_1, \dots, v_s$. Note that such vectors can have irrational coordinates, and thus we generally cannot encode this Λ. Nonetheless, we would like to compute the relations

$$M = \Big\{(x_1, \dots, x_s) \in \mathbb{Z}^s : \sum_{i=1}^{s} x_i v_i = 0\Big\}$$

between the generators when only an approximation of them is available. The challenge is to decide the required precision, which largely depends on a lower bound on $\lambda_1(\Lambda)$ (see Definition 3.21). Namely, we need to be able to decide whether an estimation that is close to zero is in fact zero.

For $t \in \mathbb{Z}_{>0}$, a *t-approximation* of $(z_1, \dots, z_n) \in \mathbb{R}^n$ is an element $(x_1, \dots, x_n) \in t^{-1}\mathbb{Z}^n$ such that $|z_j - x_j| \leq t^{-1}$ for all j. Suppose for some t that we have for all i a t-approximation w_i of v_i. We equip $\mathbb{Z}^s$ with a lattice structure given by

$$q(x) = \|x\|^2 + \Big\|\sum_{i=1}^{s} x_i w_i\Big\|^2 \cdot t^2 \in \mathbb{Z},$$

where $\|-\|$ denotes the standard norm on $\mathbb{R}^s$ and $\mathbb{R}^n$.

EXERCISE 3.56. Show that for all $x \in M$ we have $q(x) \leq (ns+1)\|x\|^2 \leq 2ns\|x\|^2$.

PROPOSITION 3.57. *Let $B = \max_i \|v_i\|$ and $\lambda = \sqrt{\lambda_1(\Lambda)}$. If M has rank k, then there exist linearly independent $u_1, \dots, u_k \in M$ such that*

$$\|u_i\|^2 \leq (n+1)^{2n+1}\left(\frac{B}{\lambda}\right)^{2n} \leq 2^{2n+1}n^{2n+1}\left(\frac{B}{\lambda}\right)^{2n}.$$

PROOF. We may assume without loss of generality that Λ is full-rank. Reorder the v_i such that $v_1, \dots, v_n$ are linearly independent. Let

$$A = \left\lfloor \left(\frac{(n+1)B}{\lambda}\right)^n \right\rfloor$$

and consider the set $S = \{-\frac{1}{2}A, -\frac{1}{2}A+1, \ldots, \frac{1}{2}A\}$ of cardinality $A+1$. Fix some $i > n$ and suppose that the map $\varphi : S^{n+1} \to \mathbb{R}^n$ given by

$$(x_1, \ldots, x_r, y) \mapsto \sum_{j=1}^{n} x_j v_j + y v_i$$

is injective. Any two images differ by a lattice vector and thus have distance at least λ. If we draw a sphere of radius $\lambda/2$ around each image, which are pair-wise disjoint, we obtain a body X with volume $(A+1)^{n+1}(\lambda/2)^n V$, where V is the volume of the unit sphere in $\mathbb{R}^n$. Note that each image is bounded in length by $\frac{1}{2}(n+1)AB$, so X is contained in the sphere of radius $\frac{1}{2}(n+1)AB + \frac{1}{2}\lambda$ around the origin. If we compare the volumes, we obtain

$$(A+1)^{n+1}(\tfrac{1}{2}\lambda)^n V \le (\tfrac{1}{2}(n+1)AB + \tfrac{1}{2}\lambda)^n V.$$

Rearranging the inequality gives

$$A+1 \le \left(\frac{(n+1)AB+\lambda}{(A+1)\lambda}\right)^n \le \left(\frac{(n+1)AB}{A\lambda}\right)^n = \left(\frac{(n+1)B}{\lambda}\right)^n,$$

which contradicts the definition of A. Hence there exist distinct $s, s' \in S^{n+1}$ with the same image under φ. If we consider their difference, we obtain $x_{i1}, \ldots, x_{in}, y_i \in \mathbb{Z}$ with $y_i \ne 0$ and $|y_i| \le A$ and $|x_{ij}| \le A$ for all $1 \le j \le n$ so that

$$\sum_{j=1}^{n} x_{ij} v_j = y_i v_i.$$

As we have this for all $i > n$ we may consider the vectors

$$\begin{aligned}
u_1 &= (x_{n+1,1}, & x_{n+1,2}, & \ \ldots, & x_{n+1,n}, & -y_{n+1}, & 0, & \ldots, & 0),\\
u_2 &= (x_{n+2,1}, & x_{n+2,2}, & \ \ldots, & x_{n+2,n}, & 0, & -y_{n+2}, & \ldots, & 0),\\
&\vdots\\
u_k &= (x_{s,1}, & x_{s,2}, & \ \ldots, & x_{s,n}, & 0, & 0, & \ldots, & -y_s).
\end{aligned}$$

Clearly they are linearly independent elements of M, and each satisfies $\|u_i\|^2 \le (n+1)A^2$. □

LEMMA 3.58. *Let $\lambda = \sqrt{\lambda_1(\Lambda)}$. Let $\omega, t \in \mathbb{Z}_{\ge 1}$ be such that*

$$t \ge (ns+1) \cdot \frac{\sqrt{\omega}}{\lambda}$$

If $x \in \mathbb{Z}^s$ satisfies $q(x) \le \omega$, then $x \in M$.

PROOF. Suppose $q(x) \le \omega$ and thus also $\|x\|^2 \le \omega$. If $x \notin M$, then

$$\begin{aligned}
\sqrt{q(x)} &> \Big\|\sum_{i=1}^{s} x_i w_i\Big\| \cdot t \ge \left(\Big\|\sum_{i=1}^{s} x_i v_i\Big\| - \Big\|\sum_{i=1}^{s} x_i (w_i - v_i)\Big\|\right) \cdot t\\
&\ge \left(\lambda - \|x\| \cdot \sum_{i=1}^{s} \|w_i - v_i\|\right) \cdot t \ge \left(\lambda - \sqrt{\omega} \cdot \left(snt^{-1}\right)\right) \cdot t\\
&= t\lambda - sn\sqrt{\omega} \ge \sqrt{\omega},
\end{aligned}$$

which is a contradiction. □

THEOREM 3.59 (Ge [**11**]). *There exists a polynomial-time algorithm that, given $t, \omega \in \mathbb{Z}_{\geq 1}$ and t-approximations $w_1, \ldots, w_s$ of vectors $v_1, \ldots, v_s \in \mathbb{R}^n$ that generate a non-zero lattice $\Lambda = \langle v_1, \ldots, v_s \rangle$ with*

$$\omega \geq 2^{2n+s+1} n^{2n+2} s \left(\frac{B}{\lambda} \right)^{2n} \quad \text{and} \quad t \geq 2ns \frac{\sqrt{\omega}}{\lambda},$$

where $B = \max_i \|v_i\|$ and $\lambda = \sqrt{\lambda_1(\Lambda)}$, computes

$$M = \Big\{ (x_1, \ldots, x_s) \in \mathbb{Z}^s : \sum_{i=1}^{s} x_i v_i = 0 \Big\}.$$

PROOF. It suffices to show that if $b_1, \ldots, b_s$ is a 2-reduced basis of $\mathbb{Z}^s$, then $\{b_i : q(b_i) \leq \omega\}$ is a basis for M, since we then can compute this basis with the LLL-algorithm. Let k be the rank of M and let $u_1, \ldots, u_k$ be as in Proposition 3.57. Then

$$q(b_i) \leq 2^{s-1} \max\{q(u_1), \ldots, q(u_k)\} \leq 2^s ns \max\{\|u_1\|^2, \ldots, \|u_k\|^2\} \leq \omega$$

for all $1 \leq i \leq k$, where the first inequality is Exercise 3.26, the second is Exercise 3.56 and the third is Proposition 3.57. Thus $b_i \in M$ by Lemma 3.58. Hence $\langle b_1, \ldots, b_k \rangle \subseteq M$ and both have the same rank, so by Exercise 3.28 they are equal. Then by Lemma 3.58 we cannot have $q(b_i) \leq \omega$ for $i > k$, so indeed $\{b_i : q(b_i) \leq \omega\}$ is a basis. □

4. Fractional ideals

In this section we move from finitely generated abelian groups to orders (see Definition 1.5). Orders in general do not have a unique prime factorization theorem like the integers do. The best we can hope for is unique prime *ideal* factorization. This observation is at the heart of the study of ideals in number rings (see Section 4.1), where we treat the ideals as analogues to elements of the number ring. In this section we will define fractional ideals, which in this analogy correspond to the elements of the ambient number field. For an ideal to factor into primes, we require the prime divisors to be *invertible*, and we will proceed to describe an algorithm to 'make ideals invertible'. As in Section 2, prime factorization of ideals is algorithmically infeasible, so we generalize the coprime basis algorithm to ideals.

DEFINITION 4.1. Let S be a commutative ring. For (additive) subgroups $I, J \subseteq S$ we write

$$I + J = \{x + y : x \in I,\ y \in J\}$$

$$I \cdot J = IJ = \Big\{ \sum_{i=1}^{n} x_i y_i : n \geq 0,\ x_1, \ldots, x_n \in I,\ y_1, \ldots, y_n \in J \Big\}$$

$$I : J = (I : J)_S = \big\{ x \in S : xJ \subseteq I \big\}.$$

We also inductively define $I^0 = (I : I)_S$ and $I^{n+1} = I^n \cdot I$ for $n \geq 0$.

Note that if $I, J \subseteq S$ are subgroups, so are $I + J$, $I \cdot J$ and $I : J$.

EXAMPLE 4.2. For $S = \mathbb{Q}$ the finitely generated subgroups are those of the form $x\mathbb{Z}$ for $x \in \mathbb{Q}$ (Exercise 4.10.a). For $x, y \in \mathbb{Q}$ we have $(x\mathbb{Z}) : (y\mathbb{Z}) = (x/y)\mathbb{Z}$ if $y \neq 0$ and $(x\mathbb{Z}) : (y\mathbb{Z}) = \mathbb{Q}$ otherwise.

LEMMA 4.3. *Let $S \subseteq A$ be commutative (sub)rings and $H, I, J \subseteq S$ subgroups. Then:*

(i) $I : J \subseteq (HI) : (HJ)$;
(ii) $I^0 = I : I$ *is a subring of* S, *the* multiplier ring *of* I;
(iii) $I^1 = I$;
(iv) $I : I = I$ *if and only if* I *is a subring of* S;
(v) *if* I *is finitely generated and* $J \cap S^* \neq \emptyset$, *then* $I : J$ *is finitely generated;*
(vi) *if* $J \cap S^* \neq \emptyset$, *then* $(I : J)_A = (I : J)_S$.

PROOF. Both (i) and (ii) follow directly from the definition.

(iii) Since $1 \in I : I$ by (ii), we have $I \subseteq (I : I) \cdot I$, while $(I : I) \cdot I \subseteq I$ by definition of division. Thus $I = (I : I) \cdot I = I^1$.

(iv) The implication ($\Rightarrow$) follows from (ii). For ($\Leftarrow$), suppose $I \subseteq S$ is a subring. Then $I \cdot I \subseteq I$ and consequently $I \subseteq I : I$. For $x \in I : I$ we have $xI \subseteq I$, so in particular $x = x \cdot 1 \in I$ and thus $I : I \subseteq I$.

(v) Let $j \in J$ be invertible. Then $j^{-1}I$ is finitely generated and hence a Noetherian $\mathbb{Z}$-module. Thus $I : J \subseteq j^{-1}I$ is finitely generated.

(vi) Clearly $(I : J)_S \subseteq (I : J)_A$ because $S \subseteq A$. Choose $j \in J \cap S^*$. If $x \in (I : J)_A$, then $x \in j^{-1}I \subseteq S$, so $x \in (I : J)_S$. ☐

DEFINITION 4.4. Let $R \subseteq S$ be commutative (sub)rings. An *invertible ideal of R within S* is an R-submodule I of S for which there exists an R-submodule J of S such that $IJ = R$. We write $\mathcal{I}_S(R)$ for the group of invertible ideals of R within S. For an invertible ideal I of R within S we write $I^{-1} = (R : I)_S$. We write $\mathrm{Q}(R)$ for the *total ring of fractions* of R, which is obtained from R by localizing the set of regular elements (Definition 1.3). A *fractional ideal* of R is a finitely generated R-submodule I of $\mathrm{Q}(R)$ such that $I\mathrm{Q}(R) = \mathrm{Q}(R)$. An *invertible ideal* of R is an invertible ideal of R within $\mathrm{Q}(R)$, and we write $\mathcal{I}(R) = \mathcal{I}_{\mathrm{Q}(R)}(R)$. If we want to stress that something is an ideal, instead of a fractional ideal, we will call it an *integral* ideal.

In most applications $\mathrm{Q}(R)$ will take the role of S in Definition 4.1, in which case we often write $I : J$ for $(I : J)_S$, and $\mathrm{Q}(R)$ will be a field.

EXERCISE 4.5. Let $R \subseteq S$ be commutative (sub)rings and let $I \in \mathcal{I}_S(R)$. Show that I is a finitely generated R-module and that I^{-1} is the unique R-submodule of S such that $II^{-1} = R$.

EXERCISE 4.6. Let $R \subseteq S$ be commutative (sub)rings and let $I, J \subseteq S$ be R-submodules and H an invertible R ideal within S. Show that $(HI) : J = H(I : J)$, $I : (HJ) = H^{-1}(I : J)$ and $I : J = HI : HJ$.

EXERCISE 4.7. Let $K \subseteq L$ be a finite degree field extension and let $R \subseteq K$ be a subring with $\mathrm{Q}(R) = K$. Show for all R-submodules $I, J \subseteq L$ that $(KI : KJ)_L \supseteq K \cdot (I : J)_L$, with equality when J is finitely generated. Give an example where equality does not hold.

EXERCISE 4.8. Let S be a commutative ring and let $R \subseteq S$ be a subring and $H, I, J \subseteq S$ be R-submodules such that $H + I = J$. Does it follow that $H : J + I : J = R$? Which inclusions hold? What if S is a number field?

4.1. Number rings. Recall that a *number field* is a field K containing the field of rational numbers $\mathbb{Q}$ such that the dimension of K over $\mathbb{Q}$ as a vector space is finite. We say an additive subgroup I of a number field K is of *full rank* if $\mathbb{Q}I = K$. A *number ring* is a ring isomorphic to a subring of a number field. An *order* is a domain whose additive group is isomorphic to $\mathbb{Z}^n$ for some $n \in \mathbb{Z}_{>0}$. Every order R is a number ring of full rank in the number field $\mathrm{Q}(R) \cong \mathbb{Q} \otimes_{\mathbb{Z}} R$. Conversely, every number field comes with a natural order.

THEOREM 4.9 (Theorem I.1 and Theorem I.2 in [**17**]). *For every number field K there exists a unique order $\mathcal{O}_K$ in K which is maximal with respect to inclusion. It is also the unique full-rank order for which every fractional ideal is invertible.* □

EXERCISE 4.10. Let $n \in \mathbb{Z}_{\geq 0}$ and let R be a subring of a number field of degree n. Show that:

a. each finitely generated subgroup $G \subseteq \mathbb{Q}^n$ can be generated by n elements;
b. for all $a \in \mathbb{Z}_{>0}$ and subgroups $H \subseteq \mathbb{Q}^n$ we have $\#(H/aH) \leq a^n$;
Hint: Assume first that H is finitely generated.
c. for all non-zero ideals $I \subseteq R$ the ring R/I is finite (Theorem 2.11 in [**28**]);
d. the ring R is Noetherian (Corollary 2.12 in [**28**]);
e. every non-zero prime ideal of R is maximal (Corollary 2.12 in [**28**]).

EXERCISE 4.11. Let R be an order. Show that every non-zero integral ideal of R is a fractional ideal. Does this hold for arbitrary commutative rings R?

EXERCISE 4.12. Let R be an order. Show that the following are equivalent:

(1) R is equal to the maximal order of $\mathrm{Q}(R)$;
(2) the sum of every two invertible ideals is again invertible;
(3) every maximal ideal of R is invertible.

EXERCISE 4.13. Let R be a number ring and let $\mathrm{Q}(R) \subseteq L$ be number fields. Show that for every $I \in \mathcal{I}_L(R)$ there exist $J \in \mathcal{I}(R)$ and $\alpha \in L^*$ such that $I = \alpha J$.

EXERCISE 4.14. Give an order R and an ideal I of R such that $I(R : I) \neq I : I$.

EXERCISE 4.15. Let K be a number field and $I \subseteq K$ be a subgroup. Then $I^2 = I$ if and only if I is a ring.
Hint: If $I^2 = I$, then I is a finitely generated ideal of the ring $\mathbb{Z} + I$.

In our algorithms we encode an order by first specifying its rank n, and then writing down its $n \times n \times n$ multiplication table on the standard basis vectors. By distributivity this completely and uniquely defines a multiplication on the order. However, not every element of $\mathbb{Z}^{n\times n\times n}$ encodes a ring structure, because it can fail for example associativity of multiplication (Exercise 4.18). We encode a number field K simply by an order R such that $\mathrm{Q}(R) = K$. We encode a finitely generated subgroup of K, in particular a fractional ideal of R, as a finitely generated subgroup of R together with a denominator. As we will see in Theorem 5.1 it is difficult to compute the maximal order of a number field.

EXERCISE 4.16. Let $\mathbb{Q}^{\mathrm{alg}}$ be some algebraic closure of $\mathbb{Q}$.

a. Show that $\#\mathbb{Q}^{\mathrm{alg}} = \#\mathbb{N}$.
b. Show that there are precisely $\#\mathbb{N}$ number fields contained in $\mathbb{Q}^{\mathrm{alg}}$.
c. Show that there are precisely $\#\mathbb{R}$ subfields of $\mathbb{Q}^{\mathrm{alg}}$.
d. The same question as (b) and (c), but counting the fields up to isomorphism.

e. Do there exist $\#\mathbb{R}$ subfields of $\mathbb{Q}^{\text{alg}}$ that are pairwise isomorphic?
f. Let $K \neq \mathbb{Q}$ be a number field. Show that there are precisely $\#\mathbb{N}$ orders and $\#\mathbb{R}$ subrings in K with field of fractions K. What holds for $K = \mathbb{Q}$?
g. Argue why it is natural to restrict the input of our algorithms to orders and number fields as opposed to general number rings.

EXERCISE 4.17 (Proposition 5.8 in [**8**]). Show that there exist polynomial-time algorithms that, given a number field K and finitely generated subgroups $I, J \subseteq K$, compute $I + J$, $I \cap J$, $I \cdot J$ and $I : J$.

EXERCISE 4.18. You may use without proof that it is possible to factor polynomials in $\mathbb{Q}[X]$ in polynomial time using the LLL-algorithm. Show that there exists a polynomial-time algorithm that, given a cube integer matrix, determines whether it encodes an order.

Let $A \subseteq B$ be commutative rings such that B is free as an A-module with $\text{rk}_A(B) = n < \infty$ (see Definition 3.7). Then the ring $\text{End}_A(B)$ of A-module endomorphisms of B is isomorphic to $\text{Mat}_n(A)$, the ring of $n \times n$-matrices with coefficients in A. This isomorphism induces a determinant $\det : \text{End}_A(B) \to A$ and trace $\text{Tr} : \text{End}_A(B) \to A$ on $\text{End}_A(B)$ which are multiplicative and A-linear respectively. We have a natural map $B \to \text{End}_A(B)$ given by $b \mapsto (x \mapsto bx)$. It induces the A-module homomorphism $\text{Tr}_{B/A} : B \to A$ when composed with Tr, which we call the *trace of B over A*. For more on traces and determinants, see Section 7 of [**27**].

EXERCISE 4.19. Show that $\text{Tr} : \text{End}_A(B) \to A$ and $\det : \text{End}_A(B) \to A$ are independent of the chosen A-basis for B.

We define the *discriminant* of B over A as $\Delta_{B/A} = \det(M) \in A$, where M is the matrix $(\text{Tr}_{B/A}(e_i e_j))_{1 \leq i,j \leq n}$ and $(e_1, \ldots, e_n)$ is a basis for B over A. This definition depends on a choice of basis for B (see Exercise 4.20). However, for $A = \mathbb{Z}$ the discriminant is uniquely determined, and we write $\Delta(B) = \Delta_{B/\mathbb{Z}}$ in this case.

EXERCISE 4.20. Prove that $\Delta_{B/A}$ is defined up to multiples by $(A^*)^2$. Conclude for $A = \mathbb{Z}$ that $\Delta_{B/A}$ is independent of choice of basis.

EXERCISE 4.21 (7-3 in [**27**]). If $C \subseteq B$ is of finite index and both are free of finite rank over $\mathbb{Z}$, then $\Delta(C) = \#(B/C)^2 \cdot \Delta(B)$.

EXERCISE 4.22. Show that there exists a polynomial-time algorithm that, given an order R, computes $\Delta(R)$. Conclude that the lengths of $\Delta(R)$ and $\#(\mathcal{O}_{Q(R)}/R)$ are polynomially bounded in terms of the representation of the order R.

EXERCISE 4.23. Suppose $d \in \mathbb{Z}$ is non-zero.
a. Show that we can write $d = \square(d)^2 \cdot \boxtimes(d)$ uniquely with $\square(d) \in \mathbb{Z}_{>0}$ and $\boxtimes(d)$ square-free, the *square part* and *square-free part* of d respectively.
b. Show that if d is not a square, then $\Delta(\mathbb{Z}[\sqrt{d}]) = 4d$.

Suppose that d is square-free and let $K = \mathbb{Q}(\sqrt{d})$.
c. Show that if $d \equiv 1 \bmod 4$, then $\mathcal{O}_K = \mathbb{Z}[(\sqrt{d}+1)/2]$.
d. Show that if $d \equiv 2, 3 \bmod 4$, then $\mathcal{O}_K = \mathbb{Z}[\sqrt{d}]$.
e. Conclude that $\Delta(\mathcal{O}_K)/\boxtimes(d) \in \{1, 4\}$.

EXERCISE 4.24. Let $n, m \in \mathbb{Z}_{>0}$. Prove that there exists a number field K such that $[K : \mathbb{Q}] = n$ and $\gcd(\Delta_K, m) = 1$.

EXERCISE 4.25. Let $A \subseteq B$ be commutative rings such that B is free of finite rank over A. Show that the map $B \to \mathrm{Hom}_A(B, A)$ given by $x \mapsto (y \mapsto \mathrm{Tr}_{B/A}(xy))$ is an isomorphism of B-modules if and only $\Delta(B/A)$ is a unit in A.

4.2. Localization. We say a commutative ring is *local* if it has a unique maximal ideal, and *semi-local* if it has only finitely many maximal ideals. For a commutative ring R and a prime ideal $\mathfrak{p} \subseteq R$, the *localization of* R *at* $\mathfrak{p}$, written $R_\mathfrak{p}$, is the ring obtained from R by inverting all elements in $R \setminus \mathfrak{p}$ (see p. 38 of [**2**]). If M is an R-module, then we call $M_\mathfrak{p} = R_\mathfrak{p} \otimes_R M$ the localization of M at $\mathfrak{p}$. We will show that an ideal of a number ring is invertible if and only if it is locally principal (Corollary 4.34).

EXAMPLE 4.26. Let $\mathfrak{p} = 2\mathbb{Z} \subseteq \mathbb{Z}$. We have $\mathbb{Z}_\mathfrak{p} = \{a/b : a, b \in \mathbb{Z}, b \,\text{odd}\}$ and $\mathfrak{p}\mathbb{Z}_\mathfrak{p}$ is its unique maximal ideal. Let R be an order. Note that R is a $\mathbb{Z}$-module, so we obtain a ring $R_\mathfrak{p} = \mathbb{Z}_\mathfrak{p} \otimes_\mathbb{Z} R$. Then all maximal ideals of $R_\mathfrak{p}$ contain $\mathfrak{p}R_\mathfrak{p}$ and thus correspond to the maximal ideals of $R_\mathfrak{p}/\mathfrak{p}R_\mathfrak{p} \cong R/2R$, which is finite. Hence $R_\mathfrak{p}$ is semi-local.

EXERCISE 4.27. Let R be a commutative ring. Show that R is a local ring if and only if $R \setminus R^*$ is an additive subgroup of R, in which case $R \setminus R^*$ is the maximal ideal.

EXERCISE 4.28. Let $R \subseteq S$ be number rings. Show that if R is semi-local, then S is semi-local.

PROPOSITION 4.29. *Let $R \subseteq S$ be commutative (sub)rings such that R is semi-local and let $I \subseteq S$ be an R-submodule. Then I is invertible within S if and only if there exists some $x \in S^*$ such that $I = xR$.*

PROOF. ($\Leftarrow$) We may take $J = x^{-1}R$.

($\Rightarrow$) Let $\mathfrak{m} \subseteq R$ be maximal. As $IJ = R \not\subseteq \mathfrak{m}$ there exist $x_\mathfrak{m} \in I$ and $y_\mathfrak{m} \in J$ such that $x_\mathfrak{m} y_\mathfrak{m} \in R \setminus \mathfrak{m}$. Since R is semi-local, there exist $\lambda_\mathfrak{m} \in R$ for each maximal $\mathfrak{m} \subseteq R$ such that for all maximal $\mathfrak{n} \subseteq R$ we have $\lambda_\mathfrak{m} \in \mathfrak{n}$ if and only if $\mathfrak{m} \neq \mathfrak{n}$: Namely, let $r_\mathfrak{n} \in \mathfrak{n} \setminus \mathfrak{m}$, which exist since the ideals are maximal, and take $\lambda_\mathfrak{m} = \prod_{\mathfrak{n} \neq \mathfrak{m}} r_\mathfrak{n}$. Consider $x = \sum_\mathfrak{m} \lambda_\mathfrak{m} x_\mathfrak{m}$ and $y = \sum_\mathfrak{m} \lambda_\mathfrak{m} y_\mathfrak{m}$. Then

$$xy = \sum_{\mathfrak{m},\mathfrak{n}} \lambda_\mathfrak{m} \lambda_\mathfrak{n} x_\mathfrak{m} y_\mathfrak{n}.$$

For all $\mathfrak{m}$ there is precisely one term not contained in $\mathfrak{m}$, which is $\lambda_\mathfrak{m}\lambda_\mathfrak{m} x_\mathfrak{m} y_\mathfrak{m}$, hence $xy \notin \mathfrak{m}$. Since $x_\mathfrak{m} y_\mathfrak{n} \in IJ = R$ we have $xy \in R$. It follows that xy is a unit of R, and x is a unit of S. Finally, $xR \subseteq I = xyI \subseteq xJI = xR$ and $I = xR$. $\square$

Recall the definition of $\mathcal{I}(R)$ from Definition 4.4.

COROLLARY 4.30. *Let R be a is semi-local commutative ring. Then $\mathcal{I}(R) \cong Q(R)^*/R^*$.* $\square$

LEMMA 4.31. *Let R be a subring of a field K. Then for every localization S of R the natural map $S \to K$ is injective, and $I = \bigcap_{\text{max. } \mathfrak{m} \subset R} I_\mathfrak{m}$ for every R-submodule $I \subseteq K$.*

PROOF. The map $S \to K$ is injective since R, and hence S, is a domain. Clearly $I \subseteq I_\mathfrak{m}$. Suppose $x \in \bigcap_\mathfrak{m} I_\mathfrak{m}$. Consider $J = (I : xR)_K \cap R$ and let $\mathfrak{m} \subseteq R$ be maximal. Since $x \in I_\mathfrak{m}$ there exists some $r \in R \setminus \mathfrak{m}$ such that $rx \in I$. Then

$r \in J$ by definition of J, so $J \not\subseteq \mathfrak{m}$. As this holds for all $\mathfrak{m}$ we conclude that $J = R$. It follows that $x \in I$ and hence $I = \bigcap_{\mathfrak{m}} I_{\mathfrak{m}}$. □

EXERCISE 4.32. Let R be a number ring and $\mathfrak{p} \subset R$ be a maximal ideal. Show that if $I_{\mathfrak{p}} \subseteq R_{\mathfrak{p}}$ is an invertible ideal over $R_{\mathfrak{p}}$, then $J = I_{\mathfrak{p}} \cap R$ is an invertible ideal over R such that $J_{\mathfrak{p}} = I_{\mathfrak{p}}$, and $J_{\mathfrak{q}} = R_{\mathfrak{q}}$ for all maximal $\mathfrak{q} \subset R$ not equal to $\mathfrak{p}$.

THEOREM 4.33 (cf. Theorem 5.3 in [**27**]). *Let $R \subseteq S$ be number rings such that $Q(R) = Q(S)$. Then we have a group isomorphism $\mathcal{I}(S) \to \bigoplus_{\text{max. } \mathfrak{p} \subset R} \mathcal{I}(S_{\mathfrak{p}})$ given by $I \mapsto (I_{\mathfrak{p}})_{\mathfrak{p}}$, with inverse $(I_{\mathfrak{p}})_{\mathfrak{p}} \mapsto \bigcap_{\mathfrak{p}} I_{\mathfrak{p}}$.*

PROOF. For any maximal $\mathfrak{q} \subset R$ let

$$\varphi_{\mathfrak{q}} : \mathcal{I}(S_{\mathfrak{q}}) \to \bigoplus_{\mathfrak{m} \subset S_{\mathfrak{q}}} \mathcal{I}((S_{\mathfrak{q}})_{\mathfrak{m}}) = \bigoplus_{\mathfrak{q} \subseteq \mathfrak{m} \subset S} \mathcal{I}(S_{\mathfrak{m}})$$

be the map as in the theorem with R replaced by $R_{\mathfrak{q}}$ and S replaced by $S_{\mathfrak{q}}$. Then the following diagram is commutative:

$$\begin{array}{ccc} \mathcal{I}(S) & \longrightarrow & \bigoplus_{\mathfrak{p} \subset R} \mathcal{I}(S_{\mathfrak{p}}) \\ & \searrow \quad \swarrow & \\ & \bigoplus_{\mathfrak{m} \subset S} \mathcal{I}(S_{\mathfrak{m}}) & \end{array}$$

Hence to show that the horizontal map is an isomorphism, it suffices to prove this for the downward maps. These maps are both injective with the appropriate left inverse by Lemma 4.31, while they are surjective by Exercise 4.32. □

COROLLARY 4.34. *Let R be a subring of a number field. Then a fractional R-ideal is invertible if and only if it is locally principal.*

PROOF. By Proposition 4.29 locally principal is equivalent to locally invertible, which in turn is equivalent to invertible by Theorem 4.33. □

LEMMA 4.35. *Let $R \subseteq S$ be number rings such that $Q(R) = Q(S)$. Then there is a surjective group homomorphism $\sigma : \mathcal{I}(R) \to \mathcal{I}(S)$ given by $I \mapsto SI$.*

PROOF. The only non-trivial part is surjectivity. Using Theorem 4.33 one verifies that the induced map $\sigma : \bigoplus_{\mathfrak{p} \subseteq R} \mathcal{I}(R_{\mathfrak{p}}) \to \bigoplus_{\mathfrak{p} \subseteq R} \mathcal{I}(S_{\mathfrak{p}})$ is the direct sum of maps $\sigma_{\mathfrak{p}} : \mathcal{I}(R_{\mathfrak{p}}) \to \mathcal{I}(S_{\mathfrak{p}})$ given by $I \mapsto SI$. Thus it suffices to prove the theorem when R is local, and hence S is semi-local by Exercise 4.28. Then the invertible ideals of S are principal by Proposition 4.29, and we may use its generator to generate an invertible R-ideal. □

EXERCISE 4.36. Let R be a number ring. Show that every fractional ideal of R is invertible if and only if every fractional ideal of R generated by at most two elements is invertible.

4.3. Blowing up. In this section we give an algorithm to make ideals invertible, i.e. given a fractional ideal I of some order R of a number field K, computing the minimal order $R \subseteq S \subseteq K$ such that SI is invertible. Borrowing terminology from algebraic geometry (cf. Proposition 7.14 in [**13**]), we will be computing a *blowup* of R at the ideal I. We give a variant on the proof of [**10**]. We warn those reading [**10**] that it contains non-standard definitions of fractional and invertible ideals.

PROPOSITION 4.37. *Let S be a commutative ring and $I \subseteq S$ a subgroup. Then:*

(1) *the* blowup *at I,*

$$\mathrm{Bl}(I) = \mathrm{Bl}_S(I) := \bigcup_{k \geq 0} (I^k : I^k)_S,$$

is a subring of S;

(2) *if $R \subseteq S$ is a subring and RI is an invertible ideal of R within S, then $\mathrm{Bl}(I) \subseteq R$;*

(3) *if RI is an invertible ideal of $R = I^n : I^n$ within S for some $n \geq 0$, then $R = \mathrm{Bl}(I)$;*

(4) *if S is a number field and I is finitely generated and non-zero, then there exists some $n \geq 0$ such that $\mathrm{Bl}(I) = I^n : I^n$, and $\mathrm{Bl}(I)$ is an order.*

PROOF. (1) It follows from Lemma 4.3 for all $k \geq 0$ that $I^k : I^k$ is a ring and that $I^k : I^k \subseteq I^{k+1} : I^{k+1}$. Hence $\mathrm{Bl}(I)$, being their union, is a ring.

(2) Note that $RI^k = (RI)^k$ is invertible within S for all $k \geq 0$. By Lemma 4.3 and Exercise 4.6 we have $I^k : I^k \subseteq (RI^k) : (RI^k) = R : R = R$ for all k, so $\mathrm{Bl}(I) \subseteq R$.

(3) This follows from (2) and the fact that $R \subseteq \mathrm{Bl}(I)$.

(4) Note that $I^k : I^k$ is both a subring and a finitely generated subgroup of S by Lemma 4.3, hence it is an order. Because $I^k : I^k \subseteq \mathcal{O}_S$, see Theorem 4.9, and $\mathcal{O}_S$ is a Noetherian $\mathbb{Z}$-module, there must exist some $n \geq 0$ such that $I^n : I^n = I^k : I^k$ for all $k \geq n$. Consequently $I^n : I^n = \mathrm{Bl}(I)$. □

EXAMPLE 4.38. Let p be a prime and $n \in \mathbb{Z}_{>1}$. Let $\pi = \sqrt[n]{p}$ and consider the orders $R = \mathbb{Z}[\pi^k : k \in \mathbb{Z}_{\geq n}] \subseteq \mathbb{Z}[\pi] = S$ and fractional ideal $I = R + R\pi$ of R. Then $I^k = R\pi^0 + R\pi^1 + \cdots + R\pi^k$ for all $k \geq 1$, and $I^{n-1} = S$. Hence $I^{n-1} : I^{n-1} = S$ by Lemma 4.3.iv. We have $I \subseteq S$ and $1 \in I$, so $SI = S$. In particular, SI is invertible over S, so $\mathrm{Bl}(I) \subseteq S$ by Proposition 4.37.2. We conclude that $\mathrm{Bl}(I) = I^{n-1} : I^{n-1} = S$. Note that S need not be a maximal order; For $n = 2$ and $p = 1 \pmod 4$ we have $S \subsetneq \mathbb{Z}[\frac{1+\pi}{2}]$.

We will show later in Theorem 4.52, in the case that I is a finitely generated non-zero subgroup of a number field K, that $\mathrm{Bl}_K(I)I$ is invertible over $\mathrm{Bl}_K(I)$. Hence in this case $\mathrm{Bl}_K(I)$ is the unique minimal subring of K over which I becomes invertible. Moreover, the proof lends itself to an algorithm to compute the blowup, which is a non-trivial statement by Exercise 4.43.

EXAMPLE 4.39. Let $S = \mathbb{Q}[X]$ and $I = \mathbb{Z} + \mathbb{Z}X$. Then $I^n \subseteq \mathbb{Z}[X]$ is the set of polynomials of degree at most n. Every non-zero element of $I^n : I^n$ needs to have degree 0, from which we deduce that $I^n : I^n = \mathbb{Z}$. In particular, $\mathrm{Bl}(I) = \mathbb{Z}$ and $\mathbb{Z}I$ is not an invertible $\mathbb{Z}$-ideal.

EXERCISE 4.40. Let S be a commutative ring and $I, J \subseteq S$ subgroups. Show that $\mathrm{Bl}(I) \cdot \mathrm{Bl}(J) \subseteq \mathrm{Bl}(IJ)$. Show that $\mathrm{Bl}(I) = I$ if and only if I is a subring of S.

EXERCISE 4.41. Let $R \subseteq S$ be a commutative subrings and and $I, J \subseteq S$ fractional R-ideals. Show that if I is invertible, then $\mathrm{Bl}(IJ) = \mathrm{Bl}(J)$.

EXERCISE 4.42. Let S be a commutative ring, $I \subseteq S$ a finitely generated non-zero subgroup and $n \in \mathbb{Z}_{\geq 0}$. Show that if RI is invertible over $R = I^n : I^n$, then $R = \mathrm{Bl}(I)$.

EXERCISE 4.43. Let $n \in \mathbb{Z}_{>1}$. Construct an order of rank n with a fractional ideal I such that $R = (I^0 : I^0) = \cdots = (I^{n-2} : I^{n-2}) \neq (I^{n-1} : I^{n-1}) = \mathrm{Bl}(I)$.

We collect some facts about generators of modules.

DEFINITION 4.44. Let R be a commutative ring and M an R-module. We define

$$\gamma_R(M) = \min\Big\{\#X : X \subseteq M,\ \sum_{x \in X} Rx = M\Big\},$$

the minimal number of generators needed to generate M as an R-module.

Note that M is finitely generated if and only if $\gamma_R(M) < \infty$. We will show how this quantity behaves with respect to quotients.

LEMMA 4.45 (Nakayama, Corollary 2.7 in [**2**]). *Let R be a commutative ring, I an ideal of R contained in every maximal ideal, and M a finitely generated R-module.*

(1) *If N is any submodule of M and $M = N + IM$, then $M = N$.*
(2) *We have $\gamma_R(M) = \gamma_{R/I}(M/IM)$.* □

EXERCISE 4.46 (cf. Corollary 5.8 in [**2**]). Let $f : R \to S$ be a morphism of commutative rings such that S is Noetherian as an R-module. Show that the map

$$\{\text{maximal } \mathfrak{m} \subset S\} \to \{\text{maximal } \mathfrak{m} \subset R\} \quad \text{given by} \quad \mathfrak{m} \mapsto f^{-1}\mathfrak{m}$$

is well-defined and has fibers of size at most $\gamma_R(S)$.

Note: We may even replace the Noetherian assumption by *integrality*.

For a local commutative ring R with unique maximal ideal $\mathfrak{m}$ write $\kappa(R) = R/\mathfrak{m}$ for the *residue field* of R.

LEMMA 4.47. *Let R be a local commutative ring with maximal ideal $\mathfrak{p}$ and A a local commutative R-algebra with maximal ideal $A\mathfrak{p}$. Let L and M be finitely generated R-modules and N a finitely generated A-module. Then:*

(1) *$\gamma_R(L \otimes_R M) = \gamma_R(L) \cdot \gamma_R(M)$;*
(2) *$\gamma_R(N) = \gamma_R(A) \cdot \gamma_A(N)$;*
(3) *$\gamma_A(A \otimes_R M) = \gamma_R(M)$.*

PROOF. Note that $\kappa(R) \otimes_R N = N/\mathfrak{p}N = N/\mathfrak{p}AN = \kappa(A) \otimes_R N$ and similarly $\kappa(R) \otimes_R A = \kappa(A)$. Applying Lemma 4.45.2 and taking the tensor product with $\kappa(R)$, we may assume that R and A are fields. Thus all modules are free and the rank corresponds with γ, in which case the results are trivial. □

LEMMA 4.48. *Let R be a local commutative ring and S a semi-local commutative R-algebra. Let $M \subseteq S$ be an R-submodule such that $M \not\subseteq \mathfrak{m}$ for all maximal ideals $\mathfrak{m} \subseteq S$. If S has at most $\#\kappa(R)$ maximal ideals, then $M \cap S^* \neq \emptyset$.*

PROOF. Note that $S \setminus S^* = \bigcup_{\mathfrak{m}} \mathfrak{m}$, where the union ranges over the maximal ideals of S, as every non-unit is contained in some proper and hence in some maximal ideal. Thus it suffices to show that $\bigcup_{\mathfrak{m}}(M \cap \mathfrak{m}) \neq M$. We will inductively prove the stronger statement that an R-module M cannot be written as a union of k proper submodules for any integer $0 \leq k \leq \#\kappa(R)$. For $k = 0$ this is trivial. Now we assume that $\bigcup_{i<k} M_i \subsetneq M$, and for the sake of contradiction that $\bigcup_{1 \leq i \leq k} M_i = M$. Choose $y \in M \setminus \bigcup_{i<k} M_i$. In particular, $y \in M_k$. Now let $x \in M \setminus M_k$. Then for all $t \in R$ we have $x + ty \in M_{i(t)}$ for some $1 \leq i(t) < k$. Since $k \leq \#\kappa(R)$ there exist $s, t \in R$ such that $s \not\equiv t$ in $\kappa(R)$ and $i(s) = i(t)$. It follows (Exercise 4.27) that $s - t \in R^*$ and thus $y = (s-t)^{-1}((x + ys) - (x + yt)) \in M_{i(t)}$, which is a contradiction. Hence $y \notin M_k$ and thus $\bigcup_{i \leq k} M_i \subsetneq M$, as was to be shown. □

LEMMA 4.49. *Let R be local commutative ring with maximal ideal $\mathfrak{p}$. Then for every $n \in \mathbb{Z}_{\geq 0}$ there exists a commutative R-algebra A which is free of finite rank over R, local with maximal ideal $A\mathfrak{p}$, and such that $\#\kappa(A) > n \cdot \mathrm{rk}_R(A)$.*

PROOF. If $\kappa(R)$ is infinite we may choose $A = R$, so assume $\kappa(R) = \mathbb{F}_q$ for some prime power q. Let $d \in \mathbb{Z}_{>0}$ and consider the field extension $\kappa(R) \subseteq \mathbb{F}_{q^d} = \mathbb{F}_q(\alpha)$ for some $\alpha \in \mathbb{F}_{q^d}$. Let $f \in R[X]$ be some monic lift of the minimal polynomial of α over $\kappa(R)$. Clearly $A = R[X]/(f)$ is free of rank d over R. Since $\mathfrak{p}$ is the unique maximal ideal of R, the maximal ideals of A correspond to those of $A/\mathfrak{p}A = \mathbb{F}_{q^d}$ by Exercise 4.46, which is a field. Hence A is local with maximal ideal $A\mathfrak{p}$. Finally, $\#\kappa(A) = \#\kappa(R)^d > n \cdot d$ for d sufficiently large, as was to be shown. □

LEMMA 4.50. *Let $R \subseteq S$ be commutative (sub)rings with R local. Let $M \subseteq S$ be an R-submodule such that $1 \in M$, and $(M^n : M^n)_S = R$ for some integer $n + 1 \geq \gamma_R\big(\bigcup_{k \geq 0} M^k\big)$. Then $M = R$.*

PROOF. Let $\mathfrak{p} \subseteq R$ be the maximal ideal. From $1 \in M$ it follows that $R = M^0 \subseteq M^1 \subseteq M^2 \subseteq \cdots$. We write $M^\infty = \bigcup_{k \geq 0} M^k$, which is an R-subalgebra of S. With $N_k = (M^k + \mathfrak{p}M^\infty)/(\mathfrak{p}M^\infty)$ we get a chain $N_0 \subseteq N_1 \subseteq N_2 \subseteq \cdots$ of $\kappa(R)$-vector spaces with limit $N_\infty = M^\infty/\mathfrak{p}M^\infty$. We have that $\dim_{\kappa(R)}(N_\infty) = \gamma_R(M^\infty) \leq n + 1$ by Lemma 4.45.2. Hence there exists a minimal k such that $N_k = N_{k+1}$, and it must satisfy $k \leq \dim_{\kappa(R)}(N_\infty) - \dim_{\kappa(R)}(N_0) \leq n$. Then $N_k = N_{k+1} = N_{k+2} = \cdots$ and thus $N_k = N_\infty$. Equivalently, we have $M^k + \mathfrak{p}M^\infty = M^\infty$, so $M^k = M^\infty$ by Lemma 4.45.1. It follows from Lemma 4.3 that $M \subseteq M^\infty = (M^\infty : M^\infty)_S = (M^n : M^n)_S = R \subseteq M$, so indeed $M = R$. □

LEMMA 4.51. *Let $R \subseteq S$ be commutative subrings, let A be a commutative R-algebra that is free as an R-module, and let $M \subseteq S$ be an R-module. Then $(M \otimes_R A : M \otimes_R A)_{S \otimes A} = (M : M)_S \otimes_R A$ as subsets of $S \otimes A$.*

PROOF. The inclusion $\supseteq$ is straight-forward. Choose a basis $\mathcal{A} \subseteq A$ of A as an R-module. Let $x \in (M \otimes_R A : M \otimes_R A)_{S \otimes A}$. Then we may uniquely write $x = \sum_{a \in \mathcal{A}} x_a \otimes a$. Then $M \otimes_R A \ni x \cdot (m \otimes 1) = \sum_{a \in \mathcal{A}} x_a m \otimes a$, so $x_a m \in M$ for all $a \in \mathcal{A}$. Hence $x_a \in (M : M)_S$ and $x \in (M : M)_S \otimes_R A$. □

THEOREM 4.52 (Dade–Taussky–Zassenhaus [**10**]). *Let K be a number field, $R \subseteq K$ a subring and $I \subseteq K$ a finitely generated non-zero R-module. Then $\mathrm{Bl}_K(I) = I^n : I^n$ for all $n \geq [K : \mathbb{Q}] - 1$. For any subring $T \subseteq K$, it holds that $\mathrm{Bl}_K(I) \subseteq T$ if and only if TI is an invertible ideal of T within K.*

Note that we may take $R = \mathbb{Z}$ in this theorem.

PROOF. Choose some $n \geq [K : \mathbb{Q}] - 1$. It suffices by Proposition 4.37 to show that $(I^n : I^n)I \in \mathcal{I}_K(I^n : I^n)$.

We first prove this under the added assumptions that $R = I^n : I^n$, that $\mathrm{Q}(R) = K$ and that R is a local ring. Let $S = R \cdot \mathcal{O}_K$ and note that it is a ring and a Noetherian R-module. In particular, S is semi-local by Exercise 4.46. Every fractional ideal of S, in particular SI, is invertible, because $\mathcal{O}_K \subseteq S$. Hence there exists by Lemma 4.35 some $J \in \mathcal{I}(R)$ such that $SI = SJ$. Then $I \in \mathcal{I}(R)$ if and only if $IJ^{-1} \in \mathcal{I}(R)$, so it suffices to show that $M = IJ^{-1}$ is invertible. Note that $SM = S$. Consider now the R-algebra A from Lemma 4.49 satisfying $\#\{\text{maximal } \mathfrak{m} \subset S\} \cdot \mathrm{rk}_R(A) < \#\kappa(A)$, and write $-_A$ for $- \otimes_R A$. Since A is free over R, we have $(M_A^n : M_A^n)_{S_A} = R_A = A$ by Lemma 4.51. As S is Noetherian, S_A is free and hence Noetherian over S. If $M_A \subseteq \mathfrak{m}$ for some maximal $\mathfrak{m} \subset S_A$, then also $M = M_A \cap S$ is contained in the maximal ideal of $\mathfrak{m} \cap S$ of S by Exercise 4.46, which is impossible since $SM = S$. From Exercise 4.46 we also obtain $\#\{\text{maximal } \mathfrak{m} \subset S_A\} \leq \mathrm{rk}_R(A) \cdot \#\{\text{maximal } \mathfrak{m} \subset S\} < \#\kappa(A)$. Thus we may apply Lemma 4.48 with R_A, S_A and M_A in the roles of R, S and M respectively. Hence we may choose some $u \in M_A \cap S_A^*$. Now let $N = u^{-1}M_A$ and $N^\infty = \bigcup_{k \geq 0} N^k$. Since $N^\infty \subseteq S_A$ is Noetherian as an R-module, we may choose some finite set of R-module generators $X \subseteq N^\infty$. By Exercise 4.10.a and Lemma 4.47.1 we have

$$\gamma_R(N^\infty) = \gamma_R\Big(\sum_{x \in X} Rx\Big) \leq \gamma_{\mathbb{Z}}\Big(\sum_{x \in X} \mathbb{Z}x\Big) \leq \dim_{\mathbb{Q}} K_A \leq \mathrm{rk}_R(A) \cdot (n+1),$$

so $\gamma_A(N^\infty) \leq n + 1$ by Lemma 4.47.2. Then by Lemma 4.50 we have $N = A$. By Lemma 4.47.3 we have $1 = \gamma_A(N) = \gamma_R(M)$, so $M = Rv$ for some $v \in S$. In particular, M is invertible, as was to be shown.

Now consider the general case. Let $S = I^n : I^n$ and $F = \mathrm{Q}(S)$. Let $x \in I$ be a non-zero element and consider $J = SIx^{-1}$. Note that $x^n J^n = SI^n = I^n$ and thus $J^n : J^n = x^n J^n : x^n J^n = I^n : I^n = S$. Since $1 \in J$, we have $J^0 \subseteq J^1 \subseteq J^2 \subseteq \cdots$. From a dimension argument it then follows that $\mathbb{Q}J^n = \mathbb{Q}J^{n+1} = \cdots$ and thus $\mathbb{Q}J^n$ is a ring. It follows from Lemma 4.3.iv and Exercise 4.7 that $\mathbb{Q}J^n = (\mathbb{Q}J^n) : (\mathbb{Q}J^n) = \mathbb{Q} \cdot (J^n : J^n) = F$, so $J \subseteq F$. Let $\mathfrak{p} \subset S$ be a maximal ideal, so that $S_\mathfrak{p}$ is local with $S_\mathfrak{p} = J_\mathfrak{p}^n : J_\mathfrak{p}^n$. Hence $J_\mathfrak{p}$ is an invertible ideal of $S_\mathfrak{p}$ by the previous case. Invertibility is a local property by Theorem 4.33, so J is an invertible S-ideal. Then $x^n J^n = (SI)^n$ is invertible over S within K. If $n = 0$, then $K = \mathbb{Q}$ and the theorem is trivial. Otherwise, SI is invertible over S within K, as was to be shown. □

COROLLARY 4.53. *There exists a polynomial-time algorithm that, given a number field K and a finitely generated non-zero subgroup $I \subseteq K$, computes* $\mathrm{Bl}(I)$.

PROOF. By Exercise 4.10.a, each subgroup $I^1, I^2, \ldots \subseteq K$ can be generated by $[K : \mathbb{Q}]$ elements. Thus we can inductively compute I^{k+1} as the image of the tensor product $I^k \otimes I \to K$, and in turn $\mathrm{Bl}(I) = I^n : I^n$ for $n = [K : \mathbb{Q}] - 1$, in polynomial time. □

EXERCISE 4.54. Let K be a number field and $I \subseteq K$ a finitely generated non-zero subgroup. Show that $\mathbb{Q}(\mathrm{Bl}(I)) = \mathbb{Q}(Ix^{-1})$ for any non-zero $x \in I$.

EXERCISE 4.55. Let $R \subseteq S \subseteq \mathrm{Q}(R)$ be number rings. Show that $\ker(\mathcal{I}(R) \to \mathcal{I}(S)) \cong (S/\mathfrak{f})^*/(R/\mathfrak{f})^*$, where $\mathfrak{f} = R : S$ is the *conductor*. Conclude that this kernel equals $\mathcal{I}(R)^{\mathrm{tors}}$, the torsion subgroup of $\mathcal{I}(R)$, when S is the maximal order.

EXERCISE 4.56. Let R be an order of a number field K. Show that the following are equivalent:
(1) There exist a finite *Boolean ring* B, i.e. a ring B such that $x^2 = x$ for all $x \in B$, a subring $A \subseteq B$ and a surjective ring homomorphism $f : \mathcal{O}_K \to B$ such that $R = f^{-1}A$.
(2) There exists an equivalence relation $\sim$ on the set $\mathrm{Hom}_{\mathrm{Ring}}(\mathcal{O}_K, \mathbb{F}_2)$ such that $R = \{x \in \mathcal{O}_K : f \sim g \Rightarrow f(x) = g(x)\}$.
(3) The group $\mathcal{I}(R)$ is torsion free.
Give an example of a non-maximal order such that $\mathcal{I}(R)$ is torsion free.

4.4. Ideals generated by two elements. Every principal fractional ideal of a number ring R is invertible, and thus its blowup is R itself. In this section we give special attention to (computing) the blowup of fractional ideals generated by two elements. Such ideals are ubiquitous: Every invertible ideal of a number ring can be generated by only two elements (Exercise 4.62).

EXERCISE 4.57. Let K be a number field and $R \subseteq K$ a subring. Show that for all finitely generated subgroups $I, J \subseteq K$ we have $\mathrm{Bl}(IJ) = \mathrm{Bl}(I) \cdot \mathrm{Bl}(J)$. Conclude that $\mathrm{Bl}(R\alpha + R\beta) = R \cdot \mathrm{Bl}(\mathbb{Z}\alpha + \mathbb{Z}\beta)$ for all $\alpha, \beta \in K^*$.

EXERCISE 4.58. Let $f \in \mathbb{Z}[X]$ be a *primitive* polynomial, i.e. a polynomial for which the only integer divisors are ± 1. Show that $(f\mathbb{Q}[X]) \cap \mathbb{Z}[X] = f\mathbb{Z}[X]$.

Hint: Gauss's lemma states that for $f, g \in \mathbb{Z}[X]$ the product fg is primitive if and only if f and g are primitive.

THEOREM 4.59. *Let $K = \mathbb{Q}(\gamma)$ be a number field of degree $n \geq 2$ and let $f = a_n X^n + \cdots + a_0 X^0 \in \mathbb{Z}[X]$ be irreducible with $f(\gamma) = 0$. Consider for $0 \leq j \leq n$ the elements*

$$p_j = a_n \gamma^j + a_{n-1}\gamma^{j-1} + \cdots + a_{n-j} \quad \text{and} \quad q_j = a_j + a_{j-1}\gamma^{-1} + \cdots + a_0 \gamma^{-j}$$

and the groups $M = \mathbb{Z} + \mathbb{Z}\gamma + \cdots + \mathbb{Z}\gamma^{n-1}$,

$$D = \mathbb{Z}p_0 + \mathbb{Z}p_1 + \cdots + \mathbb{Z}p_{n-1} \quad \text{and} \quad N = \mathbb{Z}q_0 + \mathbb{Z}q_1 + \cdots + \mathbb{Z}q_{n-1}.$$

Then

$$\mathrm{Bl}(\mathbb{Z} + \mathbb{Z}\gamma) = M : M = \mathbb{Z}[\gamma] \cap \mathbb{Z}[\gamma^{-1}] = \mathbb{Z} + N = \mathbb{Z} + D$$

and N and D are coprime invertible proper ideals of this ring, with quotients isomorphic as rings to $\mathbb{Z}/a_0\mathbb{Z}$ and $\mathbb{Z}/a_n\mathbb{Z}$ respectively, such that $N : D = \gamma \cdot \mathrm{Bl}(\mathbb{Z} + \mathbb{Z}\gamma)$.

PROOF. It follows from Theorem 4.52 that $\mathrm{Bl}(\mathbb{Z} + \mathbb{Z}\gamma) = M : M$, because $M = (\mathbb{Z} + \mathbb{Z}\gamma)^{n-1}$. We extend the definition of the a_i so that $a_i = 0$ for $i < 0$ and $i > n$, and similarly define p_i and q_i for all $i \geq 0$. It follows that $p_i = q_i = 0$ for $i \geq n$. We proceed to prove the following inclusions.

$$\mathbb{Z} + N \overset{(1)}{\subseteq} \mathbb{Z} + D \overset{(2)}{\subseteq} M : M \overset{(3)}{\subseteq} \mathbb{Z}[\gamma] \cap \mathbb{Z}[\gamma^{-1}] \overset{(4)}{\subseteq} \mathbb{Z} + N.$$

(1) Note that for $0 \leq j \leq n$ we have

$$\begin{aligned} p_{n-j} + q_j &= \gamma^{-j}(a_n \gamma^n + \cdots + a_j \gamma^j) + \gamma^{-j}(a_j \gamma^j + \cdots + a_0) \\ &= \gamma^{j-n} f(\gamma) + a_j = a_j. \end{aligned}$$

In particular $q_j \in \mathbb{Z} + D$ for all $0 \leq j \leq n$ and thus $\mathbb{Z} + N \subseteq \mathbb{Z} + D$.

(2) Let $0 \leq i, j \leq n$. If $i < n - j$, then $p_j\gamma^i \in M$. Otherwise, we have $q_{n-j}\gamma^i \in M$, and $p_j + q_{n-j} = a_{n-j} \in \mathbb{Z}$ as shown in (1), so $p_j\gamma^i \in M$. We conclude that $p_j M \subseteq M$ and $\mathbb{Z} + D \subseteq M : M$.

(3) We have $M : M \subseteq M \subseteq \mathbb{Z}[\gamma]$ since $1 \in M$. For $L = \mathbb{Z} + \mathbb{Z}\gamma^{-1} + \cdots + \mathbb{Z}(\gamma^{-1})^{n-1} = \gamma^{1-n}M$ we similarly have $M : M = L : L \subseteq \mathbb{Z}[\gamma^{-1}]$.

(4) Suppose there exists some $x \in \mathbb{Z}[\gamma] \cap \mathbb{Z}[\gamma^{-1}]$ such that $x \notin \mathbb{Z} + N$, and choose $x = \sum_{i=0}^{B} b_i\gamma^i = \sum_{i=0}^{C} c_i\gamma^{-i}$ to be such an element for which C is minimal. As $x \notin \mathbb{Z}$, we have $C > 0$. Then $g = X^C(\sum_{i=0}^{B} b_i X^i - \sum_{i=0}^{C} c_i X^{-i}) \in \mathbb{Z}[X]$. As $g(\gamma) = \gamma^C(x - x) = 0$ we conclude $g \in f\mathbb{Q}[X]$ since f is irreducible. Hence $g = fh$ for some $h \in \mathbb{Z}[X]$ by Exercise 4.58. It follows that $a_0 \mid c_C$, so we may subtract $(c_C/a_0)q_C \in \mathbb{Z} + N$ from x, violating minimality of C. This proves all inclusions.

It remains to prove the statements about N and D. Let $A = \mathrm{Bl}(\mathbb{Z}+\mathbb{Z}\gamma)$. Note that $\gamma p_j = p_{j+1} - a_{n-j-1} = q_{n-j-1}$ for all $0 \leq j < n$, so $D\gamma = N$. For $j = 0$ this gives $\gamma = (p_1 - a_{n-1})/a_n$. Because the ring $A = \mathbb{Z}+D$ has $\mathbb{Z}$-basis $1, p_1, p_2, \ldots, p_{n-1}$, we conclude $A \cap \mathbb{Z}\gamma = a_n\mathbb{Z}\gamma$. Then $A : (A + A\gamma) = \{\delta \in A : \delta\gamma \in A\}$ intersects $\mathbb{Z}$ in $a_n\mathbb{Z}$ and therefore its index in A is divisible by $|a_n|$. As $\gamma D = N \subseteq A$ we have that $D \subseteq A : (A + A\gamma)$, where we note that D has index $|a_n|$ in A by comparing $\mathbb{Z}$-bases. Hence $D = A : (A + A\gamma)$ and in particular D is an A-ideal with $A/D \cong \mathbb{Z}/a_n\mathbb{Z}$ as rings.

Analogously N is an ideal of A with $A/N \cong \mathbb{Z}/a_0\mathbb{Z}$. We have $a_j = p_{n-j} + q_j \in N + D$ for all $0 \leq j \leq n$. Since f is irreducible we have that $\mathbb{Z} = \sum_j \mathbb{Z}a_j \subseteq N + D$, so N and D are coprime. Moreover, $D \cdot (A + A\gamma) = D + N = A$, so D and consequently $N = D\gamma$ are invertible. Thus $N : D = A\gamma$. □

For more information about the ring in Theorem 4.59, see [**22**, **26**].

EXAMPLE 4.60. Let K be a number field, R a subring of K, and $\alpha, \beta \in K^*$. We will give an explicit expression for $\mathrm{Bl}(R\alpha + R\beta)$. Let $\gamma = \alpha/\beta$. By Exercise 4.57 we may write

$$\mathrm{Bl}(\alpha R + \beta R) = \mathrm{Bl}(R + \gamma R) = R \cdot \mathrm{Bl}(\mathbb{Z} + \gamma\mathbb{Z}).$$

If $\gamma \in \mathbb{Q}^*$, then $\mathbb{Z} + \gamma\mathbb{Z} \subseteq \mathbb{Q}$. Since $\mathrm{Bl}(\mathbb{Z} + \gamma\mathbb{Z})$ is an order it must equal $\mathbb{Z}$, so $\mathrm{Bl}(\alpha R + \beta R) = R$. Otherwise, we have

$$\mathrm{Bl}(\alpha R + \beta R) = R + p_1 R + \cdots + p_{n-1}R,$$

where the p_i are as in Theorem 4.59 and n is the degree of $\mathbb{Q}(\gamma)$ over $\mathbb{Q}$.

EXERCISE 4.61. Let K be a number field and $\alpha_1, \ldots, \alpha_n \in K^*$ for $n > 0$. Consider $I = \alpha_1\mathbb{Z} + \cdots + \alpha_n\mathbb{Z} \subseteq K$ and $R_i = \mathbb{Z}[\alpha_1/\alpha_i, \ldots, \alpha_n/\alpha_i]$. Show that $\mathrm{Bl}(I) = \bigcap_i R_i$.

EXERCISE 4.62. Let R be a number ring.

a. Show that every invertible ideal is generated by two elements.

b. It follows from (a) that, if R is a maximal order, then every ideal is generated by two elements. Show that the converse fails for $R = \mathbb{Z}[\sqrt{5}]$.

4.5. Coprime bases for number rings. We will proceed as in Section 2 and give a coprime basis algorithm for ideals using the blowup. Recall the notation $\mathcal{I}(R)$ for the group of invertible ideals of a commutative ring R.

DEFINITION 4.63. Let R be a number ring and let X be a set of non-zero integral ideals of R. For a number ring $R \subseteq S \subseteq \mathbb{Q}R$ we write $SX = \{SI : I \in X\}$. We write $\overline{X}_R$, or $\overline{X}$ when the ring is understood, for the closure of $X \cup \{R\}$ under addition, multiplication and integral division, i.e. $I : J$ for ideals $I, J \subset R$ for which J is invertible and $I : J \subseteq R$. A *coprime basis* for X is a set B of invertible ideals strictly contained in R which are pairwise coprime and for which X is contained in $\langle B \rangle$, the subgroup of $\mathcal{I}(R)$ generated by B. We equip the set of coprime bases of X with a partial order (cf. Exercise 2.10) where $B \leq C$ if $\langle B \rangle \subseteq \langle C \rangle$.

In particular, for a coprime basis of X to exist, the elements of X need to be invertible. Hence we cannot have a direct translation of Lemma 2.11 to the case of ideals.

EXERCISE 4.64. Let R be a number ring and X a set of non-zero integral ideals of R. Show that if $\overline{X} \subseteq \mathcal{I}(R)$, then $\overline{X}$ is closed under intersection.

EXERCISE 4.65. Let $R \subseteq S \subseteq \mathbb{Q}R$ be number rings and let X be a set of non-zero ideals of R.

a. Show that if I and J are fractional R-ideals with J invertible, then $SI : SJ = S(I : J)$, and give a counter example where J is non-invertible.

b. Show that the map $I \mapsto SI$ that maps ideals of R to ideals of S respects addition, multiplication and integral division.

c. With the notation as in Definition 4.63, conclude that $S\overline{X} \subseteq \overline{SX}$.

EXERCISE 4.66. Let R be an order and X a (possibly infinite) set of non-zero ideals of R. Show that there exists a unique number ring $R \subseteq S \subseteq K$ which is minimal with respect to inclusion such that $\overline{SX} \subseteq \mathcal{I}(S)$. Show that this S is an order.

Hint: Iteratively blow up the ideals of X and use that the index of R in $\mathcal{O}_K$ is finite.

PROPOSITION 4.67. *Let R be an order and X a (possibly infinite) set of non-zero ideals of R. Then $\overline{X} \subseteq \mathcal{I}(R)$ if and only if X has a coprime basis. If X has a coprime basis, then $\{I \in \overline{X} \mid \#\{J \in \overline{X} \mid I \subseteq J\} = 2\}$ is the minimal coprime basis (wrt. the partial order of Definition 4.63).*

PROOF. If $\overline{X} \subset \mathcal{I}(R)$, then the proof of Lemma 2.11 can be followed to show that $\{I \in \overline{X} \mid \#\{J \in \overline{X} \mid I \subseteq J\} = 2\}$ is the minimal coprime basis of X. Conversely, if X has a coprime basis B, then $\overline{X} \subseteq \langle B \rangle \subseteq \mathcal{I}(R)$. □

THEOREM 4.68. *There exists a polynomial-time algorithm that, given an order R and a finite set of non-zero ideals X of R, computes the minimal order $R \subseteq S$ for which there exists a coprime basis for SX and computes the minimal coprime basis.*

ALGORITHM 4.69. Write $\mathfrak{a}_1, \ldots, \mathfrak{a}_m$ for the elements of X not equal to R. Construct a complete simple graph G with m vertices and label the vertices $\mathfrak{a}_1, \ldots, \mathfrak{a}_m$. Set S equal to R. Then, for each i successively replace S by $\mathrm{Bl}(S\mathfrak{a}_i)$. While there are edges in G, repeat the following 5 steps:

(1) Choose an edge $\{a, b\}$ of G and let $\mathfrak{a}$ and $\mathfrak{b}$ be the labels of a and b.
(2) Compute $\mathfrak{c} = \mathfrak{a} + \mathfrak{b}$, replace S by $\mathrm{Bl}(S\mathfrak{c})$, $\mathfrak{c}$ by $S\mathfrak{c}$ and compute $\mathfrak{c}^{-1} = S : \mathfrak{c}$.
(3) Add a vertex c labeled $\mathfrak{c}$ to G and connect it to a, b and those vertices which are neighbors of both a and b.

(4) Update the labels of a and b to $\mathfrak{a}\mathfrak{c}^{-1}$ and $\mathfrak{b}\mathfrak{c}^{-1}$ respectively.
(5) For each $s \in \{a, b, c\}$, if the label of s is S, then delete s and its incident edges from G.

With L the set of labels of the remaining vertices, the required order is S with basis $\{S\mathfrak{a} \,|\, \mathfrak{a} \in L\}$ of $\overline{SX}$.

Note that after step (4), the labels of a, b and c are ideals of S, but this is not necessarily true for the remaining labels if S has become larger.

PROOF OF THEOREM 4.68. We will show that Algorithm 4.69 is correct and runs in polynomial time. First note that the output is indeed a coprime basis for SX. Observe that in each iteration S is contained in the minimal order T for which $(\overline{TX})_T \subseteq \mathcal{I}(T)$, since $\mathrm{Bl}(S\mathfrak{c})$ is the minimal order containing S in which $\mathfrak{c}$ becomes invertible. By Proposition 4.67 this means that S is indeed the minimal order for which SX has a coprime basis. That the output is the minimal coprime basis is analogous to the proof of Theorem 2.12. One similarly proves the algorithm runs in polynomial time, with one caveat: We need to remark that by Exercise 4.22 the growth of S during the algorithm is sufficiently bounded and thus all operations on the ideals can be performed in polynomial time. □

4.6. Applications of coprime bases. In this section we consider some algorithmic applications of the coprime basis algorithm for ideals.

THEOREM 4.70. *Let K be a number field and let $G \subseteq K^*$ be a subgroup. Then there exists a unique subring $S \subseteq K$, minimal with respect to inclusion, for which for all finite non-empty subsets $Y \subseteq G$ the fractional ideal SY is invertible. Moreover, this S is an order.*

PROOF. Let T be the smallest subring for which all fractional ideals $T + Tx$ with $x \in G$ are invertible, which is the smallest ring containing $\mathrm{Bl}(\mathbb{Z} + \mathbb{Z}x)$ for all $x \in G$. One easily shows that $(T+Tx)^{-1} \subseteq T$ and $Tx = (T+Tx^{-1})^{-1} : (T+Tx)^{-1}$ for all $x \in G$. Now let $T \subseteq S \subseteq K$ be the smallest subring for which all ideals in the closure $\mathcal{C}$ of $S \cdot \{(T + Tx)^{-1} \,|\, x \in G\}$ are invertible, which exists by Exercise 4.66. Write $I_x = (S + Sx)^{-1}$ for $x \in G$. Then for $Y \subseteq G$ we have

$$SY = \sum_{x \in Y} (I_{x^{-1}} : I_x) = \Big(\sum_{x \in Y} \Big(\prod_{y \in Y \setminus \{x\}} I_y \Big) I_{x^{-1}} \Big) : \Big(\prod_{y \in Y} I_y \Big),$$

where the latter numerator and denominator are in $\mathcal{C}$ and hence invertible. Clearly S is an order. It remains to show that this S is minimal.

Let U be an order such that for all finite non-empty subsets $Y \subseteq G$ the fractional ideal UY is invertible. Clearly $T \subseteq U$ since $\{1, x\}$ is a finite subset of G for all $x \in G$. Let Z be the set of fractional ideals of U of the form $UX : UY$ for finite non-empty $X, Y \subseteq G$. Then $Z \subseteq \mathcal{I}(U)$ and Z is closed under addition, multiplication and division. Since $\{(U + Ux)^{-1} \,|\, x \in G\} \subseteq Z$, we conclude that Z also contains its closure. Thus $S \subseteq U$ and S is minimal. □

EXERCISE 4.71. Show that the order S produced by Theorem 4.70 need not satisfy $\mathbb{Q}S = K$. Can you give an expression for $\mathbb{Q}S$?

THEOREM 4.72. *There exists a polynomial-time algorithm that, given an order R in a number field K and a finitely generated subgroup $G = \langle x_1, \ldots, x_n \rangle \subseteq K^*$,* *computes the order $R \subseteq S$, minimal with respect to inclusion, such that for all finite non-empty subsets $Y \subseteq G$ the fractional ideal SY is invertible.*

PROOF. We proceed as in Theorem 4.70. Compute the minimal order T for which for all i the $T + Tx_i$ are invertible, for example using Example 4.60. Note that $T + Tx_i^{-1} = x_i^{-1} \cdot (T + Tx_i)$ is also invertible, and that the $(T + Tx_i^{\pm 1})^{-1}$ are integral ideals of T. Then compute the minimal order S for which all ideals in the closure $\mathcal{C}$ of $S \cdot \{(T + Tx_i)^{-1}, (T + Tx_i^{-1})^{-1} \mid i\}$ are invertible using Theorem 4.68. We claim S is the order as required.

Clearly this S is contained in the one produced in the proof of Theorem 4.70. For the reverse inclusion it suffices to show that $S + Sx$ is invertible in S for all $x \in Y$. Write $I_y = (S + Sy)^{-1}$ for $y \in \{x_1, \ldots, x_n, x_1^{-1}, \ldots, x_n^{-1}\}$. Let $x \in Y$ and write $x = y_1 \cdots y_k$ for $y_1, \ldots, y_k \in \{x_1, \ldots, x_n, x_1^{-1}, \ldots, x_n^{-1}\}$. We have

$$S + Sx = S + \prod_{i=1}^{k} (I_{y_i^{-1}} : I_{y_i}) = \Big(\prod_{i=1}^{k} I_{y_i} + \prod_{i=1}^{k} I_{y_i^{-1}}\Big) : \Big(\prod_{i=1}^{k} I_{y_i}\Big),$$

with numerator and denominator in $\mathcal{C}$ and hence invertible. □

THEOREM 4.73. *There exists a polynomial-time algorithm that, given an order R of a number field K and fractional ideals $\mathfrak{a}_1, \ldots, \mathfrak{a}_m$ of R, computes the kernel of the map $\mathbb{Z}^m \to \mathcal{I}(\mathcal{O}_K)$ given by $(k_1, \ldots, k_m) \mapsto \prod_i (\mathcal{O}_K \mathfrak{a}_i)^{k_i}$.*

PROOF. We may write $\mathfrak{a}_i$ as a quotient $N_i : D_i$ for all i with $N_i, D_i \subseteq R$, for example with $D_i = nR$ for some $n \in \mathbb{Z}_{>0}$. Compute an order $R \subseteq S \subseteq K$ and a coprime basis $\mathfrak{c}_1, \ldots, \mathfrak{c}_n$ for $SN_1, \ldots, SN_m, SD_1, \ldots, SD_m$, as well as integers n_{ij}, d_{ij} such that $N_i = \prod_j \mathfrak{c}_j^{n_{ij}}$ and $D_i = \prod_j \mathfrak{c}_j^{d_{ij}}$. Then the kernel of $\mathbb{Z}^m \to \mathcal{I}(S)$ is the kernel of the map $\mathbb{Z}^m \to \mathbb{Z}^n$ given by $(k_1, \ldots, k_m) \mapsto (\sum_i k_i(n_{ij} - d_{ij}))_j$, which we may compute using Theorem 3.30. Finally, note that the map $\mathcal{I}(S) \to \mathcal{I}(\mathcal{O}_K)$ is injective on any subgroup generated by a coprime basis. □

In particular, we may decide whether some power-product of invertible ideals over any order R is torsion in $\mathcal{I}(R)$ (Exercise 4.55).

COROLLARY 4.74. *There exists a polynomial-time algorithm that, given elements $\alpha_1, \ldots, \alpha_m$ in a number field K and integers $k_1, \ldots, k_m \in \mathbb{Z}$, computes the kernel of the map $\mathbb{Z}^m \to K^*/\mathcal{O}_K^*$ given by $(k_1, \ldots, k_m) \mapsto \prod_i \alpha_i^{k_i} \cdot \mathcal{O}_K^*$.*

PROOF. Pick a full-rank order $R \subseteq K$ and apply Theorem 4.73 to $\alpha_1 R, \ldots, \alpha_m R$. □

4.7. Unit products. In this section we will prove the following generalization of Theorem 1.1 from the introduction.

THEOREM 4.75 (Ge [**11**]). *There exists a polynomial-time algorithm that, given elements $\alpha_1, \ldots, \alpha_m$ in a number field K, computes the kernel of the map $\mathbb{Z}^m \to K^*$ given by $(k_1, \ldots, k_m) \mapsto \prod_i \alpha_i^{k_i}$.*

Corollary 4.74 is close to being a proof of this. For this, it essentially remains to prove Theorem 4.75 for $\alpha_i \in \mathcal{O}_K^*$.

Let K be a number field and let $X = X(K)$ be the set of ring homomorphisms $K \to \mathbb{C}$. For a subgroup $G \subseteq K^*$ one obtains a subgroup

$$\Lambda_G = \Big\{ \big(\log \sigma(\alpha)\big)_{\sigma \in X} : \alpha \in G \Big\} + 2\pi \mathrm{i} \mathbb{Z}^X \subseteq \mathbb{C}^X,$$

which is independent of the chosen branch of the natural logarithm, and a group isomorphism $G \to \Lambda_G/2\pi\mathrm{i}\mathbb{Z}^X$. In our special case where $G \subseteq \mathcal{O}_K^*$, it is even a lattice (Exercise 4.77). Then, we compute the kernel of $\mathbb{Z}^m \to G$ from the map $\mathbb{Z}^m \to \Lambda_G$.

LEMMA 4.76. *Let K be a number field and let $\varepsilon \in \mathcal{O}_K$ be non-zero. Then $\varepsilon = 1$ if and only if for all $\sigma \in X$ there exists $k_\sigma \in \mathbb{Z}$ such that $|\log\sigma(\varepsilon) - k_\sigma \cdot 2\pi\mathrm{i}| < \log 2$.*

PROOF. The forward implication is trivial. Suppose now we have such $k_\sigma \in \mathbb{Z}$ for each $\sigma \in X$. Let $\sigma \in X$. With $z = \log\sigma(\varepsilon) - k_\sigma \cdot 2\pi\mathrm{i}$ we obtain

$$|\sigma(\varepsilon) - 1| = |\mathrm{e}^z - 1| = \Big|\sum_{n=1}^{\infty}\frac{z^n}{n!}\Big| \le \sum_{n=1}^{\infty}\frac{|z|^n}{n!} = \mathrm{e}^{|z|} - 1 < \mathrm{e}^{\log 2} - 1 = 1.$$

Write $N : K \to \mathbb{Q}$ for the norm function. Then $N(\varepsilon - 1)$ is an integer, because it is both integral and rational, such that

$$|N(\varepsilon - 1)| = \prod_{\sigma\in X}|\sigma(\varepsilon) - 1| < 1.$$

Thus $N(\varepsilon - 1) = 0$ and $\varepsilon = 1$. □

EXERCISE 4.77. Let K be a number field. Suppose we consider $\mathbb{C}$ as a 2-dimensional Euclidean vector space with orthonormal basis $\{1, \mathrm{i}\}$ and in turn consider the orthogonal sum $\mathbb{C}^{X(K)}$ as a Euclidean vector space. Show that for every subgroup $H \subseteq \mathcal{O}_K^*$ the subgroup $\Lambda_H \subseteq \mathbb{C}^X$ is a lattice with $\lambda_1(\Lambda_H) \ge (\log 2)^2$.

PROOF OF THEOREM 4.75. First compute using Corollary 4.74 a basis $\langle b_1, \dots, b_n\rangle$ for

$$H = \Big\{(k_1, \dots, k_m) \in \mathbb{Z}^m : \prod_i \alpha_i^{k_i} \in \mathcal{O}_K^*\Big\}.$$

Consider now the lattice

$$\Lambda_H = \sum_{i=1}^{n}\mathbb{Z}\cdot\Big(\sum_{j=1}^{m} b_{ij}\log\sigma(\alpha_j)\Big)_{\sigma\in X(K)} + 2\pi\mathrm{i}\mathbb{Z}^{X(K)},$$

to which we want to apply Theorem 3.59 to deduce the required kernel. Exercise 4.77 states that $\lambda_1(\Lambda_H) \ge (\log 2)^2$. Hence we may pick ω and t in Theorem 3.59 so that their lengths are polynomially bounded. Using the work of Schönhage [**24**] and Brent [**7**] we can compute in polynomial time a t-approximation of the elements generating Λ_H. Hence the algorithm runs in polynomial time. □

5. Computing the maximal order

This section we dedicate to the computation of the maximal order of a number field, in particular why it is difficult. We do this by proving that this problem is equally hard in some sense yet to be made precise (Definition 5.2) as computing the *radical*, written $\mathrm{rad}(n)$, of some $n \in \mathbb{Z}_{>0}$, which is the largest square-free divisor of n or equivalently the product of all primes dividing n without multiplicity. Computing radicals is considered difficult, as it seems like we need to factor n, although it is unknown whether it is actually equally hard as factorization.

We will consider rings which we encode as a finitely generated (additive) abelian group A together with a multiplication morphism $A \otimes A \to A$. This includes both orders and finite commutative rings.

Recall the definition of the the nilradical from Definition 1.3.3.

THEOREM 5.1 (cf. Theorem 1.3 in [**8**]). *The following three computational problems are equally hard:*

(1) *Given a number field K, compute $\mathcal{O}_K$.*
(2) *Given a finite ring A, compute* $\operatorname{nil}(A)$.
(3) *Given an integer $m > 0$, compute* $\operatorname{rad}(m)$.

We prove this theorem at the end of Section 5.3. As a corollary to this theorem we obtain an algorithm to compute maximal orders, albeit a rather slow one, because we have a naive algorithm to compute radicals.

5.1. Polynomial-time reductions. Recall the definition of a problem from Definition 1.2. Intuitively, a problem f is easier than a problem g if you can solve f once you know how to solve g, or more specifically to algorithms, when you can solve them in polynomial time. We can make this more formal.

DEFINITION 5.2. Let $f : I_f \to f(I_f)$ and $g : I_g \to g(I_g)$ be two problems. We say f *can be reduced to* g, or symbolically $f \preceq g$, if there exist polynomial-time algorithms $a : I_f \to I_g$ and $b : I_f \times g(I_g) \to f(I_f)$ so that $f(i) = b(i, g(a(i)))$ for all $i \in I_f$. We say f and g are *equally hard* if $f \preceq g$ and $g \preceq f$.

Effectively, this definition states that $f \preceq g$ if we can efficiently transform the input i for f into an input for g, then run g, and from its output efficiently deduce $f(i)$. Note that if there exists a polynomial-time algorithm for f, then we always have $f \preceq g$ for any non-empty problem g. This shows the concept of reductions is not very useful for problems for which we know a polynomial-time algorithm exists. The reader should be aware that there exist non-equivalent definitions of a reduction of problems, for example where it is allowed to make multiple calls to g.

EXAMPLE 5.3. The problem of computing radicals can be reduced to computing nilradicals of finite commutative rings. Namely, for $a \in \mathbb{Z}_{>0}$ we have that $\operatorname{nil}(\mathbb{Z}/a\mathbb{Z}) = \operatorname{rad}(a)\mathbb{Z}/a\mathbb{Z}$, so we can deduce $\operatorname{rad}(a)$ given $\operatorname{nil}(\mathbb{Z}/a\mathbb{Z})$.

EXERCISE 5.4 (Cf. Lemma 8.1.3 in [**15**]). Let l be prime, K a field and $a \in K$. Show that $f = X^l - a$ is irreducible in $K[X]$ if and only if a is not an l-th power in K.

Hint: Write the constant term of a factor of f in terms of a single root of f.

LEMMA 5.5. *The problem of computing radicals can be reduced to computing the maximal order of a number field.*

PROOF. Suppose we are given $a \in \mathbb{Z}_{>1}$ for which to compute $\operatorname{rad}(a)$. Choose a sufficiently small prime $l > \log_2 a$. Such an l will be at most $2\log_2 a$ by Bertrand's postulate, and $2\log_2 a$ is sufficiently small to do a naive primality test. Let $K = \mathbb{Q}(\alpha) = \mathbb{Q}[X]/(X^l - a)$, which is a field by Exercise 5.4. Compute the maximal order $\mathcal{O}$ of K. For each $0 < i < l$ express α^i in terms of a basis for $\mathcal{O}$, from which we may compute the largest integer b_i such that $\alpha^i/b_i \in \mathcal{O}$. Then $c_i = a^i/b_i^l = (\alpha^i/b_i)^l \in \mathcal{O} \cap \mathbb{Q} = \mathbb{Z}$. Now compute $\gcd\{c_i \mid 0 < i < l\}$. We claim that this equals $\operatorname{rad}(a)$.

Write $a = \prod_p p^{a_p}$ and similarly b_i and c_i. Fix some prime $p \mid a$. Then $0 < a_p \leq \log_2 a < l$. As b_i is maximal such that $a^i/b_i^l \in \mathbb{Z}$, we have that $b_{ip} = \lfloor ia_p/l \rfloor$ and that c_{ip} is the remainder of ia_p upon division by l. This remainder is positive for all i, and equals 1 for some $0 < i < l$. It follows that $\gcd\{c_i \mid 0 < i < l\} = \prod_{p \mid a} p = \operatorname{rad}(a)$, as was to be shown. □

Example 5.3 and Lemma 5.5 already prove a part of Theorem 5.1 stated in the introduction. Each of the problems in this theorem has an associated decision problem, which we will show are all equally hard as well.

Each of the problems in Theorem 5.1 has an associated decision problem, which we will show are all equally hard as well.

THEOREM 5.6 (cf. Theorem 6.12 in [**8**]). *The following three decision problems are equally hard:*

(1) *Given an order R in a number field K, decide whether $R = \mathcal{O}_K$.*
(2) *Given a finite commutative ring A, decide whether* $\mathrm{nil}(A) = 0$.
(3) *Given an integer $m > 0$, decide whether m is square-free.*

We will prove this theorem at the end of Section 5.4. As of January 23, 2026, for none of the six problems in Theorem 5.1 and Theorem 5.6 there are known polynomial-time algorithms. Note that each decision problem trivially reduces to its corresponding computational problem.

EXERCISE 5.7. Prove $3 \preceq 1$ and $3 \preceq 2$ for Theorem 5.6.
Hint: Use Exercise 4.23.

EXERCISE 5.8. Show that there exists a polynomial-time algorithm that, given $a, b \in \mathbb{Z}_{>0}$, decides whether $\mathrm{rad}(a) = \mathrm{rad}(b)$.

EXERCISE 5.9. Show that prime factorization is equally hard as decomposing a finite abelian group as a product of cyclic groups of prime-power order.

5.2. Nilradicals. In this section we will compute $\mathrm{nil}(A)$ for a finite commutative ring A given information about $\#A$, which will prove part of Theorem 5.1. We will need to treat the small primes in $\#A$ separately.

LEMMA 5.10. *Let A be a commutative ring and let $n, k \in \mathbb{Z}_{\geq 1}$. Then* $\mathrm{nil}(A/n^kA)$ *is the inverse image of* $\mathrm{nil}(A/nA)$ *under the natural map $A/n^kA \to A/nA$.*

PROOF. We have $(nA)^k \equiv 0 \bmod n^kA$, so $nA/n^kA \subseteq \mathrm{nil}(A/n^kA)$. □

LEMMA 5.11. *There exists a polynomial time algorithm that, given a prime p and a finite commutative $\mathbb{F}_p$-algebra R, computes the Frobenius morphism $x \mapsto x^p$ of R.*

PROOF. It suffices to compute x^p for each x in some additively generating set of R. Note that p can be too large to naively compute x^p with $p-1$ multiplications. Iteratively compute $x^{2^0}, x^{2^1}, x^{2^2}, \ldots, x^{2^n}$ by squaring for $n \leq \log_2 p$, and by writing p in base-2 then compute x^p in at most $\log_2 p$ multiplications. □

PROPOSITION 5.12 (cf. Algorithm 6.1 in [**8**]). *There is a polynomial-time algorithm that, given a finite commutative ring A with $\#A$ a prime power, computes the nilradical* $\mathrm{nil}(A)$.

We first state the algorithm.

ALGORITHM 5.13. Let A be a finite commutative ring of prime power order.

(1) Compute $\#A$ and find p prime and $n \in \mathbb{Z}_{\geq 0}$ such that $\#A = p^n$.
(2) Compute $B = A/pA$, which is an $\mathbb{F}_p$-algebra, and the least $t \in \mathbb{Z}_{\geq 0}$ such that $\dim_{\mathbb{F}_p} B \leq p^t$.
(3) Compute the Frobenius F of B with Lemma 5.11 and compute $K = \ker(F^t)$.

(4) Return the inverse image of K under the natural map $\pi : A \to B$.

PROOF OF PROPOSITION 5.12. We will show Algorithm 5.13 satisfies our requirements. Note that all steps in the algorithm involve only basic computations as described in Section 1.2 and Section 3. Most importantly, we use that F and thus F^t is a linear map, to compute K. Hence the algorithm runs in polynomial time. We will thus show it is correct. For each $x \in \mathrm{nil}(B)$ and $m \geq \dim_{\mathbb{F}_p} B$ we have $x^m = 0$, so $x \in \mathrm{nil}(B)$ if and only if $F^t(x) = 0$. Then $\pi^{-1}K = \mathrm{nil}(A)$ by Lemma 5.10, so the algorithm is correct. □

EXERCISE 5.14. You may assume the existence of a polynomial-time primality test, for example the AKS algorithm [**1**].

a. Show that a finite commutative ring R is local if and only if $\#R = p^k$ for some prime p and integer $k > 0$ and $\mathrm{nil}(R)$ is a maximal ideal.

b. Show that a finite commutative ring R with $\mathrm{nil}(R) = 0$ is a field if and only if there exists some prime p such that $\#R = p^k$ for some $k \in \mathbb{Z}_{>0}$ and $\dim_{\mathbb{F}_p} \ker(x \mapsto x^p - x) = 1$.

c. Conclude that there exist a polynomial-time algorithm that, given a finite commutative ring R, decides whether R is local and if so computes its maximal ideal.

Recall the definition of the trace from Section 4.1.

DEFINITION 5.15. Let $A \subseteq B$ be commutative rings such that B is free of finite rank over A as an A-module. Then we define the *trace radical* of B over A as

$$\mathrm{Trad}(B/A) = \{x \in B \mid \mathrm{Tr}_{B/A}(xB) = 0\} = \ker(x \mapsto (y \mapsto \mathrm{Tr}_{B/A}(xy))),$$

where the kernel is taken of a homomorphism of B-modules $B \to \mathrm{Hom}_A(B, A)$.

It follows trivially that $\mathrm{Trad}(B/A)$ is an ideal of B. In fact, it is the largest ideal of B contained in $\ker(\mathrm{Tr}_{B/A})$. For A a commutative ring that is finitely generated as a module over $\mathbb{Z}$ we may compute $\mathrm{Trad}(B/A)$ in polynomial time using the theory of Section 3. As the following exercise in linear algebra shows, the trace function can in some sense test nilpotency. Similarly, we hope to compute $\mathrm{nil}(B)$ from $\mathrm{Trad}(B/A)$.

EXERCISE 5.16. Let A be a field and let B be a finite-dimensional A-vector space.

a. Suppose A has characteristic 0 or p for some prime $p > \dim_A(B)$. Show that $M \in \mathrm{End}_A(B)$ is nilpotent if and only if $\mathrm{Tr}(M^n) = 0$ for all $n \in \mathbb{Z}_{>0}$.
Hint: Apply Newton's identities to the characteristic polynomial of M.

b. Suppose A has characteristic 0 or p for some prime $p > \dim_A(B)$ and that B is a commutative A-algebra. Show that $\mathrm{Trad}(B/A) = \mathrm{nil}(B)$.

c. Give a counter-example to (b) when A has prime characteristic $p \leq \dim_A(B)$.

d. What happens when we replace A by a local ring and B by a free A-module?

EXERCISE 5.17. Let $A \subseteq B$ be a commutative rings such that B is free of finite rank over A. Let $S \subseteq A$ be a multiplicative set, $\mathfrak{a} \subseteq A$ an ideal and $x \in B$.

a. Show that the image of $\mathrm{Tr}_{B/A}(x)$ in $S^{-1}A$ equals $\mathrm{Tr}_{S^{-1}B/S^{-1}A}(x)$.

b. Show that the image of $\mathrm{Tr}_{B/A}(x)$ in $A/\mathfrak{a}$ equals $\mathrm{Tr}_{(B/\mathfrak{a}B)/(A/\mathfrak{a})}(x)$.

EXERCISE 5.18. Let $A \subseteq B_1, B_2$ be commutative rings such that B_1 and B_2 are free A-modules of finite rank, and let $b = (b_1, b_2) \in B_1 \times B_2$. Show that

$$\mathrm{Tr}_{(B_1\times B_2)/A}(b) = \mathrm{Tr}_{B_1/A}(b_1) + \mathrm{Tr}_{B_2/A}(b_2) \quad \text{and}$$
$$\det{}_{(B_1\times B_2)/A}(b) = \det{}_{B_1/A}(b_1) \cdot \det{}_{B_2/A}(b_2).$$

EXERCISE 5.19. Suppose B is a free A-module and let $I, J \subseteq A$ be coprime ideals with $I \cap J = 0$. Then $B \cong (B/BI) \times (B/BJ)$ and B/BI is a free A/I-module. Show that Tr and det commute with this isomorphism when B has finite rank.

PROPOSITION 5.20. *Let m be a square-free integer and B a commutative algebra over $A = \mathbb{Z}/m\mathbb{Z}$ that is free of finite rank as a module. If $p > \mathrm{rk}_A(B)$ for all primes p dividing m, then* $\mathrm{Trad}(B/A) = \mathrm{nil}(B)$.

PROOF. We have $A \cong \prod_{p|m}(\mathbb{Z}/p\mathbb{Z})$ and $B \cong \prod_{p|m}(B/pB)$. Clearly, under these isomorphisms we have $\mathrm{nil}(B) \cong \prod_{p|m} \mathrm{nil}(B/pB)$, and from Exercise 5.19 it follows that also $\mathrm{Trad}(B/A) \cong \prod_{p|m} \mathrm{Trad}((B/pB)/(A/pA))$. Note that $p > \mathrm{rk}_A(B) = \mathrm{rk}_{A/pA}(B/pB)$ for all $p \mid m$. Hence it suffices to prove the proposition for m prime, which is Exercise 5.16. □

With this proposition, we may prove the following.

THEOREM 5.21. *There exists a polynomial-time algorithm that, given a finite commutative ring A and the integer* $\mathrm{rad}(\#A)$, *computes* $\mathrm{nil}(A)$.

ALGORITHM 5.22. Let A be a finite commutative ring and let $r = \mathrm{rad}(\#A)$ and compute $l = \lfloor \log_2(\#A) \rfloor$.

(1) Apply Algorithm 2.13 to $\{r, \#(A/rA)\} \cup \{p \mid p \le l,\ p \text{ prime}\}$ to compute a coprime basis $c_1, \ldots, c_n$.
(2) Factor $\#(A/rA) = \prod_{i\in I} c_i^{k_i}$ with I such that $k_i \ge 1$ for all $i \in I$.
(3) For each $i \in I$ compute $\mathrm{nil}(A/c_iA)$ as follows:
   (a) If $c_i \le l$, then c_i is prime and apply Algorithm 5.13.
   (b) If $c_i > l$, then A/c_iA is free over $\mathbb{Z}/c_i\mathbb{Z}$ of rank k_i and compute $\mathrm{nil}(A/c_iA)$ as $\mathrm{Trad}((A/c_iA)/(\mathbb{Z}/c_i\mathbb{Z}))$ using linear algebra.
(4) Return the inverse image of $\prod_{i\in I} \mathrm{nil}(A/c_iA)$ under the natural map $A \to \prod_{i\in I}(A/c_iA)$.

PROOF OF THEOREM 5.21. Clearly Algorithm 5.22 runs in polynomial time if it is correct. For $i \in I$ we have $\gcd(c_i, r) > 1$, so $c_i \mid r$ and c_i is square-free. If $c_i \le l$, then it is prime, so Algorithm 5.13 correctly computes $\mathrm{nil}(A/c_iA)$ by Proposition 5.12 in step (3a). Now suppose $c_i > l$. As c_i is square-free we have $A/c_iA \cong \prod_{p|c_i} A/pA$. Clearly A/pA is free over $\mathbb{Z}/p\mathbb{Z}$ since the latter is a field, and its rank must be k_i. Hence A/c_iA is free over $\mathbb{Z}/c_i\mathbb{Z}$ of rank k_i. Thus the conditions to Proposition 5.20 are satisfied and step (3b) correctly computes $\mathrm{nil}(A/c_iA)$. By Lemma 5.10 we compute $\mathrm{nil}(A)$ in step (4). □

5.3. Non-invertible ideals. By Theorem 4.52, finding larger orders within the same number field can be done by finding non-invertible ideals.

LEMMA 5.23. *Let R be a full-rank order in a number field K and let p be prime. Then $p \mid \#(\mathcal{O}_K/R)$ if and only if R has a non-invertible prime ideal containing p.*

PROOF. Every prime of R containing p is invertible if and only if every prime of R_p is invertible if and only if R_p is integrally closed if and only if $R_p = (\mathcal{O}_K)_p$ if and only if $(\mathcal{O}_K/R)_p = 1$ if and only if $p \nmid \#(\mathcal{O}_K/R)$. □

LEMMA 5.24 (cf. Theorem 1.2 in [8]). *There is a polynomial-time algorithm that, given a full-rank order R in a number field K and square-free integer d contained in all non-invertible prime ideals of R, computes $\mathcal{O}_K$.*

PROOF. For all $R \subseteq S \subseteq \mathcal{O}_K$ it holds that d is contained in all non-invertible prime ideals of S. Namely, this is equivalent to $p \mid d$ for all primes $p \mid \#(\mathcal{O}_K/S)$ by Lemma 5.23, and we have that $\#(\mathcal{O}_K/S) \mid \#(\mathcal{O}_K/R)$.

Using Algorithm 5.22 we may compute $\mathrm{rad}(R/dR)$ and in turn its inverse image $\mathfrak{a}$ in R. We compute $\mathrm{Bl}(\mathfrak{a})$, and repeat the procedure with R replaced by $\mathrm{Bl}(\mathfrak{a})$ with the same d until $R = \mathrm{Bl}(\mathfrak{a})$. Then $\mathfrak{a}$ is the product of all prime ideals containing d, so by assumption every non-invertible prime ideal of R is a factor of $\mathfrak{a}$. Because $\mathfrak{a}$ is invertible over R by Theorem 4.52, as is each factor, we have $R = \mathcal{O}_K$. □

PROPOSITION 5.25. *Computing the maximal order of a number field reduces to computing radicals of integers.*

PROOF. Let R be any order of K. Then we may compute $d = \mathrm{rad}\,|\Delta(R)|$. As $\#(\mathcal{O}_K/R) \mid \Delta(R)$ by Exercise 4.21 we get that $\mathrm{rad}(\#(\mathcal{O}_K/R)) \mid d$. Then by Lemma 5.23 every prime p for which there exists a non-invertible prime of R above p we have $p \mid d$. Since d is square-free we obtain $\mathcal{O}_K$ from Lemma 5.24. □

We can now prove the main theorem.

PROOF OF THEOREM 5.1. Example 5.3, Lemma 5.5, Theorem 5.21 and Proposition 5.25 combined prove Theorem 5.1. □

5.4. Reduced discriminants. Lemma 5.23 states that we can find the non-invertible prime ideals of an order R of a number field K precisely above rational primes dividing $\#(\mathcal{O}_K/R)$. However, algorithmically we do not have access to this index. In the search of these rational primes p, we could consider p for which $p^2 \mid \Delta(R)$, as $\Delta(R) = \#(\mathcal{O}_K/R)^2 \cdot \Delta(\mathcal{O}_K)$ by Exercise 4.21. This generally gives 'false positives', as $\Delta(\mathcal{O}_K)$ could have independent square divisors. To recover an equally strong result as Lemma 5.23 that is algorithmically accessible, we turn to the reduced discriminant.

We equip a number field K with the symmetric $\mathbb{Q}$-bilinear map $K \times K \to \mathbb{Q}$ given by $(x, y) \mapsto \mathrm{Tr}_{K/\mathbb{Q}}(xy)$. By Exercise 4.25 the hypothesis of Exercise 3.15 is satisfied and thus we have a duality as follows.

DEFINITION 5.26 (cf. Exercise 3.15). Let I be a subgroup of a number field K. We define the *trace dual* $I^\dagger$ of I as

$$I^\dagger = \{x \in K \mid \mathrm{Tr}_{K/\mathbb{Q}}(xI) \subseteq \mathbb{Z}\}.$$

We define the *reduced discriminant* $\delta(R)$ of a full-rank order R in K to be the exponent of the abelian group $R^\dagger/R$.

Since $\mathrm{Tr}_{K/\mathbb{Q}}(R) \subseteq \mathbb{Z}$ for any order R, we have $R \subseteq R^\dagger$ and thus $R^\dagger/R$ is a well-defined group. Moreover, $R^\dagger$ is a fractional R-ideal, so the quotient is in fact finite, and $\delta(R)$ is well-defined. Note that we can compute $R^\dagger$ and $\delta(R)$ in polynomial time. As is the case for the trace radical in its relation to the nilradical (Proposition 5.20), the reduced discriminant has good properties only at the 'large primes'.

EXERCISE 5.27. Let K be a number field and let $0 \subsetneq I, J \subseteq K$ be full-rank finitely generated additive subgroups. Show that if $I \subseteq J$, then $J^\dagger \subseteq I^\dagger$. Conclude that $S \subseteq R^\dagger$ for all full-rank orders R and S in K. Show that $(I^\dagger)^\dagger = I$, $(I+J)^\dagger = I^\dagger \cap J^\dagger$ and $(I \cap J)^\dagger = I^\dagger + J^\dagger$.

The following proposition motivates the name 'reduced discriminant'.

PROPOSITION 5.28. *Let R be an order. Then $|\Delta(R)| = \#(R^\dagger/R)$.*

PROOF. Let $b_1, \ldots, b_n$ be a basis of R and consider its dual basis $b_1^\dagger, \ldots, b_1^\dagger$ from Exercise 3.15. Consider then the linear map $f : \mathrm{Q}(R) \to \mathrm{Q}(R)$ that maps $b_i^\dagger$ to b_i for all i. Then $\#(R^\dagger/R) = |\det(f)|$ and

$$\Delta(R) = \det((\mathrm{Tr}(b_i b_j))_{i,j}) = \det(f) \cdot \det((\mathrm{Tr}(b_i f^{-1}(b_j)))_{i,j}) = \det(f),$$

as was to be shown. □

It follows that $\mathrm{rad}\,|\Delta(R)| = \mathrm{rad}(\delta(R))$.

LEMMA 5.29. *Let $R \subseteq S$ be full-rank orders of a number field of degree n and let $p > n$ be prime. Then $S/R \subseteq p(R^\dagger/R)$.*

PROOF. By Proposition 5.20 we have that

$$\mathrm{nil}(R/pR) = \mathrm{Trad}((R/pR)/\mathbb{F}_p) = \{x \in R \mid \mathrm{Tr}(xR) \subseteq p\mathbb{Z}\}/pR = (pR^\dagger \cap R)/pR,$$

and similarly for S. The ring homomorphism $R/pR \to S/pS$ preserves nilpotents, hence $pR^\dagger \cap R \subseteq pS^\dagger \cap S$. We then apply $(-)^\dagger$ to the inclusion $R^\dagger \cap \frac{1}{p}R \subseteq S^\dagger$ to obtain $R + pR^\dagger \supseteq S$ from Exercise 5.27. Hence $S/R \subseteq (R+pR^\dagger)/R = p(R^\dagger/R)$. □

LEMMA 5.30. *Let R be a local ring with ideals $I, J \subseteq R$. If $I + J$ is principal, then $I \subseteq J$ or $J \subseteq I$.*

PROOF. Write $\mathfrak{m} \subseteq R$ for the maximal ideal and $I + J = xR$ for some $x \in R$. Consider the $R/\mathfrak{m}$-vector space $A = xR/x\mathfrak{m}$. Note that the image of $I + J$ in A generates A. Since A is at most one-dimensional, being generated by x over $R/\mathfrak{m}$, we may assume without loss of generality that the image of I generates A. Hence $I + x\mathfrak{m} = xR$, so by Lemma 4.45.1 we have $I = xR$ and thus $J \subseteq I$. □

DEFINITION 5.31. Let R be an order and p a rational prime. We say a fractional ideal I of R is *invertible at p* if IR_p is an invertible ideal of R_p, where R_p is the localization of R at p as a $\mathbb{Z}$-algebra.

EXERCISE 5.32. Show that a fractional ideal is invertible if and only if it is invertible at all rational primes.

PROPOSITION 5.33. *Let R be an order of rank n, let p be a prime and let $a \in \mathbb{Z}_{>0}$ be such that $0 < \mathrm{ord}_p(a) < \mathrm{ord}_p(\delta(R))$. If $\delta(R)R^\dagger + aR \subseteq R$ is invertible at p, then $p \leq n$.*

PROOF. We will write $(-)_p$ for the localization at the prime p. Let $k = (\mathbb{Z}/\delta(R)\mathbb{Z})_p$ and consider the k-algebra

$$A = (R/\delta(R)R)_p = \prod_{\mathfrak{p}|p} (R/\delta(R)R)_{\mathfrak{p}} =: \prod_{\mathfrak{p}|p} A_{\mathfrak{p}},$$

where the product ranges over all prime ideals of R above p. Note that A is free over k of rank n. Since k is local, the $A_{\mathfrak{p}}$ are also free. We will show that $p \mid \mathrm{rk}_k(A_{\mathfrak{p}})$ for some $\mathfrak{p}$, from which it follows that $p \leq n$.

First we identify $\mathfrak{p}$. Let $I = \delta(R)(R^\dagger)_p$ and $J = aR_p$ and suppose $I \subseteq J$. Then $\frac{\delta(R)}{a}(R^\dagger)_p \subseteq R_p$ and $\frac{\delta(R)}{a}$ annihilates the $\mathbb{Z}_p$-module $(R^\dagger/R)_p$. Note that $(R^\dagger/R)_p$ is the p-power torsion subgroup of $R^\dagger/R$. Thus the exponent $\delta(R)$ of $R^\dagger/R$ divides $\frac{\delta(R)}{a}$ in $\mathbb{Z}_p$. But $a \notin \mathbb{Z}_p^*$ by assumption on p, so this is a contradiction. Thus $I \not\subseteq J$. By Lemma 4.31 we may choose some prime $\mathfrak{p} \mid p$ such that $I_\mathfrak{p} \not\subseteq J_\mathfrak{p}$. If we now assume that $\delta(R)R^\dagger + aR \subseteq R$ is invertible at p, then $I_\mathfrak{p} + J_\mathfrak{p}$ is invertible. Hence $J_\mathfrak{p} \subseteq I_\mathfrak{p}$ by Corollary 4.34 and Lemma 5.30.

Recall that $A_\mathfrak{q}$ is free over k for all $\mathfrak{q} \mid p$. By Exercise 5.18 the trace map $\mathrm{Tr}_{A/k} : \prod_{\mathfrak{q}|p} A_\mathfrak{q} \to k$ is given by $(x_\mathfrak{q})_\mathfrak{q} \mapsto \sum_{\mathfrak{q}|p} \mathrm{Tr}_{A_\mathfrak{q}/k}(x_\mathfrak{q})$. Note that the image of I in A is simply $\prod_{\mathfrak{q}|p} AI_\mathfrak{q}$ and similarly for J. With Exercise 5.17 we compute

$$\begin{aligned} a\,\mathrm{Tr}_{A_\mathfrak{p}/k}(A_\mathfrak{p}) &= a\,\mathrm{Tr}_{A/k}(A_\mathfrak{p}) = \mathrm{Tr}_{A/k}(AJ_\mathfrak{p}) \subseteq \mathrm{Tr}_{A/k}(AI_\mathfrak{p}) \\ &\subseteq \mathrm{Tr}_{A/k}(AI) \subseteq \mathrm{Tr}_{R_p/\mathbb{Z}_p}(I) + \delta(R)\mathbb{Z}_p \\ &= \delta(R)\,\mathrm{Tr}_{\mathbb{Q}(R)/\mathbb{Q}}((R^\dagger)_p) + \delta(R)\mathbb{Z}_p \subseteq \delta(R)\mathbb{Z}_p. \end{aligned}$$

It follows that $p \mid \frac{\delta(R)}{a} \mid \mathrm{Tr}_{A_\mathfrak{p}/k}(1) \equiv \mathrm{rk}_k(A_\mathfrak{p}) \pmod p$, as was to be shown. □

PROPOSITION 5.34. *Let R be a full-rank order of a number field K of degree n and let $p > n$ be prime. Then $p \mid \#(\mathcal{O}_K/R)$ if and only if $p^2 \mid \delta(R)$.*

PROOF. If $p \mid \#(\mathcal{O}_K/R)$, then $\mathcal{O}_K/R$ has non-trivial p-torsion, and so does $p(R^\dagger/R)$ by Lemma 5.29. It follows that $p^2 \mid \exp(R^\dagger/R) = \delta(R)$. Conversely, if $p^2 \mid \delta(R)$, then we obtain from Proposition 5.33 an ideal that is not invertible at p, hence $p \mid \#(\mathcal{O}_K/R)$ by Lemma 5.23. □

EXERCISE 5.35. Show that there exist:
a. an order R with $\mathrm{rk}(R) = 2$ and $2 \mid \#(\mathcal{O}_{\mathbb{Q}R}/R)$ and $2^2 \nmid \delta(R)$;
b. an order R with $\mathrm{rk}(R) = 2$ and $2 \nmid \#(\mathcal{O}_{\mathbb{Q}R}/R)$ and $2^2 \mid \delta(R)$.

Exercise 5.35 shows that the equivalence of Proposition 5.34 fails to hold when we drop the assumption that $p > n$. The proposition also has an algorithmic version.

EXERCISE 5.36 (cf. Theorem 6.9 in [**8**]). Show that the following problems are computationally equally hard: Given an order R of rank n, decide if there exist and if so compute

(1) an order $R \subsetneq S \subseteq \mathbb{Q}R$ such that $p \mid \#(S/R)$ for some prime $p > n$;
(2) an $a \in \mathbb{Z}_{>0}$ such that $0 < \mathrm{ord}_p(a) < \mathrm{ord}_p(\delta(R))$ for some prime $p > n$.

We can now prove the equivalence of several decision problems.

PROOF OF THEOREM 5.6. By Exercise 5.7 it suffices to reduce the problems to decide (i) whether a finite commutative ring is reduced and (ii) whether an order is maximal, to the problem of testing whether a positive integer is square-free.

(i) Suppose A is a finite commutative ring. We compute the exponent $d = \exp(A)$ and note that if $\mathrm{nil}(A)$ is to be trivial we need d to be square-free. We test this and proceed under the assumption that d is square-free. In this case $d = \mathrm{rad}(\#A)$, so we may apply Theorem 5.21 to compute $\mathrm{nil}(A)$ and verify whether it is trivial.

(ii) We first verify whether the order R has full rank. Then we may simply check whether $\delta(R)$ is square-free to conclude that either $R \neq \mathcal{O}_K$ or that the non-invertible primes of R must lie over primes $p \leq [K : \mathbb{Q}]$ by Proposition 5.34. We then apply Lemma 5.24 with d the product of all such small primes. □

EXERCISE 5.37. Recall the definition of a projective module from Exercise 3.54. Show that there exist polynomial-time algorithms for the following problems:

a. Given an integer $q > 0$ and a non-zero $\mathbb{Z}/q\mathbb{Z}$-module M, computes the maximal integer e such that M/eM is a projective $\mathbb{Z}/e\mathbb{Z}$-module.

b. Given an order R of degree n, compute an order $R \subseteq S \subseteq \mathbb{Q}R$ such that $S^\dagger$ is invertible and $S^\dagger/S$ is projective over $\mathbb{Z}/\delta(S)\mathbb{Z}$ at all primes $p > n$.

EXERCISE 5.38 (Proposition 2.5 in [**8**]). Let R be an order and I a fractional ideal of R. Show that the following are equivalent: (1) I is invertible; (2) $(R : I) : (R : I) = R$; (3) $R : (R : I) = I$ and $I : I = R$.

EXERCISE 5.39 (Proposition 2.7 in [**8**], difficult). Let R be an order in a number field. Show that the following are equivalent: (1) $R^\dagger$ is invertible; (2) for all fractional ideals I of R it holds that $I : I = R$ if and only I is invertible. We call an order with the above equivalent properties a *Gorenstein order.*

Hint: Show that $I^\dagger = R^\dagger : I$ for all I. For $(1 \Rightarrow 2)$, use Exercise 5.38.

EXERCISE 5.40. Let the notation and hypotheses be as in Theorem 4.59, denote by f' the derivative of f, and write $A = \mathrm{Bl}(\mathbb{Z} + \mathbb{Z}\gamma)$.

a (Euler; see [**29**, Proposition 3-7-12])**.** Prove that the dual (Exercise 3.15) of the $\mathbb{Q}$-basis $1, \gamma, \ldots, \gamma^{n-1}$ of K is $p_{n-1}/f'(\gamma), p_{n-2}/f'(\gamma), \ldots, p_0/f'(\gamma)$.

b. Prove that $M = D^{1-n}$ and $M^\dagger = D/f'(\gamma)$.

c. Prove that $A^\dagger = D^{1-n} \cdot M^\dagger = D \cdot M/f'(\gamma)$ and that A is a Gorenstein order (Exercise 5.39).

d. Prove that $1/f'(\gamma), \gamma/f'(\gamma), \ldots, \gamma^{n-2}/f'(\gamma), a_n\gamma^{n-1}/f'(\gamma)$ is a $\mathbb{Z}$-basis for $A^\dagger$.

e. Show that $f'(\gamma) = \sum_{i=0}^{n-1} p_i \gamma^{n-1-i} \in D \cdot M$.

EXERCISE 5.41. Let A be a finite commutative ring, and let $m \in \mathbb{Z}_{>0}$ such that $mA = 0$.

a. Prove that for some $t, b \in \mathbb{Z}_{>0}$ the following is true: for all sequences $(d_i)_{i=1}^t$ consisting of t integers $d_i \geq b$ there exists a sequence $(e_i)_{i=1}^t$ of integers satisfying $0 \leq e_i < d_i$ such that for all sequences $(f_i)_{i=1}^t$ of monic polynomials $f_i \in \mathbb{Z}[X]$ satisfying $f_i \equiv X^{d_i} - X^{e_i} \bmod m\mathbb{Z}[X]$ there exists a surjective ring homomorphism

$$\mathbb{Z}[X_1, \ldots, X_t]/(f_1(X_1), \ldots, f_t(X_t)) \to A.$$

b. Prove that there exist an order R in some number field, an ideal $I \subseteq R$, and a ring isomorphism $R/I \to A$.

c. **(open)** Can we compute a triple as in (b) in polynomial time given A?

6. Computing symbols

This section will be an application of the theory from Section 3 which is independent of the rest of this document. In algebraic number theory we find lots of 'symbols'. It is not well-defined what a symbol is, but generally they are maps which encode algebraic properties of its parameters which also satisfy some reciprocity law. In this section we will define and give algorithms for computing some of these symbols. The father of all symbols is the Legendre symbol. We will use [**19**] as a reference.

DEFINITION 6.1. Let p be an odd prime and let a be an integer coprime to p. We define the *Legendre symbol*

$$\left(\frac{a}{p}\right) = \begin{cases} +1 & a \text{ is a square in } \mathbb{Z}/p\mathbb{Z} \\ -1 & \text{otherwise} \end{cases}$$

so that $\left(\frac{a}{p}\right) \equiv a^{\frac{p-1}{2}} \bmod p$.

We can easily compute the Legendre symbol directly from the second definition in polynomial time using a square-and-multiply algorithm modulo p.

THEOREM 6.2 (Quadratic reciprocity). *Suppose p and q are distinct odd primes. Then*

$$\left(\frac{p}{q}\right)\left(\frac{q}{p}\right) = (-1)^{\frac{p-1}{2}\cdot\frac{q-1}{2}}. \quad \square$$

DEFINITION 6.3. Let b be a positive odd integer and a an integer coprime to b. In terms of the prime factorization $b = p_1^{k_1} \cdots p_n^{k_n}$ of b we define the *Jacobi symbol*

$$\left(\frac{a}{b}\right) = \left(\frac{a}{p_1}\right)^{k_1} \cdots \left(\frac{a}{p_n}\right)^{k_n},$$

where $\left(\frac{a}{p_i}\right)$ is the Legendre symbol defined previously.

Note that the Jacobi symbol extends the Legendre symbol, which justifies using the same notation for both. To compute the Jacobi symbol directly from the definition we need to be able to factor b, which is unfeasible. Quadratic reciprocity for the Jacobi symbol gives us a better method.

PROPOSITION 6.4 (Quadratic reciprocity). *Suppose a and b are coprime positive odd integers. Then*

$$\left(\frac{a}{b}\right)\left(\frac{b}{a}\right) = (-1)^{\frac{a-1}{2}\cdot\frac{b-1}{2}}.$$

PROOF. Note that for fixed b the maps $a \mapsto \left(\frac{a}{b}\right)\left(\frac{b}{a}\right)$ and $a \mapsto (-1)^{\frac{a-1}{2}\cdot\frac{b-1}{2}}$ are multiplicative. Hence it suffices to prove the proposition for a prime. Applying the same reasoning to b we have reduced to Theorem 6.2. $\square$

EXERCISE 6.5. Let p be an odd prime. Show that 2 is a square in $\mathbb{F}_p$ if and only if $p \equiv \pm 1 \bmod 8$. Conclude that $\left(\frac{2}{b}\right) = (-1)^{(b^2-1)/8}$ for all odd $b > 0$.

Hint: Write $\sqrt{2}$ in terms of 8-th roots of unity.

EXERCISE 6.6. Let $x_0, x_1, x_2, x_3 \in \mathbb{Z}_{\geq 0}$ such that x_1 is odd, x_0 and x_1 are coprime, and $x_{n+2} \equiv x_n \bmod x_{n+1}$ for $n \in \{0,1\}$. Write $x_2 = 2^u v$ for $u \in \mathbb{Z}_{\geq 0}$ and v odd. Show that:

a. $\left(\frac{x_0}{x_1}\right) = (-1)^{(x_1-1)(x_2-1)/4} \cdot \left(\frac{x_1}{x_2}\right)$ if $u = 0$.

b. $\left(\frac{x_0}{x_1}\right) = (-1)^{u\cdot\frac{x_1^2-x_3^2}{8}}(-1)^{\frac{v-1}{2}\cdot\frac{x_1-x_3}{2}} \cdot \left(\frac{x_2}{x_3}\right)$ if $u > 0$.

THEOREM 6.7. *There exists a polynomial-time algorithm that, given coprime positive odd integers a and b, computes the Jacobi symbol $\left(\frac{a}{b}\right)$.*

PROOF. We define the *gcd sequence* of positive integers x_0 and x_1 to be the sequence $(x_0, \ldots, x_N)$ where x_{n+2} is the unique integer such that $0 \leq x_{n+2} < x_{n+1}$ and $x_{n+2} \equiv x_n \bmod x_{n+1}$ and with $x_N = 0$. The proof of the Euclidean algorithm shows that this sequence contains only linearly many elements and can be computed in polynomial time. Moreover, $x_{N-1} = \gcd(x_0, x_1)$.

Compute the gcd sequence of a and b. Note that $\gcd(x_n, x_{n+1}) = \gcd(a, b) = 1$ for all $n < N$. Using Exercise 6.6 we may express $\left(\frac{a}{b}\right) = s_n \cdot \left(\frac{x_n}{x_{n+1}}\right)$ for some $s_n \in \{\pm 1\}$ iteratively for n with x_{n+1} odd. Clearly we may compute these s_n in polynomial time. As $\left(\frac{x_{N-2}}{x_{N-1}}\right) = \left(\frac{x_{N-2}}{1}\right) = 1$, we simply return s_{N-1}. □

The Jacobi symbol is defined on a subset of $\mathbb{Z}^2$. As is the theme in this document, we will 'extend' the Jacobi symbol to number rings.

REMARK 6.8. One is sometimes interested in a more general Jacobi symbol $\left(\frac{a}{b}\right)$ for coprime integers a and b, where b is not necessarily odd or positive, by defining

$$\left(\frac{a}{2}\right) = (-1)^{\frac{a^2-1}{8}} \quad \text{and} \quad \left(\frac{a}{-1}\right) = \frac{a}{|a|} = \begin{cases} +1 & \text{if } a > 0 \\ -1 & \text{if } a < 0 \end{cases}.$$

This is called the *Kronecker symbol*. The Kronecker symbol can be computed in polynomial time by writing $b = uc2^k$ where $u = \pm 1$ and c is odd and positive, and applying Theorem 6.7 to the factor $\left(\frac{a}{c}\right)$ in $\left(\frac{a}{b}\right) = \left(\frac{a}{u}\right)\left(\frac{a}{c}\right)\left(\frac{a}{2}\right)^k$.

EXERCISE 6.9. The Euclidean algorithm implied by Exercise 2.17 produces a different type of 'gcd sequence' than those used in Theorem 6.7, namely those where $|x_{n+2}| \leq |x_{n+1}|/2$ and $x_{n+2} \equiv x_n \bmod x_{n+1}$ for all n. Give a proof of Theorem 6.7 for Kronecker symbols using such gcd sequences.

6.1. Jacobi symbols in number rings. First we define the Legendre symbol for a general number ring.

DEFINITION 6.10. Let $\mathfrak{p}$ be a prime ideal in a number ring R of odd index $n_{\mathfrak{p}} = \#(R/\mathfrak{p})$ and let $a \in R$ such that $aR + \mathfrak{p} = R$. We define the *Legendre symbol*

$$\left(\frac{a}{\mathfrak{p}}\right) = \begin{cases} +1 & \text{if } a \text{ is a square in } R/\mathfrak{p} \\ -1 & \text{otherwise} \end{cases},$$

so that $\left(\frac{a}{\mathfrak{p}}\right) = a^{\frac{n_{\mathfrak{p}}-1}{2}} \bmod \mathfrak{p}$.

Extending the definition to general ideals as for the Jacobi symbol cannot be done similarly, unless R is a Dedekind domain, because it would require prime factorization of ideals. Instead, we consider the following.

DEFINITION 6.11. Suppose $\mathfrak{b}$ is an ideal of a number ring R. For a prime $\mathfrak{p}$ we define $l_{\mathfrak{p}}(\mathfrak{b}) \in \mathbb{Z}_{\geq 0}$ such that $\#(R_{\mathfrak{p}}/\mathfrak{b}_{\mathfrak{p}}) = \#(R/\mathfrak{p})^{l_{\mathfrak{p}}(\mathfrak{b})}$. For $\#(R/\mathfrak{b})$ odd and $a \in R$ with $aR + \mathfrak{b} = R$ we define the *Jacobi symbol* by

$$\left(\frac{a}{\mathfrak{b}}\right) = \prod_{\text{max.}\,\mathfrak{p} \subset R} \left(\frac{a}{\mathfrak{p}}\right)^{l_{\mathfrak{p}}(\mathfrak{b})}.$$

THEOREM 6.12 (Theorem in [**19**]). *There exists a polynomial-time algorithm that, given an order R, an ideal $\mathfrak{b}$ of R such that $\#(R/\mathfrak{b})$ is odd and $a \in R$ such that $aR + \mathfrak{b} = R$, computes the Jacobi symbol $\left(\frac{a}{\mathfrak{b}}\right)$.*

We will prove this theorem by expressing the Jacobi symbol in terms of yet another symbol.

EXERCISE 6.13. Let $\mathfrak{p}$ and $\mathfrak{b}$ be ideals in a number ring R with $\mathfrak{p}$ prime and take any composition series $0 = M_0 \subsetneq M_1 \subsetneq \cdots \subsetneq M_n = R/\mathfrak{b}$ of $R/\mathfrak{b}$ as an R-module. Show that $l_{\mathfrak{p}}(\mathfrak{b})$ equals the number of quotients M_{i+1}/M_i that are isomorphic to $R/\mathfrak{p}$ as an R-module.

6.2. Signs of automorphisms. In this section we will consider the following symbol.

DEFINITION 6.14. Let B be a finite abelian group and let $\sigma \in \mathrm{Aut}(B)$. We define (σ, B) to be the sign of σ as an element of the permutation group on B.

LEMMA 6.15. *Suppose A and C are sets and $\alpha \in \mathrm{Aut}(A)$ and $\gamma \in \mathrm{Aut}(C)$ are permutations. Write $\alpha \sqcup \gamma$ and $\alpha \times \gamma$ respectively for the induced permutation on the disjoint union $A \sqcup C$ and product $A \times C$. Then $\mathrm{sgn}(\alpha \sqcup \gamma) = \mathrm{sgn}(\alpha) \cdot \mathrm{sgn}(\gamma)$ and $\mathrm{sgn}(\alpha \times \gamma) = \mathrm{sgn}(\alpha)^{\#C} \cdot \mathrm{sgn}(\gamma)^{\#A}$.*

PROOF. That $\mathrm{sgn}(\alpha \sqcup \gamma) = \mathrm{sgn}(\alpha) \cdot \mathrm{sgn}(\gamma)$ follows from the fact that $\alpha \sqcup \gamma = \alpha\gamma$ when $\mathrm{Aut}(A)$ and $\mathrm{Aut}(C)$ are naturally embedded in $\mathrm{Aut}(A \sqcup C)$.

For the second part write $\alpha' = \alpha \times \mathrm{id}_C$ and $\gamma' = \mathrm{id}_A \times \gamma$ and note that $\alpha \times \gamma = \alpha' \cdot \gamma'$. Now α' acts as α on $\#C$ disjoint copies of A, hence by the previous $\mathrm{sgn}(\alpha') = \mathrm{sgn}(\alpha)^{\#C}$. Mutatis mutandis we obtain the same for γ', and the lemma follows from multiplicativity of the sign. □

PROPOSITION 6.16. *Suppose B is a finite abelian group and $\beta \in \mathrm{Aut}(B)$. Suppose we have an exact sequence $0 \to A \to B \to C \to 0$ such that β restricts to an automorphism α of A. Then β induces an automorphism γ of C such that the following diagram commutes*

$$\begin{array}{ccccccccc} 0 & \longrightarrow & A & \xrightarrow{f} & B & \xrightarrow{g} & C & \longrightarrow & 0 \\ & & \downarrow{\scriptstyle\alpha} & & \downarrow{\scriptstyle\beta} & & \downarrow{\scriptstyle\gamma} & & \\ 0 & \longrightarrow & A & \xrightarrow{f} & B & \xrightarrow{g} & C & \longrightarrow & 0 \end{array}$$

and if $\#C$ is odd we have $(\beta, B) = (\alpha, A) \cdot (\gamma, C)^{\#A}$.

PROOF. The map γ exists by a diagram chasing argument. Since $\#C$ is odd we may write $C = \{0\} \sqcup D \sqcup (-D)$ for some subset $D \subseteq C$. Choosing any section $D \to B$ of g (not necessarily a group homomorphism!), we may extend it to a section $h : C \to B$ in such a way that $h(-c) = -h(c)$ and $h(0) = 0$. Now the maps f and h together give a bijection of sets $A \times C \to B$. Let β' be the induced action of β on $A \times C$. By Lemma 6.15 we have that $(\alpha \times \gamma, A \times C) = (\alpha, A)^{\#C} \cdot (\gamma, C)^{\#A}$, hence to prove the proposition it suffices to show that $\sigma = (\alpha \times \gamma)^{-1} \cdot \beta'$ is an even permutation. For all $d \in D$ the action of σ restricts to $A \times \{d\}$. Note that σ commutes with -1, hence the action of σ on $A \times \{-d\}$ is isomorphic to the action on $A \times \{d\}$. Hence the restriction of σ to $A \times (C \setminus \{0\})$ is even. Finally, σ is the identity on $A \times \{0\}$, so we conclude σ is even, as was to be shown. □

The exact sequence $0 \to 2\mathbb{Z}/4\mathbb{Z} \to \mathbb{Z}/4\mathbb{Z} \to \mathbb{Z}/2\mathbb{Z} \to 0$ resists application of Proposition 6.16. If we choose the non-trivial automorphism σ given by $x \mapsto -x$ on $\mathbb{Z}/4\mathbb{Z}$ we see that its sign is -1, while the induced maps on the other terms are trivial. Hence $(\sigma, \mathbb{Z}/4\mathbb{Z})$ is not an $\mathbb{F}_2$-linear combination of $(\sigma, 2\mathbb{Z}/4\mathbb{Z})$ and $(\sigma, \mathbb{Z}/2\mathbb{Z})$.

EXERCISE 6.17. Show that for $k \in \mathbb{Z}_{\geq 2}$ and $a \in (\mathbb{Z}/2^k\mathbb{Z})^*$ we have

$$(x \mapsto ax, \mathbb{Z}/2^k\mathbb{Z}) = (-1)^{\frac{a-1}{2}}.$$

Hint: Write $\mathbb{Z}/2^k\mathbb{Z} = (\mathbb{Z}/2^k\mathbb{Z})^* \sqcup (2\mathbb{Z}/2^k\mathbb{Z})$ and show $(x \mapsto ax, (\mathbb{Z}/2^k\mathbb{Z})^*) = -1$ if and only if a generates $(\mathbb{Z}/2^k\mathbb{Z})^*$.

EXERCISE 6.18. Suppose B is a non-trivial finite abelian group. For $b \in B$ write $\lambda_b : B \to B$ for the map $x \mapsto b + x$.

a. Show that $\operatorname{sgn}(\lambda_b) = -1$ if and only if $\#(B/\langle b \rangle)$ is odd and $\#\langle b \rangle$ is even.

Let $\beta \in \operatorname{Aut}(B)$ and suppose both B and β have 2-power order.

b. Show that there exists a subgroup $C \subseteq B$ such that $\beta(C) = C$ and $\#(B/C) = 2$.

c. Let $b \in B$ such that $B = C \cup (b + C)$ with C as in (b). Show that $(\beta, B) = (\lambda_{\beta(b)-b}, C)$, and that $(\beta, B) = -1$ if and only if $C = \langle \beta(b) - b \rangle$ and $4 \mid \#B$.

We now drop the assumption that β has 2-power order.

d. Suppose $(\beta, B) = -1$. Show that $B \cong \mathbb{Z}/2^k\mathbb{Z}$ for some $k \geq 2$ or $B \cong (\mathbb{Z}/2\mathbb{Z})^2$.

e. Exhibit a polynomial-time algorithm that, given a finite abelian group A and $\alpha \in \operatorname{Aut}(A)$ such that $2 \mid \#A$, computes (α, A).

6.3. Computing signs of group automorphisms. In this section we will prove we can compute the sign of automorphisms of finite abelian groups in polynomial time. We will need an elementary lemma about determinants that mirrors Proposition 6.16.

LEMMA 6.19. *Let $\mathbb{F}$ be a field and let $0 \to A \to B \to C \to 0$ be an exact sequence of finite dimensional $\mathbb{F}$-vector spaces together with automorphism α, β and γ such that the diagram*

$$\begin{array}{ccccccccc} 0 & \longrightarrow & A & \longrightarrow & B & \longrightarrow & C & \longrightarrow & 0 \\ & & \downarrow{\scriptstyle\alpha} & & \downarrow{\scriptstyle\beta} & & \downarrow{\scriptstyle\gamma} & & \\ 0 & \longrightarrow & A & \longrightarrow & B & \longrightarrow & C & \longrightarrow & 0 \end{array}$$

commutes. Then $\det(\beta) = \det(\alpha) \cdot \det(\gamma)$. □

In terms of matrices, the above lemma simply states that for square matrices A and C and a matrix P that fits, the block matrix $B = \left(\begin{smallmatrix} A & P \\ 0 & C \end{smallmatrix}\right)$ satisfies $\det(B) = \det(A) \cdot \det(C)$.

THEOREM 6.20 (Proposition 2.3 in [**19**]). *Suppose $b \in \mathbb{Z}_{>0}$ is odd and B is a free $(\mathbb{Z}/b\mathbb{Z})$-module of finite rank. Then for all $\sigma \in \operatorname{Aut}(B)$ we have $(\sigma, B) = \left(\frac{\det(\sigma)}{b}\right)$.*

PROOF. For $B = 0$ the theorem clearly holds, so assume $b \neq 1$ and that B has rank at least 1.

First suppose b is prime and B has rank 1. Then σ is given by multiplication with $a \in (\mathbb{Z}/b\mathbb{Z})^*$. If a generates $(\mathbb{Z}/b\mathbb{Z})^*$, then the corresponding permutation fixes 0 and acts transitively on the $b - 1$ remaining elements of B, so that $(\sigma, B) = -1 = \left(\frac{a}{b}\right) = \left(\frac{\det(\sigma)}{b}\right)$. By multiplicativity of both symbols in σ, this also proves the case for prime b where a is not a generator.

Now we prove using induction the case for $\operatorname{rk} B \geq 2$. Suppose σ is given by an upper or lower triangular matrix. Then there exists a subspace $0 \subsetneq A \subsetneq B$ such that σ restricts to A. Hence we have a split exact sequence $0 \to A \to B \to C \to 0$ with $C = B/A$ and let α and γ be the induced maps on A and C respectively. Then by Proposition 6.16, the induction hypothesis and Lemma 6.19 we get

$$(\sigma, B) = (\alpha, A) \cdot (\gamma, C) = \left(\frac{\det(\alpha)}{b}\right) \cdot \left(\frac{\det(\gamma)}{b}\right) = \left(\frac{\det(\alpha)\det(\gamma)}{b}\right) = \left(\frac{\det(\sigma)}{b}\right).$$

Since every matrix can be written as a product of upper and lower triangular matrices, the case for general α follows.

Now we prove the theorem for general b with induction on the number of divisors of b. We have just proven the induction base with b prime. For b not prime we may take a divisor $1 < d < b$ of b. Let $A = dB$ and $C = B/A$ and note that they are free modules over $\mathbb{Z}/\frac{b}{d}\mathbb{Z}$ and $\mathbb{Z}/d\mathbb{Z}$ respectively. Moreover, σ induces maps α and γ on A and C respectively that make the usual diagram commute. It follows from the definition of the determinant that $\det(\alpha) \equiv \det(\sigma) \bmod \frac{b}{d}$ and $\det(\gamma) \equiv \det(\sigma) \bmod d$. Then

$$(\sigma, B) = (\alpha, A)(\gamma, C) = \left(\frac{\det(\alpha)}{b/d}\right)\left(\frac{\det(\gamma)}{d}\right) = \left(\frac{\det(\sigma)}{b/d}\right)\left(\frac{\det(\sigma)}{d}\right) = \left(\frac{\det(\sigma)}{b}\right).$$

The theorem now follows by induction. □

THEOREM 6.21 (Cf. Proposition 2.4 in [**19**]). *There exists a polynomial-time algorithm that, given an finite abelian group B and an automorphism σ of B, computes the symbol (σ, B).*

PROOF. If $2 \mid \#B$ we have Exercise 6.18, so suppose B has odd order. Using Theorem 3.52, write B as a product $\prod_{k=1}^{m}(\mathbb{Z}/n_k\mathbb{Z})$ of non-trivial cyclic groups such that $n_j \mid n_k$ for all $j > k$. Note that B fits in an exact sequence $0 \to A \to B \to C \to 0$ with $A = n_m B$ and $C = B/A$. We have $\sigma(A) = A$ and hence σ induces maps $A \to A$ and $C \to C$. Then

$$A = \prod_{k=1}^{m}(n_m\mathbb{Z}/n_k\mathbb{Z}) \cong \prod_{k=1}^{m-1}\left(\mathbb{Z}\Big/\frac{n_k}{n_m}\mathbb{Z}\right) \quad \text{and} \quad C \cong (\mathbb{Z}/n_m\mathbb{Z})^m.$$

Since C is a free $(\mathbb{Z}/n_m\mathbb{Z})$-module, we may compute (σ, C) using Theorem 6.20 in polynomial time. Note that A is a product of strictly fewer cyclic groups than B, as well as having smaller order. While $A \neq 0$ we compute (σ, A) recursively and apply Proposition 6.16 to compute (σ, B). Since m is polynomially bounded in the length of the input, there are only polynomially many recursive steps and the algorithm runs in polynomial time. □

6.4. Computing Jacobi symbols in number rings. To compute Jacobi symbols in polynomial time it now suffices to reduce to Theorem 6.21.

LEMMA 6.22. *Suppose $\mathfrak{b}$ is an ideal in a number ring R odd index $(R : \mathfrak{b})$, and suppose $a \in R$ satisfies $aR + \mathfrak{b} = R$. Then $\left(\frac{a}{\mathfrak{b}}\right) = (\alpha, R/\mathfrak{b})$ where the map α is multiplication by a.*

PROOF. First suppose $\mathfrak{b}$ is a prime ideal and choose a generator a of $(R/\mathfrak{b})^*$. Then $\left(\frac{a}{\mathfrak{b}}\right) = -1$. As a acts transitively on an even number of elements $(R/\mathfrak{b}) \setminus \{0\}$, we conclude that $(x \mapsto ax, R/\mathfrak{b}) = -1 = \left(\frac{a}{\mathfrak{b}}\right)$. The case for general a follows from multiplicativity of both symbols.

Now consider the case of general $\mathfrak{b}$. Choose some composition series $0 = M_0 \subsetneq M_1 \subsetneq \cdots \subsetneq M_n = R/\mathfrak{b}$ of $R/\mathfrak{b}$ as an R-module. Consider the exact sequence $0 \to M_i \to M_{i+1} \to M_{i+1}/M_i \to 0$ and note that $M_{i+1}/M_i \cong R/\mathfrak{p}$ as R-modules for some prime ideal $\mathfrak{p}_i$ of R. By applying Proposition 6.16 inductively we obtain $(\alpha, B) = \prod_{i=1}^{n}(\alpha, M_{i+1}/M_i) = \prod_{i=1}^{n}\left(\frac{\alpha}{\mathfrak{p}_i}\right)$. It follows from Exercise 6.13 that the latter equals $\left(\frac{\alpha}{\mathfrak{b}}\right)$. □

PROOF OF THEOREM 6.12. Compute $R/\mathfrak{b}$ and the map $\alpha : R/\mathfrak{b} \to R/\mathfrak{b}$ given by $x \mapsto ax$. Using Theorem 6.21 compute $(\alpha, R/\mathfrak{b})$, which equals $\left(\frac{a}{\mathfrak{b}}\right)$ by Lemma 6.22. □

References

[1] M. Agrawal, N. Kayal, and N. Saxena, *PRIMES is in P*, Ann. of Math. (2) **160** (2004), no. 2, 781–793, DOI 10.4007/annals.2004.160.781. MR2123939 ↑49

[2] M. F. Atiyah and I. G. Macdonald, *Introduction to commutative algebra*, Addison-Wesley Publishing Co., Reading, Mass.-London-Don Mills, Ont., 1969. MR242802 ↑9, 10, 35, 38

[3] E. Bach, G. Miller, and J. Shallit, *Sums of divisors, perfect numbers and factoring*, SIAM J. Comput. **15** (1986), no. 4, 1143–1154, DOI 10.1137/0215083. MR861378 ↑13

[4] D. J. Bernstein, *Factoring into coprimes in essentially linear time*, Journal of Algorithms **54** (2005), no. 1, 1–30, `https://www.sciencedirect.com/science/article/pii/S0196677404000732`. ↑14

[5] O. B. Berrevoets and D. M. H. van Gent, *Problemen*, Nieuw Archief voor Wiskunde **24** (2023), no. 3, 190–192, `https://www.nieuwarchief.nl/serie5/pdf/naw5-2023-24-3-190.pdf`. ↑15

[6] I. F. Blake and T. Garefalakis, *On the complexity of the discrete logarithm and Diffie-Hellman problems*, J. Complexity **20** (2004), no. 2-3, 148–170, DOI 10.1016/j.jco.2004.01.002. MR2067425 ↑17

[7] R. P. Brent, *Fast multiple-precision evaluation of elementary functions*, J. Assoc. Comput. Mach. **23** (1976), no. 2, 242–251, DOI 10.1145/321941.321944. MR395314 ↑46

[8] J. A. Buchmann and H. W. Lenstra Jr., *Approximating rings of integers in number fields* (English, with English and French summaries), J. Théor. Nombres Bordeaux **6** (1994), no. 2, 221–260, DOI 10.5802/jtnb.113. MR1360644 ↑34, 47, 48, 51, 53, 54

[9] I. Ciocanea Teodorescu, *Algorithms for finite rings*, PhD. Thesis, Leiden Univ., 2016, `https://hdl.handle.net/1887/40676`. ↑17, 24, 25, 26, 27, 28

[10] E. C. Dade, O. Taussky, and H. Zassenhaus, *On the semigroup of ideal classes in an order of an algebraic number field*, Bull. Amer. Math. Soc. **67** (1961), 305–308, DOI 10.1090/S0002-9904-1961-10594-6. MR136597 ↑37, 39

[11] G. Ge, *Algorithms related to multiplicative representations of algebraic numbers*, Ph.D. Thesis, 1993. ↑29, 31, 45

[12] J. Hadamard, *Résolution d'une question relative aux déterminants*, Bulletin des sciences mathématiques **17** (1893), 240–246. ↑20

[13] R. Hartshorne, *Algebraic geometry*, Graduate Texts in Mathematics, No. 52, Springer-Verlag, New York-Heidelberg, 1977. MR463157 ↑37

[14] J. E. Hopcroft and J. D. Ullman, *Introduction to automata theory, languages, and computation*, Addison-Wesley Series in Computer Science, Addison-Wesley Publishing Co., Reading, MA, 1979. MR645539 ↑9

[15] G. Karpilovsky, *Topics in field theory*, North-Holland Mathematics Studies, vol. 155, North-Holland Publishing Co., Amsterdam, 1989. Notas de Matemática [Mathematical Notes], 124. MR982265 ↑47

[16] S. Lang, *Linear algebra*, 3rd ed., Undergraduate Texts in Mathematics, Springer-Verlag, New York, 1987, DOI 10.1007/978-1-4757-1949-9. MR874113 ↑22

[17] S. Lang, *Algebraic number theory*, 2nd ed., Graduate Texts in Mathematics, vol. 110, Springer-Verlag, New York, 1994, DOI 10.1007/978-1-4612-0853-2. MR1282723 ↑9, 10, 33

[18] A. K. Lenstra, H. W. Lenstra Jr., and L. Lovász, *Factoring polynomials with rational coefficients*, Math. Ann. **261** (1982), no. 4, 515–534, DOI 10.1007/BF01457454. MR682664 ↑21, 22

[19] H. W. Lenstra Jr., *Computing Jacobi symbols in algebraic number fields*, Nieuw Arch. Wisk. (4) **13** (1995), no. 3, 421–426. MR1378807 ↑55, 57, 58, 59
[20] H. W. Lenstra, *Flags and lattice basis reduction*, European congress of mathematics, 2001, pp. 37–51. ↑18
[21] H. W. Lenstra Jr., *Lattices*, Algorithmic number theory: lattices, number fields, curves and cryptography, Math. Sci. Res. Inst. Publ., vol. 44, Cambridge Univ. Press, Cambridge, 2008, pp. 127–181. MR2467546 ↑20, 21, 22, 23
[22] M. M. Wood, *Rings and ideals parameterized by binary n-ic forms*, J. Lond. Math. Soc. (2) **83** (2011), no. 1, 208–231, DOI 10.1112/jlms/jdq074. MR2763952 ↑42
[23] P. Q. Nguyen and B. Vallée (eds.), *The LLL algorithm: Survey and applications*, Information Security and Cryptography, Springer-Verlag, Berlin, 2010, DOI 10.1007/978-3-642-02295-1. MR2722178 ↑21
[24] A. Schönhage, *The fundamental theorem of algebra in terms of computational complexity*, Math. Inst. Univ. Tübingen, 1982. ↑46
[25] J. H. Silverman, *An introduction to lattices, lattice reduction, and lattice-based cryptography*, Number Theory Informed by Computation (Jennifer S. Balakrishnan, Bjorn Poonen, and Akshay Venkatesh, eds.), IAS/Park City Mathematics Series, vol. 29, American Mathematical Society, Providence, RI, 2025. ↑21
[26] D. Simon, *The index of nonmonic polynomials*, Indag. Math. (N.S.) **12** (2001), no. 4, 505–517, DOI 10.1016/S0019-3577(01)80039-2. MR1908878 ↑42
[27] P. Stevenhagen, *The arithmetic of number rings*, Algorithmic number theory: lattices, number fields, curves and cryptography, Math. Sci. Res. Inst. Publ., vol. 44, Cambridge Univ. Press, Cambridge, 2008, pp. 209–266. MR2467548 ↑18, 34, 36
[28] P. Stevenhagen, *Number rings*, 2020, `https://websites.math.leidenuniv.nl/algebra/ant.pdf`. ↑33
[29] E. Weiss, *Algebraic number theory*, McGraw-Hill Book Co., Inc., New York-San Francisco-Toronto-London, 1963. MR159805 ↑54

Universiteit Leiden, Niels Bohrweg 1, 2333 CA Leiden
Email address: `d.m.h.van.gent@math.leidenuniv.nl`

IAS/Park City Mathematics Series
Volume **29**, 2025
https://doi.org/10.1090/pcms/029/00891

Counting Points on Hyperelliptic Curves over Finite Fields

David Harvey

ABSTRACT. This course is a gentle introduction to my paper "Computing zeta functions of arithmetic schemes" (2015). We walk through the algorithms of that paper in detail for the special case of a hyperelliptic curve over a prime field. In particular, we explain how to obtain "square-root time" and "average polynomial time" point counting algorithms for such a curve. The presentation is "elementary" in the sense that it does not rely on any cohomology.

CONTENTS

1. Introduction

1.1. What this course is about. This course has two main aims. The first is to provide a gentle introduction to some of the ideas behind my paper "Computing zeta functions of arithmetic schemes" [**23**]. That paper presented a number of algorithms for counting points on algebraic varieties over finite fields, or in other words, counting the number of solutions to systems of polynomial equations over finite fields. In this course we will specialise the methods of [**23**] to the case of a *hyperelliptic curve over a prime field*, i.e., a curve defined by an equation of the form $y^2 = f(x)$ over a field $\mathbf{F}_p$ where p is an odd prime. (Here "hyperelliptic" includes the case of elliptic curves; see Remark 3.1.2.) Here is a baby example of the kind of problem we consider in this course:

PROBLEM 1.1.1. Count (by hand) the number of solutions $(x, y) \in \mathbf{F}_3 \times \mathbf{F}_3$ to the equation $y^2 = x^3 + x$.

2020 *Mathematics Subject Classification.* Primary 14Dxx; Secondary 14Dxx.
Key words and phrases. Park City Mathematics Institute.

The second aim of this course is to serve as a jumping-off point for further research. Sprinkled throughout these notes are sketches of a number of unpublished point counting algorithms, and also a few other algorithms unrelated to point counting. My hope is that someone reading these notes will be sufficiently intrigued by one of these sketches to take up the challenge of fleshing out the details, writing an efficient implementation, and publishing the results.

1.2. What this course is not about. In this course I will have almost nothing to say about the many other approaches to point counting that have been developed over the last few decades. These include algorithms based on analysing the group structure of the Jacobian (see for example [**15**, §2] or [**51**]) and algorithms based on ℓ-adic cohomology [**43, 46**]. There are some links between the approach taken in this course and algorithms based on p-adic cohomology, such as [**31, 53**], but I will make only a few brief comments on the latter. One should keep in mind that it is often possible to combine information obtained from these different classes of algorithms, and the compound algorithm may well be more efficient than any of its components considered separately. Finally, I will also have nothing to say about point counting algorithms designed to run on quantum computers [**32**].

This course is also not about *applications* of point counting. There are plenty of applications in cryptography, coding theory and number theory — see for example [**10**], [**47**], and [**17**] respectively.

The literature on the above topics is vast and I cannot even begin to offer a survey here.

1.3. Outline of the course. Section 2 is a "cheat sheet" that summarises the tools we need from computer algebra, such as fast algorithms for integer and matrix multiplication. You may wish to skip this section on a first reading, and refer back occasionally when needed.

Section 3 introduces the main objects of study in this course — hyperelliptic curves and zeta functions — and works out some of their basic properties.

Section 4 develops a "trace formula" that is central to all algorithms discussed subsequently. This formula yields a congruence modulo p for the number of $\mathbf{F}_{p^r}$-rational points on a given hyperelliptic curve $y^2 = f(x)$ over $\mathbf{F}_p$, in terms of certain selected coefficients of $f^{(p-1)/2}$. Already in this section we will be able to construct point counting algorithms that significantly outperform the naive "enumeration" method.

In Section 5 we establish a *recurrence* for the coefficients of $f^{(p-1)/2}$. This recurrence feeds into the following two sections to obtain even faster point counting algorithms: in Section 6 we see how to obtain a "square-root time" point counting algorithm, whose complexity is roughly linear in $p^{1/2}$, and in Section 7 we obtain an "average polynomial time" algorithm, which reduces the complexity further to $O(\log^4 p)$, when averaged over many primes.

All of the algorithms discussed up to this point in this course have only computed the point counts modulo p. In Section 8 we develop a more powerful trace formula that yields the point counts modulo *powers* of p. Combined with the methods of Sections 5 and 6, this leads to various algorithms for computing the entire zeta function, including algorithms that run in "square-root time" and "average polynomial time". These algorithms are not worked out in detail, but are instead left as challenging exercises for the reader.

1.4. Problems. These course notes include many problems of various sorts.

A large fraction of the problems ask you to supply proofs of the various Theorems, Propositions and Lemmas. These are an integral part of the course and I strongly recommend that you attempt as many as possible.

A problem marked with the game controller icon (🎮) is an implementation problem, to be carried out in your *favourite computer algebra system* (FCAS). Examples include Sage, Magma, Maple, Mathematica and Minecraft. I try to use Sage where possible because I like to support free software. Personally, I find that one of the best ways to understand an algorithm deeply is to implement it — the machine is unforgiving and will not let you cheat. If you don't know how to use these systems, or have never done any programming before, now is an excellent time to begin learning! Some of the implementation problems are fairly substantial projects, so you might consider working in a team.

The lightning bolt icon (⚡) indicates a problem that I don't know how to solve. This does not necessarily mean that the problem is difficult or unsolved — it might only reveal some ignorance on my part, or that I haven't spent much time thinking about it yet. Two lightning bolts (⚡⚡) mean that I do not know how to solve the problem, despite having devoted considerable thought and attention to it. Again, this *may* mean that it is difficult or even impossible, but not necessarily!

1.5. Prerequisites and notation. I assume that you are familiar with basic algebraic notions such as groups, rings, fields, polynomials, the Chinese remainder theorem, and the rudiments of linear algebra over a field. We write $\#S$ for the cardinality of a set S, and $\overline{\mathbb{K}}$ for the algebraic closure of a field $\mathbb{K}$. For a ring R, we write $R[\![x]\!]$ for the ring of (formal) power series with coefficients in R, and $\operatorname{Mat}_d(R)$ for the ring of $d \times d$ matrices with entries in R.

If F is a polynomial in one variable, we write F_i for the coefficient of x^i in F. By convention $F_i = 0$ if $i < 0$ or $i > \deg F$.

The symbol p always denotes a prime. I assume that you know elementary number-theoretic facts such as Fermat's little theorem, the fact that $(x+y)^p = x^p + y^p$ in characteristic p, the structure of finite fields, and how to solve congruences modulo powers of a prime. (We will not actually use p-adic numbers anywhere, but it won't hurt to know about them!)

We write $\mathbf{F}_q$ for the field with q elements. Assuming that q is a power of an *odd* prime, we define

$$\chi_q \colon \mathbf{F}_q \to \{0, 1, -1\} \subset \mathbf{Z}$$

to be the quadratic character on $\mathbf{F}_q$, i.e.,

$$\chi_q(\beta) := \begin{cases} 0 & \beta = 0, \\ 1 & \beta \text{ is a square in } \mathbf{F}_q^*, \\ -1 & \beta \text{ is a non-square in } \mathbf{F}_q^*. \end{cases}$$

In particular, $\chi_p(\beta)$ is the same as the Legendre symbol (β/p). The identity

$$\beta^{(q-1)/2} \equiv \chi_q(\beta) \pmod{p} \qquad \text{for any } \beta \in \mathbf{F}_q \tag{1.5.1}$$

plays a central role throughout the course.

The following result is equivalent to the Prime Number Theorem; see for instance [**40**, Thm. 6.9].

LEMMA 1.5.2. *For $N \geqslant 2$ we have*

$$\sum_{p \leqslant N} \log p = N + o(N).$$

(Whenever we take sums or products over the symbol p, we always mean over primes. The symbol $o(N)$ means a function $f(N)$ such that $\lim_{N\to\infty} f(N)/N = 0$.)

We will need only the most basic notions from algebraic geometry. We write $\mathbf{A}^n_{\mathbb{K}}$ and $\mathbf{P}^n_{\mathbb{K}}$ for the affine and projective spaces of dimension n over a field $\mathbb{K}$. If X is a variety over $\mathbb{K}$ and $\mathbb{L}$ is an extension of $\mathbb{K}$, we write $X(\mathbb{L})$ for the set of $\mathbb{L}$-rational points on X.

I assume that you are comfortable with big-O notation. As a brief reminder, we write $f(n) = O(g(n))$, or occasionally $f(n) \ll g(n)$, to mean that there exists an absolute constant $C > 0$ such that $f(n) \leqslant Cg(n)$ for all n in the relevant domain. If $g(n)$ is negative or undefined for small values of n, for example, $g(n) = n \log n \log\log n$, these values of n are tacitly excluded from the domain. When we write $f(n) = O(n^{1+\varepsilon})$, this means that the bound holds for all $\varepsilon > 0$, where the implied constant may depend on ε.

1.6. Algorithms and computational complexity. I will use the terms "running time" and "complexity" interchangeably. By the running time of an algorithm, I mean the number of "bit operations" that it performs. I assume that you have some intuitive understanding of what this means, but in order to prove *theorems* about the complexity of algorithms, we need to settle on a precise computational model. In this course, the model I will always have in the back of my mind is the *deterministic multitape Turing model.* Briefly, this means a Turing machine with a fixed, finite number of tapes, and the complexity refers to the number of steps executed by the machine over the course of a computation. I do not want to get sidetracked by the technical details of this; for a formal presentation, see [**42**].

I will however mention one "gotcha" for those experienced programmers who may not have studied formal computational models before. In the multitape Turing model, one does not have available constant-time random access to elements of an array. In fact, accessing the n-th element of an array of b-bit elements requires time $O(nb)$, as the tape head has to travel all the way to the target slot.

In this course we will mainly focus on time complexity, and I will not give formal estimates for space complexity, i.e., the amount of memory used by a computation, due to lack of space. However, for certain algorithms the space complexity becomes the main bottleneck in practice, and I will point this out when relevant. We will also completely ignore the issue of parallelisation, which again is very important in practical computations.

1.7. Some thoughts about the big picture. Before going on, I would like to offer some personal reflections concerning the status of the paper [**23**] on which this course is based.

My own point counting education began in the p-adic cohomology world, mainly in the vicinity of Kedlaya's algorithm [**31**]. The development of the algorithms in [**23**] was heavily influenced by this background. Over the years, I have formed the definite impression that the approach to point counting in [**23**] is in some deep way *the same* as the p-adic cohomological approach. This is a very vague gut feeling and I do not know how to make it precise. I certainly do not know how to take an algorithm expressed in terms of p-adic cohomology and "translate" it into the setting of [**23**], or vice versa. And I do not know any cohomological interpretation of the "trace formula" [**23**, Theorem 3.1] that lies at the heart of the method. (In these notes, the closest analogue of this formula is Theorem 8.4.1.) Yet the feeling persists that the two approaches are somehow the same thing in disguise.

What is most frustrating is that each approach seems capable of doing things that the other cannot. I will give a few examples.

In one direction, the methods of [**23**] seem to lead to much more general results than the p-adic cohomology approach. For example, [**23**] presents "square root time" and "average polynomial time" point counting algorithms for completely arbitrary varieties. (For the special case of hyperelliptic curves, these results are worked out in detail in §6 and §7 of these notes.) We do not know any other way of proving these extremely general results; in the p-adic cohomology context, we can achieve results of a similar strength for only a rather limited class of varieties. Another example is the recent work of my student Madeleine Kyng, who has designed an algorithm that uses the setup in [**23**], together with some integral closure calculations, to count points on completely arbitrary curves [**35**]. Her experiments confirm that this algorithm can confidently handle curves that are apparently intractable by other known methods, including Tuitman's algorithm [**53**], which is currently the most general variant of Kedlaya's algorithm known.

In the other direction, the p-adic cohomology approach seems capable of producing better complexity bounds than [**23**] in many cases. For example, for a curve of large genus g over a small field $\mathbf{F}_p$, algorithms based on [**23**] tend to get bogged down when computing powers of the unreasonably large matrices appearing in the "trace formula". (For the corresponding phenomenon in these notes, consider the terms with large values of ℓ in (8.4.2).) Both approaches yield complexity bounds polynomial in g, but the exponent of g is usually somewhat smaller for the p-adic cohomology approach.

The complexity advantage for the p-adic cohomology algorithms is even more dramatic in higher dimensions, at least in the cases that these algorithms are applicable at all. For example, when using [**23**] to handle a quartic surface in $\mathbf{P}^3$ (a K3 surface), one needs to run the algorithm with a very high "target precision" (in these notes, the parameter λ in §8), making it essentially infeasible in practice. On the other hand, the algorithm in [**1**] needs only one or two p-adic digits, and is eminently feasible. The difference here seems to be that the p-adic cohomology algorithms have more direct access to arithmetic properties of the Frobenius action on cohomology. Another example is Lauder's deformation algorithm [**36**], which for fixed p can handle a smooth projective hypersurface of degree d in n variables in time polynomial in d^n. Using [**23**], one achieves complexity only polynomial in the rather larger quantity d^{n^2}. Again, Lauder is using properties of Frobenius that are invisible from the perspective of [**23**].

I am hopeful that one day the two approaches will be unified, and that we will have available a conceptual framework that allows us to design algorithms achieving the best of both worlds: the generality and flexibility of [**23**], and the better complexity bounds coming from p-adic cohomology.

2. A crash course on fast arithmetic

Many of the algorithms presented in this course rely on fast algorithms for arithmetic on objects such as integers, polynomials and matrices. Your computer can add, subtract and multiply integers up to about 64 bits long in constant time; these operations are hard-wired into the silicon. But if you ask your FCAS to perform arithmetic on integers larger than this, it has to be done in software, and it becomes crucial to have available algorithms whose complexity scales well for large inputs. The study of such algorithms belongs to a field known as *symbolic computation* or *computer algebra.*

This section summarises the complexity results that we need, without going into the details of the underlying algorithms. The details can be found in many sources, such as [**5**, **8**, **34**, **54**].

2.1. Integer arithmetic. We assume that integers are represented in the usual binary notation, with perhaps an extra bit to indicate the sign. (These sorts of implementation details do not affect the complexity results, but it is good to get into the habit of thinking about how to represent various mathematical objects explicitly on a Turing machine tape.)

Addition and subtraction. We can add and subtract n-bit integers in time $O(n)$, by working from the least significant bit to the most significant, propagating carries (or borrows) as we go.

Multiplication. We write $\mathsf{M}_{\mathrm{int}}(n)$ for the cost of multiplying n-bit integers. Sometimes we will write expressions like $\mathsf{M}_{\mathrm{int}}(\log p)$, where $\log p$ might not be an integer. For convenience, we interpret this to mean $\mathsf{M}_{\mathrm{int}}(\lceil \log p \rceil)$.

The classical long multiplication algorithm runs in time $O(n^2)$. Currently, the best known asymptotic bound for $\mathsf{M}_{\mathrm{int}}(n)$ is

$$\mathsf{M}_{\mathrm{int}}(n) = O(n \log n).$$

This bound is achieved via a much more complicated algorithm involving the fast Fourier transform [**28**], and is widely believed to be optimal.

PROBLEM 2.1.1. ( ) Investigate how long it takes your FCAS to multiply random n-bit integers, for $n = 10^3, 10^4, \ldots, 10^9$. How does the performance scale as a function of n?

REMARK 2.1.2. All of the computer algebra systems mentioned earlier (except possibly Minecraft) use the GMP library [**20**] under the bonnet[1] for their integer arithmetic needs. The $n \log n$ algorithm of [**28**] is probably not practical, but GMP does implement an algorithm whose theoretical complexity comes very close to $O(n \log n)$. For small n, maybe up to a few hundred bits, the running time of GMP's multiplication routine behaves like $O(n^2)$, but as n increases, the software works very hard to push the exponent down towards 1.

Division and remainder. Let u and $v > 0$ be integers with at most n bits. Suppose that we want to compute the quotient $q := \lfloor u/v \rfloor$ and corresponding remainder $r := u - qv$. The classical long division algorithm performs these tasks in time $O(n^2)$. Using a division algorithm that combines fast multiplication with Newton's method, this may be improved to

$$O(\mathsf{M}_{\mathrm{int}}(n)) = O(n \log n).$$

[1] bonnet (*noun*): metal sheet covering the engine. *U.S. English:* hood.

GCD and extended GCD. Let $u, v > 0$ be integers with at most n bits. Let g be their greatest common divisor, and let a and b be the Bezout cofactors, i.e., so that $g = au + bv$. The classical version of Euclid's algorithm computes g, a and b in time $O(n^2)$. This may be improved to

$$O(\mathsf{M}_{\mathrm{int}}(n)\log n) = O(n\log^2 n),$$

using any of several variants of the recursive "half-GCD" algorithm. In particular, we may test whether u is invertible modulo v, if and so, compute the modular inverse, in time $O(n\log^2 n)$.

PROBLEM 2.1.3. () For the same values of n as in Problem 2.1.1, investigate how long it takes your FCAS to compute the quotient (and/or the corresponding remainder) of a random $2n$-bit integer by a random n-bit integer, and how long it takes to compute the GCD (and corresponding cofactors) of two random n-bit integers. How do these running times compare to the time required to multiply a pair of n-bit integers?

Modular arithmetic. Let $m \geqslant 2$. Elements of the ring $\mathbf{Z}/m\mathbf{Z}$ may be represented by residues in the interval $0 \leqslant x < m$, and so occupy $O(\log m)$ bits of space.

We can add elements of $\mathbf{Z}/m\mathbf{Z}$ by simply adding the residues and then subtracting m if necessary. The complexity is $O(\log m)$. Similar comments apply for subtraction in $\mathbf{Z}/m\mathbf{Z}$.

To multiply two elements of $\mathbf{Z}/m\mathbf{Z}$, we simply multiply the residues, divide by m, and keep the remainder. Using the results for integer multiplication and division mentioned above, the cost of this is

$$O(\mathsf{M}_{\mathrm{int}}(\log m)) = O(\log m \log\log m).$$

To compute a quotient in $\mathbf{Z}/m\mathbf{Z}$, we can use the extended GCD algorithm to invert the divisor (if possible) and then multiply by the dividend. The cost is

$$O(\mathsf{M}_{\mathrm{int}}(\log m)\log\log m) = O(\log m\,(\log\log m)^2).$$

In some algorithms we will simply count the number of "arithmetic operations" in $\mathbf{Z}/m\mathbf{Z}$, not bothering to distinguish between additions, multiplications, and divisions. In this situation we will usually simplify matters by estimating the cost of each operation to be $O(\log^{1+\varepsilon} m)$.

REMARK 2.1.4. Again, the bound $O(\log^{1+\varepsilon} m)$ is not very realistic when $\log m$ is small; the quasi-linear behaviour only really kicks in when $\log m$ is quite large. Nevertheless, for the purpose of writing down mathematical theorems, it is convenient and traditional to express things this way. With some experience, one learns how to interpret these sorts of complexity bounds.

2.2. Polynomial arithmetic. We will mainly be interested in the case of polynomials over $\mathbf{Z}/m\mathbf{Z}$, for an integer $m \geqslant 2$. A polynomial of degree $< n$ in $(\mathbf{Z}/m\mathbf{Z})[x]$ is represented by a sequence of n coefficients, and thus occupies space $O(n\log m)$.

Multiplication. Let $\mathsf{M}_m(n)$ denote the cost of multiplying two polynomials $f, g \in (\mathbf{Z}/m\mathbf{Z})[x]$ of degree $< n$. One may reduce this problem to *integer* multiplication via the technique of *Kronecker substitution*, i.e., lift the polynomials to $F, G \in \mathbf{Z}[x]$, pack the coefficients of F and G into large integers (by evaluating at

2^c for a suitable integer $c \geqslant 1$), multiply the large integers, unpack the coefficients of $FG \in \mathbf{Z}[x]$ from the resulting integer product, and finally reduce modulo m to obtain the coefficients of $fg \in (\mathbf{Z}/m\mathbf{Z})[x]$. This leads to the estimate

$$\mathsf{M}_m(n) = O(\mathsf{M}_{\mathrm{int}}(n \log(nm))),$$

where the $\log(nm)$ term arises from estimating the size of the coefficients of FG.

If n is not too large compared to m, say $n \ll m^{O(1)}$, then this simplifies to

$$\mathsf{M}_m(n) = O(\mathsf{M}_{\mathrm{int}}(n \log m)) = O(n \log m \log(n \log m)).$$

In other words, the cost is $O(b \log b)$ where b is the total bit size of the inputs.

Remark 2.2.1. Kronecker substitution is inefficient when m is fixed and n is large: the cost becomes $O(\mathsf{M}_{\mathrm{int}}(n \log n)) = O(n \log^2 n)$. Joris van der Hoeven and I recently presented an algorithm that achieves $\mathsf{M}_m(n) = O(n \log n)$ (for fixed m) under some plausible but unproved number-theoretic hypotheses [**29**].

Problem 2.2.2. (ϟϟ) Can you design an algorithm that *provably* multiplies polynomials of degree $< n$ in $(\mathbf{Z}/m\mathbf{Z})[x]$ (for fixed m) in time $O(n \log n)$?

Division. Let $f, g \in (\mathbf{Z}/m\mathbf{Z})[x]$ have degree $< n$, with g monic. Using Newton's method combined with fast multiplication, we may compute polynomials $q, r \in (\mathbf{Z}/m\mathbf{Z})[x]$ such that $f = qg + r$ and $\deg r < \deg g$ (the quotient and remainder) in time

$$O(\mathsf{M}_m(n)).$$

2.3. Computing large powers. Let R be a commutative ring, let $u \in R$ and let $k \geqslant 1$ be an integer. To compute u^k, we will generally use the "repeated squaring" method: first recursively compute $u^{\lfloor k/2 \rfloor}$, and then recover u^k as either $(u^{\lfloor k/2 \rfloor})^2$ or $(u^{\lfloor k/2 \rfloor})^2 \cdot u$, depending on whether k is even or odd.

The total number of multiplications in R is $O(\log k)$, but the complexity analysis depends on the ring. We will need the following two cases.

Proposition 2.3.1. *Let $k \geqslant 1$ and let $u \in \mathbf{Z}/m\mathbf{Z}$. Then we may compute u^k in time*

$$O(\mathsf{M}_{\mathrm{int}}(\log m) \log k).$$

Proposition 2.3.2. *Let $k \geqslant 1$ and let $f \in (\mathbf{Z}/m\mathbf{Z})[x]$ have degree $n \geqslant 1$. Then we may compute f^k in time*

$$O(\mathsf{M}_m(nk)).$$

Notice that in the polynomial case, the cost of computing f^k is the same (up to a constant) as the cost of a *single* polynomial multiplication of degree $\deg(f^k) = nk$. This occurs because the cost of the polynomial multiplications increases according to a geometric progression as the algorithm proceeds.

2.4. Product trees. A *product tree* on a sequence of positive integers $u_1, \ldots, u_N$ is a binary tree with N leaves, whose leaves are assigned the values $u_1, \ldots, u_N$, and whose non-leaf nodes are recursively assigned the product of the values of their children. There are various ways to handle the case that N is not a power of two, and any of these are reasonable as long as the height of the tree is $O(\log N)$. An example with $N = 11$ is illustrated in Figure 2.4.1.

Starting with $u_1, \ldots, u_N$, we may compute all nodes in the product tree by repeatedly multiplying together values in adjacent nodes.

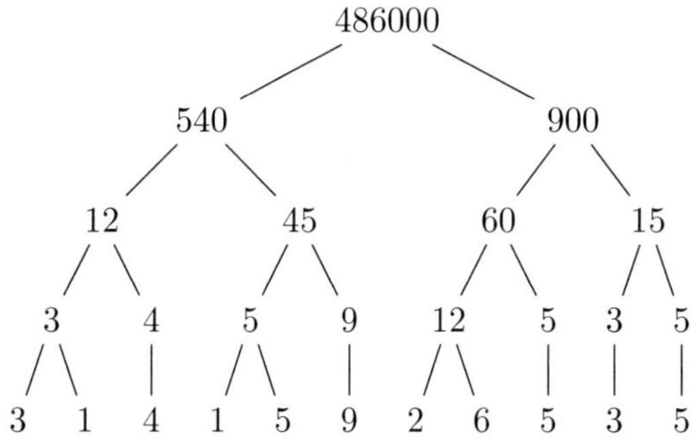

FIGURE 2.4.1. A product tree on the first 11 digits of π.

PROPOSITION 2.4.2. *The cost of computing all nodes in the product tree is*

$$O(\mathsf{M}_{\text{int}}(\log B)\log N), \qquad \textit{where } B := \prod_{j=1}^{N} u_j.$$

This follows from the fact that the total bit size of all integers in each level of the tree is $O(\log B)$, and that $\mathsf{M}_{\text{int}}(n)$ may be assumed to be super-linear, i.e., $\sum_j \mathsf{M}_{\text{int}}(n_j) \leqslant \mathsf{M}_{\text{int}}(\sum_j n_j)$.

REMARK 2.4.3. Strictly speaking, one has to be careful with the complexity bounds if many of the leaf nodes have the value 1. We will ignore this minor technicality.

The same construction may also be carried out for polynomials. The following application plays an important role in §6.

PROPOSITION 2.4.4 (Fast multipoint evaluation). *Let $m, n \geqslant 2$ be integers. Suppose that we are given as input a polynomial $Q \in (\mathbf{Z}/m\mathbf{Z})[x]$ of degree at most n, and a sequence of evaluation points $\alpha_1, \ldots, \alpha_n \in \mathbf{Z}/m\mathbf{Z}$. Then we may compute the values*

$$Q(\alpha_1), \ldots, Q(\alpha_n) \in \mathbf{Z}/m\mathbf{Z}$$

in time

$$O(\mathsf{M}_m(n)\log n).$$

Briefly, the algorithm implementing Proposition 2.4.4 runs as follows. We first use a polynomial analogue of Proposition 2.4.2 to compute a product tree on the linear polynomials

$$x - \alpha_1, \ldots, x - \alpha_n \in (\mathbf{Z}/m\mathbf{Z})[x].$$

We then compute the reduction of $Q(x)$ modulo each node in the product tree, by working recursively from the top of the tree down to the leaves, i.e., given $Q(x) \pmod{g_1(x)g_2(x)}$, we may compute $Q(x) \pmod{g_1(x)}$ and $Q(x) \pmod{g_2(x)}$ by simply dividing by each of g_1 and g_2. At the end we have $Q(x) \pmod{x - \alpha_j}$, or equvialently $Q(\alpha_j)$, for all $j = 1, \ldots, n$.

2.5. Linear algebra. Let R be a commutative ring and let $d \geqslant 1$. Consider the problem of multiplying two $d \times d$ matrices over R. The obvious algorithm for this task requires $O(d^3)$ ring operations (multiplications and additions) in R.

One of the biggest surprises in computer algebra was Strassen's discovery [**48**] that this bound can be improved to $O(d^{\log 7/\log 2})$, where $\log 7/\log 2 \approx 2.81$.

I will denote by ω any feasible *exponent of matrix multiplication*, i.e., so that $d \times d$ matrices over R may be multiplied using $O(d^\omega)$ ring operations in R. Strassen's result implies that we may take $\omega = \log 7/\log 2$.

REMARK 2.5.1. The best known exponent is currently $\omega = 2.3728596$ [**3**]. However, the result in [**3**] is stated for the case that R is a *field*, and counting field operations instead of ring operations. I suspect that this value of ω also works for an arbitrary commutative ring, but I am not an expert on these matters.

The actual running time of matrix multiplication depends on the cost of arithmetic in R. This is easy if R is a finite ring; for example, the cost of multiplication in $\mathrm{Mat}_d(\mathbf{F}_p)$ is $O(d^\omega \mathsf{M}_{\mathrm{int}}(\log p))$. But if the matrix entries are allowed to grow, such as in the case $R = \mathbf{Z}$, then the running time also depends on the size of the entries. We will defer this issue for now and revisit it in §7.

The other result we need is an estimate for the cost of computing the *inverse characteristic polynomial* of a matrix over a field.

PROPOSITION 2.5.2. *Given as input* $A \in \mathrm{Mat}_d(\mathbb{K})$, *we may compute the polynomial* $\det(I - TA) \in \mathbb{K}[T]$ *using* $O(d^\omega)$ *field operations in* $\mathbb{K}$.

This is actually quite a recent result; see [**41**].

2.6. Enumerating primes. The usual sieve of Eratosthenes is tricky to implement on a multitape Turing machine due to the lack of fast random array access. Schönhage, Grotefeld and Vetter [**45**] showed how to list the primes $p \leqslant N$ in time $O(N \log^2 N \log\log N)$ by implementing the Eratosthenes sieve using fast sorting algorithms instead. The following slightly better bound (see [**16**]) is achieved by replacing the Eratosthenes sieve with the Atkin sieve.

PROPOSITION 2.6.1. *The primes* $p \leqslant N$ *may be enumerated in time*

$$O(N \log^2 N/\log\log N).$$

3. Hyperelliptic curves and zeta functions

3.1. Hyperelliptic curves.

DEFINITION 3.1.1. Let $\mathbb{K}$ be a field of characteristic $\neq 2$ and let $g \geqslant 1$ be an integer. A *hyperelliptic curve* of genus g over $\mathbb{K}$ is the smooth algebraic curve $C/\mathbb{K}$ associated to an affine equation

$$y^2 = f(x),$$

where $f \in \mathbb{K}[x]$ is a squarefree polynomial of degree $2g+1$ or $2g+2$.

From now on, whenever we give an equation $y^2 = f(x)$ for a hyperelliptic curve of genus g, it is tacitly assumed that $f(x)$ satisfies the requirements above, i.e., it is squarefree and has degree $2g+1$ or $2g+2$.

REMARK 3.1.2. Definition 3.1.1 is slightly nonstandard when $g = 1$. A curve of genus 1 is usually not called hyperelliptic, but rather *elliptic* (assuming that it has a distinguished $\mathbb{K}$-rational point). The algorithms discussed in this course work perfectly well when $g = 1$, so it will be convenient to abuse terminology and call such a curve a "hyperelliptic curve of genus 1".

REMARK 3.1.3. When $\operatorname{char} \mathbb{K} = 2$, the general equation for a hyperelliptic curve has the form $y^2 + yh(x) = f(x)$ with $\deg h \leqslant g + 1$. We will never need to worry about fields of characteristic two in this course.

What is the best way to think about C? A rookie mistake is to homogenise the equation $y^2 = f(x)$ in the usual way to obtain a projective curve in $\mathbf{P}^2$. The problem is that this curve is usually not smooth. (To get a smooth projective model, in general one needs to embed the curve in a higher-dimensional projective space.)

A much more useful way to think about C is as a *two-to-one cover of* $\mathbf{P}^1$. Let us describe more explicitly what this means.

Suppose that

$$f(x) = f_0 + f_1 x + \cdots + f_{2g+2} x^{2g+2}, \qquad f_i \in \mathbb{K}.$$

Note that $f_{2g+2} = 0$ is allowed; this happens exactly when $\deg f = 2g + 1$.

The rational function x may be thought of as a map from C to $\mathbf{P}^1 = \mathbf{A}^1 \cup \{\infty\}$. The original equation

$$y^2 = f_0 + f_1 x + \cdots + f_{2g+2} x^{2g+2} \tag{3.1.4}$$

is really an equation for the affine patch of C consisting of those points whose x-coordinate lies in $\mathbf{A}^1$. But this patch misses all "points at infinity", i.e., the points on C whose x-coordinate is ∞. To see these points, we need to switch to a different model. Put $x = 1/u$ and $y = v/u^{g+1}$; then the equation becomes

$$v^2 = f_0 u^{2g+2} + f_1 u^{2g+1} + \cdots + f_{2g+2}. \tag{3.1.5}$$

This is an equation for a different affine patch of C, consisting of those points whose u-coordinate is not infinity, i.e., whose x-coordinate is not zero. In particular, the points where $x = \infty$ correspond exactly to the points on (3.1.5) where $u = 0$.

In summary, C may be written as a union of the two affine pieces given by (3.1.4) and (3.1.5), glued together by the relations $x = 1/u$, $y = v/u^{g+1}$.

Armed with this description, we can now say concretely what the points of $C(\mathbb{L})$ look like, for any field extension $\mathbb{L}$ of $\mathbb{K}$. Each point in $P \in C(\mathbb{L})$ has a well-defined x-coordinate $x(P) \in \mathbf{P}^1(\mathbb{L}) = \mathbb{L} \cup \{\infty\}$. The number of points in $C(\mathbb{L})$ with a specified x-coordinate is given by the following lemma. For this statement it is convenient to define

$$f(\infty) := f_{2g+2}.$$

LEMMA 3.1.6. *For any* $\alpha \in \mathbf{P}^1(\mathbb{L}) = \mathbb{L} \cup \{\infty\}$, *the number of points* $P \in C(\mathbb{L})$ *with* $x(P) = \alpha$ *is either one, two, or zero, according to whether* $f(\alpha)$ *is zero, a square in* $\mathbb{L}^*$, *or a non-square in* $\mathbb{L}^*$.

In particular:

- The points $P \in C(\mathbb{L})$ with $x(P) = 0$ correspond to solutions in $\mathbb{L}$ of
$$y^2 = f_0.$$
- The points $P \in C(\mathbb{L})$ with $x(P) = \infty$ correspond to solutions in $\mathbb{L}$ of
$$v^2 = f_{2g+2}.$$
- For any $\alpha \in \mathbb{L}^*$, the points $P \in C(\mathbb{L})$ with $x(P) = \alpha$ correspond to solutions in $\mathbb{L}$ of
$$y^2 = f(\alpha).$$
(Equivalently, these points correspond to solutions in $\mathbb{L}$ of $v^2 = \bar{f}(1/\alpha)$, where $\bar{f}(u) := u^{2g+2} f(1/u)$ is the "reversal" of $f(x)$.)

PROBLEM 3.1.7. Let C be the curve from Problem 1.1.1, now including the point(s) at infinity. Calculate $\#C(\mathbf{F}_3)$ and $\#C(\mathbf{F}_{3^2})$ (again by hand).

The covering map $x \colon C \to \mathbf{P}^1$ is ramified at exactly $2g + 2$ points, the *Weierstrass points* of C. These are exactly the points in $C(\overline{\mathbb{K}})$ whose x-coordinates satisfy $f(\alpha) = 0$. In particular, there is a Weierstrass point at $x = \infty$ if and only if $\deg f = 2g + 1$.

Finally, we make some brief comments about the squarefreeness condition in Definition 3.1.1. This condition is equivalent to requiring that $f(x)$ has no repeated roots, or alternatively that the discriminant of $f(x)$ is nonzero. (The condition "$\deg f = 2g + 1$ or $2g + 2$" may be interpreted as saying that $f(x)$ does not have repeated roots at ∞.) The purpose of the squarefreeness condition is to ensure that the curve is nonsingular. If it is not satisfied, then the curve defined by $y^2 = f(x)$ will have lower genus than expected. Nevertheless, many of the results in this course work perfectly well without this condition.

3.2. The number of points on the curve. Now suppose that $\mathbb{K}$ is a *finite* field, say $\mathbb{K} = \mathbf{F}_q$ where $q = p^a$, for p an odd prime and $a \geqslant 1$. Let $C/\mathbf{F}_q$ be a hyperelliptic curve of genus $g \geqslant 1$. In this case $C(\mathbf{F}_q)$ is obviously a *finite* set. In fact, Lemma 3.1.6 implies that $\#C(\mathbf{F}_q) \leqslant 2(q + 1)$, since there are at most two points in $C(\mathbf{F}_q)$ for each of the $q + 1$ possible values of x in $\mathbf{P}^1(\mathbf{F}_q)$. But we can do much better:

THEOREM 3.2.1 (Hasse–Weil bound). *Let $C/\mathbf{F}_q$ be a hyperelliptic curve of genus $g \geqslant 1$. Then*
$$\left|\#C(\mathbf{F}_q) - (q + 1)\right| \leqslant 2g \cdot q^{1/2}.$$

A heuristic argument for Theorem 3.2.1 might go like this: for each $\alpha \in \mathbf{P}^1(\mathbf{F}_q)$, the value of $f(\alpha)$ is "equally likely" to be a square or a non-square in $\mathbf{F}_q$, so the total number of points should behave like a random variable with mean $q + 1$ and variance proportional to q. This is probably the wrong way to think about it, but it does give roughly the right answer!

Theorem 3.2.1 also implies that

$$\left|\#C(\mathbf{F}_{q^r}) - (q^r + 1)\right| \leqslant 2g \cdot q^{r/2} \tag{3.2.2}$$

for any $r \geqslant 1$. This follows by interpreting the equation $y^2 = f(x)$ defining C as an equation over $\mathbf{F}_{q^r}$ instead of over $\mathbf{F}_q$.

We will not give a proof of Theorem 3.2.1 in this course. The result actually holds for any curve of genus g, not just for hyperelliptic curves. The theorem follows from a more general statement about the *zeta function* of the curve, discussed in the next section; see in particular Problem 3.3.8.

The main thrust of this course is to develop efficient algorithms for calculating the numbers $\#C(\mathbf{F}_{q^r})$ for $r = 1, 2, \ldots$, given as input the polynomial $f \in \mathbf{F}_q[x]$. The simplest possible algorithm for computing $\#C(\mathbf{F}_{q^r})$ is straightforward *enumeration*: loop through all $\alpha \in \mathbf{P}^1(\mathbf{F}_{q^r}) = \mathbf{F}_{q^r} \cup \{\infty\}$, and for each α compute $f(\alpha)$ and test whether it is a square in $\mathbf{F}_{q^r}$ (see Lemma 3.1.6). We will not analyse the complexity of this algorithm in detail. The main point is that the complexity grows at least as rapidly as q^r, since this is the number of x-values that must be inspected. In particular, the running time is *exponential in* r.

The enumeration algorithm can actually be a reasonable choice when the parameters are small. Various implementation tricks are available, such as using difference tables to speed up the computation of the values of $f(\alpha)$. For details see [**33**, §3] and Sutherland's `smalljac` library [**50**].

PROBLEM 3.2.3. (⚇) Let $p = 101$ and let $C/\mathbf{F}_p$ be the genus 3 curve given by

$$y^2 = x^7 - x^6 + 6x^5 - 7x^4 + 5x^3 + x^2 - x + 1.$$

Implement the enumeration algorithm in your FCAS and use it to show that

$$\#C(\mathbf{F}_p) = 153, \qquad \#C(\mathbf{F}_{p^2}) = 9935, \qquad \#C(\mathbf{F}_{p^3}) = 1029891.$$

Check that (3.2.2) holds for these values.

PROBLEM 3.2.4. (⚇) Investigate whether your FCAS has built-in functionality for computing $\#C(\mathbf{F}_{p^r})$. Do you know what algorithm it is using under the bonnet? Experimentally, how does its running time vary as a function of p, g and r?

3.3. The zeta function. Let $C/\mathbf{F}_q$ be a hyperelliptic curve of genus $g \geqslant 1$. The sequence of point counts $\#C(\mathbf{F}_{q^r})$, for $r = 1, 2, \ldots$, may be encoded into a generating function (formal power series) called the the *zeta function* of C. This is defined by the formula

$$Z_C(T) := \exp\left(\sum_{r=1}^{\infty} \frac{\#C(\mathbf{F}_{q^r})}{r} T^r\right) \in \mathbf{Q}[\![T]\!]. \tag{3.3.1}$$

Here exp denotes the usual exponential of power series, i.e.,

$$\exp U(T) := 1 + U(T) + \frac{U(T)^2}{2!} + \cdots.$$

THEOREM 3.3.2 (Weil Conjectures). *Let $C/\mathbf{F}_q$ be a hyperelliptic curve of genus $g \geqslant 1$. Then the zeta function of C is a rational function of the form*

$$Z_C(T) = \frac{L_C(T)}{(1-T)(1-qT)},$$

where $L_C(T) \in 1 + T\mathbf{Z}[T]$ is a polynomial of degree $2g$ with the following properties:

- *(Functional equation) We have*

$$L_C\left(\frac{1}{qT}\right) = (qT^2)^{-g} L_C(T).$$

- *(Riemann Hypothesis) The roots of $L_C(T)$ (in the field of complex numbers) have absolute value $q^{-1/2}$.*

Despite traditionally being known as the "Weil Conjectures", the statements in Theorem 3.3.2 are most definitely theorems! In fact, the theorem holds verbatim for an arbitrary nonsingular curve of genus g over $\mathbf{F}_q$. For a proof, see for instance [**37**] or [**47**]. (The Weil Conjectures may also be generalised far beyond the case of curves; this provided the motivating force for much of the development of 20th century algebraic geometry.)

One important implication of Theorem 3.3.2 is that knowledge of the finite object $L_C(T)$ is equivalent to knowledge of the entire infinite sequence $\#C(\mathbf{F}_{q^r})$ for $r \geqslant 1$. So another way to express the goal of this course is that we want to develop efficient algorithms for computing $L_C(T)$, given the equation $y^2 = f(x)$ as input.

REMARK 3.3.3. There is no known algorithm for computing $L_C(T)$ in polynomial time, i.e., polynomial in the amount of data needed to represent the input polynomial $f \in \mathbf{F}_q[x]$, which is $O(g \log q) = O(ag \log p)$ bits. When g is fixed, there are ℓ-adic algorithms that can compute $L_C(T)$ in time polynomial in $\log q$, but the exponent grows with g. On the other hand, when p is fixed, there are p-adic algorithms that can compute $L_C(T)$ in time polynomial in g and a. (The algorithms suggested in §8.5 also have this property.)

The following problem works out more explicitly the relationship between the point counts and the coefficients of $L_C(T)$ and $Z_C(T)$.

PROBLEM 3.3.4. Let $N_r := \#C(\mathbf{F}_{q^r})$, and put

$$Z_C(T) = 1 + \sum_{i \geqslant 1} c_i T^i, \qquad L_C(T) = 1 + \sum_{i \geqslant 1} a_i T^i,$$

where we tacitly write $a_i = 0$ for $i > 2g$.

a) Find formulas for c_1 and c_2 in terms of N_1 and N_2. Conversely, find formulas for N_1 and N_2 in terms of c_1 and c_2.
b) Same as (a) for (a_1, a_2) and (c_1, c_2).
c) More generally, show that for any $t \geqslant 1$, there exist formulas expressing the vectors $(N_1, \ldots, N_t)$, $(a_1, \ldots, a_t)$ and $(c_1, \ldots, c_t)$ in terms of each other.

Problem 3.3.4(c) implies that $L_C(T)$ is determined by the values $(N_1, \ldots, N_{2g})$. The next problem shows that we can do a little better by taking advantage of the functional equation.

PROBLEM 3.3.5. Let

$$L_C(T) = a_0 + a_1 T + \cdots + a_{2g} T^{2g},$$

where $a_0 = 1$. Show that the functional equation in Theorem 3.3.2 is equivalent to the statement that

$$a_{2g-i} = q^{g-i} a_i, \qquad 0 \leqslant i \leqslant 2g. \tag{3.3.6}$$

Conclude that knowledge of $N_1, \ldots, N_g$ (where $N_r := \#C(\mathbf{F}_{q^r})$) is enough to deduce $L_C(T)$.

Problem 3.3.5 implies that if we want to use the enumeration algorithm to compute $L_C(T)$, then we should expect the complexity to grow at least as rapidly as q^g, since this is how much work we need to do to compute $\#C(\mathbf{F}_{q^g})$. In particular, the complexity is *exponential in g*.

PROBLEM 3.3.7. Consider again the genus 3 curve from Problem 3.2.3.

a) Using the point counts found in Problem 3.2.3, compute the coefficients of T, T^2 and T^3 in $L_C(T)$.
b) Use (3.3.6) to recover the full polynomial $L_C(T)$.
c) Use $L_C(T)$ to deduce the value of $\#C(\mathbf{F}_{p^4})$.
d) Compute $\#C(\mathbf{F}_{p^4})$ directly using the enumeration algorithm, i.e., the same method as in Problem 3.2.3, and compare the results. (There are about 100 million points to count here. You will need a fairly tight implementation to finish in a reasonable amount of time.)

PROBLEM 3.3.8. Prove Theorem 3.2.1 (the Hasse–Weil bound) from Theorem 3.3.2 (the Weil Conjectures) as follows.

a) Use the Riemann Hypothesis in Theorem 3.3.2 to show that $L_C(T)$ admits a factorisation
$$L_C(T) = \prod_{i=1}^{2g}(1 - \omega_i T)$$
where $\omega_1, \ldots, \omega_{2g}$ are complex numbers with $|\omega_i| = q^{1/2}$.
b) By taking the formal logarithm of $Z_C(T)$, show that
$$\#C(\mathbf{F}_{q^r}) = q^r + 1 - \sum_{i=1}^{2g} \omega_i^r$$
for all $r \geqslant 1$. Hence deduce Theorem 3.2.1.

4. A modulo p trace formula

To keep things simple, for the rest of the course we will assume that $q = p$, i.e., we will only consider curves defined over the prime field $\mathbf{F}_p$. Recall also that p is always assumed to be odd.

The main result of this section is Theorem 4.2.7, which gives a congruence modulo p for the number of points on a hyperelliptic curve $C/\mathbf{F}_p$ with coordinates in $\mathbf{F}_{p^r}$, for any $r \geqslant 1$. This "trace formula" will play a central role in the point counting algorithms discussed later in the course.

4.1. Counting points with coordinates in $\mathbf{F}_p$. We begin with a baby version that only counts points defined over the prime field.

PROPOSITION 4.1.1. *Let $C/\mathbf{F}_p$ be a hyperelliptic curve of genus $g \geqslant 1$ defined by $y^2 = f(x)$ for some $f \in \mathbf{F}_p[x]$. Then*

$$\#C(\mathbf{F}_p) \equiv 1 - \sum_{j=1}^{g} h_{j(p-1)} \pmod{p}, \tag{4.1.2}$$

where

$$h := f^{(p-1)/2} \in \mathbf{F}_p[x].$$

Recall that $h_{j(p-1)}$ means the coefficient of $x^{j(p-1)}$ in $h(x)$. For the proof of Proposition 4.1.1, it is helpful to introduce a certain subvariety of C.

DEFINITION 4.1.3. Let $C/\mathbf{F}_p$ be a hyperelliptic curve given by $y^2 = f(x)$. We denote by $\widetilde{C}$ the subvariety of C consisting of those points for which $x \neq 0, \infty$.

PROBLEM 4.1.4. Prove Proposition 4.1.1 as follows.

a) Using (1.5.1) and Lemma 3.1.6, show that

$$\#\widetilde{C}(\mathbf{F}_p) \equiv \sum_{\alpha \in \mathbf{F}_p^*} \left(f(\alpha)^{(p-1)/2} + 1\right) \pmod{p}.$$

b) For any integer $i \geqslant 0$, show that

$$\sum_{\alpha \in \mathbf{F}_p^*} \alpha^i = \begin{cases} -1 & \text{if } i \text{ is divisible by } p-1, \\ 0 & \text{otherwise.} \end{cases}$$

c) Combining (a) and (b), show that

$$\#\widetilde{C}(\mathbf{F}_p) \equiv -1 - \sum_{j=0}^{g+1} h_{j(p-1)} \pmod{p}.$$

d) Show that

$$\#\{P \in C(\mathbf{F}_p) : x(P) = 0\} \equiv h_0 + 1 \pmod{p},$$
$$\#\{P \in C(\mathbf{F}_p) : x(P) = \infty\} \equiv h_{(g+1)(p-1)} + 1 \pmod{p}.$$

e) Deduce (4.1.2) by adding together the results of (c) and (d).

PROBLEM 4.1.5. (⌨) Compute $\#C(\mathbf{F}_p) \pmod{p}$ for the curve in Problem 3.2.3, by using your FCAS to explicitly compute $h = f^{(p-1)/2}$, and then applying Proposition 4.1.1. Notice that h is a polynomial of degree 350 over $\mathbf{F}_p$, but only three of its coefficients contribute in (4.1.2). Check that your result is consistent with the value for $\#C(\mathbf{F}_p)$ found in Problem 3.2.3.

PROBLEM 4.1.6.

a) Use the Hasse–Weil bound (Theorem 3.2.1) to show that if $p > 16g^2$, then knowledge of $\#C(\mathbf{F}_p) \pmod{p}$ determines $\#C(\mathbf{F}_p) \in \mathbf{Z}$ unambiguously.
b) For the curve in Problem 3.2.3, show that $\#C(\mathbf{F}_p) \pmod{p}$ does not provide enough information to pin down $\#C(\mathbf{F}_p)$. What possible values for $\#C(\mathbf{F}_p)$ are compatible with the Hasse–Weil bound for this curve?

4.2. Counting points with coordinates in $\mathbf{F}_{p^r}$. We now investigate how to generalise Proposition 4.1.1 to obtain a formula for $\#C(\mathbf{F}_{p^r}) \pmod{p}$ for any $r \geqslant 1$. The following problem carries out the most obvious line of attack, which is to replace the sum over $\mathbf{F}_p^*$ by a sum over $\mathbf{F}_{p^r}^*$.

PROBLEM 4.2.1. Let $C/\mathbf{F}_p$ be a hyperelliptic curve of genus $g \geqslant 1$ defined by $y^2 = f(x)$ for some $f \in \mathbf{F}_p[x]$. By mimicking the proof of Proposition 4.1.1 (see Problem 4.1.4), show that for any $r \geqslant 1$,

$$\#C(\mathbf{F}_{p^r}) \equiv 1 - \sum_{j=1}^{g} h^{(r)}_{j(p^r-1)} \pmod{p}, \tag{4.2.2}$$

where

$$h^{(r)} := f^{(p^r-1)/2} \in \mathbf{F}_p[x].$$

Unfortunately, (4.2.2) is not much use on its own because the degree of the polynomial $h^{(r)} = f^{(p^r-1)/2}$ grows exponentially in r. Except for perhaps the smallest values of r, we cannot expect to be able to explicitly compute such a large power of f. Fortunately, there is a trick that enables us to extract all the information we need from the original polynomial $h = f^{(p-1)/2}$.

DEFINITION 4.2.3. Let $C/\mathbf{F}_p$ be a hyperelliptic curve of genus $g \geqslant 1$ defined by $y^2 = f(x)$ for some $f \in \mathbf{F}_p[x]$. Define the matrix

$$A_f \in \mathrm{Mat}_g(\mathbf{F}_p)$$

by the formula

$$(A_f)_{v,u} := h_{vp-u}, \qquad 1 \leqslant u, v \leqslant g, \tag{4.2.4}$$

where

$$h := f^{(p-1)/2}.$$

The notation A_f emphasises that the matrix depends not only on the curve C, but also on the choice of equation $y^2 = f(x)$.

FIGURE 4.2.5. Illustration of the coefficients of $h = f^{(p-1)/2}$ for $p = 11$, $g = 3$, $\deg f = 2g + 2 = 8$. The circles indicate the coefficients of $1, x, \ldots, x^{40}$. Coefficients contributing to A_f are marked in black.

REMARK 4.2.6. The matrix A_f is usually known as a *Cartier–Manin matrix* associated to C; this is the matrix of the Cartier operator on the space of regular differentials on C with respect to a particular basis. It is also closely related to the *Hasse–Witt matrix*, which is the matrix of the Frobenius operator on a certain cohomology group with respect to a particular basis. The precise definitions of these objects are beyond the scope of this course. You are warned that there are some inconsistencies in the literature regarding the definition of these matrices; for a detailed discussion of this issue see **[2]**.

THEOREM 4.2.7 (Modulo p trace formula). *Let $C/\mathbf{F}_p$ be a hyperelliptic curve of genus $g \geqslant 1$ defined by $y^2 = f(x)$ for some $f \in \mathbf{F}_p[x]$, and let A_f be the associated matrix as in Definition 4.2.3. Then for any $r \geqslant 1$,*

$$\#C(\mathbf{F}_{p^r}) \equiv 1 - \mathrm{tr}(A_f^r) \pmod{p}.$$

Theorem 4.2.7 shows that to compute $\#C(\mathbf{F}_{p^r}) \pmod{p}$, it is enough to compute g^2 selected coefficients of $h = f^{(p-1)/2}$. (Proposition 4.1.1 only used a subset of g of these coefficients.) Notice that although the formula counts points in $\mathbf{F}_{p^r}$, all quantities involved in the formula, i.e., the entries of A_f, lie in $\mathbf{F}_p$.

PROBLEM 4.2.8. Check that Theorem 4.2.7 specialises to Proposition 4.1.1 in the case $r = 1$.

PROBLEM 4.2.9. Prove Theorem 4.2.7 as follows.

a) Show that for any $r \geqslant 1$,

$$f(x)^{(p^r-1)/2} = h(x)h(x^p)\cdots h(x^{p^{r-1}}),$$

where $h = f^{(p-1)/2}$ as in Definition 4.2.3.

b) Let $\widetilde{A}_f$ be the $(g+2) \times (g+2)$ matrix over $\mathbf{F}_p$ defined by

$$(\widetilde{A}_f)_{v,u} := h_{vp-u}, \qquad 0 \leqslant u, v \leqslant g+1.$$

Show by induction on $r \geqslant 1$ that

$$(\widetilde{A}_f^r)_{v,u} = (h(x)h(x^p)\cdots h(x^{p^{r-1}}))_{vp^r-u}, \qquad 0 \leqslant u, v \leqslant g+1.$$

(Hint: find an expression for the coefficient of x^{vp^r-u} in $h(x)\cdots h(x^{p^{r-1}})$ in terms of the coefficients of $h(x)\cdots h(x^{p^{r-2}})$ and $h(x)$.)

c) Show that all entries in the first and last rows of $\widetilde{A}_f$ are zero, except possibly for $(\widetilde{A}_f)_{0,0}$ and $(\widetilde{A}_f)_{g+1,g+1}$. Deduce that for all $r \geqslant 1$,

$$(A_f^r)_{v,u} = (\widetilde{A}_f^r)_{v,u}, \qquad 1 \leqslant u, v \leqslant g.$$

d) By combining (a), (b), (c) and Problem 4.2.1, complete the proof of Theorem 4.2.7.

PROBLEM 4.2.10. (🎮) For the curve in Problem 3.2.3, use your FCAS to compute $\#C(\mathbf{F}_{p^r}) \pmod p$ for $r = 1, 2, 3$, by computing $h = f^{(p-1)/2}$ (the same degree 350 polynomial that appeared in Problem 4.1.5), constructing A_f, and applying Theorem 4.2.7. Check that your answer is consistent with the point counts given in Problem 3.2.3. What happens if you try to explicitly compute the polynomials $h^{(r)}$ from Problem 4.2.1 for $r = 1, 2, 3$?

4.3. Recovering the L-polynomial modulo p. What does the trace formula (Theorem 4.2.7) tell us about the L-polynomial

$$L_C(T) = a_0 + a_1 T + \cdots + a_{2g} T^{2g}?$$

We saw in Problem 3.3.4(c) that for any $t \geqslant 1$ there exist formulas allowing us to move back and forth between $(a_1, \ldots, a_t)$ and $(N_1, \ldots, N_t)$, where $N_r := \#C(\mathbf{F}_{p^r})$. Moreover, it follows from the functional equation that $a_{g+1}, \ldots, a_{2g}$ are all zero modulo p (see Problem 3.3.5). Since the trace formula tells us how to compute $N_1, \ldots, N_g \pmod p$, it is natural to wonder whether the latter quantities determine $a_1, \ldots, a_g \pmod p$, and hence $L_C(T) \pmod p$. The answer in general is *no*, as illustrated by the following counterexample.

PROBLEM 4.3.1. (🎮) Let $p = 3$ and consider the genus 3 curves over $\mathbf{F}_p$ given by

$$C\colon\ y^2 = x^7 + 1, \qquad C'\colon\ y^2 = x^7 - x^5 + 1.$$

Use your FCAS to show that $\#C(\mathbf{F}_{p^r}) \equiv \#C'(\mathbf{F}_{p^r}) \pmod p$ for $r = 1, 2, 3$, but that $L_C(T) \not\equiv L_{C'}(T) \pmod p$.

The underlying issue is that the formulas relating $(a_1, \ldots, a_g)$ and $(N_1, \ldots, N_g)$ involve *denominators* that may be divisible by p. Some of these denominators are already clearly visible in the defining formula (3.3.1) for the zeta function.

Despite this, it turns out that there is a very elegant identity expressing $L_C(T) \pmod p$ directly in terms of A_f:

THEOREM 4.3.2. *We have*

$$L_C(T) \equiv \det(I - TA_f) \pmod{p}.$$

Theorem 4.3.2 follows quite easily from the trace formula when $p > g$, essentially because none of the relevant denominators are divisible by p. In this case knowledge of $a_1, \ldots, a_g \pmod{p}$ is equivalent to knowledge of $N_1, \ldots, N_g \pmod{p}$, and Theorem 4.3.2 is really just a restatement of the trace formula. The proof of Theorem 4.3.2 for this case is outlined in Problem 4.3.3 below.

However, when $p \leqslant g$, Theorem 4.3.2 is genuinely stronger than the trace formula; the counterexample in Problem 4.3.1 shows that A_f and $L_C(T) \pmod{p}$ contain potentially more information than the point counts $N_1, \ldots, N_g \pmod{p}$. To the best of my knowledge, the proof of Theorem 4.3.2 for $p \leqslant g$ requires cohomological methods (see [**38**]), and we will not discuss it in this course.

PROBLEM 4.3.3. This problem gives a proof of Theorem 4.3.2 in the $p > g$ case.

a) Assume temporarily that we are working over a field $\mathbb{K}$ of characteristic zero. Let $A \in \mathrm{Mat}_g(\mathbb{K})$. Show that

$$\det(I - TA) = \exp(\operatorname{tr}(\log(I - TA))). \tag{4.3.4}$$

In this formula, $\log(I - TA)$ should be interpreted as a formal power series with matrix coefficients, i.e.,

$$\log(I - TA) = -AT - \frac{1}{2}A^2T^2 - \cdots,$$

and the trace is taken term by term.

b) Now suppose that $\operatorname{char}\mathbb{K} = p$. In this case the right hand side of (4.3.4) is not even defined, since we cannot divide by p. Nevertheless, show that (4.3.4) is still correct when regarded modulo T^p, i.e., show that the coefficients of $1, T, \ldots, T^{p-1}$ on the right hand side are well-defined, and that they agree with those on the left hand side.
c) By combining (3.3.1), Theorem 4.2.7 and part (b), show that the coefficients of $1, T, \ldots, T^{p-1}$ in $(1-T)Z_C(T)$ and $\det(I - TA_f)$ agree modulo p.
d) Use (c) and Theorem 3.3.2 to prove Theorem 4.3.2 in the case that $p > g$.

PROBLEM 4.3.5. (ϟ) Is it possible to prove Theorem 4.3.2 by "elementary" (non-cohomological) means, using methods similar to the proof of Theorem 4.2.7?

PROBLEM 4.3.6. (🎮) Use your FCAS to verify Theorem 4.3.2 numerically for the curve in Problem 3.2.3, i.e., compute $\det(I - TA_f)$ and check that the result is consistent with the L-polynomial computed in Problem 3.3.7(b).

4.4. Complexity bounds. How quickly can we compute A_f from f? The most obvious approach is to expand out the power $h = f^{(p-1)/2}$ using fast polynomial arithmetic, and then extract the relevant coefficients. The following theorem indicates the complexity of this procedure.

THEOREM 4.4.1 (Computing A_f via naive expansion). *Let $C/\mathbf{F}_p$ be a hyperelliptic curve of genus $g \geqslant 1$ defined by $y^2 = f(x)$ for some $f \in \mathbf{F}_p[x]$. Then we may compute A_f in time*

$$O\big(gp\log^2(gp) + g^2\log p\big).$$

PROBLEM 4.4.2. Prove Theorem 4.4.1, by showing that the cost of computing $h = f^{(p-1)/2}$ is $O(gp \log^2(gp))$ (see Proposition 2.3.2).

(Where does the $O(g^2 \log p)$ term come from? Hint: when $g > p$, some coefficients of h appear multiple times in A_f.)

Having computed A_f, we may use Theorem 4.3.2 to deduce $L_C(T) \pmod p$. Thanks to Proposition 2.5.2, the complexity of this step is given as follows.

COROLLARY 4.4.3. *Given A_f, we may compute $L_C(T) \pmod p$ in time*

$$O(g^\omega \log^{1+\varepsilon} p).$$

Combining this corollary with Theorem 4.4.1, we obtain our first point counting algorithm that is genuinely more powerful than the enumeration method: namely, given $f \in \mathbf{F}_p[x]$ as input, we may compute $L_C(T) \pmod p$ in time

$$O\big(gp \log^2(gp) + g^\omega \log^{1+\varepsilon} p\big). \tag{4.4.4}$$

We stress that this complexity bound is *polynomial* in g. This should be contrasted with the *exponential* dependence on g in the case of the enumeration algorithm, which must inspect at least p^g values of x.

On the other hand, whereas the enumeration algorithm computes the whole polynomial $L_C(T)$, the new algorithm only recovers $L_C(T) \pmod p$. This shortcoming will be addressed later in §8. In the meantime, in §§5–7 we will focus on improving the complexity with respect to p, which is slightly worse than linear in (4.4.4).

5. Recurrences for polynomial powers

Fix some $f \in \mathbf{F}_p[x]$ defining a hyperelliptic curve $C/\mathbf{F}_p$ of genus $g \geqslant 1$. In this section we establish a *recurrence* for the coefficients of $h(x) = f(x)^{(p-1)/2}$. We then attempt to use this recurrence to compute the coefficients h_{vp-u} appearing in the A_f matrix (see Definition 4.2.3). Unfortunately, this fails due to occasional divisions by p. In §5.2 we show how to resolve this difficulty by lifting the whole computation to $\mathbf{Z}/p^\mu\mathbf{Z}$ for a suitable "precision" parameter $\mu \geqslant 1$.

Throughout this section we write

$$m := \frac{p-1}{2}, \qquad d := 2g + 2.$$

To simplify the discussion, we assume throughout that

$$f_0 \neq 0.$$

See Remark 5.1.7 for some hints on what to do if $f_0 = 0$.

5.1. A recurrence over $\mathbf{F}_p$.

PROPOSITION 5.1.1. *Let $f \in \mathbf{F}_p[x]$ and $h = f^m$ be as above. Then for any integer $k \geqslant 1$ not divisible by p, we have*

$$h_k = \frac{1}{kf_0} \sum_{j=1}^{d} (\tfrac{1}{2}j - k) f_j h_{k-j}. \tag{5.1.2}$$

In other words, (5.1.2) expresses each h_k as a linear combination of the previous d coefficients $h_{k-1}, \ldots, h_{k-d}$, provided that we can divide by kf_0 in $\mathbf{F}_p$. (Recall that by convention $h_i = 0$ for $i < 0$.)

PROBLEM 5.1.3. Prove Proposition 5.1.1 as follows. Let ∂ denote the differential operator $s \mapsto x\frac{ds}{dx}$, i.e.,

$$\partial\big(s_n x^n + \cdots + s_2 x^2 + s_1 x + s_0\big) := n s_n x^n + \cdots + 2 s_2 x^2 + s_1 x.$$

Show that $h = f^m$ satisfies the differential equation

$$f \cdot \partial h = m \cdot \partial f \cdot h. \tag{5.1.4}$$

By equating coefficients of x^k in this identity, deduce (5.1.2).

What happens if we try to use (5.1.2) to compute the coefficients h_{vp-u} appearing in A_f? To get started, we need to know the constant term h_0. It is easy to compute $h_0 = (f_0)^m \in \mathbf{F}_p$ using a fast powering algorithm (see Proposition 2.3.1); this only requires time $O(\log m \cdot \log^{1+\varepsilon} p) = O(\log^{2+\varepsilon} p)$.

We now apply the recurrence to compute successively $h_1, h_2, \ldots, h_{p-1}$. The last g such coefficients yield the first row of A_f. But after this point we hit a wall: we cannot compute h_p because that would involve dividing by zero in $\mathbf{F}_p$. Another way of saying this is that the differential equation (5.1.4) simply does not contain enough information to recover h_p from knowledge of $h_0, \ldots, h_{p-1}$. Unfortunately, this is not good enough for our purposes; for the second row of A_f we need to get to h_{2p-1}, and for the remaining rows we need to get all the way up to h_{gp-1}.

PROBLEM 5.1.5. Verify (5.1.2) numerically in your FCAS, i.e., take a small g, moderate p, and random $f \in \mathbf{F}_p[x]$, and use the recurrence to compute successively $h_0, \ldots, h_{p-1} \in \mathbf{F}_p$. Check that the coefficients produced in this way agree with a direct computation of $h = f^m$.

PROBLEM 5.1.6. Derive a recurrence for $h = f^m$ going in the opposite direction, i.e., expressing h_k in terms of $h_{k+1}, \ldots, h_{k+d}$. What conditions on k and the coefficients of $f(x)$ must be satisfied for your formula to be valid?

REMARK 5.1.7. At the outset we imposed the hypothesis $f_0 \neq 0$. To handle an input polynomial with $f_0 = 0$, there are a few possible ways to proceed.

One reasonable solution is to replace $f(x)$ by $f(x)/x$. Notice that the constant term of $f(x)/x$ is nonzero because we assumed f to be squarefree. We can easily write down a similar recurrence for $(f(x)/x)^m$, and the problem of computing selected coefficients of $f(x)^m$ is really the same as computing selected coefficients of $(f(x)/x)^m$. If anything, the latter problem is slightly easier because the degree is smaller.

Another option is to choose some $c \in \mathbf{F}_p$ such that $f(c) \neq 0$ (such c exists provided that $p > 2g + 2$), and replace $f(x)$ by $f(x + c)$. Since the equations $y^2 = f(x)$ and $y^2 = f(x + c)$ define isomorphic curves, the modified equation will lead to the same zeta function.

5.2. Lifting the coefficient ring. Following a suggestion of Bostan, Gaudry and Schost [7], our strategy for circumventing the "division by p" problem will be to *lift* the entire computation from $\mathbf{F}_p = \mathbf{Z}/p\mathbf{Z}$ to the ring $\mathbf{Z}/p^\mu\mathbf{Z}$ for some parameter $\mu \geqslant 1$ (to be chosen later).

We begin by choosing a modulo p^μ lift of f, i.e., a polynomial

$$F = F_d x^d + \cdots + F_1 x + F_0 \in (\mathbf{Z}/p^\mu\mathbf{Z})[x]$$

such that $F \equiv f \pmod p$. For example, if the coefficients of f are represented by integers from the set $\{0, 1, \ldots, p-1\}$, we can take F to be the polynomial whose coefficients are literally the same integers, but now regarded as elements of $\mathbf{Z}/p^\mu\mathbf{Z}$.

Now define

$$H := F^m \in (\mathbf{Z}/p^\mu\mathbf{Z})[x].$$

The reduction of H modulo p is simply h, so to compute the desired coefficients of h, it suffices to compute the corresponding coefficients of H modulo p.

The coefficients of H satisfy a recurrence analogous to the one satisfied by the coefficients of h:

PROPOSITION 5.2.1. *Let $F \in (\mathbf{Z}/p^\mu\mathbf{Z})[x]$ and $H = F^m$ be as above. Then for any integer $k \geqslant 1$, we have*

$$kH_k = \frac{1}{F_0} \sum_{j=1}^{d} \big((m+1)j - k\big) F_j H_{k-j}. \tag{5.2.2}$$

Keep in mind that (5.2.2) is really a congruence modulo p^μ. Also, note that our running assumption $f_0 \neq 0$ implies that $F_0 \not\equiv 0 \pmod p$, so F_0 is a unit in $\mathbf{Z}/p^\mu\mathbf{Z}$.

PROBLEM 5.2.3. Prove Proposition 5.2.1, by first showing that $H = F^m$ satisfies the differential equation

$$F \cdot \partial H = m \cdot \partial F \cdot H. \tag{5.2.4}$$

What happens if we try to use (5.2.2) to compute the coefficients of H? As before, we can compute the first term $H_0 = (F_0)^m \in \mathbf{Z}/p^\mu\mathbf{Z}$ efficiently using a fast powering algorithm. Then we try to solve (5.2.2) for $H_1, H_2, \ldots$ in turn. Whenever we encounter an index k divisible by p, the division by k introduces some *error* into our computed value for H_k. For example, our computed value for H_p will be correct modulo $p^{\mu-1}$, but not necessarily modulo p^μ. Moreover, these errors might propagate into subsequent values of H_k. It is even conceivable that the errors might become so large that (5.2.2) fails to be solvable for some large k.

The following problem asks you to carry out a "naive" analysis of the error propagation. We write $v_p(k)$ for the p-adic valuation of k, i.e., the largest integer $t \geqslant 0$ such that $p^t \mid k$.

PROBLEM 5.2.5. Let $n \geqslant 1$. Show that after solving for $H_1, \ldots, H_n$, the total number of powers of p that we have divided by is $v_p(n!)$. Deduce that if we choose

$$\mu := v_p(n!) + 1,$$

then our computed values of $H_0, \ldots, H_n \in \mathbf{Z}/p^\mu\mathbf{Z}$ will agree modulo p with the target values $h_0, \ldots, h_n \in \mathbf{F}_p$.

Problem 5.2.5 suggests that the "precision loss" in our proposed algorithm is at least *linear* in n. For example, to compute A_f we need to get up to h_{gp-1}, so we need to take

$$\mu = v_p((gp-1)!) + 1 \geqslant g.$$

Rather surprisingly, it turns out that the errors that occur in reality are much smaller than those predicted by the above discussion. Indeed, we will next show that the precision loss is only *logarithmic* in n. (I am indebted to Jan Tuitman for explaining the following argument to me. The underlying principle is similar to [**31**, Lemma 2].)

First we prove that the congruence (5.2.2) is *always* solvable for H_k, no matter how much error has accumulated, as long as all the previously computed values satisfy the recurrence.

PROPOSITION 5.2.6. *Let $n \geqslant 1$, and suppose that we are given $\widetilde{H}_0, \ldots, \widetilde{H}_{n-1} \in \mathbf{Z}/p^\mu\mathbf{Z}$ such that $\widetilde{H}_0 = H_0$ and such that*

$$k\widetilde{H}_k = \frac{1}{F_0}\sum_{j=1}^{d}((m+1)j-k)F_j\widetilde{H}_{k-j} \tag{5.2.7}$$

for all $k = 1, \ldots, n-1$. Then for $k = n$, the congruence (5.2.7) admits at least one solution $\widetilde{H}_n \in \mathbf{Z}/p^\mu\mathbf{Z}$.

PROBLEM 5.2.8. Prove Proposition 5.2.6 as follows.

a) Let
$$\widetilde{H} := \widetilde{H}_0 + \widetilde{H}_1 x + \cdots + \widetilde{H}_{n-1}x^{n-1} \in (\mathbf{Z}/p^\mu\mathbf{Z})[x].$$
Show that
$$F \cdot \partial\widetilde{H} \equiv m \cdot \partial F \cdot \widetilde{H} \pmod{x^n},$$
i.e., $\widetilde{H}$ satisfies the same differential equation as H, modulo x^n.

b) Let
$$E = 1 + E_1 x + \cdots + E_{n-1}x^{n-1} \in (\mathbf{Z}/p^\mu\mathbf{Z})[x]$$
be such that
$$E \equiv \widetilde{H}/H \pmod{x^n}.$$
(In other words, E consists of the first n coefficients of the power series expansion of $\widetilde{H}/H$. Division by H is permissible, since our running assumption $f_0 \neq 0$ implies that $H_0 = (F_0)^m$ is a unit in $\mathbf{Z}/p^\mu\mathbf{Z}$.)

Show that
$$\partial E = 0.$$

c) Let
$$\widetilde{H}' := E \cdot H \pmod{x^{n+1}}.$$
Show that $\widetilde{H}'$ satisfies the differential equation modulo x^{n+1}. Deduce that $\widetilde{H}'_n$ is a solution to (5.2.7) for $k = n$.

Proposition 5.2.6 implies that our "algorithm" never terminates: we may compute a sequence of approximations $\widetilde{H}_0, \widetilde{H}_1, \ldots \in \mathbf{Z}/p^\mu\mathbf{Z}$ that all satisfy (5.2.7), continuing happily into the sunset. (This sequence is not uniquely determined, as the congruence (5.2.7) will have more than one solution for $\widetilde{H}_k$ whenever $p \mid k$.)

The next result establishes the promised logarithmic error bound for these approximations.

PROPOSITION 5.2.9. *Let $n \geqslant 1$, and suppose that we are given $\widetilde{H}_0, \ldots, \widetilde{H}_n \in \mathbf{Z}/p^\mu\mathbf{Z}$ such that $\widetilde{H}_0 = H_0$ and such that (5.2.7) holds for $k = 1, \ldots, n$. Then*

$$\widetilde{H}_k \equiv H_k \pmod{p^{\mu - \lfloor \log_p k \rfloor}} \tag{5.2.10}$$

for all $k = 1, \ldots, n$ such that $\lfloor \log_p k \rfloor \leqslant \mu$.

PROBLEM 5.2.11. Prove Proposition 5.2.9 as follows.

a) Let
$$\widetilde{H} := \widetilde{H}_0 + \widetilde{H}_1 x + \cdots + \widetilde{H}_n x^n \in (\mathbf{Z}/p^\mu \mathbf{Z})[x],$$
and set $E := \widetilde{H}/H \pmod{x^{n+1}}$. The same argument as in parts (a) and (b) of Problem 5.2.8 shows that $\partial E = 0$. Deduce that
$$E_k \equiv 0 \pmod{p^{\mu - v_p(k)}}$$
for $k = 1, \dots, n$, whenever $v_p(k) \leqslant \mu$.

b) Conclude that (5.2.10) holds whenever $\lfloor \log_p k \rfloor \leqslant \mu$.

PROBLEM 5.2.12. () Verify Proposition 5.2.6 and Proposition 5.2.9 numerically in your FCAS. That is, for some small parameters g, p and μ, take some random $F \in (\mathbf{Z}/p^\mu \mathbf{Z})[x]$ and compute a sequence $\widetilde{H}_0, \widetilde{H}_1, \ldots \in (\mathbf{Z}/p^\mu \mathbf{Z})$ satisfying (5.2.7). Check that these values agree with the coefficients of $H = F^m$, up to the accuracy indicated in Proposition 5.2.9.

5.3. Back to point counting. Let us now see what Proposition 5.2.9 tells us about computing A_f. Recall that we need to compute h_{vp-u} for $1 \leqslant u, v \leqslant g$, which means that we must use the recurrence to compute the coefficients of $h(x)$ up to and including h_{gp-1}. This leads to the following result.

COROLLARY 5.3.1. *If we take*
$$\mu := \lfloor \log_p(gp - 1) \rfloor + 1,$$
then the approximations $\widetilde{H}_k$ are correct modulo p for $k = 0, \ldots, gp-1$. In particular,
$$h_{vp-u} \equiv \widetilde{H}_{vp-u} \pmod{p}, \qquad 1 \leqslant u, v \leqslant g.$$

PROBLEM 5.3.2. () Use your FCAS to verify Corollary 5.3.1 for the genus 3 curve from Problem 3.2.3. That is, taking $\mu := \lfloor \log_p(3p-1) \rfloor + 1 = 2$, choose a lift $F \in (\mathbf{Z}/p^2\mathbf{Z})[x]$ of f, and solve (5.2.7) for $\widetilde{H}_1, \ldots, \widetilde{H}_{3p-1} \in \mathbf{Z}/p^2\mathbf{Z}$. Check that you do correctly recover those $h_{vp-u} \in \mathbf{F}_p$ appearing in (4.2.4).

(Note that the "naive" analysis would require you to take $\mu = 3$, since you have to divide by p twice.)

After computing all of the h_{vp-u}, we may construct the matrix A_f as in (4.2.4), and then compute $L_C(T) \pmod{p}$ via Corollary 4.4.3. This leads to the following result.

THEOREM 5.3.3 (Computing A_f via recurrences). *Let $C/\mathbf{F}_p$ be a hyperelliptic curve of genus $g \geqslant 1$ defined by $y^2 = f(x)$ for some $f \in \mathbf{F}_p[x]$. Assume that $p \geqslant g$. Then we may compute A_f in time*
$$O(g^2 p \log^{1+\varepsilon} p).$$

PROBLEM 5.3.4. Prove Theorem 5.3.3 as follows.

a) Show that for each $k = 1, \ldots, gp-1$, we may compute $\widetilde{H}_k \in \mathbf{Z}/p^\mu\mathbf{Z}$ from $\widetilde{H}_{k-1}, \ldots, \widetilde{H}_{k-d}$, i.e., solve (5.2.7), in time
$$O(d\,(\mu \log p)^{1+\varepsilon}).$$
(You may need various results from §2.1.)

b) Show that the assumption $p \geqslant g$ implies that we may take $\mu = 2$.

c) Complete the proof of Theorem 5.3.3.

REMARK 5.3.5. Recall that in Theorem 4.4.1 we established the complexity bound

$$O(gp\log^2 p)$$

for computing A_f (assuming that $p \gg g$), by simply expanding out the polynomial $h = f^{(p-1)/2}$. It is interesting to compare this to the bound

$$O(g^2 p\log^{1+\varepsilon} p)$$

obtained in this section via recurrences (Theorem 5.3.3). On one hand, we have lost a factor of g. On the other hand, we have gained a factor of almost $\log p$. In practice, we have probably gained much more than this: the real problem with the algorithm in Theorem 4.4.1 is that polynomial multiplication is extremely memory-intensive. By contrast, the recurrence algorithm uses almost no memory: one only has to keep track of the last $d = O(g)$ coefficients. Experience shows that this makes an enormous difference to the overall performance in practice.

PROBLEM 5.3.6. Recall that the cost of multiplication in $\mathbf{F}_p$ is $O(\log p \log\log p)$, whereas the cost of division is $O(\log p\,(\log\log p)^2)$ (see §2.1). When solving (5.2.7), we need to divide by k. Show how to modify the algorithm to remove almost all of these divisions. (Hint: what happens when you multiply both sides by $(k-1)!$?) Conclude that the complexity bound in Theorem 5.3.3 may be improved to $O(g^2 p\log p\log\log p)$.

(This sort of optimisation is very important in practice. Divisions in $\mathbf{F}_p$ are typically *much* more expensive than multiplications, by a factor much worse than the theoretical $\log\log p$.)

PROBLEM 5.3.7. (ϟ) Recall from §5.1 that we may compute the *first row* of A_f without any divisions by p. Sutherland and I showed that it is possible to recover the entire matrix A_f from knowledge of just the first rows of the matrices corresponding to g *translates* of the original curve [**26**, §5]. Specifically, one may take the curves $y^2 = f_i(x)$ with $f_i(x) := f(x + c_i)$ for $i = 1, \dots, g$, where $c_1, \dots, c_g$ are distinct elements of $\mathbf{F}_p$. Doing things this way, one needs to solve (5.2.7) the same number of times as before, but working modulo p everywhere instead of modulo p^2. The proof in [**26**] relies on the interpretation of A_f as the matrix of a Frobenius operator on differentials. Can you give an "elementary" argument, more in the style of the proof of Theorem 4.2.7 suggested in Problem 4.2.9?

REMARK 5.3.8. I am unsure of the history of the main idea of this section, i.e., using a linear recurrence to compute selected coefficients of high index in a large power of a univariate polynomial of small degree. I learned this idea from [**7**], and the earliest reference of which I am aware is [**18**].

6. A square-root time algorithm

In the previous sections we have seen two algorithms for computing A_f whose complexity is roughly linear in p (Theorem 4.4.1 and Theorem 5.3.3). In this section we will see how to improve this to roughly linear in the *square root* of p.

6.1. A warm-up: Wilson primes. To explain the basic idea, let us first consider a simpler problem. Recall that Wilson's theorem in elementary number theory states that for any prime p,

$$(p-1)! \equiv -1 \pmod{p}.$$

We say that p is a *Wilson prime* if the above congruence holds modulo p^2. For example, 5 is a Wilson prime because $4! = 24 \equiv -1 \pmod{5^2}$. Only three Wilson primes are known: 5, 13 and 563. The search for Wilson primes has an interesting history; see [**11**] for more information.

Suppose we want to test whether a given prime p is a Wilson prime. How quickly can we compute $(p-1)! \pmod{p^2}$?

We could simply compute $(p-1)! \in \mathbf{Z}$ and then find its remainder modulo p^2, but this is grossly inefficient because $(p-1)!$ grows very rapidly with p.

A better method is to start with 1, and then multiply successively by 2, 3, and so on up to $p-1$, taking the remainder modulo p^2 after each multiplication. Using this method we never need to consider integers larger than about p^3, and the running time is essentially linear in p.

The following theorem shows that we can do much better than this.

THEOREM 6.1.1. *We may compute* $(p-1)! \pmod{p^2}$ *in time*

$$O(p^{1/2} \log^3 p).$$

To prove the theorem, let $s := p - 1$ and consider the product

$$1 \times 2 \times \cdots \times s \pmod{p^2}.$$

The idea is to split up this product into roughly $s^{1/2}$ groups of $s^{1/2}$ terms. Define

$$t := \lfloor s^{1/2} \rfloor,$$

so that

$$s = t^2 + t', \qquad 0 \leqslant t' \leqslant 2s^{1/2}.$$

Now break up the product into t groups of t terms, plus t' leftover terms, i.e.,

$$\big(1 \times \cdots \times t\big)\big((t+1) \times \cdots \times 2t\big) \cdots \big(((t-1)t+1) \times \cdots \times t^2\big)\big((t^2+1) \times \cdots \times s\big).$$

Introducing the polynomial

$$Q(k) := (k+1)(k+2)\cdots(k+t) \in (\mathbf{Z}/p^2\mathbf{Z})[k],$$

the product may be rewritten as

$$Q(0) \cdot Q(t) \cdot Q(2t) \cdots Q((t-1)t) \cdot \big((t^2+1) \times \cdots \times s\big).$$

PROBLEM 6.1.2. Prove Theorem 6.1.1 as follows.

a) Show that we may use a product tree (see §2.4) to compute $Q \in (\mathbf{Z}/p^2\mathbf{Z})[k]$ in time
$$O(p^{1/2} \log^3 p).$$
(By "compute Q", we mean compute its coefficients.)

b) Show that we may evaluate $Q(k)$ at the t points $k = 0, t, \ldots, (t-1)t$ via fast multipoint evaluation (Proposition 2.4.4) in time
$$O(p^{1/2} \log^3 p).$$

c) Show that we may multiply together the values found in (b), and the t' leftover terms, in time
$$O(p^{1/2} \log p \log\log p).$$

PROBLEM 6.1.3. (🎮) Implement the above method for computing $(p-1)! \pmod{p^2}$ in your FCAS. Can you observe the running time growing roughly like $p^{1/2}$? How big does p need to get before your implementation beats the straightforward linear-time method, i.e., multiplying successively by $2, \ldots, p-1 \pmod{p^2}$?

PROBLEM 6.1.4. The above strategy for computing factorials was first proposed by Strassen [**49**]. As an application, he described the following deterministic algorithm for integer factorisation. Consider the simplest case where we want to factor a number of the form $N = pq$, where p and q are distinct (unknown) primes. Show that we may compute $\lfloor N^{1/2} \rfloor! \pmod{N}$, and hence recover the factors p and q, in time $O(N^{1/4} \log^3 N)$. (A similar scheme was suggested by Pollard [**44**] at around the same time.)

PROBLEM 6.1.5. *Wolstenholme's theorem* states that

$$\binom{2p-1}{p-1} \equiv 1 \pmod{p^3}$$

for all primes $p \geqslant 5$. A *Wolstenholme prime* is a prime that satisfies the congruence modulo p^4. Describe an algorithm that can test whether p is a Wolstenholme prime in time $O(p^{1/2} \log^3 p)$.

(Several other characterisations of Wolstenholme primes are known; see for example [**39**]. There are only two Wolstenholme primes less than 10^{11}, namely 16843 and 2124679 [**6**]. I am not aware of any explicit mention in the literature of an $O(p^{1/2+\varepsilon})$-time algorithm for this problem, but it is hinted at in [**39**].)

REMARK 6.1.6. Bostan, Gaudry and Schost [**7**] gave an improvement to Strassen's method that reduces the complexity in Theorem 6.1.1 to

$$O(p^{1/2} \log^2 p).$$

They never actually compute the coefficients of $Q(k)$. Instead, they give a subroutine that takes as input the values of the polynomial $Q_n(k) := (k+1)\cdots(k+n)$ at n points in an arithmetic progression, and outputs the values of $Q_{2n}(k) := (k+1)\cdots(k+2n)$ at $2n$ points in a related arithmetic progression, at the cost of $O(1)$ polynomial multiplications of degree n. They apply this doubling process repeatedly to "grow" the polynomial and the set of evaluation points in parallel to reach the desired size. As far as I am aware, this is the best published complexity bound for computing $(p-1)! \pmod{p^2}$ for a single prime p.

REMARK 6.1.7. A downside of all variants of the square root time algorithm is that they use a lot of memory, and this is usually the bottleneck that limits the size of the problems that can be tackled in practice. The basic version worked out in Problem 6.1.2 uses $O(p^{1/2} \log^2 p)$ space; with some effort this can be reduced to $O(p^{1/2} \log p)$ [**55**, Lemma 2.1]. The Bostan–Gaudry–Schost variant also uses $O(p^{1/2} \log p)$ space. In all of these algorithms, it is possible to make a time-space tradeoff by adjusting the degree and the number of evaluation points. Roughly speaking, one can save a factor of $\alpha > 1$ in space by giving up a factor of α in time.

PROBLEM 6.1.8. (⚡) Can the complexity of Theorem 6.1.1 be further improved via the "factorial sieving" scheme described in [**14**, §2]? (This is also closely related to the integer factorisation algorithm in [**12**].) For example, to compute $N!$, we may split into odd and even terms, and remove a factor of 2 from each even term. Iterating this process, we arrive at a factorisation of the form

$$N! = 2^n(1 \times 3 \times \cdots \times N_0)(1 \times 3 \times \cdots \times N_1)\cdots$$

where $N_i \approx N/2^i$ for each $i \geqslant 0$. Notice that the first product has length roughly $N/2$ and the remaining products are subproducts of the first one. I would therefore expect that the entire expression could be computed using a modified square-root scheme in roughly the same time as the first product. This should save a factor of $2^{1/2}$ compared to handling $N!$ directly. Now suppose that we use the same trick to pull out factors of a few other small primes, not just the prime 2. Does this lead to an *asymptotic* improvement in the complexity of Theorem 6.1.1? My guess would be perhaps

$$O\left(\frac{p^{1/2}\log^2 p}{(\log\log p)^{1/2}}\right)?$$

6.2. Solving recurrences in square-root time. Let us now return to the situation from §5. Recall that we have a polynomial

$$F = F_0 + F_1 x + \cdots + F_d x^d \in (\mathbf{Z}/p^\mu\mathbf{Z})[x]$$

where $d := 2g + 2$, and we assume that $F_0 \not\equiv 0 \pmod p$. We are interested in computing certain coefficients of $H := F^m$ where $m := (p-1)/2$.

It will be helpful to re-express the recurrence (5.2.2) in matrix-vector form. For $k \geqslant 0$, define a row vector U_k consisting of d consecutive coefficients of H, namely

$$U_k := (H_{k-d+1}, \ldots, H_{k-1}, H_k) \in (\mathbf{Z}/p^\mu\mathbf{Z})^d.$$

Then (5.2.2) may be rewritten as

$$kU_k = \frac{1}{F_0} U_{k-1} T_k \tag{6.2.1}$$

where $T_k \in \mathrm{Mat}_d(\mathbf{Z}/p^\mu\mathbf{Z})$ is the *transition matrix* given by

$$T_k := \begin{pmatrix} 0 & \cdots & 0 & 0 & (d(m+1)-k)F_d \\ kF_0 & & & & \vdots \\ & \ddots & & & \vdots \\ & & kF_0 & & (2(m+1)-k)F_2 \\ & & & kF_0 & (m+1-k)F_1 \end{pmatrix}. \tag{6.2.2}$$

The last column of T_k implements (5.2.2) to compute kF_0H_k, and the remaining off-diagonal entries simply slide the window over from index $k-1$ to k. The "initial condition" is given by

$$U_0 = (0, \ldots, 0, H_0).$$

To compute a single coefficient H_n for a large index n, we want to chain several steps together. Using (6.2.1) repeatedly, we obtain

$$(k+1)\cdots(k+s)\cdot U_{k+s} = \frac{1}{(F_0)^s} U_k T_{k+1} \cdots T_{k+s}$$

for any $k \geqslant 0$, $s \geqslant 1$. This shows that the s-fold matrix product

$$T_{k+1}\cdots T_{k+s} \in \mathrm{Mat}_d(\mathbf{Z}/p^\mu\mathbf{Z})$$

may be viewed as a transition matrix going from index k to index $k+s$.

The most obvious algorithm for computing such an s-fold product is to simply multiply the matrices together one at a time. This has complexity growing linearly in s, and is obviously no better than executing the recurrence separately for each k. The next result shows that the complexity can be improved to roughly linear in $s^{1/2}$, by using a matrix generalisation of the square-root trick from §6.1.

PROPOSITION 6.2.3. *Given $k_0 \in \mathbf{Z}/p^\mu\mathbf{Z}$ and an integer $s \geqslant 2$ with $s \ll p^{O(1)}$, we may compute the matrix product*

$$T_{k_0+1} \cdots T_{k_0+s}$$

in time

$$O\big(d^\omega \mathsf{M}_{\text{int}}(s^{1/2}\mu \log p) \log s\big).$$

Proposition 6.2.3 may be be proved by the same strategy as Theorem 6.1.1, i.e., splitting up the product into roughly $s^{1/2}$ groups of $s^{1/2}$ terms. The following problem asks you to fill in the details.

PROBLEM 6.2.4.

a) Let $t := \lfloor s^{1/2} \rfloor$ and $t' := s - t^2$ as in §6.1, and define

$$Q(k) := T_{k+1} \cdots T_{k+t} \in \operatorname{Mat}_d((\mathbf{Z}/p^\mu\mathbf{Z})[k]).$$

Show that the entries of $Q(k)$ are polynomials in k of degree at most t.

b) Show that we may use a product tree to compute $Q(k)$, i.e., compute the coefficients of the polynomial in each entry of the matrix, in time

$$O\big(d^\omega \mathsf{M}_{\text{int}}(s^{1/2}\mu \log p) \log s\big).$$

c) Using fast multipoint evaluation, show that we may compute the values

$$Q(k_0), Q(k_0 + t), \ldots, Q(k_0 + (t-1)t) \in \operatorname{Mat}_d(\mathbf{Z}/p^\mu\mathbf{Z})$$

in time

$$O\big(d^2 \mathsf{M}_{\text{int}}(s^{1/2}\mu \log p) \log s\big).$$

d) Finish the proof of Proposition 6.2.3 by estimating the cost of multiplying together the values computed in (c) and the t' leftover matrices.

PROBLEM 6.2.5. *Kurepa's conjecture* states that there exists no odd prime p such that p divides the "left factorial" of p, which is defined to be

$$!p := 0! + 1! + \cdots + (p-1)!.$$

By setting up an appropriate recurrence involving 2×2 matrices, design an algorithm that can test Kurepa's conjecture for a single prime p in time $O(p^{1/2} \log^3 p)$. (An algorithm along these lines is mentioned briefly in [**4**].)

REMARK 6.2.6. A matrix version of the square-root trick first appeared in [**9**, §6], and several improvements were given in [**7**]. The latter improves the complexity of Proposition 6.2.3 to

$$O\big(d^\omega s^{1/2} \mathsf{M}_{\text{int}}(\mu \log p) + d^2 \mathsf{M}_{\text{int}}(s^{1/2}\mu \log p)\big).$$

6.3. Back to point counting. We now discuss how to apply Proposition 6.2.3 to the problem of computing A_f, i.e., computing $H_{vp-u} \pmod p$ for $1 \leqslant u, v \leqslant g$.

For each $v = 1, \ldots, g$, the coefficients H_{vp-u} for $u = 1, \ldots, g$ (corresponding to the entries of the v-th row of A_f) are exactly the last g entries of the vector U_{vp-1}. Therefore, we may rephrase our goal as wanting to compute the vectors

$$U_{p-1}, U_{2p-1}, \ldots, U_{gp-1} \pmod p.$$

Of course, we must still deal with the complications arising from the divisions by p that were discussed in §5. Let $\widetilde{H}_0, \widetilde{H}_1, \ldots \in \mathbf{Z}/p^\mu\mathbf{Z}$ be the sequence of approximations for H_k defined in §5, and let

$$\widetilde{U}_k := (\widetilde{H}_{k-d+1}, \ldots, \widetilde{H}_{k-1}, \widetilde{H}_k) \in (\mathbf{Z}/p^\mu\mathbf{Z})^d$$

be the corresponding vectors.

Notice that for any $k \geqslant 1$ we can compute $\widetilde{U}_k$ from $\widetilde{U}_{k-1}$ by exactly the same method explained in §5, i.e., after sliding all the entries over by one slot, we can compute the last entry of $\widetilde{U}_k$, namely $\widetilde{H}_k$, by solving (5.2.7) (see Problem 5.3.4(a)). Recall that when $k \equiv 0 \pmod p$, the resulting value of $\widetilde{H}_k$, and hence $\widetilde{U}_k$, is not uniquely determined.

On the other hand, when $k \not\equiv 0 \pmod p$, the vector $\widetilde{U}_k$ *is* uniquely determined by $\widetilde{U}_{k-1}$. In fact, from (6.2.1) we have

$$\widetilde{U}_k = \frac{1}{kF_0}\widetilde{U}_{k-1}T_k.$$

Chaining $p-1$ of these identities together, for any $i \geqslant 0$ we obtain

$$\widetilde{U}_{ip+p-1} = \frac{1}{(ip+1)\cdots(ip+p-1)(F_0)^{p-1}}\widetilde{U}_{ip}T_{ip+1}\cdots T_{ip+p-1}. \tag{6.3.1}$$

The above discussion suggests the following strategy. Starting with $\widetilde{U}_0$, we use the matrix $T_1 \cdots T_{p-1}$ to sprint to $\widetilde{U}_{p-1}$ in square-root time with the help of Proposition 6.2.3; then take a single step to $\widetilde{U}_p$; then sprint again to $\widetilde{U}_{2p-1}$ with the matrix $T_{p+1} \cdots T_{2p-1}$, and so on. This strategy may be formalised more precisely as follows.

PROPOSITION 6.3.2. *Suppose that the quantities*

$$T_{ip+1} \cdots T_{ip+p-1} \in \mathrm{Mat}_d(\mathbf{Z}/p^\mu\mathbf{Z}) \tag{6.3.3}$$

and

$$(ip+1)\cdots(ip+p-1) \in \mathbf{Z}/p^\mu\mathbf{Z} \tag{6.3.4}$$

are known for $i = 0, \ldots, g-1$. Then we may compute A_f in time

$$O\big((g^3 + \log p)(\mu \log p)^{1+\varepsilon}\big).$$

PROBLEM 6.3.5. Prove Proposition 6.3.2 as follows.

a) Show that H_0, and hence $\widetilde{U}_0 = U_0$, may be computed using $O(\log p)$ arithmetic operations in $\mathbf{Z}/p^\mu\mathbf{Z}$.
b) Using (6.3.1), show that for each i, we may compute $\widetilde{U}_{ip+p-1}$ from $\widetilde{U}_{ip}$ using $O(d^2)$ operations in $\mathbf{Z}/p^\mu\mathbf{Z}$.
c) Show that for each i, we may compute $\widetilde{U}_{(i+1)p}$ from $\widetilde{U}_{ip+p-1}$ using $O(d)$ operations in $\mathbf{Z}/p^\mu\mathbf{Z}$.

d) Finish the proof of Proposition 6.3.2.

The last remaining piece of the puzzle is to actually compute the quantities (6.3.3) and (6.3.4) for $i = 0, \dots, g-1$. The obvious thing to do is to apply Proposition 6.2.3 to handle each value of i in square-root time. But we have one more trick up our sleeves, which enables us to save a factor of $O(g)$. Recall from Problem 5.3.4(b) that if we impose the mild assumption $p \geqslant g$, then we can get away with a working precision of $\mu = 2$. This leads to the following "p-adic interpolation" optimisation.

PROPOSITION 6.3.6. *Assume that $\mu = 2$, and suppose that* (6.3.3) *and* (6.3.4) *are known for $i = 0$ and $i = 1$. Then we may compute* (6.3.3) *and* (6.3.4) *for the remaining values $i = 2, \dots, g-1$ in time*

$$O(g^3 \log^{1+\varepsilon} p).$$

PROBLEM 6.3.7. Prove Proposition 6.3.6 as follows. We only deal with (6.3.3); handling (6.3.4) is even easier. Consider the matrix

$$R(k) := T_{k+1} \cdots T_{k+p-1} \in \mathrm{Mat}_d((\mathbf{Z}/p^2\mathbf{Z})[k]).$$

Let us write this as

$$R(k) = R_0 + R_1 k + \cdots + R_{p-1} k^{p-1},$$

for matrices $R_0, \dots, R_{p-1} \in \mathrm{Mat}_d(\mathbf{Z}/p^2\mathbf{Z})$.

a) Show that given $R(0)$ and $R(p)$, we may compute $R_0 \pmod{p^2}$ and $R_1 \pmod{p}$ using $O(d^2)$ operations in $\mathbf{Z}/p^2\mathbf{Z}$.
b) Then show that for any i, we may deduce $R(ip)$ using $O(d^2)$ operations in $\mathbf{Z}/p^2\mathbf{Z}$.
c) Finish the proof of Proposition 6.3.6.

Putting everything together, we finally obtain a square-root bound for the cost of computing A_f:

THEOREM 6.3.8 (Computing A_f in square-root time). *Let $C/\mathbf{F}_p$ be a hyperelliptic curve of genus $g \geqslant 1$ defined by $y^2 = f(x)$ for some $f \in \mathbf{F}_p[x]$. Assume that $p \geqslant g^2$. Then we may compute A_f in time*

$$O(g^\omega p^{1/2} \log^3 p).$$

PROBLEM 6.3.9. Prove Theorem 6.3.8, by first using Proposition 6.2.3 to evaluate (6.3.3) for $i = 0, 1$ (and a similar method to handle (6.3.4)), and then applying Proposition 6.3.6 and Proposition 6.3.2.

REMARK 6.3.10. In the special case $g = 1$, we of course only need (6.3.3) and (6.3.4) for $i = 0$, not for $i = 1$. When $g = 2$, we need both values of i, but we need not bother with Proposition 6.3.6.

PROBLEM 6.3.11. () Implement the algorithm from Theorem 6.3.8 in your FCAS. Can you observe the running time behaving like $p^{1/2}$, and growing polynomially in g? How big does p have to get to beat the linear-time algorithms from Theorem 4.4.1 and Theorem 5.3.3?

PROBLEM 6.3.12. The products (6.3.3) for $i = 0$ and $i = 1$ are actually subproducts of the longer product $T_1 \cdots T_{2p-1}$. Show that by tweaking the proof of Proposition 6.2.3, it is possible to compute both products in about the same time as it takes to compute the longer product. Show that for large p, this trick speeds up the computation of A_f by a factor of about $2^{1/2}$.

(For a generalisation, see [**7**, §7].)

REMARK 6.3.13. Using the Bostan–Gaudry–Schost improvements mentioned in Remark 6.2.6, the complexity in Theorem 6.3.8 may be improved to

$$O\big(g^\omega p^{1/2} \log p \log \log p + g^2 p^{1/2} \log^2 p\big).$$

When p is large compared to g, this is the best complexity bound that I know for the problem of computing A_f. Note that this bound improves on [**7**, Theorem 17 (part 2)] by a factor of about $g^{3/2}$, due to the use of better p-adic error bounds (Proposition 5.2.9) and subsequent use of the interpolation trick (Proposition 6.3.6). Note also that for p sufficiently large relative to g, the second term is dominant, and this is optimal with respect to g since the output (the matrix A_f) has size $O(g^2)$. I suspect, but have not checked in detail, that the same complexity bound (up to a constant factor) could be obtained by the methods of [**21**], which works in the framework of p-adic cohomology (i.e., Kedlaya's algorithm). If this is the case, it would be interesting to compare (theoretically) the constant factors between the two algorithms. It would also be interesting to write an efficient implementation and compare its performance against the code accompanying [**21**].

7. An average polynomial time algorithm

Suppose that someone hands us a hyperelliptic curve C defined not over a finite field, but over $\mathbf{Q}$, say $y^2 = \bar{F}(x)$ for $\bar{F} \in \mathbf{Z}[x]$. Then for each prime $p \geqslant 3$, we may reduce $\bar{F}$ modulo p to obtain a polynomial $f_p \in \mathbf{F}_p[x]$, and hence an equation $y^2 = f_p(x)$. For all but finitely many p, this equation defines a hyperelliptic curve C_p over $\mathbf{F}_p$ (see §7.2), and for these "good" primes it makes sense to consider the zeta function of C_p,

$$Z_{C_p}(T) = \frac{L_{C_p}(T)}{(1-T)(1-pT)}.$$

One might guess that the L-polynomials $L_p(T) := L_{C_p}(T)$ for different primes should be completely unrelated: knowing the number of solutions to an equation modulo one prime p_1 should tell us nothing at all about the number of solutions modulo a different prime p_2. Nevertheless, when one averages over many primes, intriguing statistical regularities begin to appear. Examples include the Birch and Swinnerton-Dyer conjecture and the Sato–Tate conjecture, both of which were originally formulated for elliptic curves, and later generalised to curves of higher genus, including hyperelliptic curves.

Further discussion of these phenomena is beyond the scope of this course, but they motivate the following computational question: given a large integer N, how quickly can we compute $L_p(T)$ for all (good) primes $p \leqslant N$?

In Theorem 6.3.8 we saw that for a single prime p we can compute A_{f_p}, and hence $L_p(T) \pmod p$, in time $O(g^\omega p^{1/2} \log^3 p)$. If we do this separately for each prime $p \leqslant N$, the overall complexity is

$$O(g^\omega N^{3/2+\varepsilon}).$$

In this section we will see that it is possible to do much better! In fact, we will show that we can compute A_{f_p} for all $p \leqslant N$ (except finitely many) in time

$$O(g^\omega N \log^3 N).$$

This is incredibly fast — for fixed g, it is almost *linear* in the size of the output! Indeed, each A_{f_p} needs $O(g^2 \log p)$ bits to write down, so according to Lemma 1.5.2 the total output size is

$$\sum_{p \leqslant N} O(g^2 \log p) = O(g^2 N).$$

Another point of view is that the *average* complexity per prime $p \leqslant N$ is

$$\frac{O(g^\omega N \log^3 N)}{N / \log N} = O(g^\omega \log^4 N).$$

Loosely speaking, the time spent on each prime p is only $O(g^\omega \log^4 p)$, which is polynomial in the bit size of f_p. For this reason, the algorithm is said to run in "average polynomial time". In §8 we will see how to extend this result to compute not just $L_p(T) \pmod p$, but the entire L-polynomial $L_p(T) \in \mathbf{Z}[T]$, for all $p \leqslant N$, in average polynomial time. (Compare with Remark 3.3.3.)

All of these average polynomial time point counting algorithms depend on a marvellous device called the *accumulating remainder tree*. The accumulating remainder tree was invented by Robert Gerbicz, an applied mathematics student, in the context of a large-scale search for Wilson primes. (See §6.1 for the definition of Wilson primes.) He outlined the algorithm in a 2011 post on mersenneforum.org [**19**], and it appeared in published form a few years later [**11**]. In retrospect, it is surprising that the algorithm was not discovered decades earlier. The basic ingredients were already known by the 1980s, but somehow all the experts missed it.

7.1. A warm-up: Wilson primes. The Wilson prime example provides an excellent introduction to the main ideas behind the accumulating remainder tree. In this section we will look at this example in detail, and prove the following theorem.

THEOREM 7.1.1. *Given an integer $N \geqslant 2$, we may compute $(p-1)! \pmod{p^2}$ for all primes $p \leqslant N$ in time*

$$O(N \log^3 N).$$

To be very explicit, the quantities that we want to compute are

$$\begin{aligned}
&1 \times 2 && \pmod{3^2}, \\
&1 \times 2 \times 3 \times 4 && \pmod{5^2}, \\
&1 \times 2 \times 3 \times 4 \times 5 \times 6 && \pmod{7^2}, \\
&1 \times 2 \times 3 \times 4 \times 5 \times 6 \times 7 \times 8 \times 9 \times 10 && \pmod{11^2},
\end{aligned} \tag{7.1.2}$$

and so on, for all $p \leqslant N$.

For simplicity, we assume throughout that N is a power of two, say $N = 2^b$.

We start by defining a certain binary tree called the *modulus tree*. For $j = 1, \ldots, N$, define

$$m_j := \begin{cases} j^2 & \text{if } j \text{ is prime}, \\ 1 & \text{otherwise}. \end{cases}$$

Then the modulus tree is the product tree associated to $m_1, \ldots, m_N$ (see §2.4). An example for $N = 16$ is shown in Figure 7.1.3.

We index the levels of the tree by $\ell = 0, \ldots, b$, where $\ell = 0$ corresponds to the root and $\ell = b$ to the leaves. The nodes at level ℓ are labelled by the pairs (ℓ, j) for

$j = 1, \ldots, 2^\ell$. We write $m_{\ell,j}$ for the value in the modulus tree at the node (ℓ, j); thus for example $m_{b,j} = m_j$ for $j = 1, \ldots, N$, and $m_{0,1} = \prod_{p \leqslant N} p^2$.

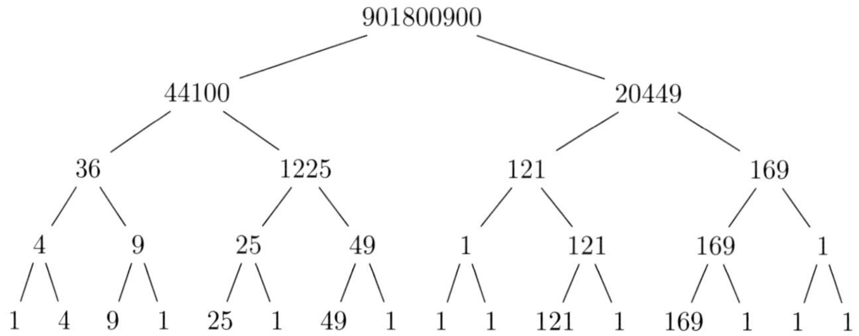

FIGURE 7.1.3. The modulus tree for $N = 16$.

LEMMA 7.1.4. *We may compute the modulus tree, i.e., all values* $m_{\ell,j}$, *in time*

$$O(N \log^2 N).$$

PROBLEM 7.1.5. Use Lemma 1.5.2 to show that the product $\prod_{p \leqslant N} p^2$ has $O(N)$ bits. Then prove Lemma 7.1.4 via Proposition 2.6.1 and Proposition 2.4.2.

We now define a second binary tree, unimaginatively named the *value tree*, to be the product tree on the values $V_j := j$ for $j = 1, \ldots, N$. We write $V_{\ell,j}$ for the value at the node (ℓ, j). For instance, the value at the root node is $V_{0,1} = N!$. An example for $N = 16$ is shown in Figure 7.1.6.

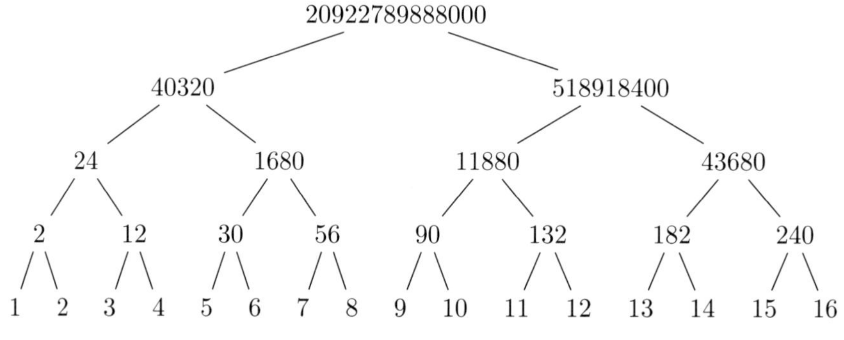

FIGURE 7.1.6. The value tree for $N = 16$.

LEMMA 7.1.7. *We may compute the value tree, i.e., all values* $V_{\ell,i}$, *in time*

$$O(N \log^3 N).$$

PROBLEM 7.1.8. Prove Lemma 7.1.7 along the same lines as the proof of Lemma 7.1.4, by first showing that the bit size of $N!$ is $O(N \log N)$.

Finally we arrive at our third and most interesting binary tree, the *accumulating remainder tree*. The value assigned to the node (ℓ, j) is defined to be

$$R_{\ell,j} := V_{\ell,1} \cdots V_{\ell,j-1} \pmod{m_{\ell,j}},$$

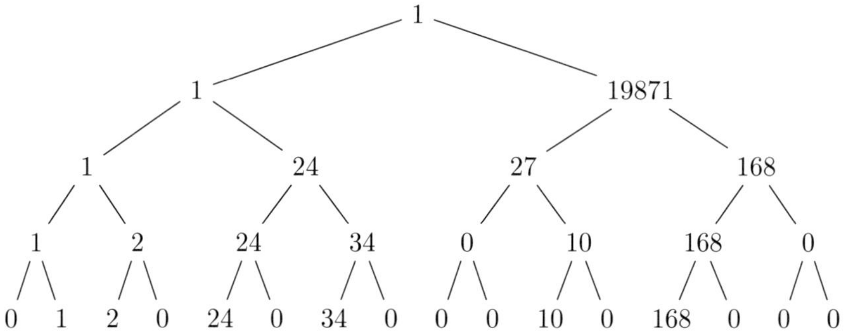

FIGURE 7.1.9. The accumulating remainder tree for $N = 16$.

thought of as an element of $\mathbf{Z}/m_{\ell,j}\mathbf{Z}$. In other words, $R_{\ell,j}$ is the product of all values strictly to the *left* of the node (ℓ, j) in the value tree, taken modulo the corresponding value $m_{\ell,j}$ in the modulus tree. Note that if $j = 1$, the product is empty so by definition $R_{\ell,j} = 1 \pmod{m_{\ell,j}}$. See Figure 7.1.9.

PROBLEM 7.1.10. Show that for any prime $p \leqslant N$,

$$R_{b,p} = (p-1)! \pmod{p^2}.$$

In other words, the values in the leaf nodes of the accumulating remainder tree are exactly the Wilson remainders. For example, in Figure 7.1.9 we can immediately see that $p = 5$ and $p = 13$ are the only Wilson primes less than 16.

The following result is the engine that drives the whole accumulating remainder tree algorithm.

LEMMA 7.1.11. *Let (ℓ, j) be a non-leaf node, i.e., with $\ell < b$. Then*

$$\begin{aligned} R_{\ell+1,2j-1} &\equiv R_{\ell,j} && \pmod{m_{\ell+1,2j-1}}, \\ R_{\ell+1,2j} &\equiv R_{\ell,j} V_{\ell+1,2j-1} && \pmod{m_{\ell+1,2j}}. \end{aligned}$$

Lemma 7.1.11 says that if we start with any non-leaf node, then its *left child* is obtained by simply reducing its value modulo the corresponding modulus tree node, and that its *right child* is obtained by first multiplying by an appropriate node from the value tree, and then reducing modulo the corresponding modulus tree node. For example, in Figure 7.1.9, starting with the value 19871 at the node $(\ell, j) = (1, 2)$, the value of its left child is $R_{2,3} = 19871 \bmod 121 = 27$, and its right child is $R_{2,4} = 19871 \cdot 11880 \bmod 169 = 168$.

COROLLARY 7.1.12. *Assume that the modulus tree and value tree are known. We may compute the remainder tree, i.e., all values $R_{\ell,j}$, in time*

$$O(N \log^3 N).$$

PROBLEM 7.1.13. Prove Lemma 7.1.11, and then deduce Corollary 7.1.12 by using Lemma 7.1.11 to show that if we know the values of the accumulating remainder tree at level ℓ for some $\ell < b$, then we may compute the values at level $\ell + 1$ in time

$$O(N \log^2 N).$$

The main theorem of this section, Theorem 7.1.1, now follows immediately by stringing together Lemma 7.1.4, Lemma 7.1.7, Corollary 7.1.12, and Problem 7.1.10.

PROBLEM 7.1.14. (🎮) Implement the algorithm described in Theorem 7.1.1 to compute $(p-1)! \pmod{p^2}$ for all $p \leqslant N$ in your FCAS. For what N is it faster than running the $O(p^{1/2+\varepsilon})$ implementation from Problem 6.1.3 separately for each $p \leqslant N$?

An algorithm similar to the one described above was used in [**11**] to carry out a large-scale search for Wilson primes. The computation expended about 1.1 million hours of CPU time (in the early 2010s), and found that there are no Wilson primes less than 2×10^{13}, apart from the three already known. The main difficulty in this project was memory usage: for $N = 2 \times 10^{13}$, just writing down the value tree, as defined above, would require about 5000 terabytes of storage! For this and other reasons, the algorithm actually used in [**11**] was considerably more complicated than the one presented in this section.

PROBLEM 7.1.15. Describe an algorithm that can find all Wolstenholme primes $p \leqslant N$ in time $O(N \log^3 N)$.

(See Problem 6.1.5 for the definition of Wolstenholme primes. The observation that this problem yields to an average polynomial time algorithm is not new — I learned of this from Robert Gerbicz many years ago. But it does not yet seem to have appeared in print.)

PROBLEM 7.1.16. Describe an algorithm that can find all counterexamples $p \leqslant N$ to Kurepa's conjecture in time $O(N \log^3 N)$. Note that some of the trees will now consist of *matrices* instead of scalars.

(See Problem 6.2.5 for the statement of Kurepa's conjecture. An average polynomial time algorithm for this problem was described in [**4**], and the authors used it to show that there are no counterexamples less than $2^{40} \approx 1.1 \times 10^{12}$. They also expressed the opinion, which I share, that Kurepa's conjecture is almost certainly false.)

PROBLEM 7.1.17. (⚡) Can the "factorial sieving" scheme in [**14**, §2] (see Problem 6.1.8) be used to improve the complexity of Theorem 7.1.1, perhaps to

$$O\left(\frac{N \log^3 N}{\log \log N}\right)?$$

PROBLEM 7.1.18. (⚡⚡) Is it possible to improve the $O(N \log^3 N)$ complexity bound for the accumulating remainder tree algorithm? I have wondered about this for many years. The bottleneck seems to be the value tree, which occupies altogether $O(N \log^2 N)$ bits of space, and really does seem to require $O(N \log^3 N)$ time to build. On the other hand, in some sense only $O(N \log N)$ bits of information from the value tree are fed into the the accumulating remainder tree, namely the remainders $V_{\ell+1,2j-1} \pmod{m_{\ell+1,2j}}$ (see Lemma 7.1.11). This suggests that when we compute the value tree, we are somehow computing too much data by a factor of $\log N$. Can we somehow use this observation to improve the overall complexity to $O(N \log^2 N)$? Or even something less ambitious like $O(N \log^3 N / \log \log N)$?

Note that for the specific case of Wilson primes, there is a lot of additional structure that might lead to a provable asymptotic speedup; see for example Problem

7.1.17, or the factorial identities described in [**11**, §3]. That would be interesting in its own right, but in this problem what I am really asking for is a generic speedup that applies to all instances of the accumulating remainder tree, including the point counting algorithms presented in the next section.

7.2. Back to point counting. Let us now return from our Wilson prime detour to the problem of point counting on hyperelliptic curves.

The setup is that $C/\mathbf{Q}$ is a hyperelliptic curve of genus $g \geqslant 1$ given by $y^2 = \bar{F}(x)$ where $\bar{F} \in \mathbf{Z}[x]$ is squarefree of degree $2g+1$ or $2g+2$. To avoid problems with recurrences later on, let us also assume that

$$\bar{F}_0 \neq 0.$$

(The $\bar{F}_0 = 0$ case may be handled along similar lines to Remark 5.1.7.) For each prime $p \geqslant 3$, we write $f_p \in \mathbf{F}_p[x]$ for the reduction of $\bar{F}$ modulo p.

DEFINITION 7.2.1. We say that a prime $p \geqslant 3$ is *admissible* for $\bar{F}$ if $p \geqslant g$, and if the polynomial $f_p \in \mathbf{F}_p[x]$ is squarefree, has the same degree as $\bar{F}$, and has nonzero constant term.

Equivalently, $p \geqslant g$ is admissible if it does not divide the leading coefficient, constant term, or discriminant of $\bar{F}$. This implies that there are only finitely many inadmissible primes for $\bar{F}$.

REMARK 7.2.2. The above definition of admissibility is nonstandard, and is customised for the requirements of the algorithms later in this section. It is closely related to the concept of *good reduction* (for a definition see [**30**, §A.9.1]), but is not the same. If p is admissible for $\bar{F}$, then C has good reduction at p, but the converse is false in general.

If p is admissible for $\bar{F}$, then the curve C_p defined by $y^2 = f_p(x)$ is a hyperelliptic curve over $\mathbf{F}_p$ of the same genus g, so it makes sense to define $L_p(T) := L_{C_p}(T)$.

Now suppose that we are given some bound $N \geqslant 3$, and we wish to compute $L_p(T) \pmod p$ for all admissible $p \leqslant N$. By Theorem 4.3.2, it suffices to compute the matrices

$$A_{f_p} \in \operatorname{Mat}_g(\mathbf{F}_p)$$

for all admissible $p \leqslant N$. In other words, we want to compute the quantities

$$(\bar{F}^{(p-1)/2})_{vp-u} \pmod p, \qquad 1 \leqslant u, v \leqslant g$$

for each p. At first glance this looks hopeless. Not only is the modulus different for each p, but we want to compute different coefficients for each p, and we are even dealing with a different polynomial $\bar{F}^{(p-1)/2}$ for each p! It is not at all obvious that there is any redundancy between the various primes that can be exploited.

To make progress, we will leverage the framework of §§5–6. Since we are only considering $p \geqslant g$, we may fix $\mu := 2$ (see Problem 5.3.4(b)). For each admissible p, let $F^{(p)} \in (\mathbf{Z}/p^2\mathbf{Z})[x]$ be the polynomial defined by

$$F^{(p)} := \bar{F} \pmod{p^2}.$$

By construction, $F^{(p)}$ is a lift of $f_p \in \mathbf{F}_p[x]$. Moreover, since p is admissible we have $(f_p)_0 \neq 0$, so everything in §§5–6 may be carried out for $F := F^{(p)}$.

REMARK 7.2.3. Notice what just happened: the single "global" polynomial $\bar{F} \in \mathbf{Z}[x]$ serves as a lift of f_p for *all* p at the same time. This is one of the key ideas of the average polynomial time approach.

Recall the transition matrix $T_k \in \mathrm{Mat}_d(\mathbf{Z}/p^2\mathbf{Z})$ defined in (6.2.2); to indicate the dependence on p, we will denote it here by $T_k^{(p)}$. According to Proposition 6.3.2 and Proposition 6.3.6, to compute A_{f_p} it suffices to compute the products

$$(7.2.4)\qquad \begin{aligned} T_1^{(p)} \cdots T_{p-1}^{(p)} &\in \mathrm{Mat}_d(\mathbf{Z}/p^2\mathbf{Z}), & 1 \cdot 2 \cdots (p-1) &\in \mathbf{Z}/p^2\mathbf{Z}, \\ T_{p+1}^{(p)} \cdots T_{2p-1}^{(p)} &\in \mathrm{Mat}_d(\mathbf{Z}/p^2\mathbf{Z}), & (p+1) \cdots (2p-1) &\in \mathbf{Z}/p^2\mathbf{Z}. \end{aligned}$$

In the present context, we need to compute these products for all admissible $p \leqslant N$ simultaneously. For example, we want to compute

$$\begin{aligned} &T_1^{(3)} T_2^{(3)} && \pmod{3^2}, && T_4^{(3)} T_5^{(3)} && \pmod{3^2}, \\ &T_1^{(5)} T_2^{(5)} T_3^{(5)} T_4^{(5)} && \pmod{5^2}, && T_6^{(5)} T_7^{(5)} T_8^{(5)} T_9^{(5)} && \pmod{5^2}, \\ &T_1^{(7)} T_2^{(7)} T_3^{(7)} T_4^{(7)} T_5^{(7)} T_6^{(7)} && \pmod{7^2}, && T_8^{(7)} T_9^{(7)} T_{10}^{(7)} T_{11}^{(7)} T_{12}^{(7)} T_{13}^{(7)} && \pmod{7^2}, \end{aligned}$$

and so on. Formally, these products look extremely similar to the products appearing in (7.1.2). However, when we attempt to throw the accumulating remainder tree at this problem, we encounter two new difficulties that did not arise in the Wilson prime context.

First, notice that the indices do not line up properly in the $i = 1$ case. For example, the product $T_6 T_7 T_8 T_9$ for $p = 5$ is not a subproduct of the corresponding product $T_8 \cdots T_{13}$ for $p = 7$.

The second and more serious problem is that *the entries of the transition matrices depend on* p! Specifically, the last column of T_k involves the quantity $m = (p-1)/2$, which is different for each p. So we cannot directly build a "value tree" in the same way that we did for the Wilson prime problem.

7.3. Introducing a generic prime. We may solve both of the problems mentioned above by employing a technique suggested in [**23**, §4.4].

Consider the ring

$$\mathbf{Z}[P]/P^2\mathbf{Z}[P],$$

which for brevity we will write as $\mathbf{Z}[P]/P^2$. This may be regarded as a ring of truncated power series in P; a typical element of $\mathbf{Z}[P]/P^2$ has the form $\gamma(P) = \gamma_0 + \gamma_1 P$ for integers γ_0 and γ_1. Multiplication in this ring is given by

$$(\gamma_0 + \gamma_1 P)(\delta_0 + \delta_1 P) = \gamma_0\delta_0 + (\gamma_0\delta_1 + \gamma_1\delta_0)P.$$

We will think of P as a "generic prime" that has not yet been specialised to an actual prime number.

For any $\gamma = \gamma_0 + \gamma_1 P \in \mathbf{Z}[P]/P^2$, it makes sense to substitute a prime number p for P to obtain an element

$$\gamma(p) := \gamma_0 + \gamma_1 p \in \mathbf{Z}/p^2\mathbf{Z}.$$

For each p, this operation is clearly a ring homomorphism

$$\mathbf{Z}[P]/P^2 \longrightarrow \mathbf{Z}/p^2\mathbf{Z},$$

i.e., for any $\gamma, \delta \in \mathbf{Z}[P]/P^2$ we have $(\gamma+\delta)(p) = \gamma(p)+\delta(p)$ and $(\gamma\cdot\delta)(p) = \gamma(p)\cdot\delta(p)$ in $\mathbf{Z}/p^2\mathbf{Z}$.

These observations extend to matrices as well. If $\gamma \in \mathrm{Mat}_d(\mathbf{Z}[P]/P^2)$, say $\gamma = \gamma_0 + \gamma_1 P$ for $\gamma_0, \gamma_1 \in \mathrm{Mat}_d(\mathbf{Z})$, then we may substitute an actual prime p for P to obtain $\gamma(p) := \gamma_0 + \gamma_1 p \in \mathrm{Mat}_d(\mathbf{Z}/p^2\mathbf{Z})$. Again, for each p this operation is a ring homomorphism $\mathrm{Mat}_d(\mathbf{Z}[P]/P^2) \to \mathrm{Mat}_d(\mathbf{Z}/p^2\mathbf{Z})$.

The idea is now to define a matrix over $\mathbf{Z}[P]/P^2$, that is completely independent of p, but that nevertheless specialises for each p to the desired transition matrix over $\mathbf{Z}/p^2\mathbf{Z}$. For each $i = 0, 1$ and $j \geqslant 1$, define

$$\bar{T}_j^i \in \operatorname{Mat}_d(\mathbf{Z}[P]/P^2)$$

to be the matrix obtained from $2 \cdot T_k$ (see (6.2.2)) by making the substitutions

$$m \mapsto \frac{P-1}{2}, \qquad k \mapsto iP + j,$$

and replacing the coefficients of F by the corresponding coefficients of $\bar{F}$. Explicitly,

$$\bar{T}_j^i(P) = \begin{pmatrix} 0 & \cdots & 0 & 0 & (d(P+1) - 2(iP+j))\bar{F}_d \\ 2(iP+j)\bar{F}_0 & & & & \vdots \\ & \ddots & & & \vdots \\ & & 2(iP+j)\bar{F}_0 & & (2(P+1) - 2(iP+j))\bar{F}_2 \\ & & & 2(iP+j)\bar{F}_0 & (P+1-2(iP+j))\bar{F}_1 \end{pmatrix}.$$

(The purpose of the extra factor of 2 is to make the entries of $\bar{T}_j^i(P)$ integral.)

PROBLEM 7.3.1. Let $i = 0, 1$. Show that for any prime $p \geqslant 3$,

$$\bar{T}_j^i(p) \equiv 2 \cdot T_{ip+j}^{(p)} \pmod{p^2}.$$

Deduce that

$$(\bar{T}_1^i \cdots \bar{T}_{p-1}^i)(p) \equiv 2^{p-1} \cdot T_{ip+1}^{(p)} \cdots T_{ip+p-1}^{(p)} \pmod{p^2}. \tag{7.3.2}$$

We are now in business! Briefly, the algorithm runs as follows. First we build a modulus tree on the squares of the primes, exactly as in §7.1. Next, for each $i = 0, 1$ we build a value tree from the matrices

$$\bar{T}_1^i, \ldots, \bar{T}_N^i \in \operatorname{Mat}_d(\mathbf{Z}[P]/P^2).$$

We then compute an accumulating remainder tree whose leaf nodes contain the matrices

$$\begin{aligned} &\bar{T}_1^i \bar{T}_2^i && \pmod{3^2}, \\ &\bar{T}_1^i \bar{T}_2^i \bar{T}_3^i \bar{T}_4^i && \pmod{5^2}, \\ &\bar{T}_1^i \bar{T}_2^i \bar{T}_3^i \bar{T}_4^i \bar{T}_5^i \bar{T}_6^i && \pmod{7^2}, \end{aligned}$$

and so on, for all $p \leqslant N$. Note that the leaf node corresponding to a given p lies in the ring

$$\operatorname{Mat}_d((\mathbf{Z}/p^2\mathbf{Z})[P]/P^2).$$

At this stage we are finally allowed to "remember" which prime p the generic P is supposed to stand for. Substituting p for P in each leaf node, we obtain the matrices

$$(\bar{T}_1^i \cdots \bar{T}_{p-1}^i)(p) \pmod{p^2}$$

as elements of $\operatorname{Mat}_d(\mathbf{Z}/p^2\mathbf{Z})$. Thanks to (7.3.2), after removing the factors of 2^{p-1} we have found all of the desired matrices in (7.2.4). The scalar quantities in (7.2.4) may be computed by an analogous (much simpler) procedure. Finally, we recover the desired A_{f_p} by applying Proposition 6.3.2 and Proposition 6.3.6 separately for each p.

7.4. Complexity analysis. In this section we investigate the complexity of the algorithm sketched above. We will prove the following theorem.

THEOREM 7.4.1 (Computing A_{f_p} in average polynomial time). *Let $C/\mathbf{Q}$ be a hyperelliptic curve of genus $g \geqslant 1$ defined by $y^2 = \bar{F}(x)$ for some $\bar{F} \in \mathbf{Z}[x]$. Let $N \geqslant 2$ and assume that*

$$\log N \gg \log \max_i |\bar{F}_i| \qquad \text{and} \qquad \log N \gg g. \tag{7.4.2}$$

Then we may compute A_{f_p} for all admissible $p \leqslant N$ in time

$$O(g^\omega N \log^3 N).$$

It is possible to drop the hypothesis on $\log \bar{F}_i$ and give a more detailed complexity bound that takes into account the size of the coefficients of $\bar{F}$, but this just complicates the analysis.

As in §7.1, for simplicity we assume that $N = 2^b$.

Compared to the Wilson prime case, the main difficulty in the complexity analysis is keeping track of the sizes of the objects (polynomials, matrices) being manipulated. The following definitions assist in managing this issue.

DEFINITION 7.4.3.

- For $\gamma = \gamma_0 + \gamma_1 P \in \mathbf{Z}[P]/P^2$, define
$$\|\gamma\| := |\gamma_0| + |\gamma_1|.$$
- For $\gamma \in \mathrm{Mat}_d(\mathbf{Z}[P]/P^2)$, define
$$\|\gamma\| := \max_{1 \leqslant i \leqslant d} \sum_{j=1}^{d} \|\gamma_{i,j}\|.$$
In other words, the norm of a matrix is defined to be the maximum of the L^1 norms of its rows.

The next problem asks you to show that these norms are both *submultiplicative.*

PROBLEM 7.4.4.

a) Let $\gamma, \delta \in \mathbf{Z}[P]/P^2$. Show that $\|\gamma\delta\| \leqslant \|\gamma\|\|\delta\|$.
b) Let $\gamma, \delta \in \mathrm{Mat}_d(\mathbf{Z}[P]/P^2)$. Show that $\|\gamma\delta\| \leqslant \|\gamma\|\|\delta\|$.

PROBLEM 7.4.5. Show that the amount of space (i.e., number of bits) required to represent a matrix $\gamma \in \mathrm{Mat}_d(\mathbf{Z}[P]/P^2)$ is

$$O(d^2 \log \|\gamma\|),$$

and that matrices $\gamma, \delta \in \mathrm{Mat}_d(\mathbf{Z}[P]/P^2)$ may be multiplied in time

$$O\big(d^\omega \mathsf{M}_{\mathrm{int}}(\log \|\gamma\| + \log \|\delta\|)\big).$$

We now define two value trees, one for each $i = 0, 1$, to be the product trees on the values

$$\bar{T}_1^i, \dots, \bar{T}_N^i \in \mathrm{Mat}_d(\mathbf{Z}[P]/P^2).$$

We write $V_{\ell,j}^i \in \mathrm{Mat}_d(\mathbf{Z}[P]/P^2)$ for the matrix in the node (ℓ, j) of the i-th value tree, for $0 \leqslant \ell \leqslant b$ and $1 \leqslant j \leqslant 2^\ell$.

LEMMA 7.4.6. *We may compute the value trees for $i = 0, 1$ in time*

$$O(d^\omega N \log^3 N).$$

PROBLEM 7.4.7. Prove Lemma 7.4.6 as follows.

a) Using the hypotheses (7.4.2), show that

$$\|\bar{T}_j^i\| = O(\log N), \qquad 1 \leqslant j \leqslant N.$$

b) Show that the total bit size of the nodes at each level of each value tree is

$$O(d^2 N \log N).$$

c) Finish the proof of Lemma 7.4.6, along the same lines as the proof of Lemma 7.1.7.

Next, we define an accumulating remainder tree for each $i = 0, 1$. In the i-th tree, the value assigned to the node (ℓ, j) is

$$R_{\ell,j}^i := V_{\ell,1}^i \cdots V_{\ell,j-1}^i \pmod{m_{\ell,j}} \qquad \in \operatorname{Mat}_d((\mathbf{Z}/m_{\ell,j}\mathbf{Z})[P]/P^2).$$

LEMMA 7.4.8. *Assume that the modulus tree and value trees are known. We may compute the accumulating remainder trees for* $i = 0, 1$ *in time*

$$O(d^\omega N \log^3 N).$$

PROBLEM 7.4.9. Prove Lemma 7.4.8 along the same lines as the proof of Corollary 7.1.12.

PROBLEM 7.4.10. Prove Theorem 7.4.1 as follows.

a) State and prove analogues of Lemma 7.4.6 and Lemma 7.4.8 to handle the scalar quantities in (7.2.4).

b) Assume that all four accumulating remainder trees are known, i.e., the matrix and scalar trees, for both $i = 0$ and $i = 1$. Use (7.3.2) to prove that for each admissible prime p we may deduce all quantities in (7.2.4) from the leaf nodes of these trees in time

$$O\big((g^2 + \log p) \log^{1+\varepsilon} p\big).$$

Show that we may then use Proposition 6.3.2 and Proposition 6.3.6 to recover A_{f_p} in time

$$O\big((g^3 + \log p) \log^{1+\varepsilon} p\big).$$

c) Show that total cost of (b) over all $p \leqslant N$ is

$$O\big((g^3 + \log N) N \log^\varepsilon N\big),$$

and use the hypotheses (7.4.2) to show that this is dominated by

$$O(g^\omega N \log^3 N).$$

d) Finish the proof of Theorem 7.4.1.

PROBLEM 7.4.11. () Implement the algorithm described above to compute A_{f_p} for all admissible $p \leqslant N$ in your FCAS. For what N is it faster than running the $O(p^{1/2+\varepsilon})$ implementation from Problem 6.3.11 separately for each $p \leqslant N$?

PROBLEM 7.4.12. Here is a sketch of another approach that yields an average polynomial time algorithm for computing $L_C(T) \pmod p$ without using the "generic prime" trick. Fix some $v = 1, \ldots, g$, and consider the vectors

$$W_k^v := (\bar{F}_{v(2k+1)-1}^k, \ldots, \bar{F}_{v(2k+1)-d}^k) \in \mathbf{Z}^d, \qquad k \geqslant 1.$$

For $k = (p-1)/2$, this vector (reduced modulo p) contains the v-th row of A_{f_p}. Show how to construct a "transition matrix" over $\mathbf{Z}$ that takes W_{k-1}^v to W_k^v, whose entries do not depend on p.

(This "diagonal recurrence" is much closer in spirit to the first published average polynomial time point counting algorithm [**22**], and the same idea was used in the first reported *implementation* of such an algorithm [**25**]. The recurrences obtained in this way are horribly complicated, and were abandoned in subsequent implementations such as [**13, 24, 26, 52**].)

REMARK 7.4.13. As mentioned previously in the context of Wilson primes, these algorithms use a huge amount of memory. One systematic method for reducing the memory usage is to replace the accumulating remainder tree by a remainder *forest*, as explained in [**25**, §4.1].

REMARK 7.4.14. The obvious algorithm for multiplying two matrices in $\mathrm{Mat}_d(\mathbf{Z})$ with n-bit entries runs in time $O(d^\omega \mathsf{M}_{\mathrm{int}}(n))$. This may be improved to

$$O(d^2 \mathsf{M}_{\mathrm{int}}(n))$$

if n is large enough compared to d, by reusing the Fourier transforms of the matrix entries [**27**]. This idea can be used to improve the main complexity bound in Theorem 7.4.1 to

$$O(g^2 N \log^3 N). \tag{7.4.15}$$

This is the best complexity bound I know of for the problem of computing A_{f_p} for all $p \leqslant N$. The dependence on g is optimal, as the output (the list of A_{F_p} matrices) has total bit size $O(g^2 N)$. The bound (7.4.15) improves on the main theorem of [**26**] by a factor of $O(g)$; the source of the improvement is that we have in effect replaced the "translated curves" trick (see Problem 5.3.7) by p-adic interpolation (Proposition 6.3.6). I am not aware of any serious attempts at implementing this algorithm, and I do not know how it would perform in practice compared to [**26**].

8. A modulo p^λ trace formula

So far, all of the point counting algorithms developed in this course have relied on Theorem 4.2.7, which yields information about point counts modulo p. The main result of this section is Theorem 8.4.1, which gives a congruence for these point counts modulo a *power* of p. Since we know *a priori* upper bounds for the point counts, this will enable us to obtain the point counts exactly in $\mathbf{Z}$.

REMARK 8.0.1. Theorem 8.4.1 may be viewed as a specialisation of the trace formula [**23**, Theorem 3.1] to the case of hyperelliptic curves over a prime field. The proofs in both cases are entirely elementary, in the sense that no cohomology is involved.

REMARK 8.0.2. The approach taken in this section is not the only way to gain information above and beyond Theorem 4.2.7. One could also use ℓ-adic methods and/or generic group methods to supplement the modulo p data; see for example the last paragraph of [**7**, §8]. We will not discuss this further in this course.

8.1. A motivating example. Recall from (1.5.1) that if $\beta \in \mathbf{F}_p$, then $\beta^{(p-1)/2} \equiv \chi_p(\beta) \pmod p$. What happens if we step back and look at things modulo p^2?

Let $\beta \in \mathbf{Z}/p^2\mathbf{Z}$, and let $\overline{\beta} \in \mathbf{F}_p$ be its reduction modulo p. Then

$$\beta^{(p-1)/2} \equiv \chi_p(\overline{\beta}) + cp \pmod{p^2} \tag{8.1.1}$$

for some integer c. For the purposes of a modulo p trace formula such as Theorem 4.2.7, the cp term is irrelevant. But if we want to construct a "modulo p^2 trace formula", the cp term is major nuisance. What we want is an analogue of the expression $\beta^{(p-1)/2}$ that yields $\chi_p(\overline{\beta}) \pmod{p^2}$, *without* the garbage cp term.

We can achieve this by playing some algebraic games with (8.1.1). Let $\beta \in \mathbf{Z}/p^2\mathbf{Z}$ and consider the expression $\beta^{3(p-1)/2}$. From (8.1.1) we have

$$\beta^{3(p-1)/2} \equiv (\chi_p(\overline{\beta}) + cp)^3 \equiv \chi_p(\overline{\beta})^3 + 3\chi_p(\overline{\beta})^2 cp \pmod{p^2}.$$

We now distinguish two cases. First, if $\chi_p(\overline{\beta}) = \pm 1$, then $\chi_p(\overline{\beta})^2 = 1$ and we get

$$\beta^{3(p-1)/2} \equiv \chi_p(\overline{\beta}) + 3cp \pmod{p^2}. \tag{8.1.2}$$

We can now *eliminate* the unwanted cp term by taking a suitable linear combination of (8.1.1) and (8.1.2):

$$\frac{1}{2}\left(3\beta^{(p-1)/2} - \beta^{3(p-1)/2}\right) \equiv \chi_p(\overline{\beta}) \pmod{p^2}. \tag{8.1.3}$$

What about the case $\chi_p(\overline{\beta}) = 0$? In this case $\overline{\beta} = 0$, so $p \mid \beta$. This implies that $\beta^{(p-1)/2} \equiv 0 \pmod{p^2}$, provided that $p \geqslant 5$. Therefore, the formula (8.1.3) holds for all β, except possibly when $p = 3$.

Equation (8.1.3) suggests that (for $p \geqslant 5$) the expression

$$\varphi(\beta^{(p-1)/2}), \qquad \text{where } \varphi(t) := \frac{1}{2}(3t - t^3), \tag{8.1.4}$$

plays the role of a "quadratic residue indicator function modulo p^2". In the next subsection we will explore how to generalise this construction to obtain an analogous indicator function modulo p^λ, and subsequently a "modulo p^λ trace formula", for any desired $\lambda \geqslant 1$.

PROBLEM 8.1.5. As mentioned above, (8.1.3) only works for $p \geqslant 5$. Find a polynomial $\varphi \in \mathbf{Q}[t]$ such that $\varphi(\beta^{(p-1)/2}) \equiv \chi_p(\overline{\beta}) \pmod{p^2}$ for all $\beta \in \mathbf{Z}/p^2\mathbf{Z}$ and all primes $p \geqslant 3$.

8.2. The indicator function.

LEMMA 8.2.1. *Let $\lambda \geqslant 1$ be an integer. There exists a unique polynomial $\varphi_\lambda \in \mathbf{Q}[t]$ of degree at most $2\lambda - 1$ such that*

$$\varphi_\lambda(t) \equiv \begin{cases} 1 & \pmod{(1-t)^\lambda}, \\ -1 & \pmod{(1+t)^\lambda}. \end{cases} \tag{8.2.2}$$

The polynomial φ_λ is odd and its coefficients lie in $\mathbf{Z}[\frac{1}{2}]$.

For example, when $\lambda = 2$ we have $\varphi_\lambda(t) = \frac{1}{2}(3t - t^3)$, which is exactly the polynomial that showed up in (8.1.4).

Note that the notation $\mathbf{Z}[\frac{1}{2}]$ means the ring generated by $\mathbf{Z}$ and $\frac{1}{2}$, i.e., the set of all rational numbers whose denominators are powers of two. "Odd" means that the polynomial has only terms of odd degree.

The existence (and uniqueness) of $\varphi_\lambda \in \mathbf{Q}[t]$ of degree at most $2\lambda - 1$ satisfying (8.2.2) follows immediately from the Chinese Remainder Theorem for $\mathbf{Q}[t]$. The oddness (oddity?) of φ_λ is also easy to deduce from (8.2.2). Proving that the coefficients of φ_λ lie in $\mathbf{Z}[\frac{1}{2}]$ requires slightly more work, and is discussed in the following problem.

PROBLEM 8.2.3. Prove Lemma 8.2.1 as follows.

a) Show that $1 - 2^{-\lambda}(1-t)^\lambda$ is divisible by $1+t$ in $\mathbf{Z}[\frac{1}{2}][t]$. Let $\eta \in \mathbf{Z}[\frac{1}{2}][t]$ be the quotient.
b) Let $\psi \in \mathbf{Q}[t]$ be the unique polynomial of degree at most $\lambda - 1$ such that $\psi \equiv \eta^\lambda \pmod{(1-t)^\lambda}$. Show that $\psi \in \mathbf{Z}[\frac{1}{2}][t]$ and that $(1+t)^\lambda \psi \equiv 1 \pmod{(1-t)^\lambda}$.
c) Show that $\varphi_\lambda(t) = \psi(t)(1+t)^\lambda - \psi(-t)(1-t)^\lambda$, and hence prove Lemma 8.2.1.

PROBLEM 8.2.4. Establish the following explicit formula for $\varphi_\lambda(t)$:

$$\varphi_\lambda(t) = \sum_{j=0}^{\lambda-1} \frac{1}{2^{\lambda+j}} \binom{\lambda+j-1}{j} \left((1-t)^j(1+t)^\lambda - (1+t)^j(1-t)^\lambda\right).$$

Now we can use $\varphi_\lambda(t)$ to construct the desired quadratic residue indicator function modulo p^λ:

COROLLARY 8.2.5. *Let $p \geqslant 2\lambda + 1$ be a prime. Let $\beta \in \mathbf{Z}/p^\lambda\mathbf{Z}$ and let $\overline{\beta}$ be its reduction modulo p. Then*

$$\varphi_\lambda(\beta^{(p-1)/2}) \equiv \chi_p(\overline{\beta}) \pmod{p^\lambda}.$$

For example, for $\lambda = 2$, Corollary 8.2.5 says exactly that the expression given in (8.1.4) is an indicator function modulo p^2, provided that $p \geqslant 5$.

PROBLEM 8.2.6. Prove Corollary 8.2.5 as follows.

a) Suppose that $\chi_p(\overline{\beta}) = 1$. From (8.2.2) we may write
$$\varphi_\lambda(t) = 1 + \psi(t)(1-t)^\lambda.$$
for some $\psi \in \mathbf{Z}[\frac{1}{2}][t]$. Use this to show that
$$\varphi_\lambda(\beta^{(p-1)/2}) \equiv 1 \pmod{p^\lambda}.$$
b) Deal with the case $\chi_p(\overline{\beta}) = -1$ along similar lines to (a).
c) Finish the proof by handling the remaining case $\chi_p(\overline{\beta}) = 0$. Where do you use the hypothesis that $p \geqslant 2\lambda + 1$?

PROBLEM 8.2.7. Write out explicitly the modulo p^3 analogue of (8.1.3) (which holds for $p \geqslant 7$).

PROBLEM 8.2.8. (🎮) Use your FCAS to compute the polynomial $\varphi_\lambda(t)$ for a few small values of λ, and check numerically that Corollary 8.2.5 holds for a few randomly chosen primes p and $\beta \in \mathbf{Z}/p^\lambda\mathbf{Z}$.

PROBLEM 8.2.9. Let $\lambda \geqslant 1$. Find a polynomial $\varphi \in \mathbf{Q}[t]$ for which an analogue of Corollary 8.2.5 holds for all $p \geqslant 3$. You will need to allow the degree of φ to be bigger than $2\lambda - 1$. What is the smallest degree you can get away with?

(The case $\lambda = 2$ was discussed in Problem 8.1.5. This problem is related to the "fudge factor" denoted by τ in [**23**, Theorem 3.1].)

8.3. Counting points with coordinates in $\mathbf{F}_p$. We now explore how to use the indicator function to write down a modulo p^λ trace formula. As a warm-up, we give a formula for counting points defined over $\mathbf{F}_p$. Recall that $\widetilde{C}$ is the variety obtained from C by removing the points with $x = 0, \infty$ (see Definition 4.1.3).

PROPOSITION 8.3.1. *Let $\lambda \geqslant 1$ and let $p \geqslant 2\lambda + 1$ be a prime. Let $C/\mathbf{F}_p$ be a hyperelliptic curve of genus $g \geqslant 1$ defined by $y^2 = f(x)$ for some $f \in \mathbf{F}_p[x]$. Let*

$$F = F_0 + F_1 x + \cdots + F_{2g+2} x^{2g+2} \in (\mathbf{Z}/p^\lambda\mathbf{Z})[x]$$

be any lift of f, i.e., so that $F \equiv f \pmod{p^\lambda}$. Let

$$H := F^{(p-1)/2} \in (\mathbf{Z}/p^\lambda\mathbf{Z})[x].$$

Then

$$\#\widetilde{C}(\mathbf{F}_p) \equiv (p-1)\left(1 + \sum_{\ell=0}^{\lambda-1} \varphi_{\lambda,2\ell+1} \sum_{j=0}^{(2\ell+1)(g+1)} H^{2\ell+1}_{j(p-1)}\right) \pmod{p^\lambda}, \tag{8.3.2}$$

where $\varphi_{\lambda,i}$ denotes the coefficient of t^i in $\varphi_\lambda(t)$.

There are three key differences between (8.3.2) and the modulo p version (4.1.2) that deserve to be highlighted.

First, whereas (4.1.2) is expressed entirely in terms of the defining polynomial $f \in \mathbf{F}_p[x]$, the modulo p^λ version (8.3.2) involves an arbitrary *lift* of that polynomial to $F \in (\mathbf{Z}/p^\lambda\mathbf{Z})[x]$.

Second, whereas (4.1.2) involves only coefficients from a single polynomial $h = f^{(p-1)/2}$, the formula (8.3.2) involves coefficients from several powers of $H = F^{(p-1)/2}$, namely H, $H^3, \ldots, H^{2\lambda-1}$. For example, in the case $g = 3$, $\deg F = 8$, $\lambda = 3$, the right hand side of (8.3.2) involves:

- the coefficients of $1, x^{p-1}, \ldots, x^{4(p-1)}$ in $H = F^{(p-1)/2}$,
- the coefficients of $1, x^{p-1}, \ldots, x^{12(p-1)}$ in $H^3 = F^{3(p-1)/2}$,
- the coefficients of $1, x^{p-1}, \ldots, x^{20(p-1)}$ in $H^5 = F^{5(p-1)/2}$.

Third, whereas (4.1.2) gives a congruence for $\#C(\mathbf{F}_p)$, the new formula (8.3.2) gives a congruence for $\#\widetilde{C}(\mathbf{F}_p)$. This is not unrelated to the presence of the "extra terms" $j = 0$ and $j = (2\ell+1)(g+1)$ that have no analogue in (4.1.2). Actually, it *is* possible to write down an explicit formula for $\#C(\mathbf{F}_p) \pmod{p^\lambda}$, but unfortunately it is not as neat as (4.1.2). For further discussion, see Problem 8.4.9.

Let us turn now to the proof of Proposition 8.3.1. We will require a few facts about the $(p-1)$-th roots of unity in $\mathbf{Z}/p^\lambda\mathbf{Z}$. Define

$$\Sigma_\lambda(\mathbf{F}_p^*) := \{\alpha \in \mathbf{Z}/p^\lambda\mathbf{Z} : \alpha^{p-1} = 1\}.$$

PROPOSITION 8.3.3. *The reduction map $\mathbf{Z}/p^\lambda\mathbf{Z} \to \mathbf{F}_p$ induces a bijection between $\Sigma_\lambda(\mathbf{F}_p^*)$ and $\mathbf{F}_p^*$.*

The element $\alpha \in \Sigma_\lambda(\mathbf{F}_p^*)$ corresponding to $\beta \in \mathbf{F}_p^*$ is called the *Teichmüller representative* (modulo p^λ) for β. For example, when $p = 5$, the Teichmüller representatives modulo p^3 for the elements $1, 2, 3, 4 \in \mathbf{F}_5^*$ are respectively $1, 57, 68, 124 \in \mathbf{Z}/5^3\mathbf{Z}$. These are exactly the fourth roots of unity in $\mathbf{Z}/5^3\mathbf{Z}$.

PROBLEM 8.3.4.

a) Prove Proposition 8.3.3. (Hint: if $\beta \in \mathbf{F}_p^*$, lift β to some $\gamma \in \mathbf{Z}/p^\lambda\mathbf{Z}$ and consider $\alpha = \gamma^{p^{\lambda-1}}$.)

b) For any integer $i \geqslant 0$, show that

$$\sum_{\alpha \in \Sigma_\lambda(\mathbf{F}_p^*)} \alpha^i = \begin{cases} p-1 & \text{if } i \text{ is divisible by } p-1, \\ 0 & \text{otherwise.} \end{cases}$$

c) Prove Proposition 8.3.1, by using part (b) and Corollary 8.2.5 to evaluate the sum

$$\sum_{\alpha \in \Sigma_\lambda(\mathbf{F}_p^*)} \left(\varphi_\lambda\big(F(\alpha)^{(p-1)/2}\big) + 1\right) \pmod{p^\lambda}$$

in two different ways.

PROBLEM 8.3.5. (🎮) Check Proposition 8.3.1 numerically in your FCAS, i.e., for some small p, g, λ and randomly chosen $f \in \mathbf{F}_p[x]$, choose a suitable lift $F \in (\mathbf{Z}/p^\lambda\mathbf{Z})[x]$ and check that (8.3.2) correctly counts the number of points in $\widetilde{C}(\mathbf{F}_p)$ modulo p^λ.

8.4. Counting points with coordinates in $\mathbf{F}_{p^r}$. Proposition 8.3.1 is not very interesting by itself, because we already know how to compute $\#C(\mathbf{F}_p)$ by computing it modulo p, at least if p is not too small (see Problem 4.1.6). What we really want is a modulo p^λ trace formula for counting points with coordinates in $\mathbf{F}_{p^r}$, i.e., a simultaneous generalisation of Theorem 4.2.7 and Proposition 8.3.1. The following result does the job.

THEOREM 8.4.1 (Modulo p^λ trace formula). *Let $\lambda \geqslant 1$ and let $p \geqslant 2\lambda+1$ be a prime. Let $C/\mathbf{F}_p$ be a hyperelliptic curve of genus $g \geqslant 1$ defined by $y^2 = f(x)$ for some $f \in \mathbf{F}_p[x]$. Let*

$$F = F_0 + F_1 x + \cdots + F_{2g+2}x^{2g+2} \in (\mathbf{Z}/p^\lambda\mathbf{Z})[x]$$

be any lift of f, i.e., so that $F \equiv f \pmod{p}$, and let

$$H := F^{(p-1)/2} \in (\mathbf{Z}/p^\lambda\mathbf{Z})[x].$$

For $\ell = 0, \ldots, \lambda-1$, let

$$\widetilde{A}_{F,\ell} \in \operatorname{Mat}_{(2\ell+1)(g+1)+1}(\mathbf{Z}/p^\lambda\mathbf{Z})$$

be the matrix defined by

$$(\widetilde{A}_{F,\ell})_{v,u} := (H^{2\ell+1})_{vp-u}, \qquad 0 \leqslant u, v \leqslant (2\ell+1)(g+1).$$

Then for any $r \geqslant 1$,

$$\#\widetilde{C}(\mathbf{F}_{p^r}) \equiv (p^r - 1)\left(1 + \sum_{\ell=0}^{\lambda-1} \varphi_{\lambda,2\ell+1} \operatorname{tr}(\widetilde{A}_{F,\ell}^r)\right) \pmod{p^\lambda}. \tag{8.4.2}$$

The main new feature in Theorem 8.4.1 is that instead of one matrix $\widetilde{A}_f$, we have to deal with λ matrices $\widetilde{A}_{F,\ell}$ for $\ell = 0, \ldots, \lambda-1$, whose size increases with ℓ.

PROBLEM 8.4.3. Check that Theorem 8.4.1 specialises to Proposition 8.3.1 in the case $r = 1$.

To prove Proposition 8.4.1, we need to be able to sum over the "(p^r-1)-th roots of unity modulo p^λ". To make sense of this, recall that $\mathbf{F}_{p^r}$ may be represented as

$$\mathbf{F}_{p^r} = \mathbf{F}_p[\theta]/z(\theta)$$

for some irreducible monic polynomial $z \in \mathbf{F}_p[\theta]$ of degree r. Let $Z \in (\mathbf{Z}/p^\lambda\mathbf{Z})[\theta]$ be any monic degree-r lift of z, i.e., so that $Z \equiv z \pmod{p}$, and define

$$W_\lambda(\mathbf{F}_{p^r}) := (\mathbf{Z}/p^\lambda\mathbf{Z})[\theta]/Z(\theta).$$

Elements of $W_\lambda(\mathbf{F}_{p^r})$ may be represented by polynomials of degree at most $r-1$ in $(\mathbf{Z}/p^\lambda\mathbf{Z})[\theta]$. The ring $W_\lambda(\mathbf{F}_{p^r})$ plays the analogous role for $\mathbf{F}_{p^r}$ as $\mathbf{Z}/p^\lambda\mathbf{Z}$ does for $\mathbf{F}_p$. (Readers familiar with Witt vectors will recognise this construction.) Now define

$$\Sigma_\lambda(\mathbf{F}_{p^r}^*) := \{\alpha \in W_\lambda(\mathbf{F}_{p^r}) : \alpha^{p^r-1} = 1\}.$$

PROPOSITION 8.4.4. *For each $r \geqslant 1$, the reduction map $W_\lambda(\mathbf{F}_{p^r}) \to \mathbf{F}_{p^r}$ induces a bijection between $\Sigma_\lambda(\mathbf{F}_{p^r}^*)$ and $\mathbf{F}_{p^r}^*$.*

The element of $\Sigma_\lambda(\mathbf{F}_{p^r}^*)$ corresponding to a given $\beta \in \mathbf{F}_{p^r}^*$ is called the Teichmüller representative (modulo p^λ) for β.

PROBLEM 8.4.5. Suppose that we represent $\mathbf{F}_{3^2}$ as $\mathbf{F}_3[\theta]/(\theta^2+1)$, and $W_2(\mathbf{F}_{3^2})$ as $(\mathbf{Z}/3^2\mathbf{Z})[\theta]/(\theta^2+1)$. Find the Teichmüller representatives for all elements of $\mathbf{F}_{3^2}^*$.

PROBLEM 8.4.6.

a) Prove Proposition 8.4.4.
b) For any integer $i \geqslant 0$, show that

$$\sum_{\alpha \in \Sigma_\lambda(\mathbf{F}_{p^r}^*)} \alpha^i = \begin{cases} p^r - 1 & \text{if } i \text{ is divisible by } p^r-1, \\ 0 & \text{otherwise.} \end{cases}$$

PROBLEM 8.4.7. Prove Theorem 8.4.1 as follows.

a) Let

$$H^{(r)}(x) := H(x)H(x^p)\cdots H(x^{p^{r-1}}) \in (\mathbf{Z}/p^\lambda\mathbf{Z})[x].$$

By evaluating the sum

$$\sum_{\alpha \in \Sigma_\lambda(\mathbf{F}_{p^r}^*)} \Big(\varphi_\lambda\big(H^{(r)}(\alpha)\big) + 1\Big) \pmod{p^\lambda}$$

in two different ways, show that

$$\#\widetilde{C}(\mathbf{F}_{p^r}) \equiv$$

$$(p^r-1)\left(1 + \sum_{\ell=0}^{\lambda-1} \varphi_{\lambda,2\ell+1} \sum_{j=0}^{(2\ell+1)(g+1)} (H^{(r)})^{2\ell+1}_{j(p^r-1)}\right) \pmod{p^\lambda}.$$

b) Using a similar argument to Problem 4.2.9(b), show that for each $\ell = 0, \ldots, \lambda-1$ we have

$$\sum_{j=0}^{(2\ell+1)(g+1)} (H^{(r)})^{2\ell+1}_{j(p^r-1)} = \operatorname{tr}(\widetilde{A}_{F,\ell}^r).$$

PROBLEM 8.4.8. (⌨) Check Theorem 8.4.1 numerically in your FCAS, i.e., for some small p, g, λ, r, and randomly chosen $f \in \mathbf{F}_p[x]$, take an arbitrary lift $F \in (\mathbf{Z}/p^\lambda\mathbf{Z})[x]$, evaluate (8.4.2), and check that the result agrees modulo p^λ with the actual number of points (obtained for example by the enumeration method).

PROBLEM 8.4.9. Notation and hypotheses as in Theorem 8.4.1. In this problem we investigate the possibility of an elegant modulo p^λ trace formula for $C(\mathbf{F}_{p^r})$ rather than $\widetilde{C}(\mathbf{F}_{p^r})$.

a) For $\ell = 0, \ldots, \lambda - 1$ define matrices
$$A_{F,\ell} \in \mathrm{Mat}_{(2\ell+1)(g+1)-1}(\mathbf{Z}/p^\lambda\mathbf{Z})$$
by the formula
$$(A_{F,\ell})_{v,u} := (H^{2\ell+1})_{vp-u}, \qquad 1 \leqslant u, v \leqslant (2\ell+1)(g+1) - 1.$$
State and prove a formula of the type
$$\#C(\mathbf{F}_{p^r}) \equiv (p^r - 1)\left(1 + \sum_{\ell=0}^{\lambda-1} \varphi_{\lambda,2\ell+1}\,\mathrm{tr}(A_{F,\ell}^r)\right) + \text{(correction terms)} \pmod{p^\lambda},$$
where the "correction terms" involve suitable linear combinations of powers of F_0 and F_{2g+2}.

b) Show that that the correction terms vanish when $r \geqslant \lambda$. In other words, when $r \geqslant \lambda$, we get a stylish trace formula for $C(\mathbf{F}_{p^r})$, just like 4.2.7.

PROBLEM 8.4.10. (ϟ) In point counting algorithms based on p-adic cohomology, the strategy is usually to construct a certain matrix over $\mathbf{Q}_p$ (the matrix of a Frobenius operator on some p-adic vector space) whose characteristic polynomial yields the L-polynomial $L_p(T)$ modulo arbitrarily large powers of p. Is there a similar formula that yields $L_p(T) \pmod{p^\lambda}$ in terms of the matrices $\widetilde{A}_{F,\ell}$ appearing in Theorem 8.4.1? (Or in terms of the matrices $A_{F,\ell}$ in Problem 8.4.9?) More generally, is there a "p-adic cohomological" interpretation of the modulo p^λ trace formulas given in this course?

8.5. Algorithms and complexity bounds. In this final section of the course, we discuss how to design algorithms that use Theorem 8.4.1 to compute not just $L_C(T) \pmod p$, but the entire L-polynomial $L_C(T) \in \mathbf{Z}[T]$. We will not give many details, but rather leave these to you as challenging exercises.

Our first task is to determine a suitable value of λ.

LEMMA 8.5.1. *Let λ be the smallest integer such that*
$$\lambda > \frac{g}{2} + \log_p(4g).$$
Then $L_C(T) \in \mathbf{Z}[T]$ may be recovered from the values
$$\#C(\mathbf{F}_p), \ldots, \#C(\mathbf{F}_{p^g}) \pmod{p^\lambda}. \tag{8.5.2}$$

PROBLEM 8.5.3. Prove Lemma 8.5.1, via (3.2.2) and Problem 3.3.5.

In principle, it is now straightforward to compute $L_C(T) \in \mathbf{Z}[T]$. Given as input $f \in \mathbf{F}_p[x]$, we lift to some $F \in (\mathbf{Z}/p^\lambda\mathbf{Z})[x]$ where λ is chosen as in Lemma

8.5.1. Notice in particular that $\lambda = O(g)$. According to Theorem 8.4.1, we can recover the quantities in (8.5.2) from the matrices

$$\widetilde{A}_{F,0}, \dots, \widetilde{A}_{F,\lambda-1}. \tag{8.5.4}$$

(We tacitly assume here that $p \geqslant 2\lambda + 1$. To handle smaller p, see Problem 8.2.9.)

PROBLEM 8.5.5. Estimate the complexity of computing $L_C(T) \in \mathbf{Z}[T]$ from knowledge of the matrices in (8.5.4). In other words, state and prove a suitable analogue of Corollary 4.4.3. You should get a complexity bound of the form

$$O(g^c \log^{1+\varepsilon} p)$$

for some constant $c > 0$.

Computing the matrices in (8.5.4) amounts to computing certain selected coefficients of $H, H^3, \dots, H^{2\lambda-1}$ modulo p^λ, where $H = F^{(p-1)/2}$, or in other words, the polynomials

$$F^{(p-1)/2}, F^{3(p-1)/2}, \dots, F^{(2\lambda-1)(p-1)/2}. \tag{8.5.6}$$

This task may be carried out by adapting any of the various strategies given earlier in the course for computing the coefficients of $h = f^{(p-1)/2}$. Let us examine each of these strategies in turn.

PROBLEM 8.5.7 (Computing $A_{F,\ell}$ via naive expansion). State and prove an analogue of Theorem 4.4.1 for computing the matrices in (8.5.4), i.e., by simply expanding out the polynomials in (8.5.6). You should get a complexity bound of the form

$$O(g^c p \log^2(gp))$$

for some constant $c > 0$.

Next we consider the recurrence strategy from §5. For each power $H^{2\ell+1}$, we may set up a suitable recurrence, and define a sequence of approximations for the coefficients of $H^{2\ell+1}$, analogous to the sequence $\widetilde{H}_k$ from §5.2. In order to deal with the "division-by-p" problem, we need to lift F further to $(\mathbf{Z}/p^\mu\mathbf{Z})[x]$ for some $\mu > \lambda$.

PROBLEM 8.5.8.

a) State and prove an analogue of Corollary 5.3.1, i.e., determine how big μ must be to ensure that our approximations for the coefficients of $H^{2\ell+1}$ are correct modulo p^λ.
b) State and prove an analogue of Problem 5.3.4(b), i.e., show that for p large enough relative to g, to correctly compute the matrices in (8.5.4) it suffices to take $\mu := \lambda + 1$. (In particular, we may take $\mu = O(g)$.)

PROBLEM 8.5.9 (Computing $A_{F,\ell}$ via recurrences). State and prove an analogue of Theorem 5.3.3 for computing the matrices in (8.5.4). You should get a complexity bound of the form

$$O(g^c p \log^{1+\varepsilon} p)$$

for some constant $c > 0$.

Next in line is the square-root algorithm.

PROBLEM 8.5.10. Generalise Proposition 6.3.6 (the interpolation trick) from the $\mu = 2$ case to arbitrary values of μ. More precisely, show that once we have computed (6.3.3) and (6.3.4) for $i = 0, \dots, \mu - 1$, then we may quickly deduce (6.3.3) and (6.3.4) for any desired value of i. (Careful: you may need to impose a lower bound on p to avoid running into difficulties during the interpolation.)

PROBLEM 8.5.11 (Computing $A_{F,\ell}$ in square-root time). State and prove an analogue of Theorem 6.3.8 for computing the matrices in (8.5.4). Taking into account Remark 6.2.6, you should get a complexity bound of the form

$$O(g^c p^{1/2} \log^2 p)$$

for some constant $c > 0$. (It should be possible to generalise Problem 6.3.12 from two subproducts to μ subproducts, in order to squeeze out another factor of $g^{1/2}$.)

REMARK 8.5.12. The complexity bound attained in the previous problem is of a similar nature to the main result of [**21**], which is based on p-adic cohomology. As mentioned in Remark 6.3.13, it would be interesting to compare the two approaches, both from a theoretical and a practical point of view.

Finally, we consider the average polynomial time algorithm. After replacing the ring $\mathbf{Z}[P]/P^2$ with $\mathbf{Z}[P]/P^\mu$, everything should run much the same way as before.

PROBLEM 8.5.13 (Computing $A_{F,\ell}$ in average polynomial time). State and prove an analogue of Theorem 7.4.1 for computing the matrices in (8.5.4), for all admissible $p \leqslant N$, for some suitably adjusted notion of "admissible". You should get a complexity bound of the form

$$O(g^c N \log^3 N)$$

for some constant $c > 0$.

PROBLEM 8.5.14. (⚡) In all of the algorithms suggested above, the powers H, $H^3, \dots, H^{2\lambda-1}$ are handled separately. Can the exponents of g in the complexity bounds be further improved by taking advantage of some redundancy between these powers?

To the best of my knowledge, no-one has yet made a serious effort to implement *any* average polynomial point counting algorithm that obtains point counts modulo p^λ for any $\lambda > 1$.

PROBLEM 8.5.15. (🎮) Implement all the algorithms discussed in this section!

Acknowledgments. Many thanks to Alex Best for his comments on a draft of these notes.

References

[1] T. G. Abbott, K. S. Kedlaya, and D. Roe, *Bounding Picard numbers of surfaces using p-adic cohomology*, Arithmetics, geometry, and coding theory (AGCT 2005), 2010, pp. 125–159. MR2856564 ↑67

[2] J. D. Achter and E. W. Howe, *Hasse-Witt and Cartier-Manin matrices: a warning and a request*, Arithmetic geometry: computation and applications, Contemp. Math., vol. 722, Amer. Math. Soc., [Providence], RI, 2019, pp. 1–18, DOI 10.1090/conm/722/14534. MR3896846 ↑79

[3] J. Alman and V. Vassilevska Williams, *A refined laser method and faster matrix multiplication*, Proceedings of the 2021 ACM-SIAM Symposium on Discrete Algorithms (SODA), [Society for Industrial and Applied Mathematics (SIAM)], Philadelphia, PA, 2021, pp. 522–539, DOI 10.1137/1.9781611976465.32. MR4262465 ↑72

[4] V. Andrejić, A. Bostan, and M. Tatarevic, *Improved algorithms for left factorial residues*, Inform. Process. Lett. **167** (2021), Paper No. 106078, 3 pp., DOI 10.1016/j.ipl.2020.106078. MR4195830 ↑91, 98

[5] D. J. Bernstein, *Fast multiplication and its applications*, Algorithmic number theory: lattices, number fields, curves and cryptography, Math. Sci. Res. Inst. Publ., vol. 44, Cambridge Univ. Press, Cambridge, 2008, pp. 325–384. MR2467550 ↑68

[6] A. R. Booker, S. Hathi, M. J. Mossinghoff, and T. S. Trudgian, *Wolstenholme and Vandiver primes*, Ramanujan J. **58** (2022), no. 3, 913–941, DOI 10.1007/s11139-021-00438-3. MR4437433 ↑89

[7] A. Bostan, P. Gaudry, and É. Schost, *Linear recurrences with polynomial coefficients and application to integer factorization and Cartier-Manin operator*, SIAM J. Comput. **36** (2007), no. 6, 1777–1806, DOI 10.1137/S0097539704443793. MR2299425 ↑83, 87, 89, 91, 94, 104

[8] R. P. Brent and P. Zimmermann, *Modern computer arithmetic*, Cambridge Monographs on Applied and Computational Mathematics, vol. 18, Cambridge University Press, Cambridge, 2011. MR2760886 ↑68

[9] D. V. Chudnovsky and G. V. Chudnovsky, *Approximations and complex multiplication according to Ramanujan*, Ramanujan revisited (Urbana-Champaign, Ill., 1987), Academic Press, Boston, MA, 1988, pp. 375–472. MR938975 ↑91

[10] H. Cohen, G. Frey, R. Avanzi, C. Doche, T. Lange, K. Nguyen, and F. Vercauteren (eds.), *Handbook of elliptic and hyperelliptic curve cryptography*, Discrete Mathematics and its Applications (Boca Raton), Chapman & Hall/CRC, Boca Raton, FL, 2006. MR2162716 ↑64

[11] E. Costa, R. Gerbicz, and D. Harvey, *A search for Wilson primes*, Math. Comp. **83** (2014), no. 290, 3071–3091, DOI 10.1090/S0025-5718-2014-02800-7. MR3246824 ↑88, 95, 98, 99

[12] E. Costa and D. Harvey, *Faster deterministic integer factorization*, Math. Comp. **83** (2014), no. 285, 339–345, DOI 10.1090/S0025-5718-2013-02707-X. MR3120593 ↑89

[13] E. Costa, D. Harvey, and A. V. Sutherland, *Counting points on smooth plane quartics*, Res. Number Theory **9** (2023), no. 1, Paper No. 1, 32 pp., DOI 10.1007/s40993-022-00397-8. MR4514545 ↑104

[14] R. Crandall, K. Dilcher, and C. Pomerance, *A search for Wieferich and Wilson primes*, Math. Comp. **66** (1997), no. 217, 433–449, DOI 10.1090/S0025-5718-97-00791-6. MR1372002 ↑89, 98

[15] N. D. Elkies, *Elliptic and modular curves over finite fields and related computational issues*, Computational perspectives on number theory (Chicago, IL, 1995), AMS/IP Stud. Adv. Math., vol. 7, Amer. Math. Soc., Providence, RI, 1998, pp. 21–76, DOI 10.1090/amsip/007/03. MR1486831 ↑64

[16] M. Farach-Colton and M.-T. Tsai, *On the complexity of computing prime tables*, Algorithms and computation, Lecture Notes in Comput. Sci., vol. 9472, Springer, Heidelberg, 2015, pp. 677–688, DOI 10.1007/978-3-662-48971-0_57. MR3489524 ↑72

[17] F. Fité, K. S. Kedlaya, V. Rotger, and A. V. Sutherland, *Sato-Tate distributions and Galois endomorphism modules in genus 2*, Compos. Math. **148** (2012), no. 5, 1390–1442, DOI 10.1112/S0010437X12000279. MR2982436 ↑64

[18] P. Flajolet and B. Salvy, *The SIGSAM challenges: Symbolic asymptotics in practice*, SIGSAM Bull. **31** (1997dec), no. 4, 36–47, DOI 10.1145/274888.274890. ↑87

[19] R. Gerbicz, 2011, `https://www.mersenneforum.org/showthread.php?p=270602`. ↑95

[20] T. Granlund, *The GNU Multiple Precision Arithmetic Library*, `http://gmplib.org/`. ↑68

[21] D. Harvey, *Kedlaya's algorithm in larger characteristic*, Int. Math. Res. Not. IMRN **22** (2007), Art. ID rnm095, 29 pp., DOI 10.1093/imrn/rnm095. MR2376210 ↑94, 112

[22] D. Harvey, *Counting points on hyperelliptic curves in average polynomial time*, Ann. of Math. (2) **179** (2014), no. 2, 783–803, DOI 10.4007/annals.2014.179.2.7. MR3152945 ↑104

[23] D. Harvey, *Computing zeta functions of arithmetic schemes*, Proc. Lond. Math. Soc. (3) **111** (2015), no. 6, 1379–1401, DOI 10.1112/plms/pdv056. MR3447797 ↑63, 66, 67, 100, 104, 106

[24] D. Harvey, M. Massierer, and A. V. Sutherland, *Computing L-series of geometrically hyperelliptic curves of genus three*, LMS J. Comput. Math. **19** (2016), no. suppl. A, 220–234, DOI 10.1112/S1461157016000383. MR3540957 ↑104

[25] D. Harvey and A. V. Sutherland, *Computing Hasse-Witt matrices of hyperelliptic curves in average polynomial time*, LMS J. Comput. Math. **17** (2014), no. suppl. A, 257–273, DOI 10.1112/S1461157014000187. MR3240808 ↑104

[26] D. Harvey and A. V. Sutherland, *Computing Hasse-Witt matrices of hyperelliptic curves in average polynomial time, II*, Frobenius distributions: Lang-Trotter and Sato-Tate conjectures, 2016, pp. 127–147, DOI 10.1090/conm/663/13352. MR3502941 ↑87, 104

[27] D. Harvey and J. van der Hoeven, *On the complexity of integer matrix multiplication*, J. Symbolic Comput. **89** (2018), 1–8, DOI 10.1016/j.jsc.2017.11.001. MR3804803 ↑104

[28] D. Harvey and J. van der Hoeven, *Integer multiplication in time* $O(n \log n)$, Ann. of Math. (2) **193** (2021), no. 2, 563–617, DOI 10.4007/annals.2021.193.2.4. MR4224716 ↑68

[29] D. Harvey and J. van der Hoeven, *Polynomial multiplication over finite fields in time* $O(n \log n)$, J. ACM **69** (2022), no. 2, Art. 12, 40 pp., DOI 10.1145/3505584. MR4433320 ↑70

[30] M. Hindry and J. H. Silverman, *Diophantine geometry: An introduction*, Graduate Texts in Mathematics, vol. 201, Springer-Verlag, New York, 2000, DOI 10.1007/978-1-4612-1210-2. MR1745599 ↑99

[31] K. S. Kedlaya, *Counting points on hyperelliptic curves using Monsky-Washnitzer cohomology*, J. Ramanujan Math. Soc. **16** (2001), no. 4, 323–338. MR1877805 ↑64, 66, 85

[32] K. S. Kedlaya, *Quantum computation of zeta functions of curves*, Comput. Complexity **15** (2006), no. 1, 1–19, DOI 10.1007/s00037-006-0204-7. MR2226067 ↑64

[33] K. S. Kedlaya and A. V. Sutherland, *Computing L-series of hyperelliptic curves*, Algorithmic number theory, 2008, pp. 312–326, Lecture Notes in Comput. Sci., vol. 5011, DOI 10.1007/978-3-540-79456-1_21. MR2467855 (2010d:11070) ↑75

[34] D. E. Knuth, *The art of computer programming. Vol. 2: Seminumerical algorithms*, Addison-Wesley, Reading, MA, 1998. Third edition [of MR0286318]. MR3077153 ↑68

[35] M. Kyng, *Computing zeta functions of algebraic curves using Harvey's trace formula*, Preprint, `arXiv:2203.02070`, 2022. ↑67

[36] A. G. B. Lauder, *Counting solutions to equations in many variables over finite fields*, Found. Comput. Math. **4** (2004), no. 3, 221–267, DOI 10.1007/s10208-003-0093-y. MR2078663 ↑67

[37] D. Lorenzini, *An invitation to arithmetic geometry*, Graduate Studies in Mathematics, vol. 9, American Mathematical Society, Providence, RI, 1996, DOI 10.1090/gsm/009. MR1376367 ↑76

[38] J. I. Manin, *The Hasse–Witt matrix of an algebraic curve*, Fifteen papers on algebra, 1965, pp. 245–264. Translated by J.W.S. Cassels. ↑81

[39] R. J. McIntosh and E. L. Roettger, *A search for Fibonacci-Wieferich and Wolstenholme primes*, Math. Comp. **76** (2007), no. 260, 2087–2094, DOI 10.1090/S0025-5718-07-01955-2. MR2336284 ↑89

[40] H. L. Montgomery and R. C. Vaughan, *Multiplicative number theory. I. Classical theory*, Cambridge Studies in Advanced Mathematics, vol. 97, Cambridge University Press, Cambridge, 2007. MR2378655 ↑65

[41] V. Neiger and C. Pernet, *Deterministic computation of the characteristic polynomial in the time of matrix multiplication*, J. Complexity **67** (2021), Paper No. 101572, 35 pp., DOI 10.1016/j.jco.2021.101572. MR4311525 ↑72

[42] C. H. Papadimitriou, *Computational complexity*, Addison-Wesley Publishing Company, Reading, MA, 1994. MR1251285 ↑66

[43] J. Pila, *Frobenius maps of abelian varieties and finding roots of unity in finite fields*, Math. Comp. **55** (1990), no. 192, 745–763, DOI 10.2307/2008445. MR1035941 ↑64

[44] J. M. Pollard, *Theorems on factorization and primality testing*, Proc. Cambridge Philos. Soc. **76** (1974), 521–528, DOI 10.1017/s0305004100049252. MR354514 ↑89

[45] A. Schönhage, A. F. W. Grotefeld, and E. Vetter, *Fast algorithms*, Bibliographisches Institut, Mannheim, 1994. A multitape Turing machine implementation. MR1290996 ↑72

[46] R. Schoof, *Elliptic curves over finite fields and the computation of square roots mod* p, Math. Comp. **44** (1985), no. 170, 483–494, DOI 10.2307/2007968. MR777280 ↑64

[47] H. Stichtenoth, *Algebraic function fields and codes*, 2nd ed., Graduate Texts in Mathematics, vol. 254, Springer-Verlag, Berlin, 2009. MR2464941 ↑64, 76

[48] V. Strassen, *Gaussian elimination is not optimal*, Numer. Math. **13** (1969), 354–356, DOI 10.1007/BF02165411. MR248973 ↑72

[49] V. Strassen, *Einige Resultate über Berechnungskomplexität*, Jber. Deutsch. Math.-Verein. **78** (1976/77), no. 1, 1–8. MR438807 ↑89

[50] A. V. Sutherland, *Smalljac software library version 4.1.3*, `https://math.mit.edu/~drew/smalljac_v4.1.3.tar`. ↑75

[51] A. V. Sutherland, *Order computations in generic groups*, ProQuest LLC, Ann Arbor, MI, 2007. Thesis (Ph.D.)–Massachusetts Institute of Technology. MR2717420 ↑64

[52] A. V. Sutherland, *Counting points on superelliptic curves in average polynomial time*, ANTS XIV—Proceedings of the Fourteenth Algorithmic Number Theory Symposium, Open Book Ser., vol. 4, Math. Sci. Publ., Berkeley, CA, 2020, pp. 403–422, DOI 10.2140/obs.2020.4.403. MR4235126 ↑104

[53] J. Tuitman, *Counting points on curves using a map to* $\mathbf{P}^1$*, II*, Finite Fields Appl. **45** (2017), 301–322, DOI 10.1016/j.ffa.2016.12.008. MR3631366 ↑64, 67

[54] J. von zur Gathen and J. Gerhard, *Modern computer algebra*, 3rd ed., Cambridge University Press, Cambridge, 2013, DOI 10.1017/CBO9781139856065. MR3087522 ↑68

[55] J. von zur Gathen and V. Shoup, *Computing Frobenius maps and factoring polynomials*, Comput. Complexity **2** (1992), no. 3, 187–224, DOI 10.1007/BF01272074. MR1220071 ↑89

School of Mathematics and Statistics, UNSW Sydney, NSW 2052, Australia
Email address: `d.harvey@unsw.edu.au`

IAS/Park City Mathematics Series
Volume **29**, 2025
https://doi.org/10.1090/pcms/029/00892

Number Field Counting, Class Group Heuristics, and Computation

Melanie Matchett Wood

Abstract. These notes give an introduction to the asymptotic counting of number fields by Galois group and the Cohen-Lenstra heuristics for the distribution of class groups of quadratic fields. We discuss the connection of the class group heuristics to random matrices and discuss the moments of the conjectured distribution and the role they play. We point out places where computation has been crucial in the development of this theory, and suggest some important directions for future computational work. We give many exercises, ranging from quite easy, to quite difficult, including longer term projects.

Any comments, corrections, typos, etc. are very welcome. Please email them to me at mmwood@math.harvard.edu

1. Counting number fields

1.1. Questions. Number fields, i.e. finite extensions of the rational numbers $\mathbb{Q}$, are the basic objects of algebraic number theory. We can ask how many there are, and there are infinitely many, and so we ask a more refined question. For a number field K, we write D_K for its *absolute discriminant*, that is, the absolute value of its discriminant.

Question 1.1.1. *Given a positive integer d, how many isomorphism classes of number fields of degree d and absolute discriminant at most X are there, asymptotically in X?*

"Asymptotically in X" means that if $N_d(X)$ is the number of isomorphism classes of number fields K of degree d over $\mathbb{Q}$ with $D_K \leq X$, we are looking to find a simple function $f(X)$ such that

$$N_d(X) = f(X) + o(f(X)).$$

It turns out in the questions considered here, usually one shows or conjectures that $f(X)$ is of the form $X^a(\log X)^b$ for some real numbers a, b. We write $g(X) = f(X) + o(h(X))$ to mean that

$$\lim_{X\to\infty} \frac{|g(X) - f(X)|}{h(X)} = 0.$$

2020 *Mathematics Subject Classification.* Primary 14Dxx; Secondary 14Dxx.
Key words and phrases. Park City Mathematics Institute.

We have asked about isomorphism classes of number fields, though it is also natural to ask about subfields of $\bar{\mathbb{Q}}$ instead. These two questions are closely related (see Exercise 1.1.3). We know classically that $N_d(X)$ is finite.

THEOREM 1.1.2 (Hermite's theorem, see, e.g., III.2.16 in [**88**]). *Given a real number X, there are finitely many isomorphism classes of number fields K with $|D_K| \leq X$.*

EXERCISE 1.1.3. *Show that $N_d(\infty)$ is infinite for each positive integer $d \geq 2$. What lower bound does your argument give for $N_d(X)$?*

There are many ways one might approach Exercise 1.1.3, and they lead in some of the different directions of work in the area. One might give an infinite list of fields, using polynomials or adjoining dth roots, but then one needs some argument to make sure that the elements of the list are actually degree d fields and distinct. One might use the theorems of class field theory, which handles these issues well but only can access abelian fields.

Given Theorem 1.1.2, one might ask why we even fix a degree d at all. First of all, fixing the degree makes the question have more accessible parts. However, in addition, once we have some of the answers, e.g. for $d = 2, 3, 4$, we can see that the various degrees have natural answers, suggesting the most natural form of the answer for all degrees would simply be the sum over the answers for each d. In fact, from the answers in low degree, we further see that the level at which we have natural answers is if we fix not only a degree but also a Galois group. More precisely, we should fix the Galois group of the Galois closure and its action of the embeddings of the number field in $\bar{\mathbb{Q}}$.

DEFINITION. A *permutation group* G is a group G, a set X, and a faithful action of G on X. We often leave X implicit in the notation. An isomorphism of two permutation groups (G, X) and (G', X') is a group isomorphism $f : G \to G'$ and a bijection $h : X \to X'$ such that for all $g \in G$ and $x \in X$, we have $h(g(x)) = f(g)(h(x))$. The permutation group is *transitive* if for all $x, y \in X$ there is a $g \in G$ such that $gx = y$ and *simply transitive* if there is always exactly 1 such g. We say $|X|$ is the *degree* of the permutation group, and that the permutation group is *finite* if it has finite degree.

EXERCISE 1.1.4. *Show that, for each n, isomorphism classes of permutation groups whose underlying set has size n are in natural bijection with conjugacy classes of subgroups of the symmetric group S_n.*

DEFINITION. For a number field K, we write $\mathrm{Gal}(K/\mathbb{Q})$ for the permutation group whose underlying group is $\mathrm{Gal}(\tilde{K}/\mathbb{Q})$, where $\tilde{K}$ is the Galois closure of K over $\mathbb{Q}$, acting on the $[K : \mathbb{Q}]$ homomorphisms $K \to \bar{\mathbb{Q}}$. More precisely, $\tau \in \mathrm{Gal}(\tilde{K}/\mathbb{Q})$ sends $i \in Hom(K, \bar{\mathbb{Q}})$ to τi.

While it might be jarring at first to write $\mathrm{Gal}(K/\mathbb{Q})$ when K is not Galois, this is a very helpful concept and notation. For example, theorems in arithmetic statistics show us that the degree 4 fields whose Galois closure has group D_4 behave quite differently from those whose Galois closure has group S_4.

EXERCISE 1.1.5. *If K is a number field, show that* $\mathrm{Gal}(\tilde{K}/\mathbb{Q})$ *acts faithfully on the $[K : \mathbb{Q}]$ homomorphisms $K \to \bar{\mathbb{Q}}$.*

EXERCISE 1.1.6. *If K is a number field, show that* $\mathrm{Gal}(K/\mathbb{Q})$ *is a transitive permutation group.*

Any group G acts on itself via left multiplication, and we call this permutation group the *regular representation.*

EXERCISE 1.1.7. *If K is Galois, show that the permutation group* $\mathrm{Gal}(K/\mathbb{Q})$ *is isomorphic to the usual Galois group with its regular representation.*

EXERCISE 1.1.8. *Show that if G is a transitive permutation group of finite degree $|G|$ (equivalently a finite simply transitive permutation group), then it is isomorphic to the underlying group in its regular representation.*

EXERCISE 1.1.9. *Show that if A is a finite abelian transitive permutation group, then A is isomorphic to the underlying group in its regular representation.*

EXERCISE 1.1.10. *Find an example of two non-isomorphic transitive permutation groups of the same degree whose underlying group is isomorphic. [Hint: the GroupNames database online, or similar, might be useful.]*

Exercise 1.1.10 shows why we keep track of the permutation group and not just the group, even though in low degrees we can often refer to *the* transitive degree d action of a particular group, e.g. the transitive degree 4 action of D_4. We can now state the most common version of the counting number fields question in arithmetic statistics.

Given a positive integer d and a transitive permutation group G of degree d, let S_G be the set of isomorphism classes of number fields K with $\mathrm{Gal}(K/\mathbb{Q}) \simeq G$. We call such number fields *G-extensions* (or if we replace $\mathbb{Q}$ with a base field K_0, we would call them G-extensions of K_0).[1] When a group G has a single transitive permutation group structure (up to isomorphism) of degree d, we often write d-ic (e.g. cubic, quartic,...) G-extensions to mean that G should be endowed with this transitive permutation group structure. Let $N_G(X)$ be the number of $K \in S_G$ with $|D_K| \leq X$.

QUESTION 1.1.11. *What is $N_G(X)$, asymptotically in X?*

Note that the question, for a given finite group G, of whether $N_G(X)$ is ever at least 1, is the Inverse Galois Problem, a famously hard and open question. However, there are many G for which we do know there exist many number fields with $\mathrm{Gal}(K/\mathbb{Q}) \simeq G$, and for some of these we can answer Question 1.1.11. Note that if we answer Question 1.1.11 for all transitive permutation groups G of a given degree d, since there are only finitely many, we can add the results to get an answer to Question 1.1.1. Indeed, if for one G of degree d we have an asymptotic of $f(X)$, to answer Question 1.1.1, it would suffice to have, for each other G of degree d, either an asymptotic for $N_G(X)$ or an upper bound of $o(f(X))$. This approach is how we know the answer to Question 1.1.1 for $d \leq 5$ [**16, 18, 38, 45**], which in each $d \leq 5$ is

$$N_d(X) = c_d X + o(X)$$

for some $c_d > 0$.

[1] It is generally better practice to define a G-extension to be an isomorphism class of pairs (K, ϕ) where ϕ is a choice of isomorphism $\mathrm{Gal}(K/\mathbb{Q}) \simeq G$. This inflates the number of automorphisms by a predictable amount, but when one needs to use an isomorphism $\mathrm{Gal}(K/\mathbb{Q}) \simeq G$, it is much more convenient to have one specified. This will arise for us as soon as we consider class groups of number fields.

EXERCISE 1.1.12. *If K is a number field with $G = \mathrm{Gal}(K/\mathbb{Q})$ and $S \subset G$ the stabilizer of an element (i.e. one of the embeddings $K \to \tilde{K}$), show that* $\mathrm{Aut}(K)$ *is isomorphic to quotient of the normalizer of S in G by S.*

EXERCISE 1.1.13. *Given a positive integer d and a permutation group G of degree d, an a number field K with* $\mathrm{Gal}(K/\mathbb{Q}) \simeq G$*, show that there are exactly $\frac{d}{|\mathrm{Aut}(K)|}$ subfields of $\bar{\mathbb{Q}}$ that are isomorphic to K.*

1.2. Approaches and results. The answer is Question 1.1.11 is known for many G. For example, we consider the permutation group S_2 in its standard permutation action on 2 elements and $N_{S_2}(X)$ counts quadratic fields. Early in algebraic number theory, we classify all quadratic fields, and show they are determined by their discriminant. A quadratic field K must contain some α satisfying an equation $\alpha^2 + a\alpha - d = 0$ for some $a, d \in \mathbb{Q}$. We can replace α with $\alpha + a/2$ and then have $a = 0$. We can replace α with an integer multiple to have $d \in \mathbb{Z}$, but then we can also divide α by an integer to have that d is square-free. So every quadratic field is $\mathbb{Q}(\sqrt{d})$ for some square-free integer $d \neq 1$. How do we know these fields are all in distinct isomorphism classes? Typically, we compute their rings of integers and their discriminants, and then we can observe they all have unique discriminants. Moreover, the discriminants that arise are exactly the set $\mathcal{F}$ of integers D such that

(1) $D \equiv 1 \pmod 4$ and D is square-free and $D \neq 1$, or
(2) $D = 4d$, where $d \equiv 2$ or $3 \pmod 4$ and d square-free.

We will now discuss how to count these integers. We have $N_{S_2}(X) = \mathcal{F} \cap [-X, X]$. We can rewrite the above conditions defining $\mathcal{F}$ as above as requiring that $D \equiv 1, 5, 8, 9, 12, 13 \pmod{16}$ and that for each odd prime p, we have $D \not\equiv 0 \pmod{p^2}$ (and that D is not 1). The problem is essentially the same in spirit as how one could count square-free integers, but the conditions at 2 require some additional care. For a square-free number n, let $B_n(X)$ be the number of non-zero integers $D \in [-X, X]$ that fail our conditions at each prime dividing n, i.e. if n is even then $D \pmod{16} \notin \{1, 5, 8, 9, 12, 13\}$ and for each odd prime $p \mid n$, we have $D \equiv 0 \pmod{p^2}$.

EXERCISE 1.2.1. *Show that for all $X \geq 2$, we have*

$$N_{S_2}(X) = -1 + \sum_{1 \leq n \leq 2\sqrt{X}} \mu(n) B_n(X),$$

where μ is the Möbius function. [Hint: consider some non-zero $D \in [-X, X]$ that fails the conditions at primes $2, p_1, \ldots, p_k$ or odd primes $p_1, \ldots, p_k$. How many times is it counted on the right-hand side?]

If X is a multiple of $4n^2$, then it is easy to count $B_n(X)$ exactly. Otherwise, we can estimate. We write $f(n, X) = g(n, X) + O(h(n, X))$ to mean that there exists a constant C such that for all $n, X \geq 1$ we have

$$|f(n, X) - g(n, X)| \leq C h(n, X).$$

EXERCISE 1.2.2. *If n is an odd square-free integer, show $B_n(X) = 2\lfloor \frac{X}{n^2} \rfloor$, and so*

$$B_n(X) = 2\frac{X}{n^2} + O(1).$$

EXERCISE 1.2.3. *If $n = 2m$, with m odd square-free, show $B_n(X) = 20\lfloor \frac{X}{4n^2} \rfloor + \epsilon$, where $0 \leq \epsilon \leq 20$, and so*

$$B_n(X) = 2\frac{5/2}{n^2}X + O(1).$$

Let $d_n = n^2$ if n is odd and $d_n = 2n^2/5$ if n is even. So we can conclude

$$N_{S_2}(X) = 2 \sum_{1 \leq n \leq 2\sqrt{X}} \mu(n)\frac{X}{d_n} + O(\sqrt{X}).$$

EXERCISE 1.2.4. *Show*

$$2\sum_{n \geq 1} \mu(n)\frac{X}{d_n} = 2X\left(1 - \frac{5}{8}\right) \prod_{\substack{2<p \\ prime}} (1 - p^{-2}) = \zeta(2)^{-1}X.$$

EXERCISE 1.2.5. *Show*

$$2 \sum_{n > 2\sqrt{X}} \mu(n)\frac{X}{d_n} = O(\sqrt{X}).$$

The exercises above combine to show

$$N_{S_2}(X) = \zeta(2)^{-1}X + O(\sqrt{X}) = \zeta(2)^{-1}X + o(X).$$

EXERCISE 1.2.6. *Plot a graph of $N_{S_2}(X)$ and $\zeta(2)^{-1}X$, and then "zoom in" by plotting*

$$\frac{N_{S_2}(X) - \zeta(2)^{-1}X}{\sqrt{X}}.$$

Does there appear to be a "secondary term," e.g. a function $f(x)$ of the form $cX^a(\log X)^b$ for reals a, b, c such that

$$N_{S_2}(X) = \zeta(2)^{-1}X + f(X) + o(f(X))?$$

Does there appear to be a an improved error term, e.g. a value $a < 1/2$ such that

$$N_{S_2}(X) = \zeta(2)^{-1}X + O(X^a)?$$

If the answers to the above questions are no, we can informally consider $O(\sqrt{X})$ as a "true error term," as opposed to the current state of knowledge. We could use the above questions to make this a precise notion, but it does not seem precision here is warranted, since the second question above ignores $\log X$ terms, and both ignore $\log\log X$ and other terms that might appear.

However, even once we consider number fields of degree 3, the situation becomes vastly more complicated. Still each cubic field can be generated by a degree 3 algebraic number α satisfying some $\alpha^3 + p\alpha + q = 0$, for $p, q \in \mathbb{Z}$, but it becomes much harder to use this to count. For one, the discriminant no longer is quite as discriminating.

EXERCISE 1.2.7. *Use a database (e.g. the LMFDB* [**80**]*) to find the smallest D such that there are two non-isomorphic cubic fields of discriminant D.*

However, even to obtain the discriminant from p, q is not something we can write down a simple formula for because we cannot write a formula for the ring of integers. (We can of course, give a formula for the discriminant of the ring $\mathbb{Z}[\alpha]$, but we do not know its index in the maximal order of K.) So we cannot say either what the discriminant of the field given by (p, q) is or when (p, q) and (p', q') give the same field. Remarkably, Shankar and Tsimerman [**97**] give a heuristic argument that gives a predicted asymptotic for S_d-extensions for each d by considering degree d polynomials, quantifying how much overcounting of fields is done, and understanding the discriminant of the maximal order in terms of local conditions. Their approach even gives a proof when $d = 3$, though we will describe an older proof below.

Even though all number fields of degree d come from degree d polynomials, the problems of overcounting and determining the discriminant represent fundamental differences between understanding the (finite) set of isomorphism classes of number fields of bounded discriminant and understanding the (infinite) set of polynomials of bounded discriminant (or even further contrast with understanding the natural finite set of polynomials with bounded coefficients). This contrast plays out in many places, from theoretical results (such as what we will see about counts of quartic D_4 and S_4 fields below), to computational challenges in making tables of number fields (see [**36, 65, 109**]).

However, another approach of parametrizing cubic fields was used by Davenport and Heilbronn in 1971 [**45**] to give the asymptotics of $N_3(X)$ as

$$N_3(X) = \frac{1}{3\zeta(3)}X + o(X).$$

Class field theory allows one to count Galois cubic fields, and Cohn in 1954 [**42**] had shown

$$N_{C_3}(X) = c_{C_3}X^{1/2} + o(X^{1/2}),$$

where we write C_3 for the cyclic group of order 3 in its regular representation (necessarily, see Exercise 1.1.9) and c_{C_3} is some explicit constant given by Cohn. From these results, if we write S_3 for the permutation group in its standard representation on a set of 3 elements, it follows that

$$N_{S_3}(X) = \frac{1}{3\zeta(3)}X + o(X).$$

Let D_4 be the dihedral group of 8 elements with its permutation action as automorphisms of a square acting on its four vertices. In 2002, Cohen, Diaz y Diaz, and Olivier [**38**] showed

$$N_{D_4}(X) = c_{D_4}X + o(X),$$

where c_{D_4} is an explicitly given constant. Their approach was to count quadratic extensions of quadratic fields, of which all quartic fields K with $\text{Gal}(K/\mathbb{Q}) \simeq D_4$ are. Bhargava [**16**] showed

$$N_{S_4}(X) = c_{S_4}X + o(X),$$

where c_{S_4} is an explicitly given constant.

It is also known that for G a transitive permutation group of degree 4 other than D_4, we have $N_G(X) = o(X)$. Thus the above two results show that asymptotically, a positive proportion (around 17%) of quartic fields, ordered by discriminant, are D_4-extensions, and a positive proportion (around 83%) of quartic fields, ordered by discriminant are S_4-extensions. If we instead consider degree n polynomials with bounded integral coefficients, asymptotically 100% have Galois group S_n. This is suggested by Hilbert's Irreducibility Theorem and was shown by van der Waerden [**106**] (see also [**19**]).

Today we know the answer to Question 1.1.11 for many more G, but many of the high-level strategies are descendants of one of these 3 approaches.

Class Field Theory. Class field theory is a powerful tool for understanding abelian extensions. In 1985, Mäki [**81**] gave asymptotics for $N_G(X)$ for each finite abelian group G in its regular representation, using the Kronecker-Weber theorem. One can also see work of Frei, Loughran, and Newton [**53**] for a more modern, class field theoretic approach to these asymptotics for abelian G. We briefly outline Cohn's proof [**42**] for C_3 extensions to give the flavor.

Let K be a number field. Let C_K be the idéle class group of K, i.e. $(\prod'_v K_v^*)/K$, where the product is over places v of K, and is restricted to elements that are units in all but finitely many places (with the restricted product topology). Let G_K be the absolute Galois group of K. Class field theory tells us that for a finite abelian group A, the Artin map $C_K \to G_K^{ab}$ induces a bijection

$$\mathrm{Sur}(G_K, A) \to \mathrm{Sur}(C_K, A)$$

(where Sur denotes continuous surjective group homomorphisms). For a finite place v the image of the local units $\mathcal{O}_v^*$ is the image of the inertia group at v of the corresponding map, and for a real place v the image of K_v^* is the image of the inertia group.

EXERCISE 1.2.8. *Let* $\mathbb{R}_{>0}^*$ *denote the positive reals (under multiplication). Show that the natural map*

$$\prod_p \mathbb{Z}_p^* \times \mathbb{R}_{>0}^* \to C_{\mathbb{Q}}$$

is an isomorphism. Conclude that for a finite abelian group A, we have a bijection

$$\mathrm{Hom}(G_{\mathbb{Q}}, A) \to \mathrm{Hom}(\prod_p \mathbb{Z}_p^*, A),$$

(where Hom *denotes continuous homomorphisms) where the image of the inertia group at p corresponds to the image of* $\mathbb{Z}_p^*$.

EXERCISE 1.2.9. *Show that* $\mathbb{Z}_p^*$ *has a continuous surjective homomorphism to* C_3 *if and only if p is* 0 *or* 1 *mod* 3, *in which case it has exactly two such homomorphisms.*

EXERCISE 1.2.10. *Show that a cyclic cubic field K has a square discriminant D, and if P is the product of the rational primes ramified in K, then if* $3 \nmid P$ *then* $D = P^2$ *and if* $3 \mid P$, *then* $D = 9P^2$. *[Hint: there are many ways to do the last part, e.g. using* [**88**, III.2.6], *the relationship between the class field theory conductor and the discriminant, or tables of local fields.]*

For a $\phi \in \mathrm{Hom}(G_{\mathbb{Q}}, C_3)$, we write D_ϕ for the absolute discriminant of the corresponding extension of $\mathbb{Q}$. Thus we can explicitly write the Dirichlet series of C_3 extensions

$$D(s) = \sum_{\phi \in \mathrm{Hom}(G_{\mathbb{Q}}, C_3)} \frac{1}{D_\phi^s} = (1 + 2 \cdot 9^{-2s}) \prod_{\substack{p \equiv 1 \pmod 3 \\ \text{prime}}} (1 + 2p^{-2s}).$$

This tells us what all cyclic cubic extensions of $\mathbb{Q}$, with their discriminants, are, and it remains a question of analytic number theory of the integers to do the asymptotic counting. This can be done by analysis of the right-most pole of $D(s)$ and an application of a Tauberian theorem such as the following.

THEOREM 1.2.11 ([**87**, Corollary p. 121]). *Let $f(s) = \sum_{n\geq 1} a_n n^{-s}$ with $a_n \geq 0$ be convergent for $\Re s > a > 0$. Assume that in the domain of convergence $f(s) = g(s)(s-a)^{-w} + h(s)$ holds, where $g(s), h(s)$ are holomorphic functions in the closed half plane $\Re s \geq a$, and $g(a) \neq 0$, and $w > 0$. Then*

$$\sum_{1\leq n\leq X} a_n = \frac{g(a)}{a\Gamma(w)} x^a (\log x)^{w-1} + o(x^a(\log x)^{w-1}).$$

EXERCISE 1.2.12. *Let χ be a non-trivial Dirichlet character of modulus 3. Show that*

$$\frac{D(s)}{L(2s,\chi)\zeta(2s)}$$

is analytic for $\Re(s) > 1/4$. Use this and Theorem 1.2.11 to conclude Cohn's asymptotic for $N_{C_3}(X)$.

EXERCISE 1.2.13. *Use Theorem 1.2.11 to give another proof that $N_{S_2}(X) = \frac{1}{\zeta(2)}X + o(X)$.*

EXERCISE 1.2.14. *Use Perron's formula (which can be used to prove Theorem 1.2.11) to analyze the size of $N_{S_2}(X) - \frac{1}{\zeta(2)}X$. How does what you find relate to what you found in Exercise 1.2.6?*

See [**115**, Section 8] for more examples and a more detailed exposition of this approach.

Parametrization and geometry of numbers. For many years, the approach of parametrization and geometry of number had only been applied to count cubic fields. However, Bhargava made significant breakthroughs in parametrizations (including the realization that resolvent rings needed to be parametrized and the construction of integral models of these) [**15, 17**] and geometry of numbers (including introducing averaging methods and methods for dealing with cusps) [**16, 18**] that allowed him to extend this approach to count quartic fields K with $\mathrm{Gal}(K/\mathbb{Q}) \simeq S_4$ and quintic fields K with $\mathrm{Gal}(K/\mathbb{Q}) \simeq S_5$. The work of Davenport-Heilbronn was also the starting place for Belabas and Fouvry [**14**] and for Bhargava and the author [**29**] to count Galois sextic S_3 extensions.

Now we briefly outline the approach of Davenport-Heilbronn. We give a parametrization originally due to Delone and Faddeev [**46**] (and refined by Gan, Gross, and Savin [**56**]), that is a cleaner version of what Davenport and Heilbronn used.

Let $\mathcal{O}_K$ be the ring of integers in a cubic number field.

EXERCISE 1.2.15. *Show that 1 generates a direct summand of $\mathcal{O}_K$ as a $\mathbb{Z}$-module.*

Let $1, W, T$ be a $\mathbb{Z}$ basis of $\mathcal{O}_K$. Since

$$WT = q + rW + sT,$$

for some $q, r, s \in \mathbb{Z}$, we can take $\omega = W - s$ and $\theta = T - r$ and have $1, \omega, \theta$ a $\mathbb{Z}$ basis of R with $\omega\theta \in \mathbb{Z}$. We call such a basis a *normalized* basis. Next, we write down a multiplication table for a normalized basis:

$$\begin{aligned} \omega\theta &= n \\ \omega^2 &= m + b\omega - a\theta \\ \theta^2 &= \ell + d\omega - c\theta, \end{aligned} \tag{1.2.16}$$

where $n, m, \ell, a, b, c, d \in \mathbb{Z}$. However, not all values of n, m, ℓ, a, b, c, d are possible.

EXERCISE 1.2.17. *Show that the associativity of multiplication in $\mathcal{O}_K$ exactly corresponds to the equations*

$$n = -ad \quad m = -ac \quad \ell = -bd. \tag{1.2.18}$$

We package these a, b, c, d into a binary cubic form $ax^3 + bx^2y + cxy^2 + dy^3$.

EXERCISE 1.2.19. *Find a binary cubic form associated to $\mathbb{Z}[\sqrt[3]{2}]$ and some normalized basis. Find another one (using a different choice of normalized basis). What is the discriminant of the binary cubic forms you found? What is the discriminant of $\mathbb{Z}[\sqrt[3]{2}]$?*

A choice of normalized basis of $\mathcal{O}_K$ is equivalent to a choice of $\mathbb{Z}$ basis of the $\mathbb{Z}$-module $\mathcal{O}_K/\mathbb{Z}$. The action of $\mathrm{GL}_2(\mathbb{Z})$ on bases of $\mathcal{O}_K/\mathbb{Z}$ gives a $\mathrm{GL}_2(\mathbb{Z})$ action on the tuples $(a, b, c, d) \in \mathbb{Z}^4$, arising above, such that the orbits are in bijection with isomorphism classes of $\mathcal{O}_K$. The action is (almost) the 4 dimensional representation of $\mathrm{GL}_2(\mathbb{Z})$ on binary cubic forms. Let $f(x, y) = ax^3 + bx^2y + cxy^2 + dy^3$. Let $g \in \mathrm{GL}_2(\mathbb{Z})$. Then we can let $\mathrm{GL}_2(\mathbb{Z})$ act on binary cubic forms via

$$(gf)(x, y) = \frac{1}{\det(g)} f((x, y)g), \tag{1.2.20}$$

where $(x, y)g$ is the multiplication of a row vector by a matrix on the right. This action exactly translates into the action on the parameters (a, b, c, d) given by action on the choice of basis of $\mathcal{O}_K/\mathbb{Z}$.

EXERCISE 1.2.21. *Check that for $g = \left(\begin{smallmatrix} 0 & 1 \\ 1 & 0 \end{smallmatrix}\right)$ the action on the set of (a, b, c, d) is as claimed. Check it for $g = \left(\begin{smallmatrix} 1 & 1 \\ 0 & 1 \end{smallmatrix}\right)$ and $g = \left(\begin{smallmatrix} -1 & 0 \\ 0 & 1 \end{smallmatrix}\right)$ as well, and conclude the action is as claimed for all of* $\mathrm{GL}_2(\mathbb{Z})$.

It turns out the discriminant of $\mathcal{O}_K$ is equal to the discriminant of the associated binary cubic forms. (These are both polynomials in a, b, c, d and it is computationally straightforward to check they are the same.) Thus it remains to understand (1)which tuples (a, b, c, d) arise, and (2) to count their $\mathrm{GL}_2(\mathbb{Z})$ orbits, asymptotically by discriminant. For (1), it turns out that this can be understood by some relatively simply criteria (the form should be irreducible over $\mathbb{Q}$, along with a condition mod p^2 for every prime p). For (2), one can count orbits by counting the (relevant) lattice points in a fundamental domain for the action, which can be done with methods from the geometry of numbers.

See also [**115**, Section 11], and [**22**] for further exposition of this approach from a modern point of view, as well as [**62**, p.12-18] for a series of exercises that provides a do-it-yourself proof of Davenport-Heilbronn's result.

Counting fields up to discriminant X is closely related to producing complete, non-redundant lists or tables of such fields, sometimes called *tabulating* fields. See [**36**, Section 9] for an exposition of algorithms for such tabulation. For general degree, they are required to essentially consider polynomials or algebraic numbers, which as we have seen above do not line up perfectly with fields, and thus require considering and computing with many more polynomials than actually produce unique fields. However, for cubic fields, this parametrization relating isomorphism classes of cubic fields to certain lattice points in a fundamental domain also is the basis of Belabas's [**11**] approach to tabulating cubic fields, which is far more efficient than what is available for general degree. It would be very interesting to see if Bhargava's parametrizations of quartic and quintic fields [**15, 17**] could be used to give an efficient approach to tabulating quartic and quintic fields (see also [**37**] for other approaches for quartic fields).

Extensions of extensions. The general approach to counting number fields which are composite extensions by counting extensions of extensions has been used by many authors. Klüners [**69**] was able to asymptotically count G extensions for many G of the form $C_2 \wr H$. Wang [**111**] and Masri, Thorne, Tsai, and Wang were able to asymptotically count G extensions for $G = S_n \times A$ where $n = 3, 4$, or 5 and A is an abelian group. There is forthcoming work of Alberts, Lemke Oliver, Wang, and the author that counts composite extensions for many more G.

Cohen, Diaz y Diaz, and Oliver were able to count quartic fields K with $\mathrm{Gal}(K/\mathbb{Q}) \simeq D_4$ fields by using the fact that these fields were all quadratic extensions of quadratic extensions, and we can enumerate quadratic extensions of any field in a relatively concrete way. We outline such an approach here.

EXERCISE 1.2.22. *Show that each quartic field K with* $\mathrm{Gal}(K/\mathbb{Q}) \simeq D_4$ *contains a unique quadratic field as a subfield.*

EXERCISE 1.2.23. *Show that any quadratic extension K of a quadratic field has* $\mathrm{Gal}(K/\mathbb{Q}) \simeq D_4, \mathbb{Z}/4\mathbb{Z}$, *or* $\mathbb{Z}/2\mathbb{Z} \times \mathbb{Z}/2\mathbb{Z}$.

We can use the class field theory approach, outlined above, to show the following:

(1) For each quadratic field F, the number $N_{F,S_2}(X)$ of quadratic extensions K of F with $\mathrm{Nm}_{F/\mathbb{Q}}(\mathrm{Disc}(K/F)) \leq X$ satisfies
$$N_{F,S_2}(X) = c_F X + o(X),$$
for some $c_F > 0$.

(2) There exists a number C such that for each quadratic field F, for all X
$$N_{F,S_2}(X) \leq C D_F^{2/3} X.$$

EXERCISE 1.2.24. *Use class field theory and the Tauberian theorem above to show statement (1) above.*

We can combine the two statements above, along with our knowledge of the quadratic extensions of $\mathbb{Q}$, to count quadratic extensions K of quadratic fields F. Recall $|\mathrm{Disc}(K/\mathbb{Q})| = \mathrm{Nm}_{F/\mathbb{Q}}(\mathrm{Disc}(K/F)) D_F^2$. If $N(X)$ is the number of quadratic extensions of quadratic fields (more precisely we count pairs (F, K), where $F/\mathbb{Q}$ is an isomorphism class of quadratic fields, and K/F is an isomorphism class of quadratic extensions), we have

$$\begin{aligned} N(X) &= \sum_{[F:\mathbb{Q}]=2} N_{F,S_2}\left(\frac{X}{D_F^2}\right) \\ &= \sum_{[F:\mathbb{Q}]=2} \left(\left(\frac{c_F}{D_F^2}\right) X + o(X)\right). \end{aligned}$$

Unfortunately, we cannot exchange the $o(X)$ (and its implicit limit statement) with an infinite sum in F. However, if we fix some Y, and only consider F with $D_F \le Y$, then we have

$$\begin{aligned} N_Y(X) :=& \sum_{\substack{[F:\mathbb{Q}]=2 \\ D_F \le Y}} N_{F,S_2}\left(\frac{X}{D_F^2}\right) \\ =& \sum_{\substack{[F:\mathbb{Q}]=2 \\ D_F \le Y}} \left(\left(\frac{c_F}{D_F^2}\right) X + o(X)\right) \\ =& \left(\sum_{\substack{[F:\mathbb{Q}]=2 \\ D_F \le Y}} \frac{c_F}{D_F^2}\right) X + o(X). \end{aligned}$$

EXERCISE 1.2.25. *Using statements (1) and (2) above, show that* $\sum_{[F:\mathbb{Q}]=2} \frac{c_F}{D_F^2}$ *converges.*

This gives a lower bound

$$\liminf_{X\to\infty} \frac{N(X)}{X} \ge \sum_{[F:\mathbb{Q}]=2} \frac{c_F}{D_F^2}.$$

However, we also have

$$\begin{aligned} N(X) \le& N_Y(X) + \sum_{\substack{[F:\mathbb{Q}]=2 \\ D_F > Y}} N_{F,S_2}\left(\frac{X}{D_F^2}\right) \\ \le& N_Y(X) + \sum_{\substack{[F:\mathbb{Q}]=2 \\ D_F > Y}} C D_F^{-4/3} X. \end{aligned}$$

EXERCISE 1.2.26. *Show* $\sum_{[F:\mathbb{Q}]=2} D_F^{-4/3}$ *converges.*

So we have

$$\limsup_{X\to\infty} \frac{N(X)}{X} \le \sum_{\substack{[F:\mathbb{Q}]=2 \\ D_F \le Y}} \frac{c_F}{D_F^2} + \sum_{\substack{[F:\mathbb{Q}]=2 \\ D_F > Y}} C D_F^{-4/3},$$

and taking the limit as $Y \to \infty$ we obtain

$$\limsup_{X\to\infty} \frac{N(X)}{X} < \sum_{\substack{[F:\mathbb{Q}]=2 \\ D_F \le Y}} \frac{c_F}{D_F^2}.$$

Combined with the lim inf statement above, we have

$$N(X) = \left(\sum_{[F:\mathbb{Q}]=2} \frac{c_F}{D_F^2}\right) X + o(X).$$

To obtain an asymptotic on $N_{D_4}(X)$, we can use the fact that the asymptotics for $\mathbb{Z}/4\mathbb{Z}$ and $\mathbb{Z}/2\mathbb{Z} \times \mathbb{Z}/2\mathbb{Z}$ quartic fields (which can be obtained by class field theory) are of smaller order, i.e.

$$N_{\mathbb{Z}/4\mathbb{Z}}(X) + N_{\mathbb{Z}/2\mathbb{Z}\times\mathbb{Z}/2\mathbb{Z}}(X) = o(X).$$

To be quite precise, we should account for how many times each D_4 quartic appears in $N(X)$.

EXERCISE 1.2.27. *Show that the map from pairs* (F, K), *where* F *is an isomorphism class of quadratic field, and* K/F *is an isomorphism class of a quadratic extension of* F *such that* $\mathrm{Gal}(K/\mathbb{Q}) \simeq D_4$, *to isomorphism classes of* D_4*-quartic extensions of* $\mathbb{Q}$ *is* $2-1$.

1.3. Conjectures and more results. For each transitive finite permutation group G, Malle [**83**] defined integers a_G, b_G such that he conjectured that there was some $c_G > 0$ (not predicted) such that

$$N_G(X) = c_G X^{1/a_G} (\log X)^{b_G}. \tag{1.3.1}$$

Though the examples mentioned above all confirm cases of Malle's conjecture, Klüners [**67**] found a counterexample in $G = C_3 \wr C_2$, for which there were more extensions than conjectured. One still has "Weak Malle," an earlier conjecture of Malle [**82**] that there is a $c_G > 0$, and for every $\epsilon > 0$ and $c'_{G,\epsilon}$ such that for all X we have

$$c_G X^{1/a_G} \leq N_G(X) \leq c'_{G,\epsilon} X^{1/a_G+\epsilon}. \tag{1.3.2}$$

Conjecture (1.3.2) has no known counterexamples and is generally expected to be true. We expect that one could add the $(\log X)^{b_G}$ term to the lower bound in (1.3.2) and the inequality still hold. Türkelli [**105**, Conjecture 6.7] has given a revised definition of b_G for which he conjectures (1.3.1) holds. See [**115**, Section 10] for further exposition of Malle's conjecture, as well as the general heuristic principle that implies the conjecture as well as related conjectures (called the Malle-Bhargava Principle).

New approaches have asymptotically counted G-extensions for other G, including work of Fouvry and Koymans [**52**] that counts nonic Heisenberg extensions, work of Klüners [**66**] for generalized quaternion groups G, and work of Koymans and Pagano [**72**] that counts G extensions for many nilpotent G.

1.4. Variations and directions. There are of course many variations one can put on Question 1.1.11. We will not give a complete literature review but try to point the reader to some recent papers from which they can learn about various directions. In the cases for which we know the answer, it is common that the answer is also given when the fields counted have a restricted signature. It is common to also be able to give asymptotics only counting fields satisfying local conditions at some finite (or even infinite) set of primes [**23, 53, 113**]. See [**115**] for an introduction to counting with local conditions, including results and conjectures in this direction.

When asymptotics are known, one can ask for improved error bounds [**13, 96**] or secondary terms. While Davenport and Heilbronn's result on counting cubic fields is a theorem, tables such as those produced by Belabas [**11**] show various slow convergence to the known asymptotic. Supported by significant numerical evidence from such tables, Roberts [**94**] was able to conjecture a precise secondary term for the counting of cubic fields (beyond the main term of Davenport and Heilbronn). Taniguchi and Thorne [**103**] and Bhargava, Shankar, and Tsimerman [**22**] independently proved Roberts conjectured secondary term. The best current result is of Bhargava, Taniguchi, and Thorne [**26**] and gives

$$N_{S_3}(X) = \frac{1}{3\zeta(3)}X + c'_{S_3}X^{5/6} + O(X^{2/3}(\log X)^{2.09}),$$

where c'_{S_3} is an explicitly given, non-zero constant.

Wang [**110**] proves the first secondary terms beyond cubic fields, in counting $S_3 \times A$ extensions, where A is an abelian odd order group. In general there are no conjectures for secondary terms, and it would be a useful computational project to make numerical computations tabulating fields that were strong enough to suggest more secondary terms, true error terms, or speed of convergence for counting G-extensions for other G. It is possible some such computations could be done with pre-existing tables of fields, at least as a starting place.

When asymptotics are not known, one can ask for lower or upper bounds [**4, 25, 73**], or to prove Weak Malle [**4, 70**]. In light of recent progress on the upper bound for degree d extensions [**9, 24, 43, 90**], as well as for G-extensions [**48**], it is very natural to ask if these ideas can be used to develop more efficient algorithm for tabulating number fields of a fixed degree or Galois group. Such improved algorithms could be very helpful to produce more exhaustive tables, and in particular to help with the questions of predicting secondary and error terms discussed above.

It would be very interesting to develop a "sampling" algorithm that might be able to numerically estimate $N_G(X)$ for some X large enough that we are not able to produce tables listing all G-extensions up to absolute discriminant X. Current tabulation algorithms consider various polynomials and some fraction of them end up producing new fields for the table. Is there a way to do some computations with random polynomials that could provide good enough estimates for use in predicting secondary or error terms?

One can replace $\mathbb{Q}$ with another number field K_0, or more generally a global field [**23, 72, 108, 121**]. One can count extensions L/K that also have a specified Galois group over some subfield K_0 of K [**5, 6**]. One can also count extensions by invariants other than the discriminant, such class the class field theory conductor for abelian extensions [**113**] , Artin conductors [**7**], or more general invariants [**95**].

2. Distribution of class groups

2.1. Motivating Questions. For a finite transitive permutation group G, one can ask as we consider the number fields in S_G, what is the distribution of class groups that arise? In particular for a finite abelian group A, we could ask what is

$$\lim_{X\to\infty} \frac{|\{K \in S_G \mid \mathrm{Cl}_K \simeq A, D_K \leq X\}|}{N_G(X)}? \tag{2.1.1}$$

More generally, for a set S of number fields, and a function f taking finite abelian groups to real numbers, we ask what are the asymptotics of

$$\frac{\sum_{\substack{K\in S\\ D_K\leq X}} f(\mathrm{Cl}_K)}{\sum_{\substack{K\in S\\ D_K\leq X}} 1}? \tag{2.1.2}$$

Note that the denominator of these questions are the number field counting questions above (or similar).

Class groups, in some sense, provide the first example of the strategy of parametrization mentioned above that Davenport and Heilbronn used to count cubic fields.

Gauss studied $\mathrm{SL}_2(\mathbb{Z})$ orbits of integral binary cubic forms $ax^2 + bxy + cy^2$ (with $a, b, c \in \mathbb{Z}$), working with Lagrange's reduction theory that relates orbits to triples (a, b, c) in a certain region of $\mathbb{Z}^3$. This allowed Gauss to, for example, tabulate and enumerate those orbits. Remarkably, Gauss found a group law on these orbits. Dedekind later realized that these orbits were related to class group of quadratic fields (and we now recognize Gauss's group law as the usual group law on class groups). This relationship is what allows us to compute class groups of quadratic fields much faster than class groups of higher degree fields. See [**86**] and the extensive tables of class groups of quadratic fields available at the LMFDB [**80**].

In tables of class group of quadratic fields, one can observe that the group $\mathbb{Z}/9\mathbb{Z}$ occurs much more often than the group $\mathbb{Z}/3\mathbb{Z} \times \mathbb{Z}/3\mathbb{Z}$ even though both groups have order 9. Cohen and Lenstra [**39**] sought to explain this and other patterns available in the empircal data on the class groups of quadratic fields, and developed heuristics that gave conjectures for the distribution of class groups of quadratic fields and real abelian fields.

Before we can describe some of Cohen and Lenstra's conjectures, we have the explain a bit more context on class groups. Almost any empirical or theoretical study of class groups of quadratic fields will show you that those of real quadratic fields behave quite differently than those of imaginary quadratic fields.

EXERCISE 2.1.3. *Using a database such as the LMFDB, look at class groups of several hundred real and imaginary quadratic fields. What is the major difference you notice?*

This difference is already apparent in Gauss's study, and Gauss conjectured (in modern terms) (1)that there are only finitely many imaginary quadratic fields with a given class group, and (2) that there are infinitely many real quadratic fields with trivial class group. The conejcture (1) was proven by Heilbronn [**61**], but (2) is still open.

EXERCISE 2.1.4. *Recall the Brauer-Siegel theorem, and see why it implies (1) above. Why doesn't it give the same conclusion for real quadratic fields?*

2.2. Genus theory of quadratic fields. Secondly, Gauss found that the 2-torsion in class groups of quadratic fields (in his binary forms language) was relatively easy to describe.

THEOREM 2.2.1. *If K is a quadratic field and N_K is its narrow class group, then $N_K[2] \simeq (\mathbb{Z}/2\mathbb{Z})^{t-1}$, where t is the number of ramified primes of $K/\mathbb{Q}$.*

When $K/\mathbb{Q}$ is imaginary quadratic, then $N_K = \mathrm{Cl}_K$ by definition.

EXERCISE 2.2.2. *When $K/\mathbb{Q}$ is real quadratic, show that $N_K \to \mathrm{Cl}_K$ is either an isomorphic or has kernel isomorphic to $\mathbb{Z}/2\mathbb{Z}$ (depending on whether the fundamental unit has negative or positive norm). Conclude that $\mathrm{Cl}_K[2] \simeq (\mathbb{Z}/2\mathbb{Z})^{t-1-\epsilon}$, where ϵ is either 1 or 0.*

While Gauss proved Theorem 2.2.1 very explicitly with binary quadratic forms, we will give a proof here using class field theory. This both gives good practice using the statements of class field theory, but also puts this phenomenon in a larger context. However, first we will note that one can see quite explicitly some potential 2-torsion. If p is a prime ramified in a quadratic field K, where $(p) = \wp^2$, then $\wp$

is either trivial or order 2 in Cl_K. There is no known simple way to produce such possible p-torsion for any primes $p > 2$.

To understand $\mathrm{Cl}_K / 2\,\mathrm{Cl}_K$ (which is abstractly isomorphic to $\mathrm{Cl}_K[2]$ by the theory of finite abelian groups), we can then try to understand $\mathrm{Sur}(\mathrm{Cl}_K, \mathbb{Z}/2\mathbb{Z})$, which via class field theory, corresponds to the set of everywhere unramified quadratic extensions of K. Similarly to understand the narrow class group, we need to understand quadratic extensions of K unramified at all finite places.

EXERCISE 2.2.3. *Let K be a number field. For a real place v let $K_v^{>0}$ be the subgroup of positive elements of K_v^*. Show that there is a unique continuous homomorphism*

$$C_K / \left(\prod_{v \text{ finite}} \mathcal{O}_v^* \times \prod_{v \text{ real}} K_v^{>0} \times \prod_{v \text{ complex}} K_v^* \right) \to N_K$$

sending, for each finite place v of K, a uniformizer $\pi_v \in K_v^*$ to the image of the corresponding prime ideal of v in N_K, and that the homomorphism is an isomorphism. Show this implies that N_K is isomorphic to the Galois group of the maximal abelian extension of K unramified at all finite places.

EXERCISE 2.2.4. *Let K be a number field. Show that there is a unique continuous homomorphism*

$$C_K / \left(\prod_{v \text{ finite}} \mathcal{O}_v^* \times \prod_{v \text{ real}} K_v^* \times \prod_{v \text{ complex}} K_v^* \right) \to \mathrm{Cl}_K$$

sending, for each finite place v of K, a uniformizer $\pi_v \in K_v^*$ to the image of the corresponding prime ideal of v in N, and that the homomorphism is an isomorphism. Show this implies that Cl_K is isomorphic to the Galois group of the maximal abelian extension of K unramified at all places.

EXERCISE 2.2.5. *If $K/\mathbb{Q}$ is quadratic, show that the action of the generator of* $\mathrm{Gal}(K/\mathbb{Q})$ *on N_K is by multiplication by -1.*

EXERCISE 2.2.6. *Show if K is a quadratic field and L/K is everywhere unramified, that $L/\mathbb{Q}$ is Galois, and* $\mathrm{Gal}(L/\mathbb{Q}) \simeq \mathbb{Z}/2\mathbb{Z} \times \mathbb{Z}/2\mathbb{Z}$. *[Hint: this may require reviewing the statements of class field theory to recall that, for K Galois, the map $G_K \to G_K$ given by conjugating by $\sigma \in \mathbb{G}_\mathbb{Q}$ corresponds, via the Artin map, to the natural action of σ, through* $\mathrm{Gal}(K/\mathbb{Q})$, *on C_K. Also, one can use an inertia group for $K/\mathbb{Q}$ to determine the Galois group from the two abelian groups of order 4.]*

PROOF OF THEOREM 2.2.1. Let $N = N_K$. We have, from the theory of finite abelian groups that $|N/2N| = |\mathrm{Hom}(N, \mathbb{Z}/2\mathbb{Z})$. If we consider the homomorphisms $N \to \mathbb{Z}/2\mathbb{Z}$, there is one trivial homomorphism, and the rest correspond, via the class field theory above, exactly to the unramified at finite places quadratic extensions L of K. We have seen that each such L is a Galois $\mathbb{Z}/2\mathbb{Z} \times \mathbb{Z}/2\mathbb{Z}$ extension of $\mathbb{Q}$.

So now we can ask, what are the $\mathbb{Z}/2\mathbb{Z} \times \mathbb{Z}/2\mathbb{Z}$ extensions L of $\mathbb{Q}$ containing K? By Galois theory, these correspond to normal subgroups S of $G_\mathbb{Q}$, contained in G_K and with quotient isomorphic to $\mathbb{Z}/2\mathbb{Z} \times \mathbb{Z}/2\mathbb{Z}$. Then these almost correspond to surjections $G_\mathbb{Q} \to \mathbb{Z}/2\mathbb{Z} \times \mathbb{Z}/2\mathbb{Z}$, whose projection onto the second coordinate gives the map $G_\mathbb{Q}/G_K \simeq \mathbb{Z}/2\mathbb{Z}$, except two surjections with the same kernel correspond to the same field. So two surjections correspond to the same field if and only if they differ by an automorphism of $\mathbb{Z}/2\mathbb{Z} \times \mathbb{Z}/2\mathbb{Z}$, and since we have fixed the projection onto the second coordinate, it must be an automorphism that preserves that projection. There are two such automorphisms, so each field L corresponds to two surjections $G_{\mathbb{Q}} \to \mathbb{Z}/2\mathbb{Z} \times \mathbb{Z}/2\mathbb{Z}$ whose projection onto the second coordinate gives the map $G_{\mathbb{Q}}/G_K \simeq \mathbb{Z}/2\mathbb{Z}$.

Given $L/\mathbb{Q}$ Galois with Galois group $\mathbb{Z}/2\mathbb{Z} \times \mathbb{Z}/2\mathbb{Z}$ and K the fixed field of $\langle(1,0)\rangle$ (as arranged above), we have L/K is unramified at all finite places if and only if the inertia groups of all finite places in $\mathrm{Gal}(L/\mathbb{Q}) = \mathbb{Z}/2\mathbb{Z}\times\mathbb{Z}/2\mathbb{Z}$ have trivial intersection with $\langle(1,0)\rangle$. The only subgroups of $\mathbb{Z}/2\mathbb{Z} \times \mathbb{Z}/2\mathbb{Z}$ with this property are the trivial subgroup, $\langle(1,1)\rangle$, and $\langle(0,1)\rangle$. If a finite place v of $\mathbb{Q}$ is unramified in K, then it follows that for L/K unramified at all finite places we must have the inertia group of v in $\mathbb{Z}/2\mathbb{Z} \times \mathbb{Z}/2\mathbb{Z}$ trivial.

From the above exercise, we have that

$$\prod_{v \text{ finite of } \mathbb{Q}} \mathcal{O}_v^* \times \mathbb{R}^* \simeq C_{\mathbb{Q}},$$

and so a continuous surjection $C_{\mathbb{Q}} \to \mathbb{Z}/2\mathbb{Z}\times\mathbb{Z}/2\mathbb{Z}$ unramified outside of a set S of places of $\mathbb{Q}$ (those ramified in K) corresponds exactly to a continuous surjection

$$\phi : \prod_{v \in S} \mathcal{O}_v^* \to \mathbb{Z}/2\mathbb{Z} \times \mathbb{Z}/2\mathbb{Z}.$$

In K, we are given, for each $v \in S$, a continuous surjection $\mathcal{O}_v^* \to \mathbb{Z}/2\mathbb{Z}$, and we see that to construct a ϕ corresponding to L as above we may lift this map to $\mathcal{O}_v^* \to \langle(1,1)\rangle$ or $\mathcal{O}_v^* \to \langle(0,1)\rangle$. If we lift all of the maps from $\mathcal{O}_v^*$ to the same subgroup, then ϕ won't be surjective, but otherwise it will be surjective. Any such surjective ϕ gives an L/K quadratic and unramified at finite places, and any L/K quadratic and unramified at finite places comes from two such surjective ϕ. Thus we have that the number of L is $\frac{2^{|S|-2}}{2} = 2^{|S|-1} - 1$, and thus $|N/2N| = |\operatorname{Hom}(N, \mathbb{Z}/2\mathbb{Z}) = 2^{t-1}$, as desired. $\square$

So from the point of view of class field theory, these unramified abelian extensions L/K are "predictable" because they come from abelian extensions of $\mathbb{Q}$. We might wonder if there are are other such unramified abelian extensions that we can produce from abelian extensions of $\mathbb{Q}$.

EXERCISE 2.2.7. *Let K be a quadratic extension. Let $E/\mathbb{Q}$ be the maximal abelian extension of $\mathbb{Q}$ such that EK/K is unramified. So EK/K is an unramified abelian extension. Show that it corresponds to $2\operatorname{Cl}_K$ via class field theory (and Galois theory). So $\operatorname{Cl}_K/2\operatorname{Cl}_K$ is the part of Cl_K that abelian extensions of $\mathbb{Q}$ tell us about. Show the analogous statement for "unramified at finite places" and the narrow class group.*

This gives us a precise sense in which $\operatorname{Cl}_K/2\operatorname{Cl}_K$ is well-understood that provably does not apply to the rest of Cl_K. In any case, as there is this "predictable" piece of the class group Cl_K of a quadratic field in the 2-Sylow subgroup, Cohen and Lenstra, in their heuristics for the distribution of class groups, only considered the odd part $\operatorname{Cl}_K^{odd}$ (the quotient of Cl_K by its 2-Sylow subgroup).

2.3. The Cohen-Lenstra heuristics. For simplicity, we will fix an odd p, and only consider $\mathrm{Cl}_{K,p}$, the Sylow p-subgroup of Cl_K. Let IQ and RQ be the sets of isomorphism classes of imaginary and real quadratic fields, respectively. Cohen and Lenstra's heuristics imply that for any "reasonable" [2] function f from the set of isomorphism classes of finite abelian p-groups to $\mathbb{R}$, that

$$\lim_{X\to\infty} \frac{\sum_{\substack{K\in \mathrm{IQ}\\ D_K\le X}} f(\mathrm{Cl}_K)}{\sum_{\substack{K\in \mathrm{IQ}\\ D_K\le X}} 1} = \frac{\sum_{\substack{A \text{fin. ab.}\\ p\text{-group}}} \frac{f(A)}{|\mathrm{Aut}(A)|}}{\sum_{\substack{A \text{fin. ab.}\\ p\text{-group}}} \frac{1}{|\mathrm{Aut}(A)|}} \tag{2.3.1}$$

and

$$\lim_{X\to\infty} \frac{\sum_{\substack{K\in \mathrm{RQ}\\ D_K\le X}} f(\mathrm{Cl}_K)}{\sum_{\substack{K\in \mathrm{RQ}\\ D_K\le X}} 1} = \frac{\sum_{\substack{A \text{fin. ab.}\\ p\text{-group}}} \frac{f(A)}{|A||\mathrm{Aut}(A)|}}{\sum_{\substack{A \text{fin. ab.}\\ p\text{-group}}} \frac{1}{|A||\mathrm{Aut}(A)|}}. \tag{2.3.2}$$

Note the denominators in (2.3.1) and (2.3.2) do not depend on f. They are simply some constant coming from the theory of finite abelian p-groups. Indeed, we have

$$\sum_{\substack{A \text{fin. ab.}\\ p\text{-group}}} \frac{1}{|\mathrm{Aut}(A)|} = \prod_{i\ge 1}(1-p^{-i})^{-1} \quad \text{and} \quad \sum_{\substack{A \text{fin. ab.}\\ p\text{-group}}} \frac{1}{|A||\mathrm{Aut}(A)|} = \prod_{i\ge 2}(1-p^{-i})^{-1}.$$

If we take f to be the characteristic function of a particular group, we see that the conjectures imply a particular p-group appears with a frequency inversely proportional to $|\mathrm{Aut}(A)|$ in the imaginary case and inversely proportional to $|A||\mathrm{Aut}(A)|$ in the real case. So in particular, this conjecture explains, in a precise way, the different frequencies with which different groups were empirically appearing in the tables. See [**115**, Section 5] and [**112**] for further exposition on these conjectures.

EXERCISE 2.3.3. *How often do the conjectures above predict that imaginary, respectively real, quadratic fields have cyclic Sylow p-subgroup?*

The heuristics of Cohen and Lenstra greatly benefited from the existence of tables of class groups of quadratic fields that could be used as evidence for the precise conjectures being made.

EXERCISE 2.3.4. *Using existing databases, make some tables of the frequency of 3-Sylow subgroups in class groups of imaginary (and/or real) quadratic fields up to some large discriminant. How do they compare to the conjectured values?*

There are no general conjectures about how quickly the limits in (2.3.1) and (2.3.2) approach their conjectured values. One could imagine conjectural secondary terms as well as conjectural error terms. Using the same approach as for counting cubic fields, we have secondary terms in one case [**22, 26, 103**] (we will discuss further below why the "same approach" applies). Not only would such a conjecture be interesting in its own right, but also if believable (say, based on evidence in certain cases and general heuristic reasoning), it might help us evaluate in other cases whether given data on class groups gives good evidence for a particular conjecture.

PROJECT 2.3.5. *Using existing databases, make some graphs of how the proportion of imaginary (and/or) real quadratic fields of absolute discriminant in $(X, 2X]$ with a given Sylow p-subgroup of class group changes with X. Do any of your graphs*

[2] No particular definition of reasonable is given, but see [**21**, Section 5.6] and [**10**, Section 7].

(or the underlying numbers) suggest a rate at which the limit is approached? Do they suggest precise secondary terms?

Note that in the above exercise if you can uniformly randomly select a sample of fields in each range, you can ensure you get values close to the actual proportions via probability (e.g. Chebyshev's inequality). For quadratic fields, since you are likely getting class groups from a table as opposed to computing them, this doesn't seem worthwhile. However, if you are in a setting where you have to *compute* each class group, then sampling will vastly increase how large of an X you may consider.

Based on computations of the sort suggested in Project 2.3.5, Lewis and Williams [**75**] conjecture approximate secondary terms for the proportion of real quadratic fields having class group with non-trivial Sylow p-subgroup. It would be interesting for these conjectures to be made more precise as in [**94**] and also to have conjectural secondary terms in other cases. (See [**64**, Section 2] for some intriguing conjectures and data along these lines.) Such conjectures, or a more general theory of how large we expect secondary or true error terms to be, can be very important in understanding if numerical data is supporting a conjecture, contradicting a conjecture, or not powerful enough to do either.

Two important motivations for Cohen-Lenstra heuristics originally were the agreement with tables and the general philosophy that objects occur in nature with frequency inversely proportional to the size of their automorphism group. See, e.g. Exercise 1.1.13 as well as the following exercise for some examples of this ubiquitous phenomenon.

EXERCISE 2.3.6. *Consider all* multiplication tables on n elements, *i.e. functions* $\{1,\dots,n\}\times\{1,\dots,n\}\to\{1,\dots,n\}$ *that are associative and have inverses and an identity. Show that a finite group* G *or order* n *is isomorphic to* $n!/|\operatorname{Aut}(G)|$ *of the groups given by these tables.*

3. A matrix model for class groups

3.1. The model. Venkatesh and Ellenberg [**108**, Section 4.1] introduced an interesting perspective on class groups of quadratic fields as follows. Let K be a quadratic field and let S be a set of primes of K that generate Cl_K. We write $\mathcal{O}_S^*$ for the S-units in the integers $\mathcal{O}_K$, and I_K^S for the abelian group of fractional ideals generated by the elements of S. Let μ_K be the group of roots of unity in K. Then

$$\mathrm{Cl}(K)=\operatorname{cok}(\mathcal{O}_S^*/\mu_K\to I_K^S), \tag{3.1.1}$$

where the map takes α to the ideal (α).

EXERCISE 3.1.2. *If* A *is a finite abelian group given as* $\operatorname{cok}(U\to V)$ *for* U,V *finitely generated abelian groups, then the Sylow* p*-subgroup* A_p *is* $\operatorname{cok}(U\otimes Z_p\to V\otimes Z_p)$.

So

$$\mathrm{Cl}(K)_p=\operatorname{cok}(\mathcal{O}_S^*/\mu_K\otimes Z_p\to I_K^S\otimes Z_p).$$

We have that I_K^S is a free abelian group of rank $|S|$, so $I_K^S\otimes Z_p\simeq\mathbb{Z}_p^{|S|}$. Dirichlet's unit theorem gives that $\mathcal{O}_S^*/\mu_K$ is a free abelian group of rank $|S|+u$, where $U=0$ if K is imaginary and $u=1$ if K is quadratic. So $\mathcal{O}_S^*/\mu_K\otimes Z_p\simeq\mathbb{Z}_p^{|S|+u}$. If we choose bases for I_K^S and $\mathcal{O}_S^*/\mu_K$ as $\mathbb{Z}$-modules (which would give specific isomorphisms as above), then the map $\mathcal{O}_S^*/\mu_K\otimes Z_p\to I_K^S\otimes Z_p$ can be written as a matrix $M \in \mathrm{Mat}_{|S|\times|S|+u}(\mathbb{Z}_p)$. (Though M has integral entries, since we are interested in $\mathrm{Cl}_{K,p}$ we take the cokernel as a $\mathbb{Z}_p$-matrix.)

EXERCISE 3.1.3. *If A is a finite abelian p-group, the isomorphism type of A can be determined from $A\otimes\mathbb{Z}/p\mathbb{Z}, A\otimes\mathbb{Z}/p^2\mathbb{Z}, A\otimes\mathbb{Z}/p^3\mathbb{Z},\dots$.*

It is then natural to consider the distribution of $\mathrm{Cl}_K \otimes\mathbb{Z}/p^k\mathbb{Z}$.

EXERCISE 3.1.4. *Let $N \in \mathrm{Mat}_{n\times m}(\mathbb{Z}_p)$ and let $\bar{N} \in \mathrm{Mat}_{n\times m}(\mathbb{Z}/p^k\mathbb{Z})$ be obtained from N by reducing each coefficient mod p^k. Show*

$$\mathrm{cok}(N : \mathbb{Z}_p^m \to \mathbb{Z}_p^n)\otimes\mathbb{Z}/p^k\mathbb{Z} \simeq \mathrm{cok}(\bar{N} : (\mathbb{Z}/p^k\mathbb{Z})^m \to (\mathbb{Z}/p^k\mathbb{Z})^n).$$

So if we let $\bar{M}$ be the reduction of M mod p^k, we are interested in the distribution of the cokernels of these $\bar{M} \in \mathrm{Mat}_{|S|\times|S|+u}(\mathbb{Z}/p^k\mathbb{Z})$, where u is either 0 or 1. Now a priori we have no idea what the distribution of these $\bar{M}$ is, but a baseline heuristic might be to guess they are approximately uniform in $\mathrm{Mat}_{n\times n+u}(\mathbb{Z}/p^k\mathbb{Z})$. Putting this together over different k, one might guess that the M are approximately distributed according to Haar measure on $\mathrm{Mat}_{n\times n+u}(\mathbb{Z}_p)$. This doesn't make perfect sense, because the n involved is not fixed, but as a heuristic we just imagine that n is large.

In any case, this leads to a natural question. What is the distribution of the cokernel of a random matrix from $\mathrm{Mat}_{n\times n+u}(\mathbb{Z}_p)$ taken with respect to Haar measure? Friedman and Washington [**55**] answered this question in the case $u = 0$. They were also motivated by the Cohen-Lenstra heuristics, but particularly the analog in which $\mathbb{Q}$ is replaced by $\mathbb{F}_q(t)$, in which one can write the class group as a cokernel of a different matrix involving the Frobenius.

Let N be a random matrix from $\mathrm{Mat}_{n\times n+u}(\mathbb{Z}_p)$ taken with respect to Haar measure. We wish to determine $\mathrm{Prob}(\mathrm{cok}(N) \simeq A)$ for each finite abelian p-group A, and we will follow the approach of [**78**] which does the analog for non-abelian profinite groups.

EXERCISE 3.1.5. *Show that if $u \geq 0$, then* $\mathrm{Prob}(\mathrm{cok}(N)$ *finite*$) = 1$.

Note that $\mathrm{Prob}(\mathrm{cok}(N) \simeq A) = |\,\mathrm{Aut}(A)|^{-1}\mathbb{E}(\#\,\mathrm{Isom}(\mathrm{cok}(N), A))$. (We use $\mathbb{E}$ to denote the expectation, or average, of a random real number.) Since $\mathrm{cok}\,N = \mathbb{Z}_p^n/N(\mathbb{Z}_p^{n+u})$, any isomorphism $\mathrm{cok}(N) \to A$ gives a distinct surjection of $\mathbb{Z}_p$-modules $\mathbb{Z}_p^n \to A$. So we consider a surjection $F : \mathbb{Z}_p^n \to A$, and ask, what is the probability that F descends to an isomorphism $\mathrm{cok}(N) \to A$? First, each column of N has to land in $\ker(F)$, and since each column is chosen independently from Haar measure on $\mathbb{Z}_p^n$, this happens with probability $[\mathbb{Z}_p^n : \ker F]^{n+u} = |A|^{n+u}$. The Haar measure on $\mathbb{Z}_p^n$ restricted to $\ker F$ is the Haar measure on $\ker F$. So the probability that F descends to an isomorphism $\mathrm{cok}(N) \to A$, conditioned on $FN = 0$ (i.e. each column of N is in $\ker F$), is the probability that $n+u$ independent elements from $\ker F \simeq \mathbb{Z}_p^n$ taken from Haar measure generate $\mathbb{Z}_p^n$. By Nakayma's lemma, this generation is equivalent to generation mod p, where the Haar measure becomes the uniform measure.

EXERCISE 3.1.6. *Show that the probability that $n+u$ uniform random elements of $(\mathbb{Z}/p\mathbb{Z})^n$ generate the group is*

$$\prod_{i=u+1}^{n+u}(1-p^{-i}).$$

EXERCISE 3.1.7. *If A is a finite abelian p-group of rank r (i.e. $A \otimes \mathbb{Z}/p\mathbb{Z} \simeq (\mathbb{Z}/p\mathbb{Z})^r$), show*

$$\#\operatorname{Sur}_{\mathbb{Z}_p}(\mathbb{Z}_p^n, A) = |A|^n \prod_{i=n-r+1}^{n} (1-p^{-i})$$

(where $\operatorname{Sur}_{\mathbb{Z}_p}$ denotes $\mathbb{Z}_p$-module surjections).

We conclude, that for A a finite abelian group, we have

(3.1.8)

$$\begin{aligned}
\operatorname{Prob}(\operatorname{cok}(N) \simeq A) &= |\operatorname{Aut}(A)|^{-1}\mathbb{E}(\#\operatorname{Isom}(\operatorname{cok}(N), A)) \\
&= |\operatorname{Aut}(A)|^{-1}|\#\operatorname{Sur}_{\mathbb{Z}_p}(\mathbb{Z}_p^n, A)||A|^{-(n+u)} \prod_{i=u+1}^{n+u} (1-p^{-i}) \\
&= |\operatorname{Aut}(A)|^{-1}|A|^n \prod_{i=n-r+1}^{n} (1-p^{-i})|A|^{-(n+u)} \prod_{i=u+1}^{n+u} (1-p^{-i}) \\
&= |\operatorname{Aut}(A)|^{-1}|A|^{-u} \prod_{i=u+1}^{n+u} (1-p^{-i}) \prod_{i=n-r+1}^{n} (1-p^{-i}).
\end{aligned}$$

We are interested in this value for a fixed A (and hence r), when n is large. So it is interesting to note that the above probabilities approach a limit as $n \to \infty$, and in particular approach

$$|\operatorname{Aut}(A)|^{-1}|A|^{-u} \prod_{i \geq u+1} (1-p^{-i}),$$

which are precisely the probabilities prediction by the Cohen-Lenstra heuristics for class groups of imaginary and real quadratic fields.

EXERCISE 3.1.9. *Find a formula for $|\operatorname{Aut}(G)|$ when G is an abelian p-group. [Hint: it is useful to choose $\lambda_1 \geq \lambda_2 \geq \cdots$ to denote the isomorphism type of the group, where $p^{i-1}G/p^iG \simeq (\mathbb{Z}/p\mathbb{Z})^{\lambda_i}$. As in the proof above, first count homomorphisms, and then restrict to counting those that are surjective.]*

3.2. Class group matrices. From the above analysis, it is tempting to make a conjecture about the distribution of the matrices M described in the last section as K varies over quadratic fields. In particular one imagines a conjecture that the M are approximately or asymptotically distributed with respect to Haar measure on p-adic matrices which would imply the Cohen-Lenstra conjecture for the distribution of class groups. There are several subtleties. One needs, for each K, to pick a set S of primes. One could take, for example, all primes of K up to norm B, where B is the Minkowski bound so that S necessarily generates Cl_K. One needs to pick bases for I^S and $\mathcal{O}_S^*/\mu_K$. Of course, there is a natural basis for I^S given by the elements of S.

QUESTION 3.2.1. *Is there a "natural" basis of $\mathcal{O}_S^*$?*

Then, one needs an equidistribution conjecture for elements in groups, where the groups themselves are changing (since $|S|$ will not be constant). Friedman and Washington [**55**, Section 2] suggest one way to do this.

Preliminary computations however, using choices as suggested above, with a basis of $\mathcal{O}_S^*$ given by the output of computer algebra computations of $\mathcal{O}_S^*$, suggest these matrices are quite far from equidistributed for Haar measure.

PROJECT 3.2.2. *Do such computations and see what the distribution of matrices arising looks like. Do they look like they are coming from Haar measure? Try other choices of setup e.g. fixing an n and then taking all K such that known bounds guarantee that the first n primes of K are sufficient to generate the class group, so one has many matrices of the same dimension.*

3.3. Universality. If the actual matrices defining Sylow p-subgroups of class groups as above are not distributed as in they are coming from Haar measure, does that mean that the Cohen-Lenstra conjectures are wrong? The answer is very much no. In fact, there is a *universality* phenomenon that says a very wide range of distributions of random matrices, asymptotically as the size of the matrices goes to infinity, have cokernel distribution approach the Cohen-Lenstra distribution. This universality is in the sense of the Central Limit Theorem.

THEOREM 3.3.1 (Central Limit Theorem). *Let $X_1, X_2, \ldots$ be independent, identically distributed random real numbers with finite mean $\mu = \mathbb{E}(X_i)$ and finite variance σ^2. Then as $n \to \infty$,*

$$\sqrt{n}\left(\frac{X_1 + \cdots + X_n}{n} - \mu\right)$$

converge in distribution to the normal distribution with mean 0 and variance σ^2.

The universality aspect of the Central Limit Theorem is that X_i can have almost any distribution. That includes the case when the X_i are normal, which is a highly symmetric distribution that we think of as analogous to Haar measure. However, the Central Limit Theorem also applies for X_i that are quite irregular. No matter what X_i you put in, the asymptotic output of this weighted averaging process is a normal distribution.

There is a morally similar result of the author for cokernels of random matrices.

THEOREM 3.3.2 ([**119**, Theorem 1.3]). *Let p be a prime, u be a non-negative integer and $\epsilon > 0$ be a real number. For each positive integer n, let $M(n)$ be a random matrix valued in $\mathrm{Mat}_{n\times n+u}(\mathbb{Z}_p)$ with independent entries. Further, for any entry $M(n)_{i,j}$ of any $M(n)$ and any $r \in \mathbb{Z}/p\mathbb{Z}$, we require that*

$$\mathbb{P}(M(n)_{i,j} \equiv r \pmod{p}) \leq 1 - \epsilon. \tag{3.3.3}$$

Then for any finite abelian p-group A,

$$\lim_{n\to\infty} \mathbb{P}(\mathrm{cok}(M(n)) \simeq A) = \frac{\prod_{i\geq 1}^{\infty}(1-p^{-i})}{|A|^u |\operatorname{Aut}(A)|}. \tag{3.3.4}$$

Theorem 3.3.2 allows great flexibility in the random matrices, in particular the matrix entries do not have to be identically distributed. The matrix entries are certainly not required to take every value mod p (and hence are not required to take every value mod p^k), but have a much weaker requirement that they not be entirely concentrated on one value mod p. It is easy to see some such condition is required, as the all 0's matrix is technically a random matrix with independent entries. The given hypothesis (3.3.3) can certainly be weakened, see [**89**, Theorem 4.1]). The independence of the entries is indeed a strong requirement, and something along those lines is of course necessary, but one could imagine weakening the hypothesis to some kind of asymptotic or approximate independence. See also [**120**, Section 3] for further exposition of this kind of universality phenomenon.

PROJECT 3.3.5. *Based on computations from 3.2.2, make an empirically supported conjecture about the distribution of the matrices M from imaginary and/or real quadratic fields. Can the methods of* [**119**] *be extended to show that matrices from your conjectured satisfy* (3.3.4)*?*

4. Moments of class group distributions

So far, the main function f we have been considering averaging in (2.1.2) (i.e. over class groups of number fields) has been the characteristic function of having a specified Sylow p-subgroup.

4.1. Definition of moments of a random finite abelian group. In fact, there is another class of important f, which are indexed by finite abelian groups B, where

$$f_B(X) := \#\operatorname{Sur}(X, B).$$

Let

$$E(f,S) := \lim_{X\to\infty} \frac{\sum_{\substack{K\in S\\ D_K\leq X}} f(\mathrm{Cl}_K)}{\sum_{\substack{K\in S\\ D_K\leq X}} 1}$$

be the empirical average of f over a family S of number fields, assuming the limit exists. (Showing that such a limit exists is quite difficult, and indeed is only known in the few cases where we know what the limit is.)

EXERCISE 4.1.1. *If B is a finite abelian group, show that the values of $E(\#\operatorname{Sur}(-,B'),S)$ over subgroups B' of B determine the value of $E(\#\operatorname{Hom}(-,B),S)$, and that the values of $E(\#\operatorname{Hom}(-,B'),S)$ over subgroups B' of B determine the value of $E(\#\operatorname{Sur}(-,B),S)$.*

EXERCISE 4.1.2. *If X is a finite abelian group of exponent p, show that $\#\operatorname{Hom}(X,(\mathbb{Z}/p\mathbb{Z})^k) = |X|^k$.*

This is reminiscent of certain functions $f_k(X) = X^k$, whose averages are important statistics of distributions of real numbers. If X is a random number, then $\mathbb{E}(X^k)$ is called the k-th moment of X or the distribution of X. These moments are often more accessible than direct information about a distribution. Yet wonderfully, knowledge of all of the moments often determines a distribution uniquely. For example, we have the following classical theorem on the moment problem.

THEOREM 4.1.3 (Carleman's condition). *Let X be a random real number such that $M_k = \mathbb{E}(X^k)$ is finite for all integers $k \geq 0$. Then if*

$$\sum_{k\geq 1} M_{2k}^{-\frac{1}{2k}} = \infty, \tag{4.1.4}$$

then there is a unique distribution for a random real number Y such that $\mathbb{E}(Y^k) = M_k$ for all $k \geq 0$.

EXERCISE 4.1.5. *Can you apply Theorem 4.1.3 for $M_k = e^k$? What about $M_k = e^{k^2}$?*

The averages of $\#\operatorname{Sur}(-, B)$ are important enough that they are called the *moments* of a distribution of finite abelian groups. The moments are indexed by finite abelian groups. If X is a random finite abelian group, then the average $\mathbb{E}(\#\operatorname{Sur}(X, B))$ is the B-moment of X, or of the distribution of X. As in Exercise 4.1.2, if we reduce a finite abelian group to a certain list of integer invariants, then the averages of $\#\operatorname{Hom}(-, B)$ are precisely the mixed moments of those integer invariants (in the classical sense of mixed moments) [**35**, Section 3.3]. Exercise 4.1.1 shows that as far as knowledge of the moments of a random variable (or even the limit of moments of a sequence of random variables) is concerned, it doesn't matter if we count homomorphisms or surjective homomorphisms. This is analogous to the relationship between the moments $\mathbb{E}(X^k)$ and the factorial moments $\mathbb{E}(X(X-1)\cdots(X-k+1))$ of a random number. One uses one version of the moments as opposed to the other generally when they make the computations and formulas simpler.

One important feature of these moments of finite abelian groups is that they determine a distribution uniquely (when they are not too large), analogously to Theorem 4.1.3.

THEOREM 4.1.6 ([**120**, Corollary 2.6]). *Let $M_A \in \mathbb{R}$ for each finite abelian group A such that there is a constant C such that $M_A \leq C|\wedge^2 A|$ for every A. Let X, Y be random finite abelian groups, If for every finite abelian group A, we have*

$$\mathbb{E}(\#\operatorname{Sur}(X, A)) = \mathbb{E}(\#\operatorname{Sur}(Y, A)) = M_A,$$

then X and Y have the same distribution, i.e. for every finite abelian group B,

$$\operatorname{Prob}(X \simeq B) = \operatorname{Prob}(Y \simeq B).$$

For our questions of distribution of class groups, we are never considering a single random group, but rather a sequence of random groups (indexed by X, where for each X we take a uniform random field with $D_K \leq X$). The averages $E(f, S)$ we consider are limits of averages. So we naturally ask when the limit moments of a sequence of random variables implies the sequence has a given limit distribution. This works well for finite abelian p-groups, but breaks down when we consider all finite abelian groups, at least for the most naive sense of limit distribution.

THEOREM 4.1.7 (see [**116**, Thm 8.3, proof of Cor 9.2]). *Let p be prime, and let $\mathcal{A}$ be the set of finite abelian p-groups. Let $M_A \in \mathbb{R}$ for each $A \in \mathcal{A}$ such that there is a constant C such that $M_A \leq C|\wedge^2 A|$ for all $A \in \mathcal{A}$. Let $Y, X_1, X_2, \ldots$ be random groups in $\mathcal{A}$. If for every $A \in \mathcal{A}$, we have*

$$\lim_{n\to\infty} \mathbb{E}(\#\operatorname{Sur}(X_n, A)) = \mathbb{E}(\#\operatorname{Sur}(Y, A)) = M_A,$$

then for every $B \in \mathcal{A}$,

$$\lim_{n\to\infty} \operatorname{Prob}(X_n \simeq B) = \operatorname{Prob}(Y \simeq B).$$

As in [**112**, Example 6.14], we can consider a random finite abelian group X, e.g. such that

$$\operatorname{Prob}(X \simeq A) = \frac{\zeta(2)^{-1}\zeta(3)^{-1}\zeta(4)^{-1}\cdots}{|A||\operatorname{Aut} A|}.$$

(There is a random group with this distribution–see e.g. [**117**, Proposition 2.1].) Then consider the random groups $X_p := X \times \mathbb{Z}/p\mathbb{Z}$ for each prime p.

EXERCISE 4.1.8. *For any finite abelian group A, show that*

$$\lim_{p\to\infty} \mathbb{E}(\#\operatorname{Sur}(X\times\mathbb{Z}/p\mathbb{Z}, A)) = \mathbb{E}(\#\operatorname{Sur}(X, A)).$$

What is $\lim_{n\to\infty}\operatorname{Prob}(X_n \simeq A)$*?*

While Theorem 4.1.6 shows that on finite abelian p-groups, limit moments contain enough information to determine a limit distribution, the converse is false.

EXERCISE 4.1.9. *Construct a sequence $Y, X_1, X_2, \dots$ of random finite abelian p-groups such that for every finite abelian p-group B we have*

$$\lim_{n\to\infty}\operatorname{Prob}(X_n \simeq B) = \operatorname{Prob}(Y \simeq B).$$

but yet for some finite abelian p-group A we have

$$\lim_{n\to\infty}\mathbb{E}(\#\operatorname{Sur}(X_n, A)) \neq \mathbb{E}(\#\operatorname{Sur}(Y, A)).$$

Show that you can do this for any Y with finite moments. Show that you can have the limit $\lim_{n\to\infty}\mathbb{E}(\#\operatorname{Sur}(X_n, A))$ *not exist.*

So, when we are taking limits of distributions of finite abelian p-groups, we see that these limiting moments are even stronger information than the limiting distributions. See [**120**, Section 2] and [**116**, Sections 1.5, 8] for further exposition on the moment problem for random groups and its applications to number theory.

PROJECT 4.1.10. *Similar to Project 2.3.5, using existing databases, make some graphs of how various moments of the class groups of imaginary (and/or) real quadratic fields of absolute discriminant converge to the predicted values. (See* [**84**] *and* [**118**, Appendix] *for some such computations.) Do any of your graphs (or the underlying numbers) suggest a rate at which the limit is approached? Do they suggest precise secondary terms?*

The importance of having conjectures for the rate of convergence or secondary terms has been discussed above for the average proportion of class groups having a fixed Sylow p-subgroup. In the case of moments, it is perhaps even more important, because convergence to moments is a more desirable target to check given that, as we will see below, the conjectural moment values are generally nice rational numbers (as opposed to the conjectural proportion values).

4.2. Relationship of moments to field counting. What about the aspect of moments that they are supposed to be easier to access? This turns out to also be the case for distributions of class groups. The only non-trivial f for which the averages (2.3.1) or (2.3.2) predicted by the Cohen-Lenstra heuristics for quadratic fields is known is $\#Sur(-, \mathbb{Z}/3\mathbb{Z})$. These averages are results of Davenport and Heilbronn [**45**], and predate the Cohen-Lenstra Heuristics. They turn out to be closely related to counting cubic fields. (See also [**68**] and [**115**, Section 3] for more discussion on the relationship between counting number fields and the Cohen-Lenstra heuristics, and [**92**] for a discussion of the relationship between both these questions and conjectures on upper bounds on class groups.)

Let A be an odd finite abelian group, and let $A \rtimes_{-1} \mathbb{Z}/2\mathbb{Z}$ be the semi-direct product where $\mathbb{Z}/2\mathbb{Z}$ acts on A by multiplication by -1.

EXERCISE 4.2.1. *Let K be a quadratic field and L/K and unramified abelian extension with* $\mathrm{Gal}(L/K) \simeq A$. *Show that $L/\mathbb{Q}$ is Galois and* $\mathrm{Gal}(L/K) \simeq A \rtimes_{-1} \mathbb{Z}/2\mathbb{Z}$. *(Don't forget that when $|A|$ is even there can be more than one group extension of $\mathbb{Z}/2\mathbb{Z}$ by A where the associated action is by -1.)*

EXERCISE 4.2.2. *Let A be an odd order finite abelian group. Let $L/\mathbb{Q}$ be Galois with Galois group $A \rtimes_{-1} \mathbb{Z}/2\mathbb{Z}$. Show that L contains a unique quadratic field K_L.*

Thus, for an odd order abelian group A, we have a correspondence between pairs (K, L) where K is an isomorphism class of quadratic fields and L/K is an isomorphism class of unramified abelian extensions of K with Galois group A and isomorphism classes of $L/\mathbb{Q}$ Galois with Galois group $A \rtimes_{-1} \mathbb{Z}/2\mathbb{Z}$ such that L is unramified over its unique quadratic subfield.

EXERCISE 4.2.3. *Show that this correspondence is* $2-1$. *(See Exercise* 1.2.27.*)*

Given K, isomorphism classes L/K of unramified abelian extensions of K with Galois group A correspond to $\mathrm{Aut}(A)$ orbits of surjections $\mathrm{Sur}(\mathrm{Cl}_K, A)$, via class field theory. Thus $\mathrm{Sur}(\mathrm{Cl}_K, A)$ has a natural meaning in terms of unramified extensions, which gives another motivation for the definition of moments above–we are averaging a count that is algebraically meaningful.

Let S be a set of quadratic fields K and A a finite abelian group A. Let T be the set of isomorphism classes of Galois $A \rtimes_{-1} \mathbb{Z}/2\mathbb{Z}$ extensions, unramified over their quadratic subfields, such that their quadratic subfield is in S. Then

$$\sum_{K \in S} \# \mathrm{Sur}(\mathrm{Cl}_K, A) = 2|\mathrm{Aut}(A)| \# T.$$

Thus the moments of class group distributions are closely related the the number field counting questions in the first section, for groups $G = A \rtimes_{-1} \mathbb{Z}/2\mathbb{Z}$ in their regular representation. The main difference is that we restrict ramification behavior (and since in class group questions we usually fix the signature of the quadratic field, this would also require an additional restriction on the counted G fields at infinity).

We can slightly shift the problem, by recognizing that since $\mathbb{Z}/2\mathbb{Z}$ is a subgroup of $A \rtimes_{-1} \mathbb{Z}/2\mathbb{Z}$, that a Galois $A \rtimes_{-1} \mathbb{Z}/2\mathbb{Z}$ extension $L/\mathbb{Q}$ also has a subfield F such that L/F is quadratic.

EXERCISE 4.2.4. *In the above situation, show that $F/\mathbb{Q}$ is not Galois, and that L is its Galois closure. In fact, there are many such subfields F, but show they are all isomorphic.*

EXERCISE 4.2.5 *In the above situation, find the Galois group* $\mathrm{Gal}(F/\mathbb{Q})$, *as a permutation group.*

EXERCISE 4.2.6. *In the above situation, show that the absolute discriminants have the property $D_F = D_{K_L}^{(|A|-1)/2}$. (Hint: it may be useful to use the fact that for a Galois extension $L/\mathbb{Q}$ with subgroup H, the discriminant* $\mathrm{Disc}\, L^H$ *of the fixed field is the Artin conductor of the permutation representation of G on H cosets* [**88**, VII:11.8].*)*

Thus one could also count $F/\mathbb{Q}$ with the Galois group above, with sufficient conditions on ramification and at infinity to ensure that the corresponding L and K_L are as desired. When $A = \mathbb{Z}/3\mathbb{Z}$, the corresponding F are precisely non-Galois cubic extensions such that no prime of $\mathbb{Q}$ ramifies completely in F.

EXERCISE 4.2.7. *Show that in the case $A = \mathbb{Z}/3\mathbb{Z}$, for a non-Galois cubic field F to have the corresponding L/K_L unramified, it is necessary and sufficient that no prime of $\mathbb{Q}$ ramifies completely in F.*

The approach of Davenport and Heilbronn [**45**] was sufficiently flexible to impose this ramification condition on cubic fields (it turns out to be a condition on the associated binary cubic form mod p^2 for each prime p, which is already the sort of condition they had to face to consider which forms actually corresponded to fields).

THEOREM 4.2.8 ([**45**, Theorem 3]). *We have*

$$\lim_{X\to\infty} \frac{\sum_{\substack{K\in \mathrm{IQ}\\ D_K\le X}} \#\operatorname{Sur}(\mathrm{Cl}_K, \mathbb{Z}/3\mathbb{Z}))}{\sum_{\substack{K\in \mathrm{IQ}\\ D_K\le X}} 1} = 1 \tag{4.2.9}$$

and

$$\lim_{X\to\infty} \frac{\sum_{\substack{K\in \mathrm{RQ}\\ D_K\le X}} \#\operatorname{Sur}(\mathrm{Cl}_K, \mathbb{Z}/3\mathbb{Z})}{\sum_{\substack{K\in \mathrm{RQ}\\ D_K\le X}} 1} = \frac{1}{3}. \tag{4.2.10}$$

Given the Cohen-Lenstra conjectures (2.3.1) and (2.3.2), this begs the question whether
(4.2.11)

$$\sum_{\substack{A \text{fin. ab.}\\ 3\text{-group}}} \frac{\#\operatorname{Sur}(A,\mathbb{Z}/3\mathbb{Z})}{|\operatorname{Aut}(A)|}\prod_{i\ge 1}(1-3^{-i}) = 1 \quad \text{and} \quad \sum_{\substack{A \text{fin. ab.}\\ 3\text{-group}}} \frac{\#\operatorname{Sur}(A,\mathbb{Z}/3\mathbb{Z})}{|A||\operatorname{Aut}(A)|}\prod_{i\ge 2}(1-3^{-i}) = \frac{1}{3}?$$

These two equalities are indeed true (recall the the work of Davenport and Heilbronn predated the work of Cohen and Lenstra). In fact, more generally we have for a prime p and finite abelian p-group B

$$\sum_{\substack{A \text{fin. ab.}\\ p\text{-group}}} \frac{\#\operatorname{Sur}(A,B)}{|\operatorname{Aut}(A)|}\prod_{i\ge 1}(1-p^{-i}) = 1 \quad \text{and} \quad \sum_{\substack{A \text{fin. ab.}\\ p\text{-group}}} \frac{\#\operatorname{Sur}(A,B)}{|A||\operatorname{Aut}(A)|}\prod_{i\ge 2}(1-p^{-i}) = \frac{1}{|B|},$$

and the Cohen-Lenstra heuristics predict the corresponding averages for the class groups of imaginary and real quadratic fields, respectively. Since we know all finite abelian p-groups, we can express the left-hand sides above as a formal expression in p. Cohen and Lenstra developed a generating function machinery that allows one to compute such group theoretic averages (see [**39**, Section 9 (C6),(C9)] and [**119**, Lemma 3.2]). However, perhaps the most direct path to the computations above is to consider the random matrix models of Section 3.

EXERCISE 4.2.12. *Let N be a random matrix from $\operatorname{Mat}_{n\times n+u}(\mathbb{Z}_p)$ (for $u \ge 0$) taken with respect to Haar measure. For a finite abelian p-group B, find $\mathbb{E}(\#\operatorname{Sur}(\operatorname{cok}(N), B)$ and show*

$$\lim_{n\to\infty} \mathbb{E}(\#\operatorname{Sur}(\operatorname{cok}(N), B) = |B|^{-u}.$$

[Hint: this is an easier version of the computation of $\mathbb{E}(\#\operatorname{Isom}(\operatorname{cok}(N), B)$ we did in Section 3.]

Thus we have

$$\lim_{n\to\infty} \sum_{\substack{A \text{ fin. ab.}\\ p\text{-group}}} \operatorname{Prob}(\operatorname{cok}(N) \simeq A) \# \operatorname{Sur}(A, B) = |B|^{-u}. \tag{4.2.13}$$

Since we know

$$\lim_{n\to\infty} \operatorname{Prob}(\operatorname{cok}(N) \simeq A) = \frac{\prod_{i\geq u+1}(1-p^{-i})}{|A|^u |\operatorname{Aut} A|}, \tag{4.2.14}$$

if we can exchange the limit in n with the sum over A, we would conclude

$$\sum_{\substack{A \text{ fin. ab.}\\ p\text{-group}}} \frac{\# \operatorname{Sur}(A, B)}{|A|^u |\operatorname{Aut} A|} \prod_{i\geq u+1} (1-p^{-i}) = |B|^{-u}. \tag{4.2.15}$$

While such an exchange is not always possible, here it can be justified. From (3.1.8), we have

(4.2.16)

$$\frac{\operatorname{Prob}(\operatorname{cok}(N) \simeq A)}{\prod_{i=u+1}^{n+u}(1-p^{-i})} \# \operatorname{Sur}(A, B) = |\operatorname{Aut}(A)|^{-1} |A|^{-u} \prod_{i=n-r+1}^{n} (1-p^{-i}) \# \operatorname{Sur}(A, B).$$

Observe that the expression on the right is increasing in n. Thus by the monotone convergence theorem, we have

$$\lim_{n\to\infty} \sum_{\substack{A \text{ fin. ab.}\\ p\text{-group}}} \frac{\operatorname{Prob}(\operatorname{cok}(N) \simeq A)}{\prod_{i=u+1}^{n+u}(1-p^{-i})} \# \operatorname{Sur}(A, B) = \sum_{\substack{A \text{ fin. ab.}\\ p\text{-group}}} \lim_{n\to\infty} \frac{\operatorname{Prob}(\operatorname{cok}(N) \simeq A)}{\prod_{i=u+1}^{n+u}(1-p^{-i})} \# \operatorname{Sur}(A, B).$$

Since $\prod_{i=u+1}^{n+u}(1-p^{-i})$ doesn't depend on A, it (along with its limit) can pull out of the sum and we have

$$\lim_{n\to\infty} \sum_{\substack{A \text{ fin. ab.}\\ p\text{-group}}} \operatorname{Prob}(\operatorname{cok}(N) \simeq A) \# \operatorname{Sur}(A, B) = \sum_{\substack{A \text{ fin. ab.}\\ p\text{-group}}} \lim_{n\to\infty} \operatorname{Prob}(\operatorname{cok}(N) \simeq A) \# \operatorname{Sur}(A, B),$$

which, with (4.2.13) and (4.2.14), implies (4.2.15).

Theorem 4.1.7 then implies that Cohen and Lenstra's conjectural distribution for Sylow p-subgroups of class groups of imaginary, respectively real, quadratic fields is the unique distribution on finite abelian p-groups with moments all 1, respectively with B-moment $|B|^{-1}$.

EXERCISE 4.2.17. *Using genus theory (and some results from analytic number theory of the integers that you might have to look up), show that*

$$\lim_{X\to\infty} \frac{\sum_{\substack{K\in \mathrm{RQ}\\ D_K \leq X}} \# \operatorname{Sur}(\mathrm{Cl}_K, \mathbb{Z}/2\mathbb{Z})}{\sum_{\substack{K\in \mathrm{RQ}\\ D_K \leq X}} 1} = \lim_{X\to\infty} \frac{\sum_{\substack{K\in \mathrm{RQ}\\ D_K \leq X}} \# \operatorname{Sur}(\mathrm{Cl}_K, \mathbb{Z}/2\mathbb{Z})}{\sum_{\substack{K\in \mathrm{RQ}\\ D_K \leq X}} 1} = \infty$$

This shows another, quantitative, way in which the 2-Sylow subgroups of class groups of quadratic fields behave differently than the (conjectural) behavior of the odd part of the class groups.

5. Cohen-Lenstra-Martinet heuristics

What about the distribution of class groups of number fields of higher degree than 2? Cohen and Lenstra [**39**] actually gave conjectures on class groups distributions for totally real G-extensions for any abelian group G, and Cohen and Martinet [**40**] gave conjectures on class groups distributions for G-extensions for any transitive permutation group G and any signature. See also [**112**] for an exposition of these conjectures as well as more direct, equivalent forms of the conjectures.

5.1. Class groups of Galois extensions. Let G be a finite simply transitive permutation group, so in particular G-extensions are Galois over $\mathbb{Q}$. We will consider class groups of G-extensions that come with an isomorphism $\mathrm{Gal}(K/\mathbb{Q}) \simeq G$, and call these pairs of a field and an isomorphism *G-Extensions*. If K is a G-Extension, then Cl_K is a G-module (i.e. a module for the group ring $\mathbb{Z}[G]$). Note that for Cl_K to be a G-module, and not just a $\mathrm{Gal}(K/\mathbb{Q})$-module, we do require the data of the isomorphism $\mathrm{Gal}(K/\mathbb{Q}) \simeq G$. It does not make any sense, as pointed out in [**10**], to consider the class groups of G-modules if we do not choose such an isomorphism.

The conjectural probabilities in (2.3.1) and (2.3.2) involve $|\mathrm{Aut}(A)|$. Part of the philosophy of Cohen-Lenstra-Martinet is what we should always consider objects with all available structure. After all, we consider the automorphisms of A as a group, not a set. So the probabilities for class groups of G-Extensions should involve $|\mathrm{Aut}_G(A)|$, the number of automorphisms of A as a G-module.

Which finite G-modules can arise?

EXERCISE 5.1.1. *Show that* $N := \sum_{g\in G} G$ *acts as zero on* Cl_K, *i.e.* $N\,\mathrm{Cl}_K = 0$.

So we let $R = \mathbb{Z}[\mathbb{G}]/N\mathbb{Z}[\mathbb{G}]$, and have that Cl_K is a finite R-module.

EXERCISE 5.1.2. *For* $G = \mathbb{Z}/3\mathbb{Z}$, *find an example of finite abelian group that can't be the class group of a* G*-extension, because it doesn't have the structure of an* R*-module.*

EXERCISE 5.1.3. *For a finite* R*-module* A, *show that* $|\mathrm{Aut}_G(A)| = |\mathrm{Aut}_R(A)|$, *where* $\mathrm{Aut}_R(A)$ *is the group of automorphisms of* A *as an* R*-module.*

What then is the analog of the $|A|^u$ factor? In the case of quadratic fields, we saw that the actual value of this factor (i.e. the value of u), was different in the different families of fields depending on their behavior at infinity. Fix a subgroup G_∞ of G. Let S_{G,G_∞} be the set of isomorphism classes of G-Extensions of $\mathbb{Q}$ (where isomorphisms are given by isomorphisms of the field that respect the identification of the Galois group with G) whose decomposition group at infinity is conjugate to G_∞ (i.e. so that complex conjugation generates a subgroup conjugate to G_∞).

EXERCISE 5.1.4. *Find the signature and the unit rank of a* $K \in S_{G,G_\infty}$.

Let p be a prime such that $p \nmid |G|$. We will restrict to Sylow p-subgroups of class groups for simplicity. Cohen and Martinet's heuristics [**40**, Hypothesis 6.6] give a conjecture on the distribution of Sylow p-subgroups class groups of G-extensions that takes quite a bit of notation to write down. However, Wang and the author [**112**, Theorems 1.1 and 4.1, Proposition 6.6] prove that it is equivalent to conjecturing that for "reasonable" functions f from the set of finite p-group

R-modules to $\mathbb{R}$ that we have

$$(5.1.5)\qquad \lim_{X\to\infty} \frac{\sum_{\substack{K\in S_{G,G_\infty}\\ D_K\leq X}} f(\mathrm{Cl}_K)}{\sum_{\substack{K\in S_{G,G_\infty}\\ D_K\leq X}} 1} = \frac{\sum_{\substack{A \text{fin. } p\text{-group}\\ R\text{-module}}} \frac{f(A)}{|A^{G_\infty}||\operatorname{Aut}_R(A)|}}{\sum_{\substack{A \text{fin. } p\text{-group}\\ R\text{-module}}} \frac{1}{|A^{G_\infty}||\operatorname{Aut}_R(A)|}}.$$

Here A^{G_∞} is the subgroup of elements of A that are fixed by every element of G_∞.

As before, the denominator of the right-hand side of (5.1.5) is a convergent sum and can be given explicitly (see [**112**, Remark 3.4], [**40**, Theorem 3.6]).

Of course, these conjectures should specialize to the conjectures of Cohen and Lenstra (2.3.1) and (2.3.2) for quadratic extensions.

EXERCISE 5.1.6. *Check* $|A^{G_\infty}|$ *is consistent with the* $|A|^u$ *terms in the quadratic conjectures* (2.3.1) *and* (2.3.2).

When $G = S_2$, then $R = \mathbb{Z}[\sigma]/(1+\sigma) = \mathbb{Z}$. So $\operatorname{Aut}(A) = \operatorname{Aut}_R(A)$. This echoes the fact, that we saw in Exercise 2.2.5, that the Galois group acts at -1 on the class group in this case, which is not really any "additional structure" on an abelian group.

EXERCISE 5.1.7. *Let* $G = \mathbb{Z}/\ell\mathbb{Z}$ *for a prime* ℓ. *What is* R *in this case? What are all the* p*-group* R*-modules?*

EXERCISE 5.1.8. *Let* $G = \mathbb{Z}/\ell\mathbb{Z}$ *for a prime* ℓ. *Find a formula for* $|\operatorname{Aut}_R(A)|$ *when* A *is a* p*-group* R*-module. [Hint: As in Exercise* 3.1.9 *it is useful denote the isomorphism type of the* R*-module by a set of partitions indexed by the primes* $\wp$ *of* R *dividing* p*, where if* λ *is the partition corresponding to* $\wp$*, then* $\wp^{i-1}A/\wp^i A \simeq (R/\wp)^{\lambda_i}$*.]*

There were some class group tables available for Cohen and Martinet to check their conjectures against, but not nearly as many as in the case of quadratic fields. Also, it is often difficult to distinguish between slow convergence to the conjectured values and convergence that is contradicting the conjectures. Malle [**84**, **85**] later was able to compute significantly more data on class groups, and was able to discover compelling evidence from his tables that for $|G|$ odd and $p = 2$ the conjecture (5.1.5) is likely incorrect. These conjectures are expected to come from roots of unity in the base field—in this case $\{\pm 1\} \subset \mathbb{Q}$. (This expected relationship to roots of unity comes also from further computations, including computations of Malle over other base fields, theoretical considerations, as well as function field results including [**1**, **50**]). The fact that $|\{\pm 1\}| = 2$ is why $p = 2$ is a problem, and when $|G|$ is even, $p = 2$ was already excluded for other reasons. Several works have addressed the question of how to modify (5.1.5) when $p = 2$ and $|G|$ is odd or analogous situations over base fields other than $\mathbb{Q}$ [**3**, **57**, **76**, **77**].

As in the case of quadratic extensions, there are certain functions f whose averages are particularly useful and more accessible. For a finite R-module B, the average of $\#\operatorname{Sur}_R(-, B)$ over a distribution is the B-moment of a distribution of R-modules, or for a random R-module X, we say that

$$E(\#\operatorname{Sur}_R(X, B))$$

is its B-moment. As in the case of conjectures for quadratic extensions, these moments have nice values over the distribution from the right-hand side of (5.1.5).

THEOREM 5.1.9 ([**112**, Theorem 6.2]). *For any finite group G, prime $p \nmid |G|$, subgroup G_∞ of G, and p-group R-module B, we have*

$$\frac{\sum_{\substack{A \text{ fin. } p\text{-group} \\ R\text{-module}}} \frac{\#\operatorname{Sur}_R(A,B))}{|A^{G_\infty}||\operatorname{Aut}_R(A)|}}{\sum_{\substack{A \text{ fin. } p\text{-group} \\ R\text{-module}}} \frac{1}{|A^{G_\infty}||\operatorname{Aut}_R(A)|}} = \frac{1}{|B^{G_\infty}|}$$

EXERCISE 5.1.10. *For $G = \mathbb{Z}/\ell\mathbb{Z}$ for a prime ℓ, prove Theorem 5.1.9 using a strategy similar to what we used to prove (4.2.15).*

Moreover, as in Theorem 4.1.7, these moments determine a unique distribution on p-group R-modules, even through a limit [**112**, Theorem 6.12].

PROJECT 5.1.11. *Replicate and extend Malle's empirical class group distribution computations for $G = \mathbb{Z}/3\mathbb{Z}, \mathbb{Z}/5\mathbb{Z}$, and/or $\mathbb{Z}/7\mathbb{Z}$ [**84**, Sections 2,3] and [**85**, Section 6.1], i.e. over a large number of appropriate fields compute the class group distribution and first few moments (see the subsection below on different notions of moments). Can you see the bad behavior at $p = 2$? At other p, do you see convergence towards the predicted value? Can you make a conjecture about a rate of convergence, second term, or error term? Can you implement a sampling technique to randomly sample fields up to a given discriminant instead of considering all fields? What is the trade-off in computation time and accuracy?*

PROJECT 5.1.12. *The same as Project 5.1.11 except for $G = \mathbb{Z}/2\mathbb{Z} \times \mathbb{Z}/2\mathbb{Z}$ (and ignore $p = 2$, but don't forget to consider different G_∞).*

PROJECT 5.1.13. *For $G = \mathbb{Z}/2\mathbb{Z} \times \mathbb{Z}/2\mathbb{Z}$, and an odd prime p, use Exercise 5.2.2 to show that for a G-Extension K, that $\mathrm{Cl}_{K,p}$ is determined by the Sylow p-subgroups of the class groups of the three quadratic subextensions (indeed, it is their product). Can this be used to show that (5.1.5) for $G = \mathbb{Z}/2\mathbb{Z} \times \mathbb{Z}/2\mathbb{Z}$ follows from (2.3.1) and (2.3.2)? Or how could you strengthen the latter so that it would?*

For many G, the conjecture (5.1.5) is probably false even at odd p, for a different reason. When asymptotically counting fields by discriminant, for many G, a positive proportion of the fields contain some fixed subfield. (See [**113**, **121**] for thorough discussion of when this happens for abelian G.) When this is the case, the class group of that fixed subfield has an outsize contribution to the entire class group average, which tends to violate the conjectures.

EXERCISE 5.1.14. *Let K be a quadratic field in which every ramified prime is 1 mod 4. Use the class field theory methods above to show that a positive proportion of $\mathbb{Z}/4\mathbb{Z}$-extensions, asymptotically by discriminant, contain K.*

Bartel and Lenstra [**10**] use this to show a concrete counterexample to (5.1.5) for $G = \mathbb{Z}/4\mathbb{Z}$ and f the characteristic function of having 3 divide the order of the group. Based on work in [**113**] and [**10**, Proposition 6.6], they propose replacing the absolute discriminant D_K in (5.1.5) with P_K, the product of the ramified primes in K.

EXERCISE 5.1.15. *Find the asymptotics of $G = \mathbb{Z}/2\mathbb{Z}$ and $G = \mathbb{Z}/3\mathbb{Z}$ fields by P_K instead of D_K.*

EXERCISE 5.1.16. *When $G = \mathbb{Z}/4\mathbb{Z}$-extensions are counted asymptotically by P_K, show that there does not exist a quadratic field that is a subfield of a positive proportion of the $\mathbb{Z}/4\mathbb{Z}$-extensions.*

PROJECT 5.1.17. *Do computations as in Project* 5.1.11 *except for* $G = \mathbb{Z}/4\mathbb{Z}$. *Do some tables by* D_K *and some by* P_K. *Can you see the difference? Also try ordering fields by* P^2Q, *where* P *is the product of partially ramified primes and* Q *is the product of totally ramified primes. (For this ordering, show the analog of Exercise* 5.1.16.*)*

The results of Liu, Zureick-Brown, and the author in [**79**] prove that in the analogous problem for class groups of (totally split at infinity) G-extensions of $\mathbb{F}_q(t)$, ordered by product of ramified primes, for Sylow p-subgroups of the class groups for $p \nmid q(q-1)|G|$, with a $q \to \infty$ limit before the $X \to \infty$ limit, the moments are as predicted by (5.1.5). This, along with the result of Wang and the author [**112**] that these moments determine a unique distribution, suggests that "roots of unity" and choice of ordering invariant are the two main issues to be fixed with the Cohen-Lenstra-Martinet heuristics at $p \nmid |G|$.

Different notions of moments.

EXERCISE 5.1.18. *Malle* [**84, 85**] *computes the classical moments of the size of the p-torsion of a random group* X, *i.e. the average of* $|X[p]|^k$. *How is this related to our* $(\mathbb{Z}/p\mathbb{Z})^j$*-moments for a random finite abelian group? [Hint: first consider the average of* $\mathrm{Hom}(-, (\mathbb{Z}/p\mathbb{Z})^k)$*]*

A p-group R-module has an underlying abelian p-group. So when p-group R-module moments (averages of $\#Sur_R(-, B)$) determine a unique distribution of p-group R-modules, they necessarily determine a unique distribution of abelian p-groups. Then (without any limits), that must uniquely determine the abelian p-group moments (averages of $\#Sur(-, B)$). However, we can go more directly from p-group R-module moments to p-group moments.

EXERCISE 5.1.19. *If* M *is a p-group R-module, and* B *is a finite abelian p group, express* $\#Sur(M, B)$ *in terms of* $\#Sur_R(M, B_i)$ *for various R-modules* B_i *that only depend on* B. *Use this to relate the two kind of moments as suggested above.*

EXERCISE 5.1.20. *When* $G = \mathbb{Z}/\ell\mathbb{Z}$, *use the two exercises above and Theorem* 5.1.9 *to find the predicted (from* (5.1.5)*) average of* $|\mathrm{Cl}_K[p]|^k$ *over G-Extensions* K.

5.2. Class groups of non-Galois extensions. Now we consider G a finite transitive permutation group that is not necessary simply transitive, with $H \subset G$ the stabilizer of an element. We will be considering G-Extensions, which are not necessarily Galois. We can still define S_{G,G_∞} as above.

EXERCISE 5.2.1. *Let* K *be a G-Extension, and* L *its Galois closure over* $\mathbb{Q}$. *Let* $H \subset G$ *be the stabilizer of an element. Show* $\mathrm{Gal}(L/K) \simeq H$, *as groups.*

EXERCISE 5.2.2. *Let* $L/\mathbb{Q}$ *be a Galois extension with Galois group* G. *Let* H *be a subgroup of* G *and* K *be the fixed field* L^H. *Show that the inclusion map*

$$\mathrm{Cl}_K \to \mathrm{Cl}_L^H$$

is an isomorphism on Sylow p-subgroups for any $p \nmid |H|$. *[Hint: consider to norm map* $\mathrm{Cl}_L \to \mathrm{Cl}_K$ *as well.]*

Even though Cohen and Martinet do not make a general conjecture for non-Galois extensions, for $p \nmid |G|$, (5.1.5) implies a conjecture on the distribution of

$\mathrm{Cl}_{K,p}$ by pushing forward the distribution from (5.1.5) along taking H-invariants. Said another way, one can consider functions $f(A)$ that only depend on A^H in (5.1.5) and have a conjecture of the distribution of $\mathrm{Cl}_{K,p}$ for G-Extensions, with the minor caveat that (5.1.5) orders fields by the discriminant of the Galois closure and not the discriminant of the G-Extensions. (We say this is minor not because there is any clear way to go from one such average to the other, but rather because it seems there is no strong a priori reason to prefer one ordering of the fields over the other.) Explicit formulas for this pushforward are given in [**112**, Theorem 8.14], and they are indeed quite similar to (2.3.1), (2.3.2), (5.1.5). We will now describe some aspects of those formulas.

Absolutely irreducible permutation groups. A finite permutation group G on X naturally gives an $|X|$-dimensional representation of G over $\mathbb{C}$, where G acts on the basis vectors via its action on X. The associated representation always contains one copy of the trivial representation (spanned by the sum of the basis vectors), and let V_G be the quotient of this representation by that 1-dimensional trivial representation. We say G is *absolutely irreducible* if V_G is. (The *absolutely* emphasizes that we considered the representation over $\mathbb{C}$ and not $\mathbb{Q}$).

For example, if $G = S_n$ in its usual degree n permutation representation, then G is absolutely irreducible.

Let G be an absolutely irreducible finite permutation group, p a prime not diving $|G|$, and G_∞ a subgroup of G. Let $u+1$ be the number of cycles in the action of G_∞ on the underlying set. Then we have that Cohen and Martinet's conjecture (5.1.5) implies that for "reasonable" f on the set of finite abelian p-groups, we have

$$\lim_{X\to\infty} \frac{\sum_{\substack{K\in S_{G,G_\infty}\\ D_{\tilde{K}}\le X}} f(\mathrm{Cl}_K)}{\sum_{\substack{K\in S_{G,G_\infty}\\ D_{\tilde{K}}\le X}} 1} = \frac{\sum_{\substack{A \text{fin. ab.}\\ p\text{-group}}} \frac{f(A)}{|A|^u|\operatorname{Aut}(A)|}}{\sum_{\substack{A \text{fin. ab.}\\ p\text{-group}}} \frac{1}{|A|^u|\operatorname{Aut}(A)|}}. \tag{5.2.3}$$

This implication follows from [**112**, Theorem 8.14, Proposition 8.16]. (In this deduction, the absolute irreducibility implies $k = 2$ and $a_{\Gamma/\Gamma'} = \phi_2$; so, $\langle\phi_2, a_\Gamma\rangle = \dim\phi_2 = \dim a_{\Gamma/\Gamma'}$ and $\langle\phi_2, a_{\Gamma/\Gamma'}\rangle = 1$ and $\langle\chi_K,\phi_2\rangle = \langle \mathrm{Ind}_{\Gamma_\infty}^{\Gamma}\mathbb{C},\phi_2\rangle = \langle\mathbb{C}, \mathrm{Res}_{\Gamma_\infty}^{\Gamma}\phi_2\rangle = \langle\mathbb{C}, V_G\rangle_{\Gamma_\infty} = u$ and $h_i = \dim a_{\Gamma/\Gamma'}$.)

We have already seen in (4.2.15) the moments of the distributions above, and we know from Theorem 4.1.7, that they determine a unique distribution.

The same caveats about avoiding $p = 2$ (because of the roots of unity in $\mathbb{Q}$) and counting by P_K instead of D_k apply to (5.2.3) as they did to (5.1.5).

EXERCISE 5.2.4. *Show that for $K \in S_{G,G_\infty}$ that u (the number of cycles in the action of G_∞ on the underlying set, minus 1) is the rank of the group of units of K.*

PROJECT 5.2.5. *Do computations as in* 5.1.11 *for $G = S_3, S_4$, and/or S_5 in their standard permutation representations, for $p \nmid |G|$. Malle has done some computations for $G = S_3$ in* [**84**, Section 5] *and* [**85**, Section 5], *and you will likely want to use Belabas's CUBIC program* [**12**] *for S_3. For quartic and quintic fields, you could do some computations with existing databases, or build on the suggested direction above to find new field enumeration algorithms for S_4 and S_5 fields based on Bhargava's parametrizations.*

Beyond absolutely irreducible permutation groups. A takeaway from the above conjecture is that we conjecture when G is absolutely irreducible, then there is no "additional structure" on the class groups of G-Extensions (at least for Sylow p-subgroups for $p \nmid 2|G|$). However, other non-Galois G-extensions do have extra structure on their class groups.

Consider the case when $G = D_4$ with permutation action as automorphisms of a square acting on its four vertices. If K is a G-Extension, then $\mathrm{Aut}(K) = \mathbb{Z}/2\mathbb{Z}$. Of course, $\mathrm{Aut}(K)$ acts on Cl_K, so even though K is not Galois, Cl_K has additional structure.

However, also consider the following example from [**112**, Example 8.20]. Let $G = A_5$ in the transitive degree 10 action with $H = \langle (123), (12)(45) \rangle$ (as abstract groups $H \simeq S_3$).

EXERCISE 5.2.6. *For the G above, show that a G-extension K has no non-trivial automorphisms (e.g. by using Exercise* 1.1.12*).*

Yet, if K is a G-Extension, for this $G = A_5$, and p a prime > 5, then $\mathrm{Cl}_{K,p}$ is naturally a module for a ring larger than $\mathbb{Z}$. Indeed, this happens when G is not absolutely irreducible [**112**, Lemma 8.4,Proposition 8.16].

Let H be a subgroup of a finite group G. Let $e = \frac{1}{|H|}\sum_{h \in H} h$ be an element of the group algebra $\mathbb{Q}[G]$. We can consider the $\mathbb{Z}$-algebra $e\mathbb{Z}[G]e$. (Note $e\mathbb{Z}[G]e$ is additively and multiplicatively closed in $\mathbb{Q}[G]$ though we usually would not call it a subalgebra because it does not have the same unit.)

EXERCISE 5.2.7. *Show that*

$$e\mathbb{Z}[G]e \otimes_{\mathbb{Z}} \mathbb{Q} \simeq \mathbb{Q}[H\backslash G/H],$$

where $\mathbb{Q}[H\backslash G/H]$ is the Hecke algebra of a finite group.

EXERCISE 5.2.8. *Let A be a finite abelian G-module of order prime to $|G|$. Show that $e\mathbb{Z}[G]e$ acts on the invariants A^H making A^H into a module for $e\mathbb{Z}[G]e$. [Hint:for $g \in G$ and $a \in A^H$, let $ege = egea$, where the action all happens in A and the $|H|^{-1}$ acts by an integer equivalent to it mod $|A|$.]*

EXERCISE 5.2.9. *Let L be a Galois field with Galois group G. Let K be the fixed field of H. Let p be a prime not dividing $|G|$. Where N is defined as in Section* 5.1*, show that $e\mathbb{Z}[G]e/Ne\mathbb{Z}[G]e$ acts on the Sylow p-subgroup of Cl_K.*

See [**112**, Section 8] for a more complete explanation of this $\mathfrak{o} := e\mathbb{Z}[G]e/Ne\mathbb{Z}[G]e$ action on $\mathrm{Cl}_{K,p}$, and in particular the proof that $\mathfrak{o}$ is not $\mathbb{Z}$ when G is not absolutely irreducible [**112**, Proposition 8.16], and the precise formula for the pushforward of the conjectural Galois distribution on class groups from (5.1.5) along $A \mapsto A^H$, which can be given in terms of $|\mathrm{Aut}_{\mathfrak{o}}(B)|$ [**112**, Theorem 8.14].

PROJECT 5.2.10. *Do computations for $G = D_4$ in its transitive degree 4 permutation representation, i.e. on quartic D_4 fields, as in Project* 5.1.11*. Order fields by discriminant, product of ramified primes, and also the Artin conductor of the 2-dimensional irreducible representation of D_4 (see [**7**] for asymptotic counting of D_4-extensions by this invariant). How do these different orders change the invariants? Note that a positive proportion of D_4-extensions, when counted by discriminant, include a given quadratic field (which we can see from our asymptotic count above).*

There is one further interesting feature in the non-Galois case, which is that predictions can be made for some $p \mid |G|$. See [**112**, Section 7-8] for further discussion of how this works, which is beyond the scope of these notes.

Previous computations [**41**] suggested that conjectures on the 2-Sylow subgroup of class groups of non-Galois cubic fields might be wrong, which may have been because these computations were just seeing a large secondary or true error term, or may have been an indication that the conjectures were wrong, which might be coming from the roots of unity in $\mathbb{Q}$. While Bhargava has proved one average [**16**] on the 2-Sylow subgroup of class groups of cubic fields, it is still quite open what the general distribution is.

PROJECT 5.2.11. *Do computations for $G = S_3$ in its standard representation and $p = 2$, as in Project* 5.1.11. *Do the same for $G = A_4$ and/or S_4 and $p = 3$. Do the values look like they approach the values in* (5.2.3) *and moments as in* (4.2.15) *(which are the values predicted by Cohen and Martinet, even for these particular cases where $p \mid |G|$)?*

PROJECT 5.2.12. *Do computations for $G = A_5$ and/or S_5 in its standard representation and $p = 3$, as in Project* 5.1.11. *Do the same for $G = A_4$ and/or S_4 and $p = 3$. In these cases, the conjectures of Cohen and Martinet make no prediction. Can you guess a pattern from the empirical values?*

5.3. Generalizations and known results. There are of course more generalizations of the questions of class group distribution. We do not attempt to give a complete literature review, but give pointers to examples of recent work if various directions. One can replace the base field $\mathbb{Q}$ with another number field K_0, or even a global field such as a finite extension of $\mathbb{F}_q(t)$. For $G = \mathbb{Z}/p\mathbb{Z}$ for a prime p, Gerth [**58, 59**] has extended the conjectures to include conjectures on the Sylow p-subgroup.

Even within the scope of all these generalizations, there are few known class group averages. Datskovsky and Wright [**44**] found the average of $\mathrm{Sur}(\mathrm{Cl}_K, \mathbb{Z}/3\mathbb{Z})$ over quadratic extensions of a general global field of characteristic not 2 or 3. Bhargava [**16**] found the average of $\mathrm{Sur}(\mathrm{Cl}_K, \mathbb{Z}/2\mathbb{Z})$ over imaginary, respectively real, non-Galois cubic fields K. The work of Bhargava, Shankar, and Wang [**23**] address the analog of this average over general global fields. For G any transitive permutation 2-group containing a transposition, Lemke Oliver, Wang, and the author [**91**] found the average of $\mathrm{Sur}(\mathrm{Cl}_K, \mathbb{Z}/3\mathbb{Z})$ over G-extensions K/K_0 for any base number field K_0. We also find the $\mathbb{Z}/3\mathbb{Z}$-moments of certain relative class groups of these G-extensions. The method in [**91**] can be used to give these averages for many other G of the form $\mathbb{Z}/2\mathbb{Z} \wr H$, see [**91**, Section 8]. When $G = \mathbb{Z}/4\mathbb{Z}$ and f is bounded and only depends on the class group of the quadratic subfield of the cyclic quartic of a quartic G-extension K, Bartel and Lenstra [**10**] showed that the average of $f(\mathrm{Cl}_K)$ over G-extensions K is computable in finite time to arbitrary precision. These are the known averages on the part of the class group of order prime to the degree of the extensions being averaged over. However, Alex Smith [**100, 101**] has groundbreaking work that has proven the entirety of Gerth's conjecture on the distribution of $2\,\mathrm{Cl}_{K,2} = \mathrm{Cl}_{K,2} / \mathrm{Cl}_K[2]$ for quadratic fields, as well as Gerth's analogous conjecture on $\mathrm{Cl}_{K,p}$ over cyclic p-extension K, improving earlier results [**51, 71**]. When the base field is $\mathbb{F}_q(t)$ and one introduces an additional limit $q \to \infty$ into the question, there are many results [**1, 2, 50, 79, 117**] and recently these function field theorems have played a large role in guiding conjectures in the area.

One can also ask about refinements of class groups, such as ray class groups [**34**, **47**, **107**] or Arakelov class groups [**10**], or pointed groups that specify the element in the class group corresponding to certain ideals [**117**]. One can restrict the fields averaged over by imposing local conditions at finite primes [**27**,**28**,**107**,**117**], or enlarge to consider class groups of non-maximal orders [**27**, **28**]. One can also impose global conditions on the fields [**20**, **63**, **98**, **99**, **102**, **114**], such as requiring their ring of integers be be generated by a single element (as an $\mathbb{Z}$-algebra). When one cannot obtain exact asymptotics for moments of class group distributions, one may ask for upper bounds [**8**, **49**, **54**, **60**, **74**, **93**, **104**].

5.4. Unramified extensions. Most of the above mentioned results access Cl_K through class field theory as $\mathrm{Gal}(K^{un,ab}/K)$, the Galois group of the maximal unramified, abelian extension of K. The averages are proven by counting unramified extensions. From this point of view, it is natural to ask more generally about the distribution of the non-abelian (profinite) groups $\mathrm{Gal}(K^{un}/K)$, where K^{un} is the maximal unramified extension of K. As in the abelian case, we have moments, now indexed by finite groups G, and the Gth moment is the average of $\#\mathrm{Sur}(-, G)$, where Sur denotes continuous surjections. The Gth moment of $\mathrm{Gal}(K^{un}/K)$, as K ranges uniformly over a family of number fields, is then $|\operatorname{Aut}(G)|$ times the average number of unramified G-extensions of K (by Galois theory).

Liu, Zureick-Brown, and the author [**79**, Conjecture 1.3] have given conjectures on the distribution of the part of $\mathrm{Gal}(K^{un}/K)$ prime to $2|\Gamma|$ (as K ranges over totally real Γ-extensions), and in particular on the Gth moment for any finite group G of order relatively prime to $2|\Gamma|$. These conjectures generalize the conjecture of Boston, Bush, and Hajir [**31**–**33**] on the maximal pro-p quotient of $\mathrm{Gal}(K^{un}/K)$ (the p-class tower group) as K ranges over quadratic fields. The conjectures of Boston, Bush, and Hajir were based on extensive computation and numerical evidence in the number field case. The general conjectures of [**79**] are based on theoretical reasoning about the presentation of $\mathrm{Gal}(K^{un}/K)$ and on $q \to \infty$ theorems in the function field case. It would be very interesting the have more empirical evidence in the number field case, in particular on whether the moments are approaching their predicted values. See [**79**, Section 1.5] for a detailed discussion of the most interesting and accessible moments that would make useful computational projects to explore experimentally. See also [**118**] for conjectures on G-moments of $\mathrm{Gal}(K^{un}/K)$ as K ranges over quadratic fields and $|G|$ has *even* order. These conjectures are based on function field $q \to \infty$ theorems, but the paper includes an appendix in which computations are done to give empirical number field evidence. Boston and Bush [**30**] also have significant computations on the maximal pro-2 quotient of $\mathrm{Gal}(K^{un}/K)$ as K ranges over cyclic cubic fields, which is beyond the scope of the conjectures of [**79**].

References

[1] J. D. Achter, *The distribution of class groups of function fields*, J. Pure Appl. Algebra **204** (2006), no. 2, 316–333, DOI 10.1016/j.jpaa.2005.04.003. MR2184814 ↑145, 150

[2] J. D. Achter, *Results of Cohen-Lenstra type for quadratic function fields*, Computational arithmetic geometry, Contemp. Math., vol. 463, Amer. Math. Soc., Providence, RI, 2008, pp. 1–7, DOI 10.1090/conm/463/09041. MR2459984 ↑150

[3] M. Adam and G. Malle, *A class group heuristic based on the distribution of 1-eigenspaces in matrix groups*, J. Number Theory **149** (2015), 225–235, DOI 10.1016/j.jnt.2014.10.018. MR3296009 ↑145

[4] B. Alberts, *The weak form of Malle's conjecture and solvable groups*, Res. Number Theory **6** (2020), no. 1, Paper No. 10, 23, DOI 10.1007/s40993-019-0185-7. MR4047213 ↑129

[5] B. Alberts, *Statistics of the first Galois cohomology group: a refinement of Malle's conjecture*, Algebra Number Theory **15** (2021), no. 10, 2513–2569, DOI 10.2140/ant.2021.15.2513. MR4377858 ↑129

[6] B. Alberts and E. O'Dorney, *Harmonic analysis and statistics of the first Galois cohomology group*, Res. Math. Sci. **8** (2021), no. 3, Paper No. 50, 16, DOI 10.1007/s40687-021-00283-2. MR4298097 ↑129

[7] S. A. Altuğ, A. Shankar, I. Varma, and K. H. Wilson, *The number of D_4-fields ordered by conductor*, J. Eur. Math. Soc. (JEMS) **23** (2021), no. 8, 2733–2785, DOI 10.4171/jems/1070. MR4269426 ↑129, 149

[8] C. An, *ℓ-torsion in class groups of certain families of D_4-quartic fields* (English, with English and French summaries), J. Théor. Nombres Bordeaux **32** (2020), no. 1, 1–23, DOI 10.5802/jtnb.1109. MR4158576 ↑151

[9] T. C. Anderson, A. Gafni, K. Hughes, R. J. Lemke Oliver, D. Lowry-Duda, F. Thorne, J. Wang, and R. Zhang, *Improved bounds on number fields of small degree*, Discrete Anal. (2024), Paper No. 19, 24. MR4844557 ↑129

[10] A. Bartel and H. W. Lenstra Jr., *On class groups of random number fields*, Proc. Lond. Math. Soc. (3) **121** (2020), no. 4, 927–953, DOI 10.1112/plms.12343. MR4105790 ↑133, 144, 146, 150, 151

[11] K. Belabas, *A fast algorithm to compute cubic fields*, Math. Comp. **66** (1997), no. 219, 1213–1237, DOI 10.1090/S0025-5718-97-00846-6. MR1415795 ↑126, 128

[12] K. Belabas, *CUBIC*, Accessed July 24, 2022, `https://www.math.u-bordeaux.fr/~kbelabas/research/cubic.html`. ↑148

[13] K. Belabas, M. Bhargava, and C. Pomerance, *Error estimates for the Davenport-Heilbronn theorems*, Duke Math. J. **153** (2010), no. 1, 173–210, DOI 10.1215/00127094-2010-007. MR2641942 ↑128

[14] K. Belabas and É. Fouvry, *Discriminants cubiques et progressions arithmétiques* (French, with English and French summaries), Int. J. Number Theory **6** (2010), no. 7, 1491–1529, DOI 10.1142/S1793042110003605. MR2740719 ↑124

[15] M. Bhargava, *Higher composition laws. III. The parametrization of quartic rings*, Ann. of Math. (2) **159** (2004), no. 3, 1329–1360, DOI 10.4007/annals.2004.159.1329. MR2113024 ↑124, 126

[16] M. Bhargava, *The density of discriminants of quartic rings and fields*, Ann. of Math. (2) **162** (2005), no. 2, 1031–1063, DOI 10.4007/annals.2005.162.1031. MR2183288 ↑119, 122, 124, 150

[17] M. Bhargava, *Higher composition laws. IV. The parametrization of quintic rings*, Ann. of Math. (2) **167** (2008), no. 1, 53–94, DOI 10.4007/annals.2008.167.53. MR2373152 ↑124, 126

[18] M. Bhargava, *The density of discriminants of quintic rings and fields*, Ann. of Math. (2) **172** (2010), no. 3, 1559–1591, DOI 10.4007/annals.2010.172.1559. MR2745272 ↑119, 124

[19] M. Bhargava, *Galois groups of random integer polynomials and van der Waerden's conjecture*, Ann. of Math. (2) **201** (2025), no. 2, 339–377, DOI 10.4007/annals.2025.201.2.1. MR4878222 ↑122

[20] M. Bhargava, J. Hanke, and A. Shankar, *The mean number of 2-torsion elements in the class groups of n-monogenized cubic fields* (October 2020). ↑151

[21] M. Bhargava, D. M. Kane, H. W. Lenstra Jr., B. Poonen, and E. Rains, *Modeling the distribution of ranks, Selmer groups, and Shafarevich-Tate groups of elliptic curves*, Camb. J. Math. **3** (2015), no. 3, 275–321, DOI 10.4310/CJM.2015.v3.n3.a1. MR3393023 ↑133

[22] M. Bhargava, A. Shankar, and J. Tsimerman, *On the Davenport-Heilbronn theorems and second order terms*, Invent. Math. **193** (2013), no. 2, 439–499, DOI 10.1007/s00222-012-0433-0. MR3090184 ↑125, 128, 133

[23] M. Bhargava, A. Shankar, and X. Wang, *Geometry-of-numbers methods over global fields I: Prehomogeneous vector spaces*, Preprint, `arXiv:1512.03035`, December 2015. ↑128, 129, 150

[24] M. Bhargava, A. Shankar, and X. Wang, *An improvement on Schmidt's bound on the number of number fields of bounded discriminant and small degree*, Forum Math. Sigma **10** (2022), Paper No. e86, 13, DOI 10.1017/fms.2022.71. MR4493242 ↑129
[25] M. Bhargava, A. Shankar, and X. Wang, *Squarefree values of polynomial discriminants I*, Invent. Math. **228** (2022), no. 3, 1037–1073, DOI 10.1007/s00222-022-01098-w. MR4419629 ↑129
[26] M. Bhargava, T. Taniguchi, and F. Thorne, *Improved error estimates for the Davenport-Heilbronn theorems*, Math. Ann. **389** (2024), no. 4, 3471–3512, DOI 10.1007/s00208-023-02684-w. MR4768704 ↑129, 133
[27] M. Bhargava and I. Varma, *On the mean number of 2-torsion elements in the class groups, narrow class groups, and ideal groups of cubic orders and fields*, Duke Math. J. **164** (2015), no. 10, 1911–1933, DOI 10.1215/00127094-3120636. MR3369305 ↑151
[28] M. Bhargava and I. Varma, *The mean number of 3-torsion elements in the class groups and ideal groups of quadratic orders*, Proc. Lond. Math. Soc. (3) **112** (2016), no. 2, 235–266, DOI 10.1112/plms/pdv062. MR3471250 ↑151
[29] M. Bhargava and M. M. Wood, *The density of discriminants of S_3-sextic number fields*, Proc. Amer. Math. Soc. **136** (2008), no. 5, 1581–1587, DOI 10.1090/S0002-9939-07-09171-X. MR2373587 ↑124
[30] N. Boston and M. R. Bush, *Heuristics for 2-class towers of cyclic cubic fields*, Exp. Math. **33** (2024), no. 1, 111–122, DOI 10.1080/10586458.2021.2011492. MR4713866 ↑151
[31] N. Boston, M. R. Bush, and F. Hajir, *Heuristics for p-class towers of imaginary quadratic fields*, Math. Ann. **368** (2017), no. 1-2, 633–669, DOI 10.1007/s00208-016-1449-3. MR3651585 ↑151
[32] N. Boston, M. R. Bush, and F. Hajir, *Heuristics for p-class towers of real quadratic fields*, J. Inst. Math. Jussieu **20** (2021), no. 4, 1429–1452, DOI 10.1017/S1474748019000641. MR4293801 ↑151
[33] N. Boston and M. M. Wood, *Non-abelian Cohen-Lenstra heuristics over function fields*, Compos. Math. **153** (2017), no. 7, 1372–1390, DOI 10.1112/S0010437X17007102. MR3705261 ↑151
[34] B. Breen, I. Varma, and J. Voight, *On unit signatures and narrow class groups of odd degree Abelian number fields*, Trans. Amer. Math. Soc. Ser. B **10** (2023), 86–128, DOI 10.1090/btran/90. With an appendix by Noam Elkies. MR4544138 ↑151
[35] J. Clancy, N. Kaplan, T. Leake, S. Payne, and M. M. Wood, *On a Cohen-Lenstra heuristic for Jacobians of random graphs*, J. Algebraic Combin. **42** (2015), no. 3, 701–723, DOI 10.1007/s10801-015-0598-x. MR3403177 ↑139
[36] H. Cohen, *Advanced topics in computational number theory*, Graduate Texts in Mathematics, vol. 193, Springer-Verlag, New York, 2000, DOI 10.1007/978-1-4419-8489-0. MR1728313 ↑122, 125
[37] H. Cohen, F. Diaz y Diaz, and M. Olivier, *Constructing complete tables of quartic fields using Kummer theory*, Math. Comp. **72** (2003), no. 242, 941–951, DOI 10.1090/S0025-5718-02-01452-7. MR1954977 ↑126
[38] H. Cohen, F. Diaz y Diaz, and M. Olivier, *Enumerating quartic dihedral extensions of $\mathbb{Q}$*, Compositio Math. **133** (2002), no. 1, 65–93, DOI 10.1023/A:1016310902973. MR1918290 ↑119, 122
[39] H. Cohen and H. W. Lenstra Jr., *Heuristics on class groups of number fields*, Number theory, Noordwijkerhout 1983 (Noordwijkerhout, 1983), Lecture Notes in Math., vol. 1068, Springer, Berlin, 1984, pp. 33–62, DOI 10.1007/BFb0099440. MR756082 ↑130, 142, 144
[40] H. Cohen and J. Martinet, *Étude heuristique des groupes de classes des corps de nombres* (French), J. Reine Angew. Math. **404** (1990), 39–76, DOI 10.1515/crll.1990.404.39. MR1037430 ↑144, 145
[41] H. Cohen and J. Martinet, *Heuristics on class groups: some good primes are not too good*, Math. Comp. **63** (1994), no. 207, 329–334, DOI 10.2307/2153578. MR1226813 ↑150
[42] H. Cohn, *The density of abelian cubic fields*, Proc. Amer. Math. Soc. **5** (1954), 476–477, DOI 10.2307/2031963. MR64076 ↑122, 123
[43] J.-M. Couveignes, *Enumerating number fields*, Ann. of Math. (2) **192** (2020), no. 2, 487–497, DOI 10.4007/annals.2020.192.2.4. MR4151082 ↑129
[44] B. Datskovsky and D. J. Wright, *Density of discriminants of cubic extensions*, J. Reine Angew. Math. **386** (1988), 116–138, DOI 10.1515/crll.1988.386.116. MR936994 ↑150

[45] H. Davenport and H. Heilbronn, *On the density of discriminants of cubic fields. II*, Proc. Roy. Soc. London Ser. A **322** (1971), no. 1551, 405–420, DOI 10.1098/rspa.1971.0075. MR491593 ↑119, 122, 140, 142

[46] B. N. Delone and D. K. Faddeev, *The theory of irrationalities of the third degree*, Translations of Mathematical Monographs, Vol. 10, American Mathematical Society, Providence, RI, 1964. MR160744 ↑124

[47] D. S. Dummit and J. Voight, *The 2-Selmer group of a number field and heuristics for narrow class groups and signature ranks of units*, Proc. Lond. Math. Soc. (3) **117** (2018), no. 4, 682–726, DOI 10.1112/plms.12143. MR3873132 ↑151

[48] E. P. Dummit, *Counting G-extensions by discriminant*, Math. Res. Lett. **25** (2018), no. 4, 1151–1172, DOI 10.4310/MRL.2018.v25.n4.a5. MR3882158 ↑129

[49] J. Ellenberg, L. B. Pierce, and M. M. Wood, *On ℓ-torsion in class groups of number fields*, Algebra Number Theory **11** (2017), no. 8, 1739–1778, DOI 10.2140/ant.2017.11.1739. MR3720931 ↑151

[50] J. S. Ellenberg, A. Venkatesh, and C. Westerland, *Homological stability for Hurwitz spaces and the Cohen-Lenstra conjecture over function fields*, Ann. of Math. (2) **183** (2016), no. 3, 729–786, DOI 10.4007/annals.2016.183.3.1. MR3488737 ↑145, 150

[51] É. Fouvry and J. Klüners, *On the 4-rank of class groups of quadratic number fields*, Invent. Math. **167** (2007), no. 3, 455–513, DOI 10.1007/s00222-006-0021-2. MR2276261 ↑150

[52] É. Fouvry and P. Koymans, *Malle's conjecture for nonic Heisenberg extensions*, Preprint, `arXiv:2102.09465`, February 2021. ↑128

[53] C. Frei, D. Loughran, and R. Newton, *The Hasse norm principle for Abelian extensions*, Amer. J. Math. **140** (2018), no. 6, 1639–1685, DOI 10.1353/ajm.2018.0048. MR3884640 ↑123, 128

[54] C. Frei and M. Widmer, *Averages and higher moments for the ℓ-torsion in class groups*, Math. Ann. **379** (2021), no. 3-4, 1205–1229, DOI 10.1007/s00208-020-02121-2. MR4238263 ↑151

[55] E. Friedman and L. C. Washington, *On the distribution of divisor class groups of curves over a finite field*, Théorie des nombres (Quebec, PQ, 1987), de Gruyter, Berlin, 1989, pp. 227–239. MR1024565 ↑135, 136

[56] W. T. Gan, B. Gross, and G. Savin, *Fourier coefficients of modular forms on G_2*, Duke Math. J. **115** (2002), no. 1, 105–169, DOI 10.1215/S0012-7094-02-11514-2. MR1932327 ↑124

[57] D. Garton, *Random matrices, the Cohen-Lenstra heuristics, and roots of unity*, Algebra Number Theory **9** (2015), no. 1, 149–171, DOI 10.2140/ant.2015.9.149. MR3317763 ↑145

[58] F. Gerth III, *Densities for ranks of certain parts of p-class groups*, Proc. Amer. Math. Soc. **99** (1987), no. 1, 1–8, DOI 10.2307/2046260. MR866419 ↑150

[59] F. Gerth III, *Extension of conjectures of Cohen and Lenstra*, Exposition. Math. **5** (1987), no. 2, 181–184. MR887792 ↑150

[60] D. R. Heath-Brown and L. B. Pierce, *Averages and moments associated to class numbers of imaginary quadratic fields*, Compos. Math. **153** (2017), no. 11, 2287–2309, DOI 10.1112/S0010437X1700728X. MR3692746 ↑151

[61] H. Heilbronn, *On the class-number in imaginary quadratic fields*, The Quarterly Journal of Mathematics **os-5** (1934), no. 1, 150–160. ↑130

[62] W. Ho and A. Shankar, *Problem set for 2014 Arizona Winter School*, 2014. [Online; accessed 21 July 2022], `https://swc-math.github.io/aws/2014/2014HoShankarProblems.pdf`. ↑125

[63] W. Ho, A. Shankar, and I. Varma, *Odd degree number fields with odd class number*, Duke Math. J. **167** (2018), no. 5, 995–1047, DOI 10.1215/00127094-2017-0050. MR3782066 ↑151

[64] B. Hough, *Equidistribution of bounded torsion CM points*, J. Anal. Math. **138** (2019), no. 2, 765–797, DOI 10.1007/s11854-019-0044-4. MR3996057 ↑134

[65] J. W. Jones and D. P. Roberts, *A database of number fields*, LMS J. Comput. Math. **17** (2014), no. 1, 595–618, DOI 10.1112/S1461157014000424. MR3356048 ↑122

[66] J. Klüners, *Über die Berechnung von Automorphismen und Teilkörpern algebraischer Zahlkörper*, Ph.D. Thesis, Berlin, 1997. ↑128

[67] J. Klüners, *A counterexample to Malle's conjecture on the asymptotics of discriminants*, Comptes Rendus Mathématique. Académie des Sciences. Paris **340** (2005), no. 6, 411–414. MR2135320 ↑128

[68] J. Klüners, *Asymptotics of number fields and the Cohen–Lenstra heuristics*, Journal de Théorie des Nombres de Bordeaux **18** (2006), no. 3, 607–615. ↑140

[69] J. Klüners, *The distribution of number fields with wreath products as Galois groups*, International Journal of Number Theory **8** (2012), no. 3, 845–858. MR2904935 ↑126

[70] J. Klüners and G. Malle, *Counting nilpotent Galois extensions*, J. Reine Angew. Math. **572** (2004), 1–26, DOI 10.1515/crll.2004.050. MR2076117 ↑129

[71] J. Klys, *The distribution of p-torsion in degree p cyclic fields*, Algebra Number Theory **14** (2020), no. 4, 815–854, DOI 10.2140/ant.2020.14.815. MR4114057 ↑150

[72] P. Koymans and C. Pagano, *On Malle's conjecture for nilpotent groups*, Trans. Amer. Math. Soc. Ser. B **10** (2023), 310–354, DOI 10.1090/btran/140. MR4555793 ↑128, 129

[73] A. Landesman, R. J. Lemke Oliver, and F. Thorne, *Improved lower bounds for the number of fields with alternating Galois group*, Bull. Lond. Math. Soc. **53** (2021), no. 4, 1159–1173, DOI 10.1112/blms.12490. MR4311826 ↑129

[74] R. J. Lemke Oliver, J. Thorner, and A. Zaman, *An approximate form of Artin's holomorphy conjecture and non-vanishing of Artin L-functions*, Invent. Math. **235** (2024), no. 3, 893–971, DOI 10.1007/s00222-023-01232-2. MR4701881 ↑151

[75] C. Lewis and C. Williams, *Numerical secondary terms in a Cohen-Lenstra conjecture on real quadratic fields*, Involve **12** (2019), no. 2, 221–233, DOI 10.2140/involve.2019.12.221. MR3864215 ↑134

[76] M. Lipnowski, W. Sawin, and J. Tsimerman, *Cohen-Lenstra heuristics and bilinear pairings in the presence of roots of unity*, Preprint, `arXiv:2007.12533`, July 2020. ↑145

[77] Y. Liu, *Non-abelian Cohen–Lenstra Heuristics in the presence of roots of unity*, Preprint, `arXiv:2202.09471`, February 2022. ↑145

[78] Y. Liu and M. M. Wood, *The free group on n generators modulo $n+u$ random relations as n goes to infinity*, J. Reine Angew. Math. **762** (2020), 123–166, DOI 10.1515/crelle-2018-0025. MR4195658 ↑135

[79] Y. Liu, M. M. Wood, and D. Zureick-Brown, *A predicted distribution for Galois groups of maximal unramified extensions*, Invent. Math. **237** (2024), no. 1, 49–116, DOI 10.1007/s00222-024-01257-1. MR4756988 ↑147, 150, 151

[80] T. LMFDB Collaboration, *The L-functions and modular forms database*, 2022. [Online; accessed 21 July 2022], `http://www.lmfdb.org`. ↑121, 130

[81] S. Mäki, *On the density of abelian number fields*, Ann. Acad. Sci. Fenn. Ser. A I Math. Dissertationes **54** (1985), 104. MR791087 ↑123

[82] G. Malle, *On the distribution of Galois groups*, J. Number Theory **92** (2002), no. 2, 315–329, DOI 10.1006/jnth.2001.2713. MR1884706 ↑128

[83] G. Malle, *On the distribution of Galois groups. II*, Experiment. Math. **13** (2004), no. 2, 129–135. MR2068887 ↑128

[84] G. Malle, *Cohen-Lenstra heuristic and roots of unity*, J. Number Theory **128** (2008), no. 10, 2823–2835, DOI 10.1016/j.jnt.2008.01.002. MR2441080 ↑140, 145, 146, 147, 148

[85] G. Malle, *On the distribution of class groups of number fields*, Experiment. Math. **19** (2010), no. 4, 465–474, DOI 10.1080/10586458.2010.10390636. MR2778658 ↑145, 146, 147, 148

[86] A. S. Mosunov and M. J. Jacobson Jr., *Unconditional class group tabulation of imaginary quadratic fields to $|\Delta| < 2^{40}$*, Math. Comp. **85** (2016), no. 300, 1983–2009, DOI 10.1090/mcom3050. MR3471116 ↑130

[87] W. Narkiewicz, *Number theory*, World Scientific Publishing Co., Singapore; distributed by Heyden & Son, Inc., Philadelphia, PA, 1983. Translated from the Polish by S. Kanemitsu. MR748136 ↑124

[88] J. Neukirch, *Algebraic number theory*, Grundlehren der mathematischen Wissenschaften [Fundamental Principles of Mathematical Sciences], vol. 322, Springer-Verlag, Berlin, 1999. Translated from the 1992 German original and with a note by Norbert Schappacher; With a foreword by G. Harder, DOI 10.1007/978-3-662-03983-0. MR1697859 ↑118, 123, 141

[89] H. H. Nguyen and M. M. Wood, *Random integral matrices: universality of surjectivity and the cokernel*, Invent. Math. **228** (2022), no. 1, 1–76, DOI 10.1007/s00222-021-01082-w. MR4392456 ↑137

[90] R. J. Lemke Oliver and F. Thorne, *Upper bounds on number fields of given degree and bounded discriminant*, Duke Math. J. **171** (2022), no. 15, 3077–3087, DOI 10.1215/00127094-2022-0046. MR4497223 ↑129

[91] R. J. L. Oliver, J. Wang, and M. M. Wood, *The average size of 3-torsion in class groups of 2-extensions*, Forum of Mathematics, Pi, posted on September 15, 2025, DOI 10.1017/S2050508625000009. ↑150
[92] L. B. Pierce, C. L. Turnage-Butterbaugh, and M. M. Wood, *On a conjecture for ℓ-torsion in class groups of number fields: from the perspective of moments*, Math. Res. Lett. **28** (2021), no. 2, 575–621, DOI 10.4310/MRL.2021.v28.n2.a9. MR4258981 ↑140
[93] L. B. Pierce, C. L. Turnage-Butterbaugh, and M. M. Wood, *An effective Chebotarev density theorem for families of number fields, with an application to ℓ-torsion in class groups*, Invent. Math. **219** (2020), no. 2, 701–778, DOI 10.1007/s00222-019-00915-z. MR4054810 ↑151
[94] D. P. Roberts, *Density of cubic field discriminants*, Math. Comp. **70** (2001), no. 236, 1699–1705, DOI 10.1090/S0025-5718-00-01291-6. MR1836927 ↑128, 134
[95] A. Shankar and F. Thorne, *On the asymptotics of cubic fields ordered by general invariants*, Comment. Math. Helv. **99** (2024), no. 4, 769–797, DOI 10.4171/cmh/576. MR4815165 ↑129
[96] A. Shankar and J. Tsimerman, *Counting S_5-fields with a power saving error term*, Forum Math. Sigma **2** (2014), Paper No. e13, 8, DOI 10.1017/fms.2014.10. MR3264252 ↑128
[97] A. Shankar and J. Tsimerman, *Heuristics for the asymptotics of the number of S_n-number fields*, J. Lond. Math. Soc. (2) **107** (2023), no. 5, 1613–1634, DOI 10.1112/jlms.12720. MR4585297 ↑122
[98] A. Siad, *Monogenic fields with odd class number Part I: Odd degree*, arXiv.2011.08834 (November 2020). ↑151
[99] A. Siad, *Monogenic fields with odd class number Part II: Even degree*, arXiv.2011.08842 (November 2020). ↑151
[100] A. Smith, *The distribution of ℓ^∞-Selmer groups in degree ℓ twist families*, Preprint, arXiv:2207.05674, July 2022. ↑150
[101] A. Smith, *The distribution of fixed point Selmer groups in twist families*, Preprint, arXiv:2207.05143, July 2022. ↑150
[102] A. A. Swaminathan, *A new parametrization for ideal classes in rings defined by binary forms, and applications*, J. Reine Angew. Math. **798** (2023), 143–191, DOI 10.1515/crelle-2023-0006. MR4579703 ↑151
[103] T. Taniguchi and F. Thorne, *Secondary terms in counting functions for cubic fields*, Duke Math. J. **162** (2013), no. 13, 2451–2508, DOI 10.1215/00127094-2371752. MR3127806 ↑128, 133
[104] J. Thorner and A. Zaman, *A zero density estimate for Dedekind zeta functions*, Int. Math. Res. Not. IMRN **8** (2023), 6739–6761, DOI 10.1093/imrn/rnac015. MR4574386 ↑151
[105] S. Türkelli, *Connected components of Hurwitz schemes and Malle's conjecture*, J. Number Theory **155** (2015), 163–201, DOI 10.1016/j.jnt.2015.03.005. MR3349443 ↑128
[106] B. L. van der Waerden, *Die Seltenheit der reduziblen Gleichungen und der Gleichungen mit Affekt* (German), Monatsh. Math. Phys. **43** (1936), no. 1, 133–147, DOI 10.1007/BF01707594. MR1550517 ↑122
[107] I. Varma, *The mean number of 3-torsion elements in ray class groups of quadratic fields*, Israel J. Math. **252** (2022), no. 1, 149–185, DOI 10.1007/s11856-022-2346-y. MR4526829 ↑151
[108] A. Venkatesh and J. S. Ellenberg, *Statistics of number fields and function fields*, Proceedings of the International Congress of Mathematicians. Volume II, Hindustan Book Agency, New Delhi, 2010, pp. 383–402. MR2827801 ↑129, 134
[109] J. Voight, *Enumeration of totally real number fields of bounded root discriminant*, Algorithmic number theory, Lecture Notes in Comput. Sci., vol. 5011, Springer, Berlin, 2008, pp. 268–281, DOI 10.1007/978-3-540-79456-1_18. MR2467853 ↑122
[110] J. Wang, *Secondary term of asymptotic distribution of $S_3 \times A$-extensions over $\mathbb{Q}$*, Preprint, arXiv:1710.10693, October 2017. ↑129
[111] J. Wang, *Malle's conjecture for $S_n \times A$ for $n = 3, 4, 5$*, Compos. Math. **157** (2021), no. 1, 83–121, DOI 10.1112/s0010437x20007587. MR4219215 ↑126
[112] W. Wang and M. M. Wood, *Moments and interpretations of the Cohen-Lenstra-Martinet heuristics*, Comment. Math. Helv. **96** (2021), no. 2, 339–387, DOI 10.4171/cmh/514. MR4277275 ↑133, 139, 144, 145, 146, 147, 148, 149, 150
[113] M. M. Wood, *On the probabilities of local behaviors in abelian field extensions*, Compos. Math. **146** (2010), no. 1, 102–128, DOI 10.1112/S0010437X0900431X. MR2581243 ↑128, 129, 146

[114] M. M. Wood, *Parametrization of ideal classes in rings associated to binary forms*, J. Reine Angew. Math. **689** (2014), 169–199, DOI 10.1515/crelle-2012-0058. MR3187931 ↑151
[115] M. M. Wood, *Asymptotics for number fields and class groups*, Directions in Number Theory: Proceedings of the 2014 WIN3 Workshop, July 2016. ↑124, 125, 128, 133, 140
[116] M. M. Wood, *The distribution of sandpile groups of random graphs*, J. Amer. Math. Soc. **30** (2017), no. 4, 915–958, DOI 10.1090/jams/866. MR3671933 ↑139, 140
[117] M. M. Wood, *Cohen-Lenstra heuristics and local conditions*, Res. Number Theory **4** (2018), no. 4, Paper No. 41, 22, DOI 10.1007/s40993-018-0134-x. MR3858473 ↑139, 150, 151
[118] M. M. Wood, *Nonabelian Cohen-Lenstra moments*, Duke Math. J. **168** (2019), no. 3, 377–427, DOI 10.1215/00127094-2018-0037. With an appendix by the author and Philip Matchett Wood. MR3909900 ↑140, 151
[119] M. M. Wood, *Random integral matrices and the Cohen-Lenstra heuristics*, Amer. J. Math. **141** (2019), no. 2, 383–398, DOI 10.1353/ajm.2019.0008. MR3928040 ↑137, 138, 142
[120] M. M. Wood, *Probability theory for random groups arising in number theory*, ICM—International Congress of Mathematicians. Vol. 6. Sections 12–14, EMS Press, Berlin, 2023, pp. 4476–4508. MR4680411 ↑138, 139, 140
[121] D. J. Wright, *Distribution of discriminants of abelian extensions*, Proc. London Math. Soc. (3) **58** (1989), no. 1, 17–50, DOI 10.1112/plms/s3-58.1.17. MR969545 ↑129, 146

Department of Mathematics, Harvard University, Science Center Room 325, 1 Oxford Street, Cambridge, Massachusetts 02138 USA
Email address: `mmwood@math.harvard.edu`

IAS/Park City Mathematics Series
Volume **29**, 2025
https://doi.org/10.1090/pcms/029/00893

An Introduction to Lattices, Lattice Reduction, and Lattice-Based Cryptography

Joseph H. Silverman

ABSTRACT. A lattice is a discrete subgroup of $\mathbb{R}^n$. We will discuss the theory of lattices and describe how they have been used to construct practical public key cryptosystems and digital signature schemes that are, at least presently, secure against attacks by quantum computers.

CONTENTS

2020 *Mathematics Subject Classification.* Primary 94A60; Secondary 11Hxx.
Key words and phrases. Lattice, shortest vector problem, closest vector problem, public key cryptosystem, digital signature, lattice-based cryptography.

Acknowledgments

The author thanks Mingjie Chen and Timo Keller for their helpful suggestions and corrections to the original draft of these notes, Jonathan Love for his detailed referee report that improved the exposition and corrected some mistakes, and Mingjie Chen for her assistance during the PCMI graduate summer school. The author also thanks the organizers for inviting him to deliver these lectures, and PCMI for hosting the graduate summer school.

Prerequisites. These notes assume that the reader has familiarity with linear algebra, group theory, and ring theory at the level of a typical undergraduate course in these subjects. We make occasional use of concepts from point-set topology. We use the notation $A := B$ to mean that the quantity A is defined to be the quantity B.

1. Lattices and Hard Lattice Problems

1.1. Lattices: Definitions, Notation, and Basic Properties.

DEFINITION 1.1.1. A *lattice* L is a discrete[1] subgroup of $\mathbb{R}^n$.

DEFINITION 1.1.2. Let $S := \{\mathbf{v}_1, \dots, \mathbf{v}_k\}$ be a set of vectors in $\mathbb{R}^n$. The *$\mathbb{Z}$-linear span of S* is the set

$$\operatorname{Span}_{\mathbb{Z}}(S) := \{a_1\mathbf{v}_1 + a_2\mathbf{v}_2 + \cdots + a_k\mathbf{v}_k : a_1, a_2, \dots, a_k \in \mathbb{Z}\}.$$

This should not be confused with the *$\mathbb{R}$-linear span of S*, which is defined similarly by

$$\operatorname{Span}_{\mathbb{R}}(S) := \{a_1\mathbf{v}_1 + a_2\mathbf{v}_2 + \cdots + a_k\mathbf{v}_k : a_1, a_2, \dots, a_k \in \mathbb{R}\},$$

but now we allow the scalars to be arbitrary real numbers, rather than restricting them to be integers.

PROPOSITION 1.1.3. *Let L be a lattice in $\mathbb{R}^n$. Then there exist vectors $\mathbf{v}_1, \dots, \mathbf{v}_k$ for some $0 \le k \le n$ such that $\mathbf{v}_1, \dots, \mathbf{v}_k$ are $\mathbb{R}$-linearly independent and satisfy*

$$L = \operatorname{Span}_{\mathbb{Z}}(\mathbf{v}_1, \dots, \mathbf{v}_k).$$

The set $\{\mathbf{v}_1, \dots, \mathbf{v}_k\}$ is a basis *for L, and k is the* rank *or* dimension *of L.*

PROOF. Left for the reader. □

Convention: In these lectures we work almost exclusively with lattices of rank n in $\mathbb{R}^n$, i.e., with lattices of maximal rank. So unless we specify the contrary, we use the convention that[2]

$$\boxed{\textit{lattice} := \textit{lattice of maximal rank.}}$$

Thus when we say that L is a lattice, we will generally mean that L is the $\mathbb{Z}$-linear span of an $\mathbb{R}$-basis $\{\mathbf{v}_1, \dots, \mathbf{v}_n\}$ of $\mathbb{R}^n$.
Important Note: A lattice will have many bases, but some bases are "better" than others.[3] This distinction is a major theme of these lectures.

DEFINITION 1.1.4. Let L be a lattice with basis $\mathcal{B} := \{\mathbf{v}_1, \dots, \mathbf{v}_n\}$. The *fundamental domain for L associated to $\mathcal{B}$* is the set[4]

$$\mathcal{F}(\mathcal{B}) := \{t_1\mathbf{v}_1 + t_2\mathbf{v}_2 + \cdots + t_n\mathbf{v}_n : 0 \le t_i < 1\}.$$

The *determinant* (or *volume*) *of L* is[5]

$$\operatorname{Det}(L) := \operatorname{Volume}(\mathcal{F}(\mathcal{B})).$$

See Figure 1.1.5 for an example of a fundamental domain for a lattice in $\mathbb{R}^2$.

---

[1]A subset of S of $\mathbb{R}^n$ is discrete if each point of S is contained in an open ball that doesn't contain any other points of S.

[2]Lattices of non-maximal rank will come up in our proof that the LLL algroithm runs in polynomial time; see Section 2.7 and Exercise 2.8.2.

[3]Or, as Orwell's Napoleon might say, "All bases are created equal, but some are more equal than others."

[4]More properly, we should say that $\mathcal{F}(\mathcal{B})$ is a fundamental domain for the action of L on $\mathbb{R}^n$; cf. Definition 1.7.2.

[5]Although we may sometimes refer to the "volume" of L, we really mean the volume of the torus $\mathbb{R}^n/L$, which would more properly be called the co-volume of L. People also frequently work with the *discriminant of L*, which is the square of the volume; i.e., $\operatorname{Disc}(L) = \operatorname{Det}(L)^2$

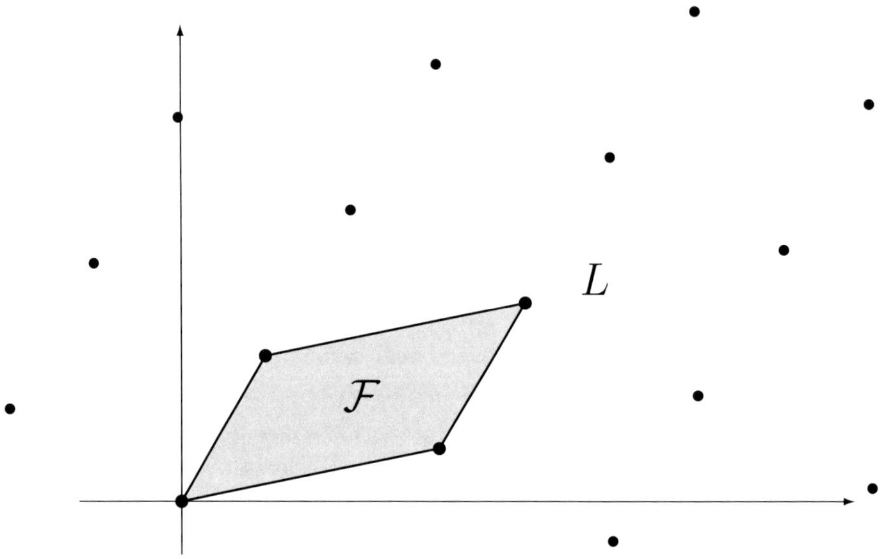

FIGURE 1.1.5. A 2-dimensional lattice L with fundamental domain $\mathcal{F}$

PROPOSITION 1.1.6. *Let $L \subset \mathbb{R}^n$ be a lattice, and let*

$$\mathcal{B} := \{\mathbf{v}_1, \ldots, \mathbf{v}_n\} \quad \textit{and} \quad \mathcal{B}' := \{\mathbf{v}'_1, \ldots, \mathbf{v}'_n\}$$

be bases for L.

(1) *Let*

$$\mathcal{M}(\mathcal{B}) := \big(\mathbf{v}_1 \mid \mathbf{v}_2 \mid \cdots \mid \mathbf{v}_n\big) \in \operatorname{Mat}_{n\times n}(\mathbb{R})$$

be the matrix whose columns are the vectors $\mathbf{v}_1, \ldots, \mathbf{v}_n$. Then

$$\operatorname{Det}(L) = \big|\det \mathcal{M}(\mathcal{B})\big|.$$

(2) *There is a matrix $A \in \mathrm{SL}_n(\mathbb{Z})$ that transforms $\mathcal{B}$ to $\mathcal{B}'$,*

$$\mathcal{M}(\mathcal{B}') = A\mathcal{M}(\mathcal{B}).$$

(3) *We have*

$$\operatorname{Volume}\big(\mathcal{F}(\mathcal{B})\big) = \operatorname{Volume}\big(\mathcal{F}(\mathcal{B}')\big).$$

In particular, $\operatorname{Det}(L)$ *depends only on L.*

PROOF. Left for the reader; see Exercise 1.8.1. □

1.2. Hard Lattice Problems. The following is one of the most important and fundamental computational problems associated to a lattice.

DEFINITION 1.2.1. Let $L \subset \mathbb{R}^n$ be a lattice. The **Shortest Vector Problem (SVP)** is the problem of finding a shortest non-zero vector in L. In other words, the SVP for L is to find a vector $\mathbf{v}_0 \in L$ satisfying

$$\|\mathbf{v}_0\| = \min_{\mathbf{v}\in(L\smallsetminus\mathbf{0})} \|\mathbf{v}\|.$$

Closely related to the SVP is a second important computational lattice problem.

DEFINITION 1.2.2. Let $L \subset \mathbb{R}^n$ be a lattice. The **Closest Vector Problem (CVP)** is the problem of finding, for a given target vector $\mathbf{t} \in \mathbb{R}^n$, a vector in L that is closest to $\mathbf{t}$. In other words, the CVP for L with target vector $\mathbf{t}$ is to find a vector $\mathbf{v}_0 \in L$ satisfying

$$\|\mathbf{v}_0 - \mathbf{t}\| = \min_{\mathbf{v} \in L} \|\mathbf{v} - \mathbf{t}\|.$$

In full generality, the exact version of both the SVP and the CVP appear to be very difficult, although for many applications it suffices to find an approximate solution, as in the following formulation.

DEFINITION 1.2.3. Let $L \subset \mathbb{R}^n$ be a lattice. The **Approximate Closest Vector Problem (apprCVP)** is the problem of finding, for a given target vector $\mathbf{t} \in \mathbb{R}^n$, a vector in L that is "reasonably close" to $\mathbf{t}$. More precisely, for $\kappa \geq 1$, the apprCVP$_\kappa$ is to find a vector $\mathbf{v}_0 \in L$ satisfying

$$\|\mathbf{v}_0 - \mathbf{t}\| \leq \kappa \min_{\mathbf{v} \in L} \|\mathbf{v} - \mathbf{t}\|.$$

1.3. Using a Basis to Try to Solve the CVP: Babai's Algorithm. Given a lattice L and a basis $\mathcal{B} := \{\mathbf{v}_1, \ldots, \mathbf{v}_n\}$ for L, Babai's Algorithm uses the basis to try to solve the CVP.

ALGORITHM 1.3.1 (**Babai's Algorithm: Intuitive Description**).

Input: *A basis* $\mathcal{B} := \{\mathbf{v}_1, \ldots, \mathbf{v}_n\}$ *for a lattice* $L \subset \mathbb{R}^n$.
A target vector $\mathbf{t} \in \mathbb{R}^n$.

Compute: *A lattice vector* $\tilde{\mathbf{v}}$ *so that the translated fundamental domain* $\mathcal{F}(\mathcal{B}) + \tilde{\mathbf{v}}$ *contains the target point* $\mathbf{t}$.

Output: *A vertex of the closure of* $\mathcal{F}(\mathcal{B}) + \tilde{\mathbf{v}}$ *that is fairly close to* $\mathbf{t}$.

See Figure 1.3.3 for an illustration in $\mathbb{R}^2$. Our high-level description of Babai's algorithm can be translated into mathematical formulas as follows:[6]

ALGORITHM 1.3.2 (**Babai's Algorithm: Precise Formulation**).

Input: *A basis* $\mathcal{B} := \{\mathbf{v}_1, \ldots, \mathbf{v}_n\}$ *for a lattice* $L \subset \mathbb{R}^n$.
A target vector $\mathbf{t} \in \mathbb{R}^n$.

Compute: $\alpha_1, \ldots, \alpha_n \in \mathbb{R}$ *so that* $\mathbf{t} = \alpha_1\mathbf{v}_1 + \cdots + \alpha_n\mathbf{v}_n$.
$\mathbf{v}_0 \leftarrow \lfloor \alpha_1 \rceil \mathbf{v}_1 + \cdots + \lfloor \alpha_n \rceil \mathbf{v}_n \in L$.

Output: $\mathbf{v}_0$

Babai's algorithm seems like a very reasonable way to try to solve the apprCVP, and if you believe Figure 1.3.3, it should work quite well in practice. Unfortunately, Figure 1.3.3 is misleading, because we have used a fundamental domain that is fairly close to rectangular. (It is also misleading because pictures in 2 or 3 dimensions fail to convey the complexity of high-dimensional lattices.) In other words, the vectors in the basis $\mathcal{B}$ used to create the fundamental domain $\mathcal{F}(\mathcal{B})$ are reasonably close to orthogonal to one another.

What happens if we choose some other basis whose elements are not so orthogonal? Figure 1.3.4 illustrates a lattice with two bases, a "good" somewhat orthogonal basis $\{\mathbf{v}_1, \mathbf{v}_2\}$ and a "bad" highly non-orthogonal basis $\{\mathbf{w}_1, \mathbf{w}_2\}$. If we try to solve the CVP using the bad basis, Figure 1.3.5 shows what can go wrong. The closest lattice point in the fundamental domain is not the lattice point that

---

[6]We write $\lfloor \alpha \rceil$ for the integer k that is closest to α; i.e., the unique $k \in \mathbb{Z}$ satisfying $k - \frac{1}{2} \leq \alpha < k + \frac{1}{2}$.

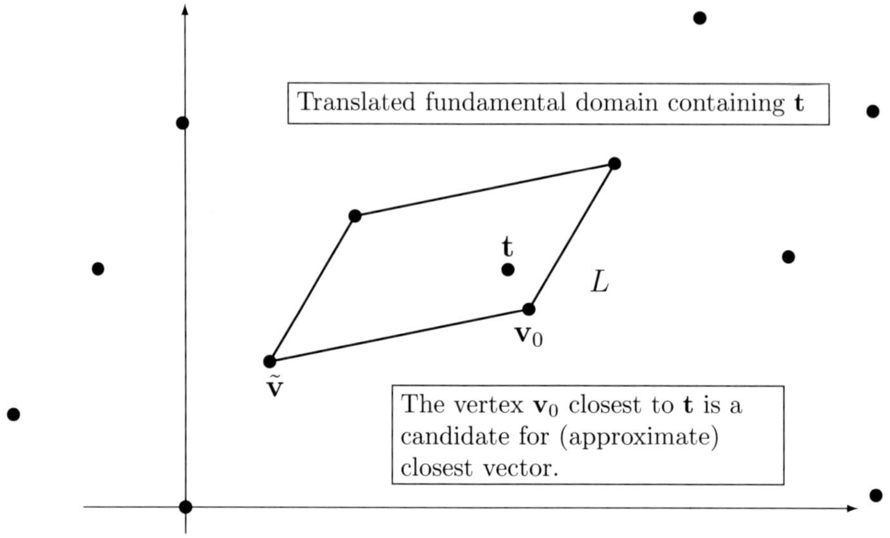

FIGURE 1.3.3. Babai's Algorithm: Using a Basis to Try to Solve the Closest Vector Problem

is closest to the target. And as the dimension goes up, the failure of a bad basis to solve apprCVP increases exponentially, in the sense that the distance from the vector that we find to the target vector is exponentially larger (as a function of $\dim(L)$) than the distance to the actual closest vector.

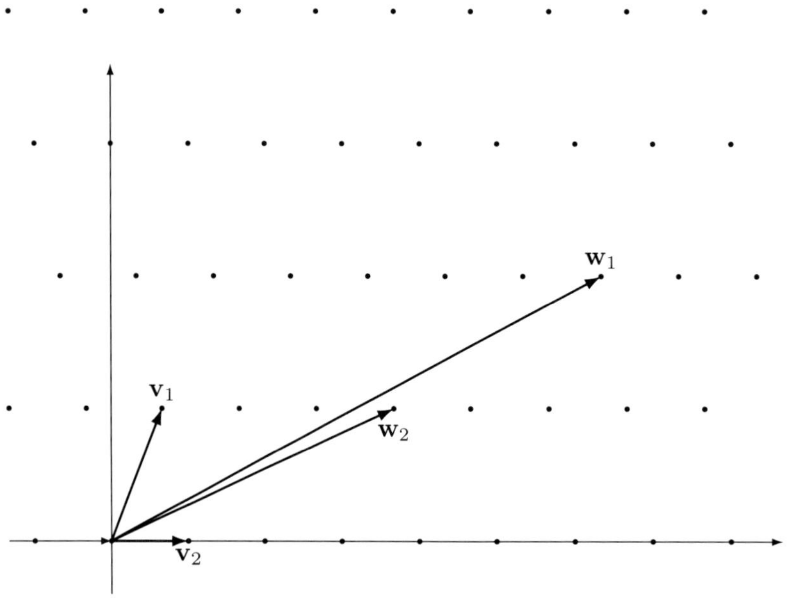

FIGURE 1.3.4. A "good" basis $\{\mathbf{v}_1, \mathbf{v}_2\}$ and a "bad" basis $\{\mathbf{w}_1, \mathbf{w}_2\}$

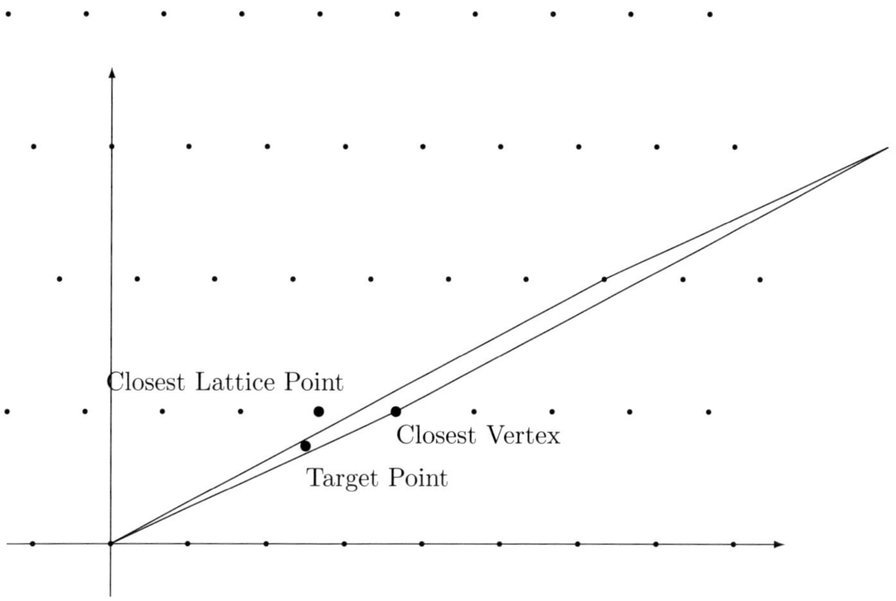

FIGURE 1.3.5. Babai's Algorithm Fails with a "Bad" Basis

REMARK 1.3.6. In Algorithm 1.3.2 we select the coefficients of the $\mathbf{v}_i$ in $\mathbf{v}_0$ simultaneously. Babai also described a slightly more complicated algorithm in which the coefficients of the output vector $\mathbf{v}_0$ are selected iteratively, using the earlier selections to potentially improve the later choices. See Exercise 2.8.7 for details.

1.4. How to Distinguish Good Bases from Bad Bases. The best possible basis would be one in which the vectors are pairwise orthogonal (perpendicular) to one another. It is not common for a lattice to have an orthogonal basis, although if it does, then its determinant is simply the product of the lengths of the basis vectors.[7] However, even for non-orthogonal bases we get an upper bound.

THEOREM 1.4.1 (Hadamard's Inequality). *Let* $\mathbf{v}_1, \ldots, \mathbf{v}_n$ *be any basis for a lattice* L. *Then*

$$\operatorname{Det}(L) \le \|\mathbf{v}_1\| \cdot \|\mathbf{v}_2\| \cdots \|\mathbf{v}_n\|,$$

with equality if and only if $\mathbf{v}_1, \ldots, \mathbf{v}_n$ *are pairwise orthogonal.*

PROOF. For $1 \le k \le n-1$, let $\mathcal{H}_k$ be the $\mathbb{R}$-linear subspace spanned by the vectors $\mathbf{v}_1, \ldots, \mathbf{v}_k$, i.e.,

$$\mathcal{H}_k := \{t_1\mathbf{v}_1 + t_2\mathbf{v}_2 + \cdots + t_k\mathbf{v}_k : t_i \in \mathbb{R}\};$$

and let $\theta_k \in [0, \frac{1}{2}\pi]$ be the angle between $\mathcal{H}_k$ and $\mathbf{v}_{k+1}$. Then

$$\operatorname{Det}(L) := \operatorname{Volume} \mathcal{F}(\mathbf{v}_1, \ldots, \mathbf{v}_n) = \prod_{k=1}^{n} \|\mathbf{v}_k\| \cdot \prod_{k=1}^{n-1} \sin(\theta_k).$$

[7]Think rectangular box. We also note that in this special situation, Babai's algorithm solves the CVP; see Exercise 1.8.5.

Since $|\sin(\theta)| \le 1$, we immediately get Hadamard's inequality; and there is equality if and only if every $\theta_k = \frac{1}{2}\pi$, which is equivalent to pairwise orthogonality of $\mathbf{v}_1, \ldots, \mathbf{v}_n$. □

Hadamard's inequality is an equality if and only if the basis vectors are orthogonal to one another. The extent to which it is not an equality measures the extent to which the basis is non-orthogonal. A famous theorem of Minkowski says that every lattice has at least one basis that is reasonably orthogonal, where the amount of non-orthogonality is bounded solely in terms of the dimension.[8]

THEOREM 1.4.2 (Minkowski). *Fix $n \ge 1$. There is a constant γ so that every lattice L of dimension n has a basis $\mathbf{v}_1, \ldots, \mathbf{v}_n$ satisfying*

$$\|\mathbf{v}_1\| \cdot \|\mathbf{v}_2\| \cdots \|\mathbf{v}_k\| \le \gamma^{k/2} \operatorname{Det}(L)^{k/n} \quad \text{for all } 1 \le k \le n. \tag{1.4.3}$$

In particular, the inequality (1.4.3) *is true for the constant*[9]

$$\gamma = \frac{4}{\pi}\Gamma\left(\frac{1}{2}n+1\right)^{2/n} \approx \frac{2n}{\pi e}.$$

PROOF. See Section 1.7 and Exercise 1.8.8. □

DEFINITION 1.4.4. The *first minimum of L*, denoted $\lambda_1(L)$, is the length of shortest non-zero vector in L,

$$\lambda_1(L) := \inf_{\mathbf{v}\in(L\smallsetminus\mathbf{0})} \|\mathbf{v}\|. \tag{1.4.5}$$

More generally, the *k'th successive minimum of L*, denoted $\lambda_k(L)$, is the smallest real number λ such that L contains k vectors that are $\mathbb{R}$-linearly independent and have length at most λ; i.e.,

$$\lambda_k(L) := \inf\Bigl\{\lambda > 0 : \dim \operatorname{Span}_{\mathbb{R}}\bigl\{\mathbf{v} \in L : \|\mathbf{v}\| \le \lambda\bigr\} \ge k\Bigr\}.$$

With this notation, Theorem 1.4.2 implies that

$$\lambda_1(L)\lambda_2(L)\cdots\lambda_k(L) \le \gamma^{k/2}\operatorname{Det}(L)^{k/n}.$$

And the SVP may be succinctly stated as that of finding a vector $\mathbf{v}_0 \in L$ that satisfies $\|\mathbf{v}_0\| = \lambda_1(L)$.

DEFINITION 1.4.6. Theorem 1.4.2 says in particular that there is a γ so that for every n-dimensional lattice L, the shortest non-zero vector in L has length at most $\gamma^{1/2}\operatorname{Det}(L)^{1/n}$. The smallest such γ is denoted γ_n and is called *Hermite's constant*; i.e.,

$$\gamma_n := \sup_{\text{Lattices } L\subset\mathbb{R}^n} \lambda_1(L)^2 \cdot \operatorname{Det}(L)^{-2/n}.$$

---

[8]Note that we cannot hope for an inequality of this sort to hold for every basis $\mathcal{B}$, since $\operatorname{Det}(L)$ is independent of $\mathcal{B}$, but we can make the product $\prod_i \|\mathbf{v}_i\|$ arbitrarily large by replacing $\mathbf{v}_1$ with $\mathbf{v}_1 + C\mathbf{v}_2$ and letting $C \in \mathbb{Z}$ go to ∞.

[9]Here $\Gamma(z) = \int_0^\infty t^{z-1}e^{-t}\,dt$ is the gamma function. The approximation, which is valid for large n, comes from Stirling's formula $\Gamma(z+1) \approx (z/e)^z \cdot \sqrt{2\pi z}$. As we will see in Sections 1.6 and 1.7, the gamma function appears because the volume of a unit ball in $\mathbb{R}^n$ is $\pi^{n/2}/\Gamma(\frac{1}{2}n+1)$.

REMARK 1.4.7. Theorem 1.4.2 says that $\gamma_n \lesssim 2n/\pi e$. A refined estimate due to Blichfeldt [**6**] says that

$$\gamma_n \leq \frac{2}{\pi}\Gamma\left(\frac{1}{2}n+2\right)^{2/n} \approx \frac{n}{\pi e}.$$

The exact value of γ_n is known only for $n \leq 8$ and $n = 24$:

n	1	2	3	4	5	6	7	8	24
γ_n^n	1	$\frac{4}{3}$	2	4	8	$\frac{64}{3}$	64	2^8	2^{48}

As we will see in Section 1.6, for a "random" lattice of dimension n we expect that Theorem 1.4.2 should be true with the even smaller constant $\gamma \approx n/2\pi e$.

1.5. Theory and Practice and a Smidgeon of History. Lattices and the SVP and CVP have been intensively studied for more than 100 years, both as intrinsic mathematical problems and for applications in pure and applied mathematics, physics and cryptography. The theoretical study of lattices is often called the

Geometry of Numbers,

a name bestowed on it by Minkowski in his 1910 book *Geometrie der Zahlen.* Standard references for this subject include [**7**, **9**, **17**, **33**, **37**].

The practical process of finding short(est) or close(st) vectors in lattices is referred to as the theory of

Lattice Reduction.

The shortest vector problem (SVP) and the closest vector problems (CVP) are closely related, since an algorithm that solves the SVP in dimension $n+1$ can generally be used to solve the CVP in dimension n; see Exercise 1.8.11. If the dimension of the lattice L is large, then both the SVP and the CVP seem to be very hard. Indeed, in full generality, the CVP is known to be NP-hard, and the SVP is NP-hard under a randomized reduction hypothesis; see for example [**28**].

The best lattice reduction methods currently known are based on the LLL Algorithm of Lenstra, Lenstra, and Lovász [**24**], originally described in *Mathematische Annalen* **261** (1982), 515–534. The LLL algorithm and some of its variants will be the featured actors in Lecture 2. They are able to find moderately short (or close) lattice vectors in polynomial time, which suffices for many applications. However, finding very short or very close vectors currently requires time that is exponential in the dimension n, making SVP and CVP suitable candidates for cryptographic applications.

REMARK 1.5.1. It is worth noting that current lattice reduction algorithms such as LLL are highly sequential. Thus they are not distributable, although they are somewhat parallelizable. Further, there are currently no known quantum algorithms that solve general instances of (approximate) SVP or CVP in anything faster than exponential time.

1.6. The Gaussian Heuristic. Let $L \subset \mathbb{R}^n$ be a "random" lattice, and let $\mathbf{t} \in \mathbb{R}^n$ be a "random" target point. How close might we expect the closest vector of L to be to $\mathbf{t}$? The following heuristic gives an answer that tends to work reasonably well in practice. One might try to rigorously justify this heuristic by defining some sort of probability measure on the space of lattices and targets, but we will not pursue this in these notes.

HEURISTIC 1.6.1 (Gaussian Heuristic). *Let n be large.*

SVP: *Let $L \subset \mathbb{R}^n$ be a random lattice. Then we expect that the smallest non-zero vector in L will satisfy*

$$\lambda_1(L) := \min_{\mathbf{v}\in(L\smallsetminus\mathbf{0})} \|\mathbf{v}\| \approx \sqrt{\frac{n}{2\pi e}}\,\mathrm{Det}(L)^{1/n}.$$

CVP: *Let $L \subset \mathbb{R}^n$ be a random lattice, and let $\mathbf{t} \in \mathbb{R}^n$ be a random target point. Then we expect*

$$\min_{\mathbf{v}\in L} \|\mathbf{v}-\mathbf{t}\| \approx \sqrt{\frac{n}{2\pi e}}\,\mathrm{Det}(L)^{1/n}.$$

HEURISTIC "PROOF". The proofs of the SVP and CVP versions are similar, so we do the latter. Let

$$\mathbb{B}_R^n(\mathbf{t}) := \bigl\{\mathbf{x}\in\mathbb{R}^n : \|\mathbf{x}-\mathbf{t}\| \le R\bigr\}$$

be the ball of radius R centered at $\mathbf{t}$. The volume of $\mathbb{B}_R^n(\mathbf{t})$ is given by

$$\mathrm{Volume}\bigl(\mathbb{B}_R^n(\mathbf{t})\bigr) = \mathrm{Volume}\bigl(\mathbb{B}_1(\mathbf{0})\bigr)R^n = \frac{\pi^{n/2}}{\Gamma\bigl(\frac{n}{2}+1\bigr)}R^n \approx \left(\frac{2\pi e}{n}\right)^{n/2} R^n.$$

Here the first equality is true by homogeneity, the second equality is a standard calculus exercise using the Gamma function, and the third (approximate) equality is valid for large n using Stirling's formula

$$\Gamma(z+1) \approx \left(\frac{z}{e}\right)^z \cdot \sqrt{2\pi z} \quad \text{as} \quad z \to \infty.$$

On the other hand, we know the following facts:

- A fundamental domain $\mathcal{F}$ for L has volume $\mathrm{Det}(L)$.
- Each translated fundamental domain $\mathcal{F}+\mathbf{v}$ with $\mathbf{v}\in L$ contains exactly one point of L.
- The translated fundamental domains cover $\mathbb{R}^n$.

We thus expect that if we take a random (compact, convex) blob in $\mathbb{R}^n$ whose volume significantly exceeds $\mathrm{Det}(L)$, then that blob is likely to contain a point of L; but if the blob has volume significantly smaller than $\mathrm{Det}(L)$, then it probably won't contain a point of L.

A short computation shows that

$$\mathrm{Volume}\bigl(\mathbb{B}_R^n(\mathbf{t})\bigr) \approx \mathrm{Det}(L) \quad\Longleftrightarrow\quad R \approx \sqrt{n/2\pi e}\,\mathrm{Det}(L)^{1/n},$$

which leads us to expect:

$$\begin{aligned} R > \sqrt{n/2\pi e}\,\mathrm{Det}(L)^{1/n} &\quad\Longrightarrow\quad L\cap\mathbb{B}_R^n(\mathbf{t}) \ne \emptyset;\\ R < \sqrt{n/2\pi e}\,\mathrm{Det}(L)^{1/n} &\quad\Longrightarrow\quad L\cap\mathbb{B}_R^n(\mathbf{t}) = \emptyset. \end{aligned}$$

Hence the solution to our CVP problem

$$\min\bigl\{\|\mathbf{v}-\mathbf{t}\| : \mathbf{v}\in L\bigr\}$$

is likely to be roughly equal to $\sqrt{n/2\pi e}\,\mathrm{Det}(L)^{1/n}$. □

REMARK 1.6.2. Blichfeldt's upper bound for Hermite's constant (Remark 1.4.7) gives a guaranteed bound for the length $\lambda_1(L)$ of a shortest non-zero vector in L, while the Gaussian heuristic gives a likely value for $\lambda_1(L)$ in a random lattice. It is instructive to see how they compare.

$$\frac{\begin{pmatrix}\text{Upper bound for } \lambda_1(L) \text{ guaran-}\\ \text{teed by Hermite and Blichfeldt}\end{pmatrix}}{\begin{pmatrix}\text{Expected length of } \lambda_1(L) \text{ sug-}\\ \text{gested by the Gaussian heuristic}\end{pmatrix}} = \frac{\sqrt{\frac{2}{\pi}} \cdot \Gamma\left(2+\frac{n}{2}\right)^{1/n} \operatorname{Det}(L)^{1/n}}{\sqrt{\frac{1}{\pi}} \cdot \Gamma\left(1+\frac{n}{2}\right)^{1/n} \operatorname{Det}(L)^{1/n}} \approx \sqrt{2}.$$

Thus the Gaussian heuristic suggests that the Blichfeldt–Hermite upper bound, which is guaranteed to work for all lattices, is only about 1.4 times larger than what we'd expect to need for most lattices.

1.7. Fundamental Domains: The Good, the Bad, and the Voronoi. Let L be a lattice with a given basis $\mathcal{B}$. We defined the associated fundamental domain $\mathcal{F}(\mathcal{B})$ to be the half-open parallelopiped spanned by the vectors in $\mathcal{B}$; see Definition 1.1.4. An important property of $\mathcal{F}(\mathcal{B})$ is that the natural map $\mathcal{F}(\mathcal{B}) \to \mathbb{R}^n/L$ is a bijection; i.e., every vector in L can be written uniquely as the sum of an element of L and an element of $\mathcal{F}(\mathcal{B})$.

DEFINITION 1.7.1. A (*closed*) *half-plane* in $\mathbb{R}^n$ is a set of the form

$$\{(x_1,\dots,x_n) \in \mathbb{R}^n : a_0 + a_1x_1 + \cdots + a_nx_n \le 0\}$$

for some $a_0,\dots,a_n \in \mathbb{R}$ with $a_1,\dots,a_n$ not all 0. A (*closed*) *polytope* in $\mathbb{R}^n$ is a bounded subset $\mathcal{P} \subset \mathbb{R}^n$ that is the intersection of finitely many (closed) half-planes.

DEFINITION 1.7.2. Let $L \subset \mathbb{R}^n$ be a lattice. A (closed polytope) *fundamental domain for L* is a closed polytope $\mathcal{P} \subset \mathbb{R}^n$ such that $\mathcal{P}$ and its interior $\mathcal{P}^\circ$ have the following properties:

(1) $\mathcal{P}^\circ \longrightarrow \mathbb{R}^n/L$ is injective. (2) $\mathcal{P} \longrightarrow \mathbb{R}^n/L$ is surjective.

A lattice L has many different fundamental domains having many different shapes. But they all have one thing in common.

PROPOSITION 1.7.3. *Let $L \subset \mathbb{R}^n$ be a lattice. All closed polytope fundamental domains for L have the same volume. This volume is called the* determinant of L *and is denoted* $\operatorname{Det}(L)$.

PROOF. Left for the reader; see Exercise 1.8.7. □

The closure of the fundamental domain $\mathcal{F}(\mathcal{B})$ associated to a basis is a closed polytope fundamental domain in the sense of Definition 1.7.2. We now describe another sort of closed polytope fundamental domain whose very definition encodes the closest vector problem.

DEFINITION 1.7.4. Let $L \subset \mathbb{R}^n$ be a lattice, and let $\mathbf{v} \in L$. The (*closed*) *Voronoi cell around* $\mathbf{v}$ is the set

$$\mathcal{V}(\mathbf{v}) := \Big\{\mathbf{t} \in \mathbb{R}^n : \|\mathbf{t}-\mathbf{v}\| \le \|\mathbf{t}-\mathbf{w}\| \text{ for all } \mathbf{w} \in (L \smallsetminus \mathbf{v})\Big\}.$$

We note that the interior points $\mathbf{t} \in \mathcal{V}(\mathbf{v})^\circ$ of the Voronoi cell are exactly the target points $\mathbf{t} \in \mathbb{R}^n$ such that the CVP for $(L,\mathbf{t})$ has $\mathbf{v}$ as its unique solution, while target points on the boundary $\mathbf{t} \in \partial\mathcal{V}(\mathbf{v})$ have the property that the CVP for $(L,\mathbf{t})$ has more than one solution.

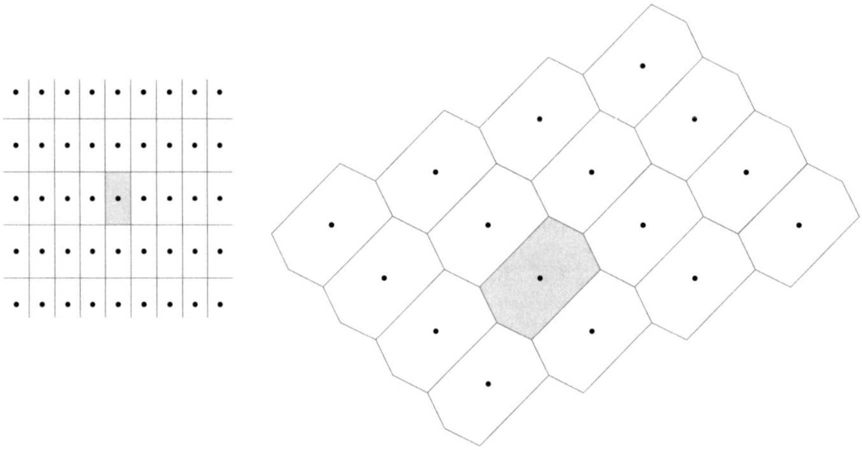

FIGURE 1.7.6. Two lattices and their Voronoi cells $\mathcal{V}(\mathbf{v})$, with one cell shaded

REMARK 1.7.5. The boundary of a Voronoi cell is a union of finitely many $(n-1)$-dimensional polytopes lying in hyperplanes in $\mathbb{R}^n$. More concretely, let $\mathbf{v}, \mathbf{w} \in L$ be distinct lattice points whose Voronoi cells share a common boundary. The set of points that are equidistant from $\mathbf{v}$ and $\mathbf{w}$ form the hyperplane $H(\mathbf{v}, \mathbf{w})$ in $\mathbb{R}^n$ that perpendicularly bisects the line segment from $\mathbf{v}_1$ to $\mathbf{v}_2$, and the common boundary of the two Voronoi cells is a subset of this hyperplane,

$$\mathcal{V}(\mathbf{v}) \cap \mathcal{V}(\mathbf{w}) \subset H(\mathbf{v}, \mathbf{w}).$$

Figure 1.7.6 illustrates the Voronoi cells for two different lattices in $\mathbb{R}^2$. If you imagine the line segment connecting lattice points in adjacent cells, you will see that the line segment is perpendicular to the common boundary of the cells.

PROPOSITION 1.7.7. *Let $L \subset \mathbb{R}^n$ be a lattice, and let $\mathbf{v} \in L$. The Voronoi cell $\mathcal{V}(\mathbf{v})$ containing $\mathbf{v}$ is a fundamental domain for L.*

PROOF. (We remark that the proof of Proposition 1.7.7 uses some basic point-set topology.) First suppose that $\mathbf{t}_1, \mathbf{t}_2 \in \mathcal{V}(\mathbf{v})^\circ$ have the same image in $\mathbb{R}^n/L$. Our goal is to prove that $\mathbf{t}_1 = \mathbf{t}_2$, which will show that $\mathcal{V}(\mathbf{v})^\circ \to \mathbb{R}^n/L$ is injective. The assumption says that

$$\mathbf{t}_1 - \mathbf{t}_2 = \mathbf{u} \quad \text{for some } \mathbf{u} \in L.$$

We suppose that $\mathbf{u} \neq \mathbf{0}$ and use the following calculation to derive a contradiction:

$$\begin{aligned}
\|\mathbf{t}_1 - \mathbf{v}\| &< \min_{\mathbf{w} \in (L \smallsetminus \mathbf{v})} \|\mathbf{t}_1 - \mathbf{w}\| && \text{since } \mathbf{t}_1 \in \mathcal{V}(\mathbf{v})^\circ, \\
&\le \big\|\mathbf{t}_1 - (\mathbf{u} + \mathbf{v})\big\| && \text{since we are assuming that } \mathbf{u} \neq 0, \\
& && \text{so we can take } \mathbf{w} = \mathbf{u} + \mathbf{v}, \\
&= \|\mathbf{t}_2 - \mathbf{v}\| && \text{since } \mathbf{t}_1 - \mathbf{t}_2 = \mathbf{u}, \\
&= \min_{\mathbf{w} \in L} \|\mathbf{t}_2 - \mathbf{w}\| && \text{since } \mathbf{t}_2 \in \mathcal{V}(\mathbf{v})^\circ, \\
&= \min_{\mathbf{w} \in L} \big\|\mathbf{t}_1 - (\mathbf{u} + \mathbf{w})\big\| && \text{since } \mathbf{t}_1 - \mathbf{t}_2 = \mathbf{u},
\end{aligned}$$

$$\begin{aligned}
&= \min_{\mathbf{w}\in L} \|\mathbf{t}_1 - \mathbf{w}\| && \text{since } \mathbf{u} \in L, \text{ so min is the same,} \\
&= \|\mathbf{t}_1 - \mathbf{v}\| && \text{since } \mathbf{t}_1 \in \mathcal{V}(\mathbf{v})^\circ.
\end{aligned}$$

This strict inequality is a contradiction, which proves that $\mathbf{u} = \mathbf{0}$, and thus $\mathbf{t}_1 = \mathbf{t}_2$.

It remains to show that $\mathcal{V}(\mathbf{v}) \to \mathbb{R}^n/L$ is surjective. Let $\widehat{\mathbf{t}} \in \mathbb{R}^n/L$ be an arbitrary point, and let $\widehat{U} \subset \mathbb{R}^n/L$ be an arbitrary (small) open neighborhood of $\widehat{\mathbf{t}}$. Since the map $\mathbb{R}^n \to \mathbb{R}^n/L$ is a local isomorphism (this uses the fact that L is discrete), we can find an open set $U \subset \mathbb{R}^n$ and a point $\mathbf{t} \in \mathbb{R}^n$ such that $\mathbf{t} \mapsto \widehat{\mathbf{t}}$ and such that the map $U \to \widehat{U}$ is a bijection.

As explained in Remark 1.7.5, the boundaries of the Voronoi cells in $\mathbb{R}^n$ are pieces of hyperplanes, so they cannot cover the open subset U. Hence there is a point $\mathbf{t}' \in U$ lying in the interior $\mathcal{V}(\mathbf{w})^\circ$ of some Voronoi cell $\mathcal{V}(\mathbf{w})$. But then

$$\mathbf{t}' - \mathbf{w} + \mathbf{v} \in \mathcal{V}(\mathbf{v})^\circ,$$

and hence

$$\widehat{\mathbf{t}}' := (\text{image of } \mathbf{t}' \text{ in } \mathbb{R}^n/L) \in \widehat{U}.$$

To recapitulate, we have shown that for every $\widehat{\mathbf{t}} \in \mathbb{R}^n/L$ and every open neighborhood $\widehat{U} \subset \mathbb{R}^n/L$ of $\widehat{\mathbf{t}}$, the image of $\mathcal{V}(\mathbf{0})^\circ$ in $\mathbb{R}^n/L$ has a point in common with $\widehat{U}$. Hence the image of $\mathcal{V}(\mathbf{0})^\circ$ is dense in $\mathbb{R}^n/L$, so its closure is all of $\mathbb{R}^n/L$. □

We can use Voronoi cells to prove Theorem 1.4.2 for $k = 1$, thereby proving a fundamental upper bound for the shortest non-zero vectors in a lattice.

Proof of Theorem 1.4.2 for $k = 1$. We are given a lattice $L \subset \mathbb{R}^n$, and we want to show that there is a vector $\mathbf{v}_1 \in L$ satisfying

$$\|\mathbf{v}_1\| \le \frac{2}{\sqrt{\pi}}\Gamma\left(\frac{1}{2}n + 1\right)^{1/n} \operatorname{Det}(L)^{1/n}.$$

We let

$$\lambda := \lambda_1(L) := \min\bigl\{\|\mathbf{w}\| : \mathbf{w} \in L,\ \mathbf{w} \ne 0\bigr\}$$

be the length of the shortest non-zero vectors in L. We claim that the open ball

$$B_{\lambda/2}(\mathbf{0}) := \bigl\{\mathbf{t} \in \mathbb{R}^n : \|\mathbf{t}\| < \lambda/2\bigr\}$$

is entirely contained within the Voronoi cell $\mathcal{V}(\mathbf{0})$ around $\mathbf{0}$. To see why, suppose that

$$\mathbf{t} \in B_{\lambda/2}(\mathbf{0}) \cap \mathcal{V}(\mathbf{v}) \quad \text{for some } \mathbf{v} \in L.$$

Then

$$\begin{aligned}
\|\mathbf{v}\| &\le \|\mathbf{t}\| + \|\mathbf{t} - \mathbf{v}\| && \text{triangle inequality,} \\
&= \|\mathbf{t}\| + \min_{\mathbf{w}\in L} \|\mathbf{t} - \mathbf{w}\| && \text{since } \mathbf{t} \in \mathcal{V}(\mathbf{v}), \\
&\le \|\mathbf{t}\| + \|\mathbf{t}\| && \text{taking } \mathbf{w} = \mathbf{0}, \\
&< \lambda && \text{since } \mathbf{t} \in B_{\lambda/2}(\mathbf{0}).
\end{aligned}$$

This strict inequality and the definition of λ tell us that $\mathbf{v} = \mathbf{0}$. Hence the ball $B_{\lambda/2}(\mathbf{0})$ is contained in the Voronoi cell $\mathcal{V}(\mathbf{0})$.

To recapitulate, we have proven that

$$B_{\lambda/2}(\mathbf{0}) \subseteq \mathcal{V}(\mathbf{0}). \tag{1.7.8}$$

Taking volumes yields

$$\begin{aligned} \operatorname{Det}(L) &= \operatorname{Volume}(\mathcal{V}(\mathbf{0})) && \text{from Proposition 1.7.3,} \\ &\geq \operatorname{Volume}(B_{\lambda/2}(\mathbf{0})) && \text{from (1.7.8),} \\ &= (\lambda/2)^n \operatorname{Volume}(B_1(\mathbf{0})). \end{aligned}$$

Solving for λ yields the desired inequality,

$$\min_{\mathbf{w}\in(L\smallsetminus\mathbf{0})} \|\mathbf{w}\| := \lambda \leq \underbrace{2\operatorname{Volume}(B_1(\mathbf{0}))^{1/n}}_{\text{This quantity depends only on } n.} \operatorname{Det}(L)^{1/n}.$$

Hence if we set

$$\gamma := 2\operatorname{Volume}(B_1(\mathbf{0}))^{1/n} = \frac{4}{\pi}\Gamma\left(\frac{1}{2}n+1\right)^{2/n} \approx \frac{2n}{\pi e},$$

then we have proven that every lattice $L \subset \mathbb{R}^n$ contains a non-zero vector of length at most $\gamma \operatorname{Det}(L)^{1/n}$. □

1.8. Exercises for Lecture 1.

Exercise 1.8.1. Prove Proposition 1.1.3, which says that every discrete subgroup of $\mathbb{R}^n$, has a finite $\mathbb{Z}$-basis, i.e., every lattice (regardless of its rank) in $\mathbb{R}^n$, has a finite $\mathbb{Z}$-basis.

Note: For the remaining exercises in this section, all lattices are assumed to have maximal rank.

Exercise 1.8.2. Prove Proposition 1.1.6 which says that the value of $\operatorname{Det}(L)$ does not depend on the choice of a basis for L.

Exercise 1.8.3. Let $L \subset \mathbb{R}^n$ be a lattice, let $\mathcal{B} = \{\mathbf{v}_1, \dots, \mathbf{v}_n\}$ be a basis for L, let $A \in \operatorname{Mat}_{n\times n}(\mathbb{Z})$ be a matrix with $\det(A) \neq 0$, and let L' be the lattice spanned by the column vectors of $\mathcal{M}(\mathcal{B})A$.

(1) Prove that L' is a sublattice of L and satisfies

$$L/L' \cong \mathbb{Z}^n/A\mathbb{Z}^n \quad \text{and} \quad (L : L') = |\det A|.$$

(2) Prove that every sublattice $L' \subseteq L$ of finite index arises in this way.

Exercise 1.8.4. Write a computer program implementing Babai's algorithm and use it to attempt to solve the following CVP problems.

(1) The lattice $L \subset \mathbb{R}^4$ is spanned by

$$\mathbf{v}_1 = (1,-1,0,1),\ \mathbf{v}_2 = (2,-1,-1,1), \mathbf{v}_3 = (2,-1,1,1),\ \mathbf{v}_4 = (1,-1,2,0),$$

and the target vector $\mathbf{t} = (11, 23, 9, 15)$.

(2) The lattice $L' \subset \mathbb{R}^4$ is spanned by

$$\mathbf{v}_1' = (-1,3,-4,-1),\ \mathbf{v}_2' = (-2,2,-1,-1),$$
$$\mathbf{v}_3' = (-6,12,-19,-5),\ \mathbf{v}_4' = (-9,16,-26,-7).$$

and the target vector $\mathbf{t} = (11, 23, 9, 15)$.

(3) Let $\mathbf{w} \in L$ and $\mathbf{w}' \in L'$ be the lattice vectors in (1) and (2) that Babai's algorithm says are close to the target vector $\mathbf{t}$. Compute the distances

$$\|\mathbf{w}-\mathbf{t}\| \quad \text{and} \quad \|\mathbf{w}'-\mathbf{t}\|.$$

(4) Prove that the lattice L and L' are in fact the same lattice, and use (3) to deduce that Babai's algorithm using the good basis in (1) gives a better solution to the CVP for L and $\mathbf{t}$ than using the bad basis in (2).

EXERCISE 1.8.5. Let L be a lattice, and let $\mathcal{B} := \{\mathbf{v}_1, \ldots, \mathbf{v}_n\}$ be a basis for L.

(1) Suppose that the vectors in $\mathcal{B}$ are pairwise orthogonal; i.e., $\mathbf{v}_i \cdot \mathbf{v}_j = 0$ for all $i \neq j$. Prove that Babai's algorithm solves the CVP for every target vector.
(2) Prove the converse; i.e., if Babai's algorithm solves the CVP for every target vector, then $\mathcal{B}$ is an orthogonal basis.

EXERCISE 1.8.6. Compare Blichfeldt's upper bound for Hermite's constant with the actual values for $1 \leq n \leq 8$ and $n = 24$, as listed in Remark 1.4.7.

EXERCISE 1.8.7. Prove Proposition 1.7.3, which says that all fundamental domains for a lattice L have the same volume.

EXERCISE 1.8.8. Prove Theorem 1.4.2 for all k. (This is a challenging problem!)

EXERCISE 1.8.9. Let $L \subset \mathbb{R}^n$ be a lattice, and let $\mathcal{B} := \{\mathbf{v}_1, \ldots, \mathbf{v}_n\}$ be a basis for L. The *orthogonality defect* of $\mathcal{B}$ is the ratio

$$\mathsf{OD}(\mathcal{B}) := \frac{\|\mathbf{v}_1\| \cdot \|\mathbf{v}_2\| \cdots \|\mathbf{v}_n\|}{\operatorname{Det}(L)}.$$

Hadamard's inequality (Theorem 1.4.1) says that $\mathsf{OD}(\mathcal{B}) \geq 1$, with equality if and only if $\mathcal{B}$ is an orthogonal basis, so the size of $\mathsf{OD}(\mathcal{B})$ is a measure of the non-orthogonality of the basis. We define

$$\kappa_n := \sup_{L \subset \mathbb{R}^n} \inf_{\mathcal{B} \text{ basis for } L} \mathsf{OD}(\mathcal{B}).$$

(1) Prove that κ_n is finite. What explicit uppper bound for κ_n is provided by Minkowski's theorem (Theorem 1.4.2)?
(2) Compute the orthogonality defects of the two bases for the lattice $L = L'$ in Exercise 1.8.4. How do they compare with the estimate for κ_4 in (1)?

EXERCISE 1.8.10. Let $\mathbb{B}^n_R \subset \mathbb{R}^n$ be a ball of radius R.

(1) Prove that the volume of $\mathbb{B}^n_R$ is given by the formula

$$\operatorname{Volume}(\mathbb{B}^n_R) := \operatorname{Volume}\big(\{\mathbf{x} \in \mathbb{R}^n : \|\mathbf{x}\| \leq R\}\big) = \frac{\pi^{n/2}}{\Gamma\big(\frac{1}{2}n + 1\big)} R^n,$$

where Γ is the *Gamma function*

$$\Gamma(z) := \int_0^\infty t^z e^{-t} \frac{dt}{t}.$$

(2) Prove that the Gamma function satisfies $\Gamma(z+1) = z\Gamma(z)$, so in particular if m is an integer, then $\Gamma(m+1) = m!$.
(3) Use

$$(\textbf{Stirling'sFormula}) \qquad \Gamma(z+1) \approx \left(\frac{z}{e}\right)^z \cdot \sqrt{2\pi z} \text{ as } z \to \infty,$$

to prove that

$$\operatorname{Volume}(\mathbb{B}^n_R) \approx \left(\frac{2\pi e}{n}\right)^{n/2} R^n.$$

EXERCISE 1.8.11. Let L be a lattice of dimension n, and let $\mathbf{t} \in \mathbb{R}^n$ be a target vector. Suppose that you have an algorithm that solves SVP in dimension $n+1$. This exercise explains how to use your SVP solver to try to solve the given CVP.

For each $\delta > 0$, we form a new lattice L_δ in dimension $n+1$ by setting

$$\mathbf{v}'_1 := (\mathbf{v}_1, 0),\ \ldots,\ \mathbf{v}'_n := (\mathbf{v}_n, 0),\ \mathbf{t}' := (\mathbf{t}, \delta)$$

and taking the span

$$L_\delta := \operatorname{Span}_{\mathbb{Z}}(\mathbf{v}'_1, \ldots, \mathbf{v}'_n, \mathbf{t}') \subset \mathbb{R}^{n+1}.$$

(1) Prove that $\lambda_1(L_\delta) \to 0$ as $\delta \to 0$.
(2) Prove that either $\lambda_1(L_\delta) \to \lambda_1(L)$ as $\delta \to \infty$.
(3) Explain why it often happens that there a value of δ so that the smallest non-zero vector in L_δ has the form

$$m_1\mathbf{v}'_1 + \cdots + m_n\mathbf{v}'_n \pm \mathbf{t}'.$$

If this occurs, explain why $\mp(m_1\mathbf{v}_1 + \cdots + m_n\mathbf{v}_n)$ is a point in L that is close to $\mathbf{t}$. (Sometimes, but not always, it will be a closest such lattice point.)
(4) Suppose that the SVP solver applied to L_δ returns the vector

$$m_1\mathbf{v}'_1 + \cdots + m_n\mathbf{v}'_n + m_0\mathbf{t}' \text{ with } |m_0| \neq 1.$$

If $m_0 = 0$, should we increase or decrease δ before trying again? Same question if $|m_0| \geq 2$.
(5) Prove that $\operatorname{Det}(L_\delta) = \operatorname{Det}(L) \cdot \delta$.
(6) Use the Gaussian heuristic (Heuristic 1.6.1) to show that a good starting value for δ is a solution to the equation

$$\frac{n+1}{2\pi e} \cdot \operatorname{Det}(L)^{2/(n+1)} \cdot \delta^{2/(n+1)} = \frac{n}{2\pi e} \cdot \operatorname{Det}(L)^{2/n} + \delta^2.$$

(7) Let $\mathbf{v}_1 = (1,0)$ and $\mathbf{v}_3 = (0,3)$, and consider the lattice and target point

$$L = \operatorname{Span}_{\mathbb{Z}}(\mathbf{v}_1, \mathbf{v}_2) \subset \mathbb{R}^2 \quad \text{and} \quad \mathbf{t} = \left(\frac{3}{5}, \frac{3}{2}\right).$$

Let $\delta > 0$. Prove that every vector in $\mathbf{w}' \in L_\delta$ of the form

$$\mathbf{w}' = m_1\mathbf{v}'_1 + m_2\mathbf{v}'_2 \pm \mathbf{t}'$$

satisfies $\|\mathbf{w}'\| > \|\mathbf{v}'_1\|$. Conclude that as δ decreases, the SVP solver as in (4) returns vectors with $|m_0| \geq 2$ for a while, and then it jumps to $m_0 = 0$, so there is no δ that gives $m_0 = \pm 1$. (The author thanks Jonathan Love for noting this phenomenon and providing an example.)

2. Lattice Reduction: The Practical Problem of Solving Hard Lattice Problems

2.1. Introduction. We recall the two fundamental lattice problems:

Shortest Vector Problem (SVP) Find a shortest non-zero vector in L. **Closest Vector Problem** (CVP) Find a vector in L closest to a target $\mathbf{t}$.

As explained in Exercise 1.8.11, an algorithm to solve the SVP can generally be adapted to solve the CVP in one lower dimension, and we have seen that Babai's method (Section 1.3) can solve apprCVP if it is given a good basis; i.e., a basis whose vectors are reasonably orthogonal. In this lecture we thus concentrate on the computational problem of constructing a good basis for a given lattice.

Suppose that L is a lattice and that we are given a basis $\mathcal{B} = \{\mathbf{v}_1, \ldots, \mathbf{v}_n\}$ for L. The Gram–Schmidt algorithm, which we now recall, transforms $\mathcal{B}$ into an orthogonal basis $\mathcal{B}^* = \{\mathbf{v}_1^*, \ldots, \mathbf{v}_n^*\}$ for $\mathbb{R}^n$.

ALGORITHM 2.1.1 (**Gram–Schmidt Algorithm**).

$$\mathbf{v}_1^* \leftarrow \mathbf{v}_1 \quad \textit{and} \quad \mathbf{v}_k^* \leftarrow \mathbf{v}_k - \sum_{i=1}^{k-1} \frac{\mathbf{v}_k \cdot \mathbf{v}_i^*}{\|\mathbf{v}_i^*\|^2} \mathbf{v}_i^* \quad \textit{for } 2 \le k \le n.$$

Intuition : $\mathbf{v}_k^* \leftarrow$ Projection of $\mathbf{v}_k$ onto $\operatorname{Span}_{\mathbb{R}}(\mathbf{v}_1, \ldots, \mathbf{v}_{k-1})^{\perp}$.

Thus $\mathbf{v}_i^*$ is a vector created using $\mathbf{v}_i$ and making the largest possible angle with the span of the previous vectors.

Unfortunately, the vectors $\mathbf{v}_1^*, \ldots, \mathbf{v}_n^*$ will probably not be in L, due to those pesky $\|\mathbf{v}_i^*\|^2$ factors in the denominators. If we want to do a Gram–Schmidt type of process and end up with vectors in L, we can try rounding the coefficients.

ALGORITHM 2.1.2 (**Gram–Schmidt Algorithm with Rounding**).

$$\mathbf{v}_1^* \leftarrow \mathbf{v}_1 \quad \textit{and} \quad \mathbf{v}_k^* \leftarrow \mathbf{v}_k - \sum_{i=1}^{k-1} \left\lfloor \frac{\mathbf{v}_k \cdot \mathbf{v}_i^*}{\|\mathbf{v}_i^*\|^2} \right\rceil \mathbf{v}_i^* \quad \textit{for } 2 \le k \le n.$$

Algorithm 2.1.2 creates a basis $\{\mathbf{v}_1^*, \ldots, \mathbf{v}_n^*\}$ for L, and this new basis might be more orthogonal than the original basis. On the other hand, it might not.

It is worth observing that the output from Algorithm 2.1.2 depends on the order in which the original basis vectors are fed into the algorithm. In particular, it's usually best to take $\mathbf{v}_1$ to be the smallest of the original vectors. More generally, sorting the original basis by size and then feeding the vectors from small to large into the algorithm is likely to give a better output than feeding them in large to small; see Exercise 2.8.5.

The LLL algorithm intertwines Algorithm 2.1.2 (Gram–Schmidt with Rounding) with the operation of swapping pairs of input vectors, leading to a process in which one moves up and down the list of basis vectors as they're being modified. Remarkably, with appropriate parameters, one can prove that the LLL algorithm is fast (polynomial time) and that it produces a basis that is pretty good. Further, there is an extension of LLL called LLL-BKZ in which one modifies larger blocks instead of swapping pairs. LLL-BKZ can produce better bases than LLL, albeit at the cost of added processing time.

In this lecture we describe the LLL algorithm, sketch the proof that it produces a reasonably good basis in polynomial time, and explain the idea behind LLL-BKZ. We start with dimension 2, where lattice reduction is very fast.

2.2. Lattice Reduction in Dimension 2. The following method of lattice reduction for 2-dimensional lattices is essentially due to Gauss.

> **Gauss Lattice Reduction in Dimension 2: Intuition**
> Alternately subtract multiples of one basis vector from the other until no further improvement is possible.

This leads to the following algorithm.

ALGORITHM 2.2.1 (**Lattice Reduction in Dimension 2**).

[1] Input: A basis $\{\mathbf{v}_1, \mathbf{v}_2\}$ for a lattice $L \subset \mathbb{R}^2$
[2] If $\|\mathbf{v}_2\| < \|\mathbf{v}_1\|$, then swap $\mathbf{v}_1$ and $\mathbf{v}_2$. [Swap Step]
[3] $m \leftarrow \mathbf{v}_1 \cdot \mathbf{v}_2 / \|\mathbf{v}_1\|^2$.
[4] If $\lfloor m \rceil = 0$
[5]   Output $\{\mathbf{v}_1, \mathbf{v}_2\}$
[6] Else
[7] $\mathbf{v}_2 \leftarrow \mathbf{v}_2 - \lfloor m \rceil \mathbf{v}_1$ [Size Reduction Step]
[8]   Go To Step 2
[9] End If

REMARK 2.2.2. If we remove the closest integer sign in Step 7 of Algorithm 2.2.1, then $\mathbf{v}_2 - m\mathbf{v}_1$ is actually orthogonal to $\mathbf{v}_1$. So Step 7 does the best that it can, subject to the requirement that the new $\mathbf{v}_2$ must be in L.

THEOREM 2.2.3. *The output* $\{\mathbf{v}_1, \mathbf{v}_2\}$ *from Algorithm* 2.2.1 *has the following properties:*

(1) $\mathbf{v}_1$ *is a solution to the SVP for* L.
(2) *The angle between* $\mathbf{v}_1$ *and* $\mathbf{v}_2$ *satisfies* $\dfrac{\pi}{3} \le \theta(\mathbf{v}_1, \mathbf{v}_2) \le \dfrac{2\pi}{3}$.

PROOF. Left for the reader; see Exerise 2.8.1. □

Size reduction (Step 7 in Algorithm 2.2.1) is illustrated in Figure 2.2.4. In this picture, the quantity $m = \mathbf{v}_1 \cdot \mathbf{v}_2 / \|\mathbf{v}_1\|^2$ is the exact factor that makes

$$\mathbf{v}_2 - m\mathbf{v}_1 \text{ perpendicular to } \mathbf{v}_1,$$

so

$$\mathbf{v}_2^* := \mathbf{v}_2 - m\mathbf{v}_1 = \text{the projection of } \mathbf{v}_2 \text{ onto } \mathbf{v}_1^\perp.$$

Rounding m to the nearest integer, we have

$$\lfloor m \rceil \ne 0 \quad \Longrightarrow \quad \mathbf{v}_2 - \lfloor m \rceil \mathbf{v}_1 \text{ is more orthogonal to } \mathbf{v}_1 \text{ than is } \mathbf{v}_2.$$

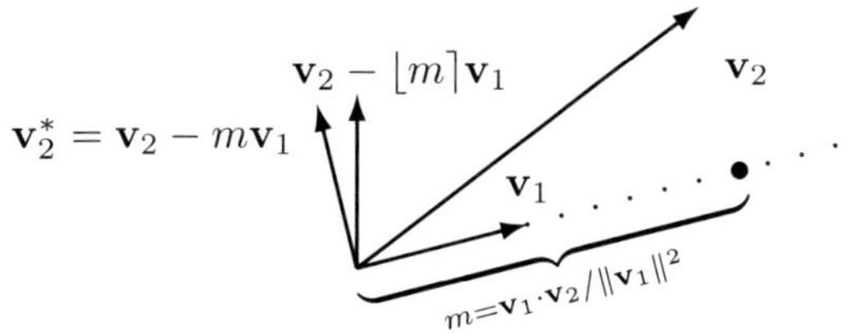

FIGURE 2.2.4. Size Reduction for Dimension 2 Lattice Reduction

2.3. The Size, Quasi-Orthogonality, and Lovász Conditions. If some coefficient in the Gram–Schmidt process satisfies

$$\frac{|\mathbf{v}_i \cdot \mathbf{v}_j^*|}{\|\mathbf{v}_j^*\|^2} > \frac{1}{2},$$

then replacing $\mathbf{v}_i$ by

$$\mathbf{v}_i - a\mathbf{v}_j \quad \text{for an appropriate } a \in \mathbb{Z}$$

makes the coefficient smaller, thereby increasing the angle. So Algorithm 2.1.2 (Gram–Schmidt with Rounding) helps until the basis has the following property.

DEFINITION 2.3.1. An ordered basis $\mathbf{v}_1, \ldots, \mathbf{v}_n$ for L satisfies the *Size Condition* if the output from the Gram–Schmidt algorithm (Algorithm 2.1.1) satisfies:

(2.3.2) SIZE CONDITION: $$\frac{|\mathbf{v}_i \cdot \mathbf{v}_j^*|}{\|\mathbf{v}_j^*\|^2} \le \frac{1}{2} \quad \text{for all } 1 \le j < i \le n.$$

On the other hand, we want the basis vectors to be at least somewhat orthogonal to one another. So we might ask that they also have the following property.

DEFINITION 2.3.3. An ordered basis $\mathbf{v}_1, \ldots, \mathbf{v}_n$ for L satisfies the *Quasi-Orthogonality Condition* if the output from the Gram–Schmidt algorithm (Algorithm 2.1.1) satisfies:

(2.3.4) QUASI-ORTHOGONALITY CONDITION: $$\|\mathbf{v}_{i+1}^*\| \ge \frac{\sqrt{3}}{2}\|\mathbf{v}_i^*\| \quad \text{for all } 1 \le i \le n-1.$$

An elaboration of the proof of Theorem 1.4.2 can be used to prove the following result.

THEOREM 2.3.5. *Every lattice has a basis satisfying both the Size Condition* (2.3.2) *and the Quasi-Orthogonality Condition* (2.3.4).

PROOF. Left for the reader; see Exercise 2.8.3. □

Unfortunately, the best known algorithms to find such a basis take time that is exponential in the dimension. The innovation of the LLL algorithm is to relax the Quasi-Orthogonality Condition.

DEFINITION 2.3.6. An ordered basis $\mathbf{v}_1, \ldots, \mathbf{v}_n$ for L satisfies the *Lovász Condition* if the output from the Gram–Schmidt algorithm (Algorithm 2.1.1) satisfies:

(2.3.7) LOVÁSZ CONDITION: $$\|\mathbf{v}_{i+1}^*\|^2 \ge \left(\frac{3}{4} - \frac{|\mathbf{v}_{i+1} \cdot \mathbf{v}_i^*|^2}{\|\mathbf{v}_i^*\|^4}\right)\|\mathbf{v}_i^*\|^2.$$

Although this looks complicated, it has a natural geometrical description. The Lovász Condition says that

$$\left\| \text{Projection of } \mathbf{v}_{i+1} \text{ onto } \operatorname{Span}_{\mathbb{R}}(\mathbf{v}_1, \dots, \mathbf{v}_{i-1})^{\perp} \right\|$$
$$\geq \frac{3}{4} \cdot \left\| \text{Projection of } \mathbf{v}_i \text{ onto } \operatorname{Span}_{\mathbb{R}}(\mathbf{v}_1, \dots, \mathbf{v}_{i-1})^{\perp} \right\|.$$

DEFINITION 2.3.8. An ordered basis for L satisfying both the Size Condition (2.3.2) and the Lovász Condition (2.3.7) is called an *LLL-Reduced Basis* for L.

THEOREM 2.3.9. *Let* $\{\mathbf{v}_1, \dots, \mathbf{v}_n\}$ *be an LLL-reduced basis for* L.

(1) $\|\mathbf{v}_1\| \leq 2^{(n-1)/2} \lambda_1(L)$.

(2) $\displaystyle\prod_{i=1}^{n} \|\mathbf{v}_i\| \leq 2^{n(n-1)/4} \operatorname{Det}(L)$.

PROOF. See Section 2.6 for a sketch of the proof. □

REMARK 2.3.10. How does an LLL-reduced basis stack up agianst the highly reduced bases that we expect from Theorem 1.4.2 or from the Gaussian heuristic? The earlier results say that every lattice L has a basis $\mathbf{v}_1, \dots, \mathbf{v}_n$ satisfying

$$\prod_{i=1}^{n} \|\mathbf{v}_i\| \lessapprox n^{n/2} \operatorname{Det}(L) \asymp 2^{\frac{1}{2} n \log_2 n} \operatorname{Det}(L).$$

So an LLL-reduced basis more-or-less has an extra n in the exponent. This is typical. All known lattice reduction algorithms either take exponential time, or else they give vectors whose norms are exponentially worse than optimal in terms of the dimension.

2.4. The Basic LLL Algorithm. The LLL lattice reduction algorithm is presented in Figure 2.4.2. We have opted to give a non-optimized version that has the advantage of being easier to understand and analyze.

THEOREM 2.4.3 (Lenstra, Lenstra, Lovász [**24**]). *Algorithm* 2.4.1 *in Figure* 2.4.2 *finds an LLL-reduced basis for* L *and has running time that is polynomial-time in* $n := \dim(L)$. *More precisely if the initial basis is* $\{\mathbf{v}_1, \dots, \mathbf{v}_n\}$ *and if we let* $B := \max \|\mathbf{v}_i\|$, *then the Swap Step (Step* 14*) is executed at most* $O(n^2 \log B)$ *times.*

PROOF. See Section 2.7 for a sketch of the proof. □

REMARK 2.4.4. Observations for the LLL Algorithm (Algorithm 2.4.1).

- At the end of the loop in Steps 4–8, the vectors $\mathbf{v}_1, \dots, \mathbf{v}_k$ satisfy the size condition (2.3.2).
- At Step 11, we already know that $\mathbf{v}_1, \dots, \mathbf{v}_{k-1}$ satisfy the Lovász condition, so we just need to check if the Lovász condition is true when we adjoin $\mathbf{v}_k$ to the list. If it is, we can go on to $\mathbf{v}_{k+1}$; but if not, then in some sense $\mathbf{v}_k$ is "better" than $\mathbf{v}_{k-1}$, so we swap them. But that means that we have to reconsider the new list of values $\mathbf{v}_1, \dots, \mathbf{v}_{k-1}$.
- The big outer k-loop from Steps 3 to 17 ends when $k = n$, at which point the list of vectors $\mathbf{v}_1, \dots, \mathbf{v}_n$ has been size reduced and satisfies the Lovász condition, so it forms an LLL-reduced basis.

ALGORITHM 2.4.1.

[1] Input: A basis $\{\mathbf{v}_1, \ldots, \mathbf{v}_n\}$ for a lattice L

[2] $k \leftarrow 2$

[3] Loop while $k \le n$

[4] Loop Down $j = k-1, k-2, \ldots, 2, 1$

[5] $\mathbf{v}_1^*, \ldots, \mathbf{v}_k^* \leftarrow \mathbf{v}_1, \ldots, \mathbf{v}_k$ [Gram–Schmidt (2.1.1)]

[6] $\mu_{k,j} \leftarrow (\mathbf{v}_k \cdot \mathbf{v}_j^*)/\|\mathbf{v}_j^*\|^2$

[7] $\mathbf{v}_k \leftarrow \mathbf{v}_k - \lfloor \mu_{k,j} \rceil \mathbf{v}_j$ [Size Reduction]

[8] End j Loop

[9] $\mathbf{v}_1^*, \ldots, \mathbf{v}_k^* \leftarrow \mathbf{v}_1, \ldots, \mathbf{v}_k$ [Gram–Schmidt (2.1.1)]

[10] $\mu_{k,k-1} \leftarrow (\mathbf{v}_k \cdot \mathbf{v}_{k-1}^*)/\|\mathbf{v}_{k-1}^*\|^2$

[11] If $\|\mathbf{v}_k^*\|^2 \ge \left(\frac{3}{4} - \mu_{k,k-1}^2\right) \|\mathbf{v}_{k-1}^*\|^2$ [Lovász Condition]

[12] $k \leftarrow k+1$

[13] Else

[14] Swap $\mathbf{v}_{k-1}$ and $\mathbf{v}_k$ [Swap Step]

[15] Set $k \leftarrow \max(k-1, 2)$

[16] End If

[17] End k Loop

[18] Output: The LLL-reduced basis $\{\mathbf{v}_1, \ldots, \mathbf{v}_n\}$

FIGURE 2.4.2. The LLL Lattice Reduction Algorithm (Unoptimized)

- We reiterate that Step 12 brings us a step closer to the end of the algorithm by incrementing k, but that there is a countervailing force in Step 15, the Swap Step, pulls us away from our final destination by decrementing k.
- Later in Section 2.7 we prove that Steps 14 and 15 are executed only finitely many times, and that the number of executions is bounded by a polynomial in n. Thus LLL runs in time that is polynomial in n.

REMARK 2.4.5. The LLL algorithm finds an LLL-reduced basis, so Theorem 2.3.9 tells us that LLL always finds a non-zero vector $\mathbf{v}$ in L that is no more than $2^{(n-2)/2}$ times as long as the shortest non-zero vector. In practice, LLL generally does better than this. But also in practice, if n is large, then $\|\mathbf{v}\|/\lambda_1(L)$ will be quite large.

2.5. Variants and Improvements to LLL. Many methods of improving LLL have been proposed over the years. Often they sacrifice provable polynomial time performance for improved operation on most lattices. One of the most important, which we briefly discuss, replaces the Swap Step with a more complicated procedure.

DEFINITION 2.5.1. A basis $\{\mathbf{v}_1, \ldots, \mathbf{v}_n\}$ is said to be a *Korkine–Zolotareff (KZ) Reduced Basis* if it satisfies both the SIZE CONDITION and the following:

For all $1 \le i \le n$, the vector $\mathbf{v}_i^*$ is the shortest non-zero vector in the projection of L onto $\mathrm{Span}_{\mathbb{R}}(\mathbf{v}_i^*, \ldots, \mathbf{v}_n^*)$.

There are no known subexponential algorithm for creating a KZ-reduced basis for all of L. Indeed, the first vector in a KZ-reduced basis for L is a shortest non-zero vector in L. So instead we use KZ-reduced bases for small sublattices of L.

THEOREM 2.5.2 (LLL-BKZ Algorithm with Blocksize β). (Schnorr–Euchner [**35, 36**]) *Rather than swapping* $\mathbf{v}_k$ *and* $\mathbf{v}_{k-1}$ *in Step* 14 *of LLL* (*Algorithm* 2.4.1), *we instead take the sublattice spanned by a block of vectors* $\mathbf{v}_i, \mathbf{v}_{i+1}, \dots, \mathbf{v}_{i+\beta-1}$ *and replace these vectors with a KZ-reduced basis for the sublattice. The LLL-BKZ algorithm terminates in no more than* $O(\beta^{a\beta}n^b)$ *steps* (*for some small constants* a *and* b) *and finds a non-zero vector* $\mathbf{v} \in L$ *satisfying*

$$\|\mathbf{v}\| \le \left(\frac{\beta}{\pi e}\right)^{\frac{n-1}{\beta-1}} \lambda_1(L).$$

PROOF. We omit the proof. □

REMARK 2.5.3. Thus LLL-BKZ solves apprSVP to within an approximation factor roughly equal to $\beta^{n/\beta}$, as compared to the approximation factor $2^{n/2}$ guaranteed by LLL, but at the cost of a running time that is exponential in β. For example, if we want to solve apprSVP with an approximation factor n^d for some $d > 0$, then we need to take $\beta^{n/\beta} \approx n^d$, which forces $\beta \approx n/d$ and a running time that is exponential in n.

REMARK 2.5.4. As a practical matter, it is currently possible to run BKZ-LLL with blocksizes in the 50 to 100 range, but definitely impractical to use blocksize 500. There has been a considerable amount of research, both theoretical and experimental, devoted to estimating both the time required to run BKZ-LLL with blocksize β and the expected approximation factor for the shortest non-zero vector $\mathbf{v}$ returned by the algorithm. We give some examples of the sorts of results that appear in the literature, and point the reader to papers such as [**2, 8, 31, 34, 38**] for further information.

First, how long does it take to run BKZ-β, which is our name for BKZ-LLL run using blocksize β? Conservative estimates currently suggest that running BKZ-β takes at least $2^{0.292\beta}$ operations on a classical computer and at least $2^{0.265\beta}$ operations a quantum computer.[10] Second, how good is the output from BKZ-β? This is normally measured in terms of the Hermite root factor.

DEFINITION 2.5.5. The *Hermite root factor for BKZ-β applied to the n-dimensional lattice L* is the quantity

$$\delta(\text{BKZ-}\beta, L) := \left\{ \frac{\left(\begin{matrix}\text{the length } \|\mathbf{v}\| \text{ of the shortest}\\ \text{non-zero vector found by BKZ-}\beta\end{matrix}\right)}{\operatorname{Det}(L)^{1/n}} \right\}^{1/(n-1)}.$$

Experiments and a geometric series heuristic suggest that for a fixed blocksize $\beta \ge 50$ and a random lattice $L \subset \mathbb{R}^n$ with n significantly larger than β, a reasonable

[10]We will not make the notion of "operation" precise, but we note that in 2022, the world's fastest computers operate at roughly 1 exaFLOPS, where FLOPS stands for "floating point operation per second" and "exa" means 10^{18}. So in terms of floating point operations, the current fastest computers in the world can perform roughly $2^{59.8}$ operations per second. This explains why BKZ-β with $\beta = 50$ is feasible, but $\beta = 500$ is not.

estimate for the Hermite root factor for BKZ-β is given by

$$\delta(\text{BKZ-}\beta, L) \approx \left(\frac{\beta}{2\pi e} \cdot (\beta\pi)^{1/\beta}\right)^{1/(2\beta-2)}.$$

An interesting alternative method for estimating $\delta(\text{BKZ-}\beta, L)$ is the BKZ-simulator of Chen and Nguyen [**8**, Algorithm 2]. The input to their simulator is the list of the logarithms of the lengths of the lattice basis vectors converted into Hermite normal form.[11] However, they assume that the lattice is random in some suitable, but not precisely specified, sense. As the preceding material indicates, there is much room for further research on the operating characteristics of lattice reduction algorithms.

REMARK 2.5.6. Many lattice-based cryptographic algorithms rely on hiding one or more small vectors in a lattice. What this means is that the lattice L has a non-zero vector $\mathbf{v}$ whose length $\|\mathbf{v}\|$ is significantly smaller than the shortest non-zero length predicted by the Gaussian heuristic described in Section 1.6. For example, suppose that the quantities

$$\overbrace{\lambda_1(L) := \min_{\mathbf{v}\in(L\smallsetminus\mathbf{0})} \|\mathbf{v}\|}^{\text{actual shortest vector length}} \text{ and } \overbrace{\gamma(L) := \sqrt{n/2\pi e}\cdot(\operatorname{Det} L)^{1/n}}^{\text{expected shortest vector length}} \text{ satisfy } \lambda_1(L) \lessapprox n^{-1/2}\gamma(L).$$

It turns out that if $\lambda_1(L)$ is exponentially smaller than $\gamma(L)$, then lattice reduction algorithms such as LLL-BKZ have a fairly easy time solving the SVP; but if $\lambda_1(L)$ is only polynomially smaller than $\gamma(L)$, then it takes exponential time to solve SVP. Heuristic arguments combined with experiments suggest that for a random lattice with a hidden small vector, BKZ-β will find the small vector if its Hermite root factor satisfies either:

$$\lambda_1(L) \le 3\sqrt{n/2\pi e}\cdot\delta(\text{BKZ-}\beta, L)^{-n}\cdot\operatorname{Det}(L)^{1/n}, \qquad [\mathbf{16}, \text{Section 2.2.1}];$$

$$\lambda_1(L) \le \sqrt{n/\beta}\cdot\delta(\text{BKZ-}\beta, L)^{-(n-2\beta)}\cdot\operatorname{Det}(L)^{1/n}, \qquad [\mathbf{3},\mathbf{4}].$$

As these very different estimates suggest, we still have a long way to go in understanding the behavior of lattice reduction algorithms for lattices containing a vector that is moderately shorter than expected.

2.6. LLL Bases Are Nice: Proof Sketch.

PROOF SKETCH OF THEOREM 2.3.9. We prove the second estimate, and leave the first for you; see Exercise 2.8.4. So our goal is to prove that an LLL-reduced basis satisfies

$$\prod_{i=1}^{n}\|\mathbf{v}_i\| \le 2^{n(n-1)/4}\operatorname{Det}(L).$$

The Lovász condition and $|\mu_{i,i-1}| \le \frac{1}{2}$ give

$$\|\mathbf{v}_i^*\|^2 \ge \left(\frac{3}{4} - \mu_{i,i-1}^2\right)\|\mathbf{v}_{i-1}^*\|^2 \ge \frac{1}{2}\|\mathbf{v}_{i-1}^*\|^2.$$

Applying repeatedly yields the following useful inequality for $j < i$:

$$\|\mathbf{v}_j^*\|^2 \le 2^{i-j}\|\mathbf{v}_i^*\|^2.$$

---

[11] An invertible square matrix is in column Hermite normal form if it is lower triangular and the diagonal entries satisfy certain size comparison properties.

We now compute

$$
\begin{aligned}
\|\mathbf{v}_i\|^2 &= \left\| \mathbf{v}_i^* + \sum_{j=1}^{i-1} \mu_{i,j}\mathbf{v}_j^* \right\|^2 && \text{from Gram–Schmidt,} \\
&= \|\mathbf{v}_i^*\|^2 + \sum_{j=1}^{i-1} \mu_{i,j}^2 \|\mathbf{v}_j^*\|^2 && \text{since } \mathbf{v}_1^*, \dots, \mathbf{v}_n^* \text{ are orthogonal,} \\
&\le \|\mathbf{v}_i^*\|^2 + \sum_{j=1}^{i-1} \frac{2^{i-j}\|\mathbf{v}_i^*\|^2}{4} && \text{since } |\mu_{i,j}| \le \tfrac{1}{2} \text{ and } \|\mathbf{v}_j^*\|^2 \le 2^{i-j}\|\mathbf{v}_i^*\|^2, \\
&= \frac{1+2^{i-1}}{2}\|\mathbf{v}_i^*\|^2 \\
&\le 2^{i-1}\|\mathbf{v}_i^*\|^2.
\end{aligned}
$$

Multiplying over $1 \le i \le n$ yields

$$
\prod_{i=1}^{n} \|\mathbf{v}_i\|^2 \le \prod_{i=1}^{n} 2^{i-1}\|\mathbf{v}_i^*\|^2 = 2^{n(n-1)/2}\left(\prod_{i=1}^{n} \|\mathbf{v}_i^*\|\right)^2 = 2^{n(n-1)/2}\operatorname{Det}(L)^2.
$$

□

2.7. LLL Runs in Polynomial Time: Proof Sketch.

PROOF SKETCH OF THEOREM 2.4.3. Our goal is to prove that LLL terminates in time that is polynomial in the dimension n. For ease of exposition, we assume that $L \subseteq \mathbb{Z}^n$.

The idea of the proof is to define a function

$$
D : \{\text{bases } \mathbf{v}_1, \dots, \mathbf{v}_n \text{ for } L\} \longrightarrow \mathbb{R}_{\ge 1}
$$

that measures the "complexity" of a basis and to prove that D has the following property for the Lovász swap step:

$$
\begin{pmatrix}\text{Lovász condi-}\\ \text{tion is false}\end{pmatrix} \Longrightarrow D\begin{pmatrix}\text{basis after}\\ \text{swap step}\end{pmatrix} \le \frac{\sqrt{3}}{2} \cdot D\begin{pmatrix}\text{basis before}\\ \text{swap step}\end{pmatrix}. \tag{2.7.1}
$$

Applying this repeatedly, we conclude that

$$
D\begin{pmatrix}\text{basis after swap step has}\\ \text{been executed } k \text{ times}\end{pmatrix} \le \left(\frac{\sqrt{3}}{2}\right)^k D\begin{pmatrix}\text{original}\\ \text{basis}\end{pmatrix}.
$$

But $D(\mathcal{B}) \ge 1$ for any basis, so

$$
\begin{pmatrix}\text{number of times that}\\ \text{the swap step is executed}\end{pmatrix} \le \frac{\log D(\text{original basis})}{\log(2/\sqrt{3}\,)}.
$$

It remains to define a function D satisfying (2.7.1) and to prove that D of the original basis is $O(n^2 \log B)$.

DEFINITION 2.7.2. For any basis $\mathcal{B} := \{\mathbf{v}_1, \dots, \mathbf{v}_n\}$ of L, we define a sequence of sublattices

$$
L_\ell := \operatorname{Span}_{\mathbb{Z}}\{\mathbf{v}_1, \dots, \mathbf{v}_\ell\}, \quad 1 \le \ell \le n.
$$

We note that L_ℓ is a lattice of rank ℓ sitting in $\mathbb{Z}^n$. It follows that its absolute determinant satisfies

$$\operatorname{Det}(L_\ell)^2 \in \mathbb{N}, \quad \text{and hence in particular that} \quad \operatorname{Det}(L_\ell) \geq 1.$$

(See Exercise 2.8.2.) We define the *complexity of the ordered basis* $\mathcal{B}$ to be the quantity

$$D(\mathcal{B}) := \prod_{\ell=1}^{n} \operatorname{Det}(L_\ell).$$

Let $\mathbf{v}_1^*, \ldots, \mathbf{v}_n^*$ be the orthogonal Gram–Schmidt basis of $\mathbb{R}^n$ associated to $\mathcal{B}$; i.e., the output of applying Algorithm 2.1.1 to $\mathcal{B}$. Then

$$\operatorname{Det}(L_\ell) = \prod_{i=1}^{\ell} \|\mathbf{v}_i^*\|, \quad \text{and hence} \quad D(\mathcal{B}) = \prod_{i=1}^{n} \|\mathbf{v}_i^*\|^{n+1-i}.$$

Since each $\operatorname{Det}(L_\ell) \geq 1$, we find that

$$1 \leq D(\mathcal{B}) \leq \Bigl(\max_{1\leq i\leq n} \|\mathbf{v}_i^*\|\Bigr)^{\frac{n(n+1)}{2}} \leq \Bigl(\max_{\mathbf{v}\in\mathcal{B}} \|\mathbf{v}\|\Bigr)^{\frac{n(n+1)}{2}} = B^{\frac{n(n+1)}{2}}.$$

Hence in order to complete the proof that LLL terminates in at most $O(n^2 \log B)$ steps, we need to prove the Lovász swap property described in (2.7.1).

We start with the assumption:

The Lovász Condition is false at iteration k.

In other words, we assume that

$$\|\mathbf{v}_k^*\| < \sqrt{\frac{3}{4} - \mu_{k,k-1}^2} \cdot \|\mathbf{v}_{k-1}^*\|. \tag{2.7.3}$$

When we perform the swap step, i.e., when we swap $\mathbf{v}_k$ and $\mathbf{v}_{k-1}$, the basis of L_{k-1} changes from

$$\mathsf{OldBasis} = \{\mathbf{v}_1, \ldots, \mathbf{v}_{k-2}, \mathbf{v}_{k-1}\} \qquad \text{to} \qquad \mathsf{NewBasis} = \{\mathbf{v}_1, \ldots, \mathbf{v}_{k-2}, \mathbf{v}_k\}.$$

After this change, the Gram–Schmidt vectors $\mathbf{v}_1^*, \ldots, \mathbf{v}_{k-2}^*$ do not change, but the new $\mathbf{v}_{k-1}^*$ is related to the old $\mathbf{v}_{k-1}^*$ by

$$(\mathbf{v}_{k-1}^*)^{\text{new}} = \mu_{k,k-1}(\mathbf{v}_{k-1}^*)^{\text{old}} + (\mathbf{v}_k^*)^{\text{old}} \tag{2.7.4}$$

This and the negation of the Lovász condition imply that

$$\begin{aligned} &\bigl\|\mu_{k,k-1}(\mathbf{v}_{k-1}^*)^{\text{old}} + (\mathbf{v}_k^*)^{\text{old}}\bigr\|^2 \\ &\quad = \mu_{k,k-1}^2\bigl\|(\mathbf{v}_{k-1}^*)^{\text{old}}\bigr\|^2 + \bigl\|(\mathbf{v}_k^*)^{\text{old}}\bigr\|^2 \\ &\qquad\qquad \text{since } (\mathbf{v}_{k-1}^*)^{\text{old}} \text{ and } (\mathbf{v}_k^*)^{\text{old}} \text{ are orthogonal,} \\ &\quad < \mu_{k,k-1}^2\bigl\|(\mathbf{v}_{k-1}^*)^{\text{old}}\bigr\|^2 + \Bigl(\frac{3}{4} - \mu_{k,k-1}^2\Bigr)\bigl\|(\mathbf{v}_{k-1}^*)^{\text{old}}\bigr\|^2 \\ &\qquad\qquad \text{from (2.7.3),} \\ &\quad = \frac{3}{4}\bigl\|(\mathbf{v}_{k-1}^*)^{\text{old}}\bigr\|^2. \end{aligned} \tag{2.7.5}$$

This allows us to estimate how the determinant changes. Thus

$$\begin{aligned}
\operatorname{Det}(L_{k-1}^{\text{new}}) &= \|\mathbf{v}_1^*\| \cdot \|\mathbf{v}_2^*\| \cdots \|\mathbf{v}_{k-2}^*\| \cdot \left\|(\mathbf{v}_{k-1}^*)^{\text{new}}\right\| \\
&= \|\mathbf{v}_1^*\| \cdot \|\mathbf{v}_2^*\| \cdots \|\mathbf{v}_{k-2}^*\| \cdot \left\|(\mathbf{v}_{k-1}^*)^{\text{old}}\right\| \cdot \frac{\left\|(\mathbf{v}_{k-1}^*)^{\text{new}}\right\|}{\left\|(\mathbf{v}_{k-1}^*)^{\text{old}}\right\|} \\
&= \operatorname{Det}(L_{k-1}^{\text{old}}) \cdot \frac{\left\|(\mathbf{v}_{k-1}^*)^{\text{new}}\right\|}{\left\|(\mathbf{v}_{k-1}^*)^{\text{old}}\right\|} \\
&= \operatorname{Det}(L_{k-1}^{\text{old}}) \cdot \frac{\left\|\mu_{k,k-1}(\mathbf{v}_{k-1}^*)^{\text{old}} + (\mathbf{v}_k^*)^{\text{old}}\right\|}{\left\|(\mathbf{v}_{k-1}^*)^{\text{old}}\right\|} \qquad \text{from (2.7.4)}, \\
&< \operatorname{Det}(L_{k-1}^{\text{old}}) \cdot \frac{\sqrt{3}}{2} \qquad \text{from (2.7.5)}.
\end{aligned}$$

The swap only affects the basis of L_{k-1}; the bases of all of the other L_i are unchanged. Hence the complexity of the new swapped basis for L is related to the complexity of the old unswapped basis by

$$\begin{aligned}
D(\mathcal{B}^{\text{new}}) &= \Bigl(\prod_{i \neq k-1} \operatorname{Det}(L_i^{\text{old}})\Bigr) \cdot \operatorname{Det}(L_{k-1}^{\text{new}}) \\
&\le \Bigl(\prod_{i \neq k-1} \operatorname{Det}(L_i^{\text{old}})\Bigr) \cdot \frac{\sqrt{3}}{2} \operatorname{Det}(L_{k-1}^{\text{old}}) \\
&= \frac{\sqrt{3}}{2} D(\mathcal{B}^{\text{old}}).
\end{aligned}$$

This completes the proof of (2.7.1), and with it, the proof of Theorem 2.4.3. □

2.8. Exercises for Lecture 2.

EXERCISE 2.8.1. Prove Theorem 2.2.3, which says that the basis of a 2-dimensional lattice computed by the algorithm described in Algorithm 2.2.1 solves SVP and gives a second basis vector that is fairly orthogonal to the first basis vector .

EXERCISE 2.8.2. Let $\mathcal{B} := \{\mathbf{v}_1, \ldots, \mathbf{v}_\ell\} \subset \mathbb{R}^n$ be a linearly independent set of vectors, and let

$$L_{\mathcal{B}} := \operatorname{Span}_{\mathbb{Z}}(\mathbf{v}_1, \ldots, \mathbf{v}_\ell)$$

be the lattice that they span. N.B. We are allowing $\ell < n$. In particular, the fundamental domain

$$\mathcal{F}(\mathcal{B}) := \{t_1\mathbf{v}_1 + \cdots + t_\ell\mathbf{v}_\ell : 0 \le t_i < 1\}$$

is an ℓ-dimensional paralllelopiped in $\mathbb{R}^n$, and $\operatorname{Det}(L_{\mathcal{B}})$ is the ℓ-dimensional volume of $\mathcal{F}(\mathcal{B})$.

(1) Let $\mathbf{v}_1^*, \ldots, \mathbf{v}_\ell^*$ be the orthgonal vectors that are the output of Gram-Schmidt. Prove that

$$\operatorname{Det}(L_{\mathcal{B}}) = \prod_{i=1}^{\ell} \|\mathbf{v}_i^*\|.$$

(2) Let A be the n-by-ℓ matrix whose columns are $\mathbf{v}_1, \ldots, \mathbf{v}_\ell$, and let ${}^tA$ be the transpose of A. Prove that

$$\operatorname{Det}(L_{\mathcal{B}})^2 = \det({}^tAA) = \det\Bigl((\mathbf{v}_i \cdot \mathbf{v}_j)_{1 \le i,j \le \ell}\Bigr).$$

(3) Deduce that if $L_{\mathcal{B}} \subseteq \mathbb{Z}^n$, i.e., if $\mathbf{v}_1, \ldots, \mathbf{v}_\ell \in \mathbb{Z}^n$, then $\operatorname{Det}(L_{\mathcal{B}}) \ge 1$.

EXERCISE 2.8.3. Prove Theorem 2.3.5, which says that every lattice has a basis satisfying both the Size Condition (2.3.2) and the Quasi-Orthogonality Condition (2.3.4). (This is a challenging problem.)

EXERCISE 2.8.4. Let $\mathbf{v}_1, \dots, \mathbf{v}_n$ be an LLL-reduced basis for L. Prove the first part of Theorem 2.3.9:

$$\|\mathbf{v}_1\| \le 2^{(n-1)/2}\lambda_1(L).$$

EXERCISE 2.8.5. Consider the Gram–Schmidt With Rounding algorithm (2.1.2). We suggested that feeding it a list of vectors in small-to-large order is likely to give a better output than large-to-small.

(1) Justify this claim rigorously for lattices in $\mathbb{R}^2$.
(2) Consider the lattice L in $\mathbb{R}^3$ whose basis $\mathcal{B}$ consists fo the three vectors

$$\mathbf{v}_1 := (9, 3, -11),\ \mathbf{v}_2 := (1, -2, 3),\ \mathbf{v}_3 := (-190, -87, 140).$$

Feed this basis into the Gram–Schmidt With Rounding algorithm using all six permutations of the three basis vectors and compare the outputs in various ways. For example, compare the shortest vector produced, the value of the product $\prod \|\mathbf{v}_i^*\|$, and/or the complexity D of the output basis.

EXERCISE 2.8.6. Write a computer program implementing the LLL algorithm and do some of the following experiments.

(1) Choose random vectors $\mathbf{v}_1, \dots, \mathbf{v}_{30}$ in $\mathbb{Z}^{30}$, say with coefficients randomly selected in the range $[-100, 100]$, and let L be the lattice that they span. Run LLL on this basis for L and compute the orthogonality defect (Exercise 1.8.9) of the output. What happens if you use the same basis, but input the basis vectors in a different order?
(2) As in (1), but after running LLL, permute the output vectors and then run them through LLL again? Do you get a better basis, as measured by the orthogonality defect?
(3) Repeat (1) and (2) for different random $\mathbf{v}_1, \dots, \mathbf{v}_{30}$ and study the extent to which the results are different.
(4) Repeat (1) and (2) for a different dimension and/or vectors whose coordinates are in chosen in some other range, and compare the results.

EXERCISE 2.8.7. Algorithm 2.8.8 in Figure 2.8.9 describes Babai's nearest hyperplane algorithm [**5**]. It tends to give better results than the simpler "round coefficients to nearest integer" version described in Algorithm 1.3.1.

(1) Show that Algorithm 2.8.8 has the following geoemtric interpretation: First find an element of the translated lattice $L - \mathbf{t}$ whose nth coordinate in the basis $\{\mathbf{v}_1^*, \dots, \mathbf{v}_n^*\}$ is minimal. Then leave the nth coordinate fixed and apply the same process to the remaining $n-1$ coordinates. Etc.
(2) Prove that the output $\mathbf{s}$ from Algorithm 2.8.8 has the form

$$\mathbf{s} = s_1\mathbf{v}_1^* + \cdots + s_n\mathbf{v}_n^* \quad \text{with } |s_i| \le \frac{1}{2} \text{ for all } i.$$

(3) Prove that the output $\mathbf{s}$ from Algorithm 2.8.8 satisfies

$$\|\mathbf{s}\| \le \sqrt{n+1} \cdot \left(\max_{1 \le j \le i \le n} \frac{\|\mathbf{v}_j^*\|}{\|\mathbf{v}_i^*\|} \right) \cdot \min_{\mathbf{v} \in L} \|\mathbf{t} - \mathbf{v}\|.$$

It thus solves the approximate CVP with the indicated approximation factor. (This is a challenging problem.)

ALGORITHM 2.8.8 (**Babai's Nearest Hyperplane Algorithm**).

Input: A basis $\{\mathbf{v}_1, \ldots, \mathbf{v}_n\}$ for a lattice L and a target vector $\mathbf{t}$
$\mathbf{v}_1^*, \ldots, \mathbf{v}_n^* \leftarrow \mathbf{v}_1, \ldots, \mathbf{v}_n$ [Gram–Schmidt (2.1.1)]
$\mathbf{s} \leftarrow -\mathbf{t}$
Loop $i = 1, 2, \ldots, n$
  $m \leftarrow \lfloor (\mathbf{s} \cdot \mathbf{v}_{n-i+1}^*)/\|\mathbf{v}_{n-i+1}^*\|^2 \rceil$
  $\mathbf{s} \leftarrow \mathbf{s} - m\mathbf{v}_{n-i+1}$
End i Loop
Output: $\mathbf{s}$

FIGURE 2.8.9. Babai's Nearest Hyperplane Algorithm

3. Public Key Cryptography 101: A Brief Introduction

We start with the *Fundamental Problem of Cryptography.* Two people, say Alice and Bob[12], want to exchange information so that the eavesdropper Eve cannot read their messages.

3.1. Cryptography in the (pre-1970s) Dark Ages. For thousands of years, cryptography worked as follows:

Private Key Cryptography — A Single Shared Key:
Step 1: Alice and Bob meet and share a Secret Key.
Step 2: Alice and Bob go their separate ways.
Step 3: Bob writes a message, which is called the Plaintext, that he wants to send to Alice.
Step 4: Bob uses the Secret Key and a publicly available Encryption Algorithm to encrypt his message. The encrypted message is called the Ciphertext.
Step 5: Bob sends the Ciphertext to Alice.
Step 6: Alice uses the Secret Key and a publicly available Decryption Algorithm to decrypt Bob's message.
Step 7: Eve might intercept the Ciphertext, but without the Secret Key, she cannot decrypt it to recover the Plaintext.

REMARK 3.1.1. A fundamental problem with this scenario is that Bob and Alice cannot send messages until after they've met and exchanged a Secret Key. What if they've never met and have no secure way to exchange the key?

EXAMPLE 3.1.2. Alice is an internet company, and Bob wants to send Alice his credit card number.

3.2. Public Key Cryptography to the Rescue. Is there some way for Alice and Bob to securely communicate even if they've never met, and even if Eve is privy to every piece of information that they exchange? Seems unlikely, but in the mid-1970s Diffie and Hellman [**10**] proposed creating cryptosystems that use two

[12]In cryptography, Alice and Bob feature in an eternal stuggle for security against the eavesdropper Eve, the opponent Oscar, the malicious attacker Mallory, and an assortment of other evildoers!

keys, a Private Key that Alice keeps secret and a Public Key that she publishes.[13] Here's how a two-key system might work:

Public Key Cryptography — A Tale of Two Keys:

Step 1: Alice creates a two-piece key consisting of a secret Private Key and a Public Key.

Step 2: Alice publishes her Public Key.

Step 3: Bob choose a Plaintext that he wants to send to Alice.

Step 4: Bob uses Alice's Public Key and a publicly available Encryption Algorithm to encrypt his message.

Step 5: Bob sends the resulting Ciphertext to Alice.

Step 6: Alice uses her secret Private Key and a publicly available Decryption Algorithm to decrypt Bob's message.

Step 7: Eve can intercept the Ciphertext, and she know's Alice's Public Key, but without knowing Alice's secret Private Key, she cannot decrypt the Ciphertext to recover the Plaintext.

This all sounds great, but Diffie and Hellman were not able to propose an explicit example of such a

Public Key Cryptosystem,

although they did propose a closely related key exchange system.

REMARK 3.2.1. In these notes we will continue to refer to *private key cryptosystems* in which there is only one key and *public key cryptosystems* in which there are two keys. However, we note that they are formally known as *symmetric cryptosystems* and *asymmetric cryptosystems*, since the former have a single key that is used for both encryption and decryption, while the latter have separate keys for these purposes.

3.3. A Mathematical Formulation. Before discussing some specific examples of public key cryptosystems, let's reformulate everything in mathematical terms. Keys and texts are elements chosen from certain sets, while encryption and decryption are functions between these sets:

$$\left[\begin{array}{ccc} \{\text{Public Keys}\} \times \{\text{Plain Texts}\} & \xrightarrow{\textsf{Encrypt}} & \{\text{Cipher Texts}\} \\ \{\text{Private Keys}\} \times \{\text{Cipher Texts}\} & \xrightarrow{\textsf{Decrypt}} & \{\text{Plain Texts}\} \end{array}\right]$$

Suppose that $(\textsc{PubKey}, \textsc{PrivKey})$ is a valid public/private key pair. Then for all plaintexts $\textsc{Msg}$ (messgaes) we want

$$\textsf{Decrypt}\Big(\textsc{PrivKey}, \textsf{Encrypt}(\textsc{PubKey}, \textsc{Msg})\Big) = \textsc{Msg}.$$

Further, even though Eve knows the Public Key and the Ciphertext, we insist that this information does not allow her to recover the Plaintext. Only Alice, who knows the Private Key, can do that.

---

[13]The concept of public key encryption was originally discovered by James Ellis in 1969 while working at the British Government Communications Headquarters (GCHQ), but his discoveries were classified by the British government and were not declassified and released until 1997, after his death. Two other researchers at GCHQ, Malcolm Williamson and Clifford Cocks, discovered the Diffie–Hellman key exchange algorithm and the RSA public key encryption system, respectively, before the rediscovery and public dissemination by Diffie, Hellman, Rivest, Shamir, and Adleman, but again the work Williamson and Cocks was not published until much later.

We may thus view Decrypt as a function that inverts Encrypt, but in order to evaluate Decrypt, an extra piece of information, the Private Key, is needed. Here is an abstract formulation of this idea.

DEFINITION 3.3.1. A *Trap Door Function* is an invertible function $f : X \to Y$ having the following properties:[14]

(1) $f(x)$ is easy to compute.
(2) $f^{-1}(y)$ is hard to compute.
(3) Knowledge of an extra piece of information, the "trapdoor", makes $f^{-1}(y)$ easy to compute.

3.4. A Menagerie of Functions that are Ostensibly Hard to Invert. We describe four mathematical problems that can be used to build trapdoor functions, assuming that the underlying mathematical problems are actually hard. For each problem, we indicate the public and private keys used by an associated public key cryptosystem, but we postpone the description of their encyption and decryption functions until Sections 3.5, 4.2, and 4.5.

3.4.1. *The Integer Factorization Problem (IFP).* Given two large prime numbers p and q and an exponent e, the exponentiation function

$$\mathbb{Z}/pq\mathbb{Z} \longrightarrow \mathbb{Z}/pq\mathbb{Z}, \quad x \longmapsto x^e \bmod pq, \tag{3.4.1}$$

is easy to compute, but hard(?) to invert unless you know p and q.

REMARK 3.4.2. We're cheating a bit, since inverting (3.4.1) isn't really the Integer Factorization Problem. We might call the

Taking Roots Modulo pq Problem.

However, for appropriate choices of p, q, and e, the fastest known algorithm for finding roots modulo pq is to first solve for p and q.

The IFP is used to build the[15]

$$\begin{bmatrix} \textbf{RSA Public Key Cryptosystem.} \\ \text{Public Key} := (pq, e), \qquad \text{Private Key} := (p, q). \end{bmatrix}$$

3.4.2. *The Discrete Logarithm Problem (DLP).* Let p be a large prime number, and let $g \in \mathbb{F}_p^*$. The powering function

$$\mathbb{Z}/(p-1)\mathbb{Z} \longrightarrow \mathbb{F}_p^*, \quad k \longmapsto g^k \bmod p,$$

is easy to compute, but hard to invert.

The DLP is used to build the

$$\begin{bmatrix} \textbf{Elgamal Public Key Cryptosystem.} \\ \text{Public Key} := (p, g, g^k) \qquad \text{Private Key} := k. \end{bmatrix}$$

---

[14] In these notes we will be somewhat informal about the words "easy" and "hard". Roughly speaking, a function whose input is n bits is *easy to compute* if it can be computed in time that is polynomial in n, and preferably not much more than linear in n. It is *hard to compute* if it takes time that is exponential, or at least super-polynomial, in n to compute. Depending on the desired security level, a quantity is currently considered hard to compute if its computation takes at least 2^{80}, or 2^{160}, or 2^{320} operations.

[15] For added effciency, it suffices for the RSA private key to be an integer d that satisfies the congruence $de \equiv 1 \pmod{pq - p - q + 1}$.

REMARK 3.4.3. Again we've cheated. Elgamal is actually based on the

Diffie–Hellman Problem: Given g, g^a, g^b, compute g^{ab}.

But again, the fastest general method known for solving the Diffie–Hellman Problem is to compute a or b, i.e., to solve the DLP.

3.4.3. *The Elliptic Curve Discrete Logarithm Problem (ECDLP).* Similar to the DLP, but the multiplicative group $\mathbb{F}_p^*$ is replacéd by the group of points $E(\mathbb{F}_p)$ on an elliptic curve. Thus let p be a large prime number, let $E/\mathbb{F}_p$ be an elliptic curve, and let $Q \in E(\mathbb{F}_p)$ be a point on E having large order N. Then the multiplication function

$$\mathbb{Z}/N\mathbb{Z} \longrightarrow E(\mathbb{F}_p), \qquad k \longrightarrow kQ,$$

is easy to compute, but hard to invert.

The ECDLP is used to build the

Elliptic Curve Elgamal Public Key Cryptosystem.
Public Key := (p, E, Q, kQ) Private Key := k.

REMARK 3.4.4. Why bother using elliptic curves, since additon on $E(\mathbb{F}_p)$ takes a lot more effort than multiplication in $\mathbb{F}_p^*$? The answer is that the best known algorithms to solve the ECDLP are much slower than those for the IFP or the DLP. This means that keys and ciphertexts using ECDLP-based cryptosystems are smaller. Thus elliptic curve cryptography is part of the never-ending battle between contradictory cryptographic goals:

- ⋆ Be maximally efficient!
- ⋆ Be maximally secure!

EXAMPLE 3.4.5. The bar codes on airline boarding passes have a limited number of bits, so they may use an ECDLP-based digital signature. Similarly, blockchain applications may need to store and transmit millions of digital signatures, so they often use ECDLP-based signatures in order to save space and bandwidth.

3.4.4. *The Closest Vector Problem (CVP).* Let L be a lattice and let

$$\mathcal{B}^{\text{bad}} := \{\mathbf{w}_1, \ldots, \mathbf{w}_n\}$$

be a bad basis for L. Then the function

$$\begin{aligned} \{0,1\}^n &\longrightarrow \mathbb{R}^n, \\ (\epsilon_1, \ldots, \epsilon_n) &\longmapsto \epsilon_1\mathbf{w}_1 + \cdots + \epsilon_n\mathbf{w}_n + (\text{small random vector}) \end{aligned}$$

is easy to compute, but hard to invert.

The CVP is used to build various cryptosystems, including the

GGH and NTRU Public Key Cryptosystems.
Public Key := a bad basis $\mathcal{B}^{\text{bad}}$ Private Key := a good basis $\mathcal{B}^{\text{good}}$.

Cryptosystems built from the CVP are discussed in more detail in Lecture 4.

3.5. From Trapdoor Functions to Public Key Cryptosystems.

3.5.1. *From IFP to RSA.* The **RSA Cryptosystem** invented by Rivest, Shamir, and Adelman, works as follows:

- Private Key: (p,q). Public Key: (pq,e).
- Plaintext: A number M mod pq.
- Ciphertext: The number $C :\equiv M^e \pmod{pq}$.
- Decryption: Compute

$$C^d \bmod pq \text{ where } de \equiv 1 \pmod{pq - p - q + 1}.$$

3.5.2. *From DLP to Elgamal.* The **Elgamal DLP-Based Cryptosystem** introduces randomness, a topic to which we shall return.

- Private Key: k Public Key: (p,g,g^k), where $g \in \mathbb{F}_p^*$.
- Plaintext: A number M mod p.
- Ciphertext: Choose a random $R \bmod p-1$. The ciphtertext is the pair of values

$$C_1 :\equiv g^R \pmod{p} \quad \text{and} \quad C_2 :\equiv M \cdot (g^k)^R \pmod{p}.$$

- Decryption: Compute $(C_1^k)^{-1} \cdot C_2 \bmod p$.

3.5.3. *From ECDLP to Elliptic Elgamal.* The **Elliptic Curve Elgamal Cryptosystem** works similarly, with the multiplicative group $\mathbb{F}_p^*$ replaced by the group of points on an elliptic curve $E(\mathbb{F}_p)$. There are also proposals for quantum secure public key cryptosystems based on maps (isogenies) between elliptic curves; see the material by Kristen Lauter in this volume.

3.5.4. *From CVP to GGH and NTRU.* It's relatively straightforward to create a public key cryptosystem from the CVP, where the public key is a bad basis and the private key is a good basis. However, lattice reduction algorithms such as LLL-BKZ make such systems insecure unless the dimension, and thus the key size, is quite large. The use of lattices having additional structure leads to practical lattice-based cryptosystems. Lectures 4 and 5 are devoted to an introduction to lattice-based cryptography.

3.6. Digital Signatures. Digital Signatures provide a way for Bob to electronically sign a digital document. The secure operation of the internet and its attendant businesses relies at least as much on digital signatures as it does on public key cryptosystems.

EXAMPLE 3.6.1. Bob is Microsoft or Apple sending an update for Alice's computer. Should Alice install it? Only if she can verify that it came from Bob.

3.6.1. *Signatures in the Dark Ages.* Bob's written signature on a document affirmed that he created the document or was willing to abide by its terms. For example:

Bobolink Bank of Boston
Pay to **Alice Adams** $100.00
BOB ROBERTS

The bank can verify Bob's signature by comparing it to a copy that they have on file. But suppose that Bob wants to sign a computer file and send it to Alice over the internet. How can he do that so that she is able verify his digital signature on the file?

3.6.2. *Digital Signature Schemes: Mathematical Description.* Just as with a public key cryptosystem, a

Digital Signature Scheme

consists of two functions:

$$\left[\begin{array}{ccc} \{\text{Private Keys}\}\times\{\text{Digital Docs}\} & \xrightarrow{\textsf{Sign}} & \{\text{Signatures}\} \\ \{\text{Public Keys}\}\times\{\text{Signatures}\}\times\{\text{Digital Docs}\} & \xrightarrow{\textsf{Verify}} & \{\text{Yes, No}\} \end{array}\right]$$

These functions are required to have the following properties:

- If (PubKey,PrivKey) is a valid public/private key pair, then for all documents Doc and potential signatures Sig we have

$$\mathsf{Verify}(\text{PubKey}, \text{Sig}, \text{Doc}) = \text{Yes} \iff \text{Sig} = \mathsf{Sign}(\text{PrivKey}, \text{Doc}).$$

- Given only PubKey and Doc, it is difficult to create a signature Sig satisfying

$$\mathsf{Verify}(\text{PubKey}, \text{Sig}, \text{Doc}) = \text{Yes}.$$

- A list (transcript) of valid signatures

$$(\text{Sig}_i, \text{Doc}_i), \quad i = 1, 2, 3, \dots, n,$$

  for the public key PubKey should not reveal useful information about the associated private key PrivKey, nor should it allow Eve to sign any documents other than those already in the list.

3.6.3. *Examples of Digital Signature Schemes.*

RSA Signatures

- Public Key: (pq, e).
- Private Key: d satisfying $de \equiv 1 \pmod{pq - p - q + 1}$.
- Document: A number $D \bmod pq$.
- Signing: The number $S :\equiv D^d \pmod{pq}$.
- Verifying: Signature is valid if $S^e \equiv D \pmod{pq}$.

Elgamal Signatures

- Public Parameters: (p, g), where $g \in \mathbb{F}_p^*$.
- Private Key: k
- Public Key: g^k.
- Document: A number $D \bmod p$.
- Signing: Choose a random $R \bmod p-1$. The signature is a pair of integers (S_1, S_2) satisfying

$$S_1 :\equiv g^R \pmod{p}, \qquad S_2 :\equiv (D - k \cdot S_1) \cdot R^{-1} \pmod{p-1}.$$

- Verifying: Signature is valid if $(g^k)^{S_1} \cdot {S_1}^{S_2} \equiv g^D \pmod{p}$.

Elliptic Elgamal Signatures

- Public Parameters: (p, E, x, Q), where $Q \in E(\mathbb{F}_p)$ is a point of (large prime) order q, and where x is the x-coordinate function on a Weierstrass equation for E.
- Private Key: $k \bmod q$
- Public Key: kQ
- Document: A number $d \bmod q$.

- Signing: Choose a random $r \bmod q$. The signature is a pair of integers (s_1, s_2) satisfying
$$s_1 \equiv x(rQ) \bmod q, \qquad s_2 \equiv (d + ks_1)r^{-1} \bmod q.$$
- Verifying: Signature is valid if $x\Big(ds_2^{-1}Q + s_1 s_2^{-1}(kQ)\Big) \equiv s_1 \pmod{q}$.

Lattice-Based Signatures

- Lecture 5 gives an introduction to lattice-based digital signature schemes.

3.7. Cryptographically Secure Hash Functions. If Bob's document Doc is large, he could break it into pieces
$$\mathrm{Doc}_1, \mathrm{Doc}_2, \mathrm{Doc}_3, \ldots$$
and sign each piece. But that's inefficient, as well as leading to security issues. Instead, Bob signs a cryptographically secure hash of his document.

Definition 3.7.1. Intuitively, a *Hash Function* takes an arbitrary length input and creates a fixed length, deterministic, but unpredictable and random looking, output:
$$\left\{\begin{matrix}\text{arbitrary length}\\ \text{bit strings}\end{matrix}\right\} \xrightarrow{\mathsf{Hash}} \left\{\begin{matrix}\text{bit strings}\\ \text{of length } b\end{matrix}\right\}.$$
The function Hash should have (at least) the following properties:

- For $D \in \{0,1\}^*$, computing $\mathsf{Hash}(D)$ is **very** fast.[16]
- Given $H \in \{0,1\}^b$, it is very hard to find even one
$$D \in \{0,1\}^* \quad \text{satisfying} \quad \mathsf{Hash}(D) = H.$$
- It is very hard to find distinct
$$D_1, D_2 \in \{0,1\}^* \quad \text{satisfying} \quad \mathsf{Hash}(D_1) = \mathsf{Hash}(D_2).$$
This property is called *Collision Resistance.*
- Altering even one bit of D changes $\mathsf{Hash}(D)$ unpredictably. We might formalize this as:
$$\mathrm{Prob}\left(\begin{matrix}\text{the } j\text{th bits of } \mathsf{Hash}(D)\\ \text{and } \mathsf{Hash}(D') \text{ are equal}\end{matrix} \;\middle|\; \begin{matrix}D \text{ and } D' \text{ differ}\\ \text{in exactly one bit}\end{matrix}\right) = 50\%.$$

3.8. Random Numbers in Cryptography. For RSA, Bob and Alice need to choose *random* prime numbers. Elgamal uses *random* numbers to encrypt and sign. And even a deterministic cryptosystem such as RSA tends to have security problems unless some randomness is introduced.

Example 3.8.1 (Plaintext Padding). The following is only semi-realistic. When Bob wants to send the message M to Alice, rather than encrypting M directly, he chooses a random string R and instead sends Alice the concatenated string
$$M' := R \,\|\, (R \,\mathtt{xor}\, M).$$
Then, even if Eve guesses part of the message, she cannot use that knowledge to help with decryption, since the bits of M have been scrambled by R. Alice recovers the actual plaintext by first computing M', and then computing
$$(R \,\mathtt{xor}\, M) \,\mathtt{xor}\, R = M.$$

[16]The notation $\{0,1\}^*$ is shorthand for the set of all finite binary sequences. More formally, it is the set of maps $a : \mathbb{N} \to \{0,1\}$ with the property that $\{n \in \mathbb{N} : a_n = 1\}$ is finite.

There are sources of bits that people believe are truely random:

- Quantum phenomena, such as radioactive decay.
- Micro-changes in temperature.

Such sources can be used in practice, but they tend to be inefficient.

What Bob and Alice need is a *Psuedo-Random Number Generator* (PRNG). For example, they might use a function

$$\mathsf{Rand} : \{0,1\}^N \longrightarrow \{0,1\}^N$$

and iterate it starting from a seed value σ_0. This gives a sequence of values

$$\sigma_1 := \mathsf{Rand}(\sigma_0), \quad \sigma_2 := \mathsf{Rand}(\sigma_1), \quad \sigma_3 := \mathsf{Rand}(\sigma_2), \ldots$$

and they would like this sequence to be "indistinguishable" from a sequence of values chosen randomly and uniformly from $\{0,1\}^N$.

Another way to create a pseudo-random number generator is to use a hash function and a seed value σ_0, and compute the list of pseudo-random numbers

$$\sigma_1 := \mathsf{Hash}(\sigma_0 \,\|\, 1), \quad \sigma_2 := \mathsf{Hash}(\sigma_0 \,\|\, 2), \quad \sigma_3 := \mathsf{Hash}(\sigma_0 \,\|\, 3), \ldots .$$

Of course, there is still the problem of generating a random seed σ_0 value. And it must be noted that there have been real-world security issues arising from insufficiently random seed values![17]

REMARK 3.8.2. Don't be fooled by the relatively innocuous definitons in Sections 3.7 and 3.8. Creating cryptographically secure hash functions and pseudo-random number generators is hard! The US National Institute of Standards and Technology (NIST) ran an open competition from 2007 to 2012 to develop an efficient and secure cryptographic hash function. The winner was publised in 2015 and is called *Secure Hash Function-3* (SHA-3), although an earlier SHA-2 is also still widely used.

3.9. How Hard are Hard Problems? So just how hard are famous "hard problems" such as the IFP, DLP, ECDLP, and CVP?

Honest Answer: No one knows!!! In the sense that we don't have a proof that any of these problems are hard.

Practical Answer: How hard are they to solve using existing algorithms on existing computers?

Problem	Steps Required to Solve the Problem	Key/Ciphertext Size to be Secure
IFP	$\approx \exp(\sqrt[3]{\log pq})$ steps	2000 to 4000 bits
DLP	$\approx \exp(\sqrt[3]{\log p})$ steps	2000 to 4000 bits
ECDLP	$\approx \sqrt{p}$ steps	300 to 400 bits
CVP	$\approx C^{\dim L}$ steps	2000 to 4000 bits

[17]Lenstra et al. [**25**] gathered about 6 million RSA keys from various sources on the internet and computed their pariwise gcds. They found that about 0.2% of the pairs had a common prime factor, which enabled them to factor those keys. They were able to do this because some significant portion (maybe all) of the seeds used to generate the primes had insufficiently many bits of randomness.

That's all great. Even 4000 bits isn't much. But you probably noticed the caveat:

| **Existing** algorithms on **existing** computers. |

3.10. Quantum Computers and Cryptography.

DEFINITION 3.10.1. A *quantum computer* is a machine in which computation on bits (0's and 1's) is replaced by computation on *qubits*.

In the popular literature, a qubit is a bit that can take on every real value between 0 and 1. Slightly more precisely, a qubit is described by a complex number representing a superposition of 0 and 1 states with certain probabilities. A quantum computer with n qubits can "perform" a simultaneous computation on 2^n states, achieving an exponential speedup over a classical computer for certain tasks.

At present, the largest quantum computers built have around 100 qubits. But governments and businesses are investing huge sums of money to build larger ones. One might make an analogy:

- First airplane flight — 1903 — flew 852 feet.
- WW I — 1914–18 — airplanes ubiquitous.
- WW-II — 1939–45 — jets flying 500+ MPH.

Here's the bombshell paper that started all the fuss:

| *Polynomial-Time Algorithms for Prime Factorization and Discrete Logarithms on a Quantum Computer*, Peter W. Shor. Proc. 35th Annual Symposium on Foundations of Computer Science, Santa Fe, NM, Nov. 20–22, 1994 |

Shor's quantum algorithms solve the IFP and DLP (and ECDLP) in more-or-less *quadratic* time, so a full-scale quantum computer would compromise the security of most classical public key systems. This has sparked much current research on public key systems that cannot (as far as we know) be broken by a quantum computer. These include lattice-based systems discussed in Sections 4 and 5, isogeny-based systems discussed in Kristen Lauter's chapter in this volume, and systems based on coding theory. People refer to work in this field as:

| **Post-Quantum Cryptography** |

3.10.1. *What, Me Worry?* If large-scale working quantum computers are decades off, why worry about them now?

- Infrastructure change is slow, time-consuming, and expensive.
- Even if post-quantum cryptosystems are not used now, we should build them into systems so that we can start using them with the flip of an (electronic) switch.
- How long do you want to protect your secrets? Your legal documents? Your blockchains? If for 10 years, then you should *probably* encrypt and sign using PQC. If 50 years, then you *definitely* should.

NIST, which estimates that by 2030 it will cost roughly \$1 billion to build a quantum computer that can break 2048-bit RSA, is in the final stages of running a multi-year open competition to select and standardize post-quantum cryptosystems.

NIST announced the Round 3 candidates in July 2020.[18] It consists of 4 public key cryptosystems and 3 digital signatures, together with 8 additional alternate candidates.

3.11. Code Makers Versus Code Breakers, Or, Cryptanalysts Are Very Clever! History is littered with the invention of "unbreakable" cryptosystems that were broken. So before you start touting your own brilliant new cryptosystem, here are a few lessons that the author has painfully learned over the years:

- Cryptanalysts, those pesky folks who break cryptosystems, are *very very* clever people.
- Cryptanalysts don't play by your rules, they set their own rules. They'll try to break your algorithm, they'll try to break your software implementation, they'll try to break the hardware that you're using!
- Cryptanalysts attack the weakest part of your system or its implementation, frequently via a method that you hadn't considered.
- If you modify a cryptosystem to make it more efficient, 99 times out of 100 you'll end up compromising its security.
- It's not enough for your system to be based on a problem whose hardest case is hard. Instead you need "most" of the instances of your problem to be hard.
- Even if most instances of your problem are hard, you still need to be able to distinguish, and avoid, the instances that are not hard. And *a priori*, it's often not clear which instances are easy (or easier).
- It is important to be able to prove, under reasonable assumptions, that an oracle that decrypts or signs is also capable of solving a hard mathematical problem. These are called *security reduction proofs*. However, they come with two important caveats:
  - What constitutes a "reasonable assumption"? This is a subtle question, and systems that were supposedly secure because of a security reduction proof have turned out to be insecure due to issues with the underlying assumptions.
  - Since we don't yet know how to prove that cryptographically useful math problems are hard, what a security reduction proof demonstrates is that breaking the cryptosystem is equivalent to solving a mathematical problem that has been studied intensively for a many years. Thus the cryptographic security underlying today's widely deployed cryptographic constructions is ultimately based on the following principle:

Proof via lack of known algorithms despite much effort

Here are a few examples of non-obvious weaknesses in otherwise ostensibly hard mathematical problems and secure implementations, where we note that an integer is said to be *smooth* if it is a product of small primes.

- For most primes p and q, one expects the IFP for pq to be hard. But it is easier if $p-1$ or $q-1$ is sufficiently smooth.

[18]Added in proof: On July 7, 2022, NIST selected one public key cryptosystem and three digital signature schemes to be used as post-quantum systems. See `https://csrc.nist.gov/projects/post-quantum-cryptography/selected-algorithms-2022`.

- For most primes p and most generators $g \in \mathbb{F}_p^*$, one expects the DLP to be hard. But it is easier if $p - 1$ is sufficiently smooth.
- For most primes p, most elliptic curves $E/\mathbb{F}_p$, and most points $Q \in E(\mathbb{F}_p)$, one expects the ECDLP to be hard. But letting N denote the order of Q in $\#E(\mathbb{F}_p)$, the ECDLP is:
  - Easier if N is sufficiently smooth.
  - Easier if there is a small f such that $p^f \equiv 1 \pmod{N}$.
  - Extremely easy if $\#E(\mathbb{F}_p) = p$.
- Let (pq, e) be an RSA key. If Eve can steal (or guess) half the bits of p, she can recover the other bits using lattice reduction!
- A long transcript of classical lattice-based signatures can be used to reconstruct the fundamental domain of a good basis. And it turns out that in some cases, the transcript doesn't even have to be all that long!
- Software and hardware implementations of cryptosystems, especially if optimized to avoid unnecessary calculations, can often be compromised by measuring their power consumption or execution time. These are called *power analysis and timing analysis attacks.* To see why they might succeed, note that flipping a bit takes longer and uses more energy than leaving it unchanged.

3.12. Exercises for Lecture 3.

EXERCISE 3.12.1.

(1) Verify that RSA decryption (Section 3.5.1) works.
(2) Verify that Elgamal decryption (Section 3.5.2) works.
(3) Verify that RSA signature verification (Section 3.6.3) works.
(4) Verify that Elgamal signature verification (Section 3.6.3) works.

EXERCISE 3.12.2.

(1) Prove that an oracle that solves the Discrete Logarithm Problem (Section 3.4.2) can be used to solve the Diffie–Hellman problem (Remark 3.4.3)
(2) Prove that an oracle that solves the Diffie–Hellman problem (Remark 3.4.3) can be used to break the Elgamal cryptosystem (Section 3.5.2).
(3) Prove that an oracle that decrypts Elgamal ciphertexts (Section 3.5.2) can be used to solve the Diffie–Hellman problem (Remark 3.4.3).
(4) It is an open question as to whether an oracle that solves the Diffie–Hellman problem can be used to solve the Discrete Logarithm Problem. Try to say something about this question.

EXERCISE 3.12.3. It should be noted that Bob chooses R randomly, and that Alice does not know M or R but otherwise knows which computations Bob performed (including the functions Hash1 and Hash2).

It's probably also best to specify that M, R, and the outputs of Hash1 and Hash2 are all the same number of bits. (Alice must know the lengths of X and Y in order to know where to split the concatenation; this assumption would also help avoid the need to introduce padding before taking xor of strings of different lengths, which is typically necessary in OAEP)

Bob wants to send the message M to Alice. He first chooses a random R and computes the following quantities:[19]

$$X \leftarrow M \mathtt{xor} \mathsf{Hash}_1(R) \quad \text{and} \quad Y \leftarrow R \mathtt{xor} \mathsf{Hash}_2(X).$$

He encrypts and sends Alice the concatenation $M' = X \parallel Y$.

(1) Show how Alice can recover M from M'; i.e., show that Alice can recover M from X and Y.

(2) Assuming that the hash functions are appropriately secure, explain why Eve needs to know every bit of X in order to recover R from Y, and she needs to know every bit of R to recover M from X.

This is called *optimal asymmetric encryption padding.*

EXERCISE 3.12.4. It is asserted in Sections 3.4.1 and 3.4.2 that raising a given quantity to a given power is easy. However, when implementing systems such as RSA or Elgamal, one needs to compute a^n in a ring $\mathbb{Z}/m\mathbb{Z}$, where m, a, n may have hundreds of digits.

(1) If one naively sets $a_0 \leftarrow 1$ and $a_{i+1} \leftarrow a \cdot a_i$, how many multiplications are required to compute a^n?

(2) Consider the following *square-and-multiply* algorithm:
   (a) Write n in binary as $n = n_0 + 2 \cdot n_1 + 4 \cdot n_2 + \cdots + 2^t \cdot n_t$, where $n_i \in \{0, 1\}$.
   (b) Set $b_0 \leftarrow a$. For $i = 1, 2, \ldots, t$, compute $b_i \leftarrow b_{i-1}^2$.
   (c) Set $a_0 \leftarrow 1$. For $i = 1, 2, \ldots, t$, compute $a_i \leftarrow \begin{cases} a_{i-1} & \text{if } n_{i-1} = 0, \\ a_{i-1} \cdot b_{i-1} & \text{if } n_{i-1} = 1. \end{cases}$

   Prove that $a_t = a^n$, and that the computation requires at most $2\log_2(n)$ multiplications and (roughly) $\log_2(n)$ storage.

(3) Read about improvements to the square-and-multiply algorithm including: (a) Implementations requiring only $O(1)$ storage. (b) Improved performance using precomputation and/or windowing methods. (c) Using a binary expansion of n with coefficients in $\{-1, 0, 1\}$. This is especially helpful for elliptic curve cryptography, where the inversion operation $P \to -P$ is very fast.

EXERCISE 3.12.5 (Shanks's babystep-giantstep algorithm). This exercise describes a method for solving the DLP. Let $g \in \mathbb{F}_p^*$ have order N, and let $h \in \mathbb{F}_p^*$. The goal is to find an exponent k so that $g^k = h$. Consider the following algorithm:[20]

- Set $n \leftarrow 1 + \lfloor \sqrt{N} \rfloor$.
- Compute $u \leftarrow g^n$. (If n is large, see Exercise 3.12.4.)
- Create two lists:

$$\begin{aligned} &\text{List 1:} \quad 1, g, g^2, g^3, \ldots, g^n \\ &\text{List 2:} \quad h, hu, hu^2, hu^3, \ldots, hu^n \end{aligned}$$

- Find an element that's in both lists, say $g^i = hu^j$.
- Output $i - nj \pmod{N}$.

[19]Alice does not know M or R, but she does know the hash functions Hash_1 and Hash_2. We further assume that M and the outputs of the hash functions have the same number of bits.

[20]The powers of g are the "babysteps," the powers of u are the "giantsteps."

(1) Prove that the two lists are guaranteed to have an element in common.
(2) Prove that the output is an integer k satisfying $g^k = h$.
(3) Prove that the algorithm requires $O(\sqrt{N}))$ multiplications and $O(\sqrt{N})$ storage.[21]
(4) Read about Pollard's rho method, which also solves the DLP in $O(\sqrt{N})$ steps, but requires only $O(1)$ storage.

EXERCISE 3.12.6 (The Pohlig-Hellman algorithm). This exercise illustrates a method for solving the DLP when the order of the element is highly composite. Let $g \in \mathbb{F}_p^*$ have order N, and let $h \in \mathbb{F}_p^*$. The goal is to find an exponent k so that $g^k = h$. For ease of exposition, we suppose that

$$N = q_1 q_2 \quad \text{with } q_1 \text{ and } q_2 \text{ distinct primes of roughly the same size.}$$

- Let

$$g_1 \leftarrow g^{N/q_1}, \qquad g_2 \leftarrow g^{N/q_2}, \qquad h_1 \leftarrow h^{N/q_1}, \qquad h_2 \leftarrow h^{N/q_2}.$$

- Solve the DLP for the pairs (g_1, h_1) and (g_2, h_2), i.e., find exponents k_1 and k_2 so that

$$g_1^{k_1} = h_1 \quad \text{and} \quad g_2^{k_2} = h_2.$$

- Solve the simultaneous congruences

$$k \equiv k_1 \pmod{q_1} \quad \text{and} \quad k \equiv k_2 \pmod{q_2}.$$

(1) Prove that the output k from this algorithm satisfies $g^k = h$.
(2) Exercise 3.12.5 describes an algorithm which says that if g has order N, then we can solve the DLP for the pair (g, h) in time that is proportional to $N^{1/2}$. Prove that if we use Exercise 3.12.5 to solve each of the subsidiary DLPs in the above algorithm, then the time to solve the original DLP is proportional to $N^{1/4}$, which represents a substantial savings.
(3) Generalize to the case that N is a product $q_1^{e_1} q_2^{e_2} \cdots q_t^{e_t}$ of many distinct primes raised to powers. If it roughly takes time $T(n)$ to solve the DLP for an element of order n, prove that your algorithm solves the DLP in time that is roughly the sum $\sum_{i=1}^{t} T(q_i^{e_i})$.
(4) Let q be a prime, and as in (3), suppose that it takes time $T(n)$ to solve the DLP for pairs (g, h) if g has order n. Let $N = q^e$ be a power of a prime q, and suppose that g has order N. Devise an algorithm to solve the DLP for (g, h) that takes time roughly $eT(q)$.

EXERCISE 3.12.7. A *smooth number* is an integer whose factorization uses only small primes; more precisely, an integer is B-smooth if its factorization is a product of prime powers using only primes less than B. Many factorization and DLP algorithms exploit smooth numbers.

(1) Read about the quadratic sieve factorization method, and how it depends on finding numbers x and y so that $x^2 - y^2 \bmod N$ is smooth.
(2) Read about the number field sieve, a generalization of the quadratic sieve that is currently the fastest known general factorization method.
(3) Read about Pollard's $p-1$ factorization algorithm, which efficiently finds prime factors p of N if $p-1$ is smooth.

---

[21] We note that there are sorting algorithms that allow one to find a common element in the two lists in roughly $O(\log N)$ steps, which is negligible.

(4) Read about Lenstra's elliptic curve factorization algorithm, which is similar to Pollard's $p-1$ algorithm, but uses elliptic curve groups $E(\mathbb{F}_p)$ in place of the multiplicative group $\mathbb{F}_p^*$.
(5) Read about the index calculus, which is an algorithm to solve the DLP in $\mathbb{F}_p^*$ that relies on finding numbers of the form g^i and $g^j h$ whose lifts to integers in the interval $[1, p-1]$ are smooth.

EXERCISE 3.12.8. Read about hash functions and pseudo-random number generators. Try to implement one of them.

4. Lattice-Based Public Key Cryptosystems

4.1. Early Days and the Ajtai-Dwork Lattice-Based Cryptosystem. In 1995 Ajtai and Dwork [**1**] described a lattice-based public key cryptosystem whose security relies on the difficulty of solving CVP in a certain set of lattices $\mathcal{L}_{\mathrm{AD}}$. They proved that breaking their system for a random lattice of dimension m in $\mathcal{L}_{\mathrm{AD}}$ is as difficult as solving SVP for *all* lattices in $\mathcal{L}_{\mathrm{AD}}$ of dimension n, where $n < m$ depends on m. This *average case-worst case equivalence* was a theoretical cryptographic milestone, but unfortunately the Ajtai-Dwork cryptosystem is impractical. Inspired by the work of Ajtai and Dwork, Goldreich–Goldwasser–Halevi [**15**] proposed a somewhat more practical lattice-based cryptosystem, while at the same time and working indpendently, Hoffstein–Pipher–Silverman [**20**] proposed an even more efficient system that they dubbed NTRU. In this chapter we describe the GGH and NTRU public key cryptosystems.

It is interesting to note that initially the major advantage of these systems was their speed, since lattice-based systems can be 10 to 100 times faster than RSA and ECC at equivalent security levels. However, as computers became faster, the speed differential became less important. Then in an unexpected turn of events, interest in lattice-based systems flourished due both to their ostensible resistance to quantum attacks and their adaptability to applications such as Gentry's homomorphic encryption scheme [**13**].[22]

4.2. The GGH Public Key Cryptosystem. The *GGH lattice-based cryptosystem* of Goldreich, Goldwasser, and Halevi [**15**] is described in Table 4.2.1.

- Note that the ciphertext vector $\mathbf{e}$ is not in the lattice L, but it is close to a lattice vector, since $\mathbf{r}$ is small.
- Bob uses the bad public basis $\mathcal{B}^{\mathrm{bad}}$ and a random small vector $\mathbf{r}$ to create the ciphertext.
- Alice uses Babai's method (Section 1.3) with her good basis $\mathcal{B}^{\mathrm{good}}$ to solve CVP. In this way she finds a vector $\mathbf{v} \in L$ that is close to the ciphertext. Since $\mathbf{r}$ is small, when Alice writes $\mathbf{v}$ in terms of $\mathcal{B}^{\mathrm{bad}}$, with high probability it will be equal to

$$\mathbf{v} = \epsilon_1 \mathbf{w}_1 + \cdots + \epsilon_n \mathbf{w}_n.$$

Form this Alice can read off $(\epsilon_1, \ldots, \epsilon_n)$, which is the plaintext.

[22]Homomorphic encryption is a fascinating area that we will not have time to discuss. Using a homomorphic encryption scheme, Alice can have Bob run an encrypted algorithm on encrypted input data to produce encrypted output. Current schemes are impractically slow for most applications, but research continues to improve the efficiency.

$$\begin{aligned}
\text{Private Key} &:= \text{a good basis } \mathcal{B}^{\text{good}} \text{ for a lattice } L \\
&:= \{\mathbf{v}_1, \ldots, \mathbf{v}_n\}. \\
\text{Public Key} &:= \text{a bad basis } \mathcal{B}^{\text{bad}} \text{ for the lattice } L \\
&:= \{\mathbf{w}_1, \ldots, \mathbf{w}_n\}. \\
\text{Plaintext} &:= \text{a binary vector } (\epsilon_1, \ldots, \epsilon_n), \text{ i.e., } \epsilon_i \in \{0,1\}. \\
\text{Ciphertext} &:= \epsilon_1 \mathbf{w}_1 + \cdots + \epsilon_n \mathbf{w}_n + \mathbf{r}, \\
&\qquad\qquad \text{where } \mathbf{r} \text{ is a small random vector.}
\end{aligned}$$

TABLE 4.2.1. The GGH Lattice-Based Cryptosystem

4.3. GGH versus LLL: A Battle for Supremacy! The security of GGH comes down to the question of just how well LLL and its variants solve CVP; or alternatively, how long it takes LLL-BKZ to find a basis that is good enough to solve CVP using Babai's method.

Experimentally, one finds that if $n = \dim(\mathcal{L}) < 100$, then LLL easily finds a basis that's good enough to break GGH. And even up to $n \approx 200$, variants of LLL-BKZ will break GGH. On the other hand, GGH may well be secure for $500 < n < 1000$; and barring some major breakthrough in lattice-reduction methods, it is likely to be secure for quite a while for (say) $2000 < n < 4000$. However, we must then deal with the following unfortunate fact. If we implement GGH using a lattice $L \subset \mathbb{R}^n$, then the GGH public key consists of n vectors in $\mathbb{R}^n$, and hence

$$\text{Size of GGH Public Key} = O(n^2) \text{ bits.}$$

For example, if we take $n = 1000$, and if we assume that each coordinate of each vector in the GGH basis is an 8 bit integer, then a GGH public key has length roughly 8-megabits. This is quite a bit larger than standard 4000 bit RSA keys or 256 bit ECC keys!

The NTRU Public Key Cryptosystem, which we describe next, solves this problem by using a special type of lattice that has bases that can be described using roughly $\frac{1}{2}n\log_2(n)$ bits, leading to key sizes that are comparable to RSA keys.

4.4. Convolution Products and Polynomial Quotient Rings. We start with some mathematical tools that we will need.

DEFINITION 4.4.1. The *convolution product* of two vectors[23]

$$\mathbf{a} := (a_0, a_1, \ldots, a_{N-1}) \quad \text{and} \quad \mathbf{b} := (b_0, b_1, \ldots, b_{N-1})$$

is the vector

$$\mathbf{c} := \mathbf{a} \star \mathbf{b} \quad \text{with} \quad c_k := \sum_{i+j \equiv k \pmod{N}} a_i b_j. \tag{4.4.2}$$

EXAMPLE 4.4.3. Let $\mathbf{a} := (1,2,3)$ and $\mathbf{b} := (4,-5,6)$. Then

$$\mathbf{a} \star \mathbf{b} = (1 \cdot 4 + 2 \cdot 6 - 3 \cdot 5,\ -1 \cdot 5 + 2 \cdot 4 + 3 \cdot 6,\ 1 \cdot 6 - 2 \cdot 5 + 3 \cdot 4) = (1, 21, 8).$$

[23]We use the word *vector* loosely to describe an n-tuple of elements chosen from some commutative ring such as $\mathbb{Z}$ or $\mathbb{Q}$ or $\mathbb{R}$ or $\mathbb{F}_q$.

PROPOSITION 4.4.4. *Vector addition and convolution product make the set of vectors into a ring. In particular,*

$$\begin{aligned} (\mathbf{a} \star \mathbf{b}) \star \mathbf{c} &= \mathbf{a} \star (\mathbf{b} \star \mathbf{c}) && \textit{Associative Law,} \\ \mathbf{a} \star (\mathbf{b} + \mathbf{c}) &= \mathbf{a} \star \mathbf{b} + \mathbf{a} \star \mathbf{c} && \textit{Distributive Law,} \\ \mathbf{a} \star \mathbf{b} &= \mathbf{b} \star \mathbf{a} && \textit{Commutative Law.} \end{aligned}$$

PROOF. We leave the proof to the reader; see Exercise 4.10.2. □

REMARK 4.4.5. There is an alternative way to describe the convolution product ring that is often useful. The idea is to identify vectors and polynomials via the association

$$\mathbf{a} := (a_0, \dots, a_{N-1}) \quad \longleftrightarrow \quad \mathbf{a}(X) := a_0 + a_1 X + \cdots + a_{N-1} X^{N-1}.$$

We view these polynomials as living in the quotient ring

$$\mathcal{R} := \mathbb{Z}[X]/(X^N - 1),$$

so when we multiply polynomials in the quotient ring, we use the rule $X^N = 1$ to write the result as a polynomial of degree at most $N-1$. With this identification, we have the equivalence

$$\mathbf{c} = \mathbf{a} \star \mathbf{b} \quad \Longleftrightarrow \quad \mathbf{c}(X) \equiv \mathbf{a}(X)\mathbf{b}(X) \pmod{X^N - 1},$$

a calculation that we leave to the reader; see Exercise 4.10.3. In other words, this identification defines a ring isomorphism

$$\begin{array}{ccc} (\mathbb{Z}^N, +, \star) & \longrightarrow & (\mathcal{R}, +, \cdot), \\ \mathbf{a} & \longmapsto & \mathbf{a}(X). \end{array}$$

All of the preceding material works if we replace $\mathbb{Z}$ by another ring such as $\mathbb{Z}/q\mathbb{Z}$. We will make frequent use of the ring

$$\mathcal{R}_q := (\mathbb{Z}/q\mathbb{Z})[X]/(X^N - 1),$$

where we note that the natural map $\mathcal{R} \to \mathcal{R}_q$ obtained by reducing coefficients modulo q is a ring homomorphism.

PROPOSITION 4.4.6. *Let q be a prime. Then there are "many" polynomials $\mathbf{a}(X) \in \mathcal{R}_q$ that have an inverse*[24] *in $\mathcal{R}_q$, i.e., such that there exists a polynomial*

$$\mathbf{a}(X)^{-1} \in \mathcal{R} \quad \textit{satisfying} \quad \mathbf{a}(X)\mathbf{a}(X)^{-1} \equiv 1 \pmod{q}.$$

More precisely, we have

$$\mathbf{a}(X) \in \mathcal{R}_q^* \quad \Longleftrightarrow \quad \gcd_{(\mathbb{Z}/q\mathbb{Z})[X]}\big(\mathbf{a}(X), X^N - 1\big) = 1.$$

PROOF. Suppose first that $\mathbf{a}(X) \in \mathcal{R}_q^*$, so there is some $\mathbf{b}(X) \in \mathcal{R}_q$ satisfying

$$\mathbf{a}(X)\mathbf{b}(X) = 1 \quad \text{in } \mathcal{R}_q.$$

Viewing $\mathbf{a}(X)$ and $\mathbf{b}(X)$ as polynomials in $\mathbb{Z}[X]$, this is the same as saying that

$$\mathbf{a}(X)\mathbf{b}(X) \equiv 1 \pmod{X^N - 1, q},$$

so we can find polynomials $\mathbf{c}(X), \mathbf{d}(X) \in \mathbb{Z}[X]$ so that

$$\mathbf{a}(X)\mathbf{b}(X) - 1 = (X^N - 1)\mathbf{c}(X) + q\mathbf{d}(X).$$

---

[24] See Exercise 4.10.4 for a further explanation of why polynomials usually have inverses in $\mathcal{R}_q$.

Reducing modulo q, we see that $\mathbf{a}(X)$ and $X^N - 1$ can have no non-trivial common factors in $(\mathbb{Z}/q\mathbb{Z})[X]$, so their gcd is 1.

Conversely, suppose that the gcd of $\mathbf{a}(X)$ and $X^N - 1$ in $(\mathbb{Z}/q\mathbb{Z})[X]$ is 1. The ring $(\mathbb{Z}/q\mathbb{Z})[X]$ is Euclidean. (This is where we use the assumption that q is prime, since it implies that $\mathbb{Z}/q\mathbb{Z}$ is a field.) The extended Euclidean algorithm, which is both effective and extremely efficient, tells us that there are elements $\mathbf{b}(X), \mathbf{c}(X) \in (\mathbb{Z}/q\mathbb{Z})[X]$ satisfying

$$\mathbf{a}(X)\mathbf{b}(X) + (X^N - 1)\mathbf{c}(X) = 1 \quad \text{in } (\mathbb{Z}/q\mathbb{Z})[X].$$

Reducing modulo $X^N - 1$, we find that $\mathbf{a}(X)\mathbf{b}(X) = 1$ in $\mathcal{R}_q$, so $\mathbf{a}(X) \in \mathcal{R}_q^*$. □

4.5. NTRUEncrypt: The NTRU Public Key Cryptosystem.

Public Parameters: Alice and Bob agree on three public parameters

$$N, \ p, \ q,$$

with N prime and with $\gcd(p, q) = 1$.[25] It is also important that q be significantly larger than p.

Key Creation: Alice chooses random polynomials $\mathbf{f}, \mathbf{g} \in \mathcal{R}$ with small coefficients.[26] She computes the inverses of $\mathbf{f}$ modulo q and modulo p,

$$\mathbf{F}_q :\equiv \mathbf{f}^{-1} \pmod{q} \qquad \text{and} \qquad \mathbf{F}_p :\equiv \mathbf{f}^{-1} \pmod{p}.$$

(If either inverse doesn't exist, she discards this $\mathbf{f}$ and chooses a new one.) Alice computes the product

$$\mathbf{h} :\equiv \mathbf{g} \cdot \mathbf{F}_q \pmod{q}.$$

In other words, she computes $\mathbf{h} := \mathbf{g} \cdot \mathbf{f}^{-1}$ in $\mathcal{R}_q$. Then Alice's public/private key pair is[27]

$$\text{Public Key} := \mathbf{h}, \qquad \text{Private Key} := \mathbf{f}.$$

Encryption: Bob's plaintext is an element $\mathbf{m} \in \mathcal{R}_p$, or more precisely, a polynomial with integer coefficients in the interval $(-p/2, p/2]$. Bob also chooses a random polynomial $\mathbf{r}$ with small integer coefficients. Bob uses these quantities to compute his ciphertext

$$\text{Ciphertext} := \mathbf{e} :\equiv p \cdot \mathbf{r} \cdot \mathbf{h} + \mathbf{m} \pmod{q}.$$

Decryption: Alice computes

$$\mathbf{a} :\equiv \mathbf{e} \cdot \mathbf{f} \pmod{q}.$$

This computation is done in the ring $\mathcal{R}_q$, and then Alice lifts the coefficients of

$$\mathbf{a} := a_0 + a_1 X + \cdots + a_{N-1} X^{N-1}$$

---

[25]In practice, as explained in Exericse 4.10.4, it is also often convenient to take p and q to be primes that are generators for $(\mathbb{Z}/N\mathbb{Z})^*$, or at least that have large order in that ring, since then the inverses $\mathbf{F}_q$ and $\mathbf{F}_p$ will exist, either certainly or with high probability, provided $\mathbf{f}(1) \neq 0$.

[26]Typically, "small" polynomials such as $\mathbf{f}$ and $\mathbf{g}$ and $\mathbf{r}$ have coefficients chosen from $\{-1, 0, 1\}$, with roughly an equal number of 1s and -1s; see Exercise 4.10.6 for further details.

[27]For efficiency, Alice might also store $\mathbf{F}_p$, but if space is an issue, she can always recompute it from $\mathbf{f}$ and p.

so that they are integers satisfying

$$-\frac{1}{2}q < a_i \le \frac{1}{2}q.$$

Alice then reduces the coefficients modulo p and computes the product

$$\mathbf{F}_p \cdot \mathbf{a} \bmod p \quad \text{in the ring } \mathcal{R}_p.$$

This quantity will equal the plaintext m.

PROPOSITION 4.5.1. *If the parameters are chosen appropriately, then NTRU decryption gives the plaintext; i.e., NTRU works as advertised.*

PROOF. The first decryption step yields the following polynomial as output:

Computation (mod q)	Reason It Works
$\mathbf{a} \equiv \mathbf{e} \cdot \mathbf{f}$	
$\equiv (p \cdot \mathbf{r} \cdot \mathbf{h} + \mathbf{m}) \cdot \mathbf{f}$	$\mathbf{e} \equiv p \cdot \mathbf{r} \cdot \mathbf{h} + \mathbf{m}$
$\equiv p \cdot \mathbf{r} \cdot \mathbf{g} + \mathbf{m} \cdot \mathbf{f}$	$\mathbf{h} \cdot \mathbf{f} \equiv \mathbf{g} \cdot \mathbf{F}_q \cdot \mathbf{f} \equiv \mathbf{g}$

The coefficients of $\mathbf{r}, \mathbf{g}, \mathbf{m}, \mathbf{f}$ are *small*, and as noted earlier, typically chosen so that the coefficients of $\mathbf{r}, \mathbf{g}, \mathbf{f}$ more-or-less sum to 0. With appropriately chosen parameters, it is then quite likely that the coefficients of

$$p \cdot \mathbf{r} \cdot \mathbf{g} + \mathbf{m} \cdot \mathbf{f}$$

will lie in an interval in the interval between $-\frac{1}{2}q$ and $\frac{1}{2}q$. So when we first compute $\mathbf{a}$ modulo q, and then lift its coefficients to be in this interval, it is likely that the polynomial

$$\mathbf{a} \quad \text{equals} \quad p \cdot \mathbf{r} \cdot \mathbf{g} + \mathbf{m} \cdot \mathbf{f} \quad \text{exactly } \mathcal{R}, \text{ and not merely modulo in } \mathcal{R}_q.$$

Now multiply $\mathbf{a}$ by $\mathbf{F}_p$ and reduce modulo p to get

$$\begin{aligned} \mathbf{F}_p \cdot \mathbf{a} &= \mathbf{F}_p \cdot (p \cdot \mathbf{r} \cdot \mathbf{g} + \mathbf{m} \cdot \mathbf{f}) \\ &\equiv \mathbf{F}_p \cdot \mathbf{m} \cdot \mathbf{f} \pmod{p} \\ &\equiv \mathbf{m} \pmod{p} \qquad \text{since } \mathbf{F}_p \cdot \mathbf{f} \equiv 1 \pmod{p}. \end{aligned}$$

This shows that decryption recovers the plaintext. □

4.6. NTRU and Lattice Problems.

DEFINITION 4.6.1. The *Convolution Modular Lattice* $L_{\mathbf{h}}$ associated to the N-dimensional vector $\mathbf{h}$ and the modulus q is the $2N$-dimensional lattice whose basis is given by the rows of the following matrix:

$$L_{\mathbf{h}} := \operatorname{RowSpan}_{\mathbb{Z}} \left(\begin{array}{cccc|cccc} 1 & 0 & \cdots & 0 & h_0 & h_1 & \cdots & h_{N-1} \\ 0 & 1 & \cdots & 0 & h_{N-1} & h_0 & \cdots & h_{N-2} \\ \vdots & \vdots & \ddots & \vdots & \vdots & \vdots & \ddots & \vdots \\ 0 & 0 & \cdots & 1 & h_1 & h_2 & \cdots & h_0 \\ \hline 0 & 0 & \cdots & 0 & q & 0 & \cdots & 0 \\ 0 & 0 & \cdots & 0 & 0 & q & \cdots & 0 \\ \vdots & \vdots & \ddots & \vdots & \vdots & \vdots & \ddots & \vdots \\ 0 & 0 & \cdots & 0 & 0 & 0 & \cdots & q \end{array} \right) \tag{4.6.2}$$

We now given an alternative, very convenient, way to describe $L_{\mathbf{h}}$.

PROPOSITION 4.6.3. *If we identify* $\mathbb{Z}^{2N}$ *with two copies of the convolution product ring* $(\mathbb{Z}^N, +, \star)$, *then*

$$L_{\mathbf{h}} = \{(\mathbf{a}, \mathbf{b}) \in \mathbb{Z}^{2N} : \mathbf{a} \star \mathbf{h} \equiv \mathbf{b} \pmod{q}\}.$$

PROOF. Suppose that $(\mathbf{a}, \mathbf{b})$ satisfies

$$\mathbf{a} \star \mathbf{h} \equiv \mathbf{b} \pmod{q}.$$

This implies that

$$\mathbf{u} := \frac{\mathbf{b} - \mathbf{a} \star \mathbf{h}}{q} \in \mathbb{Z}^N$$

has coordinates in $\mathbb{Z}$. Then we see that

$$[a_0, \ldots, a_{N-1}, b_0, \ldots, b_{N-1}]$$

is equal to the matrix product

$$[a_0, \ldots, a_{N-1}, u_0, \ldots, u_{N-1}] \begin{pmatrix} 1 & \cdots & 0 & h_0 & \cdots & h_{N-1} \\ \vdots & \ddots & \vdots & \vdots & \ddots & \vdots \\ 0 & \cdots & 1 & h_1 & \cdots & h_0 \\ 0 & \cdots & 0 & q & \cdots & 0 \\ \vdots & \ddots & \vdots & \vdots & \ddots & \vdots \\ 0 & \cdots & 0 & 0 & \cdots & q \end{pmatrix},$$

which shows that

$$[a_0, \ldots, a_{N-1}, b_0, \ldots, b_{N-1}] \in L_{\mathbf{h}}.$$

We leave the opposite inclusion for the reader; see Exercise 4.10.5. □

REMARK 4.6.4. An immediate consequence of Proposition 4.6.3 is that if $[\mathbf{a}, \mathbf{b}] \in L_{\mathbf{h}}$, then its *rotations*

(4.6.5) $$[\mathbf{e}_i \star \mathbf{a}, \mathbf{e}_i \star \mathbf{b}] \text{ are also in } L_{\mathbf{h}} \text{ for all } 0 \le i < N,$$

where $\mathbf{e}_i \in \mathbb{Z}^N$ is the usual basis factor with a 1 in the ith coordinate and 0s elsewhere. Alternatively, if we identify $\mathbf{a}$ with the polynomial $\mathbf{z}(X)$ as in Remark 4.4.5, then $\mathbf{e}_i \star \mathbf{a}$ is the product $X^i\mathbf{a}(X)$ in the polynomial quotient ring $\mathcal{R}$, and similarly for $\mathbf{e} \star \mathbf{b}$.

4.7. Recovering an NTRU Private Key via an SVP Problem. An NTRU public/private key pair satisfies

(4.7.1) $$\mathbf{f} \star \mathbf{h} \equiv \mathbf{g} \pmod{q} \quad \text{where } \mathbf{f} \text{ and } \mathbf{g} \text{ are small;}$$

i.e., $\mathbf{f}$ and $\mathbf{g}$ are vectors in $\mathbb{Z}^N$ having small coordinates. Alternatively, we may view $\mathbf{f}, \mathbf{g} \in \mathbb{Z}[X]$ as being polynomials of degree at most $N-1$ having small coefficients. Proposition 4.6.3 and (4.7.1) tell us that the lattice $L_{\mathbf{h}}$ contains the short vector

$$[\mathbf{f}, \mathbf{g}] := [f_0, f_1, \ldots, f_{N-1}, g_0, g_1, \ldots, g_{N-1}].$$

For practical choices of parameters, the vector $[\mathbf{f}, \mathbf{g}]$ and its rotations (4.6.5) are likely to be the shortest non-zero vectors in $L_{\mathbf{h}}$, so solving SVP in $L_{\mathbf{h}}$ essentially recovers the private key $\mathbf{f}$. On the other hand, the public basis for $L_{\mathbf{h}}$ provided by the rows of the matrix (4.6.2) is not a good basis, since the coefficients of $\mathbf{h}$ are more-or-less randomly distributed between $-\frac{1}{2}q$ and $\frac{1}{2}q$, so these basis vectors are significantly longer than the short vector $[\mathbf{f}, \mathbf{g}]$.

4.8. Recovering an NTRU Plaintext via a CVP Problem. An NTRU ciphertext $\mathbf{e}$ has the form

$$\mathbf{e} := p\mathbf{r} \star \mathbf{h} + \mathbf{m} \pmod{q} \quad \text{where } \mathbf{r} \text{ and } \mathbf{m} \text{ are small.}$$

We rewrite this relation in vector form as

$$\begin{aligned}[\mathbf{0}, \mathbf{e}] &= [\mathbf{0},\ p\mathbf{r} \star \mathbf{h} + \mathbf{m} \mod q] \\ &\equiv [\mathbf{r},\ \mathbf{r} \star (p\mathbf{h}) \mod q] + [-\mathbf{r}, \mathbf{m}].\end{aligned}$$

The vector $[\mathbf{r},\ \mathbf{r}\star(p\mathbf{h}) \mod q]$ is in the convolution modular lattice $L_{p\mathbf{h}}$ obtained by using $p\mathbf{h}$ in place of $\mathbf{h}$. Further, the vector $[-\mathbf{r}, \mathbf{m}]$ is quite short. Hence recovering the plaintext $\mathbf{m}$ from the ciphertext $\mathbf{e}$ is equivalent to finding the vector in $L_{p\mathbf{h}}$ that is closest to the vector $[\mathbf{0}, \mathbf{e}]$. It is then a question of estimating how hard it is to solve this CVP problem, and just as we noted in Section 4.7 we expect this CVP problem to be hard because the public basis for $L_{p\mathbf{h}}$ is a bad basis.

4.9. NTRU Operating Characteristics and Variants. Table 4.9.1 gives a rough comparison of the operating characteristics for naive implementations of RSA, ECC, and NTRUEncrypt; i.e., with no implementation tricks. But there are many ways to improve the speed, including:

- Using a small RSA encryption exponent gives encryption speed $O(n^2)$.
- One can do precomputation and use windowing techniques to make ECC encryption and decryption faster, although the former requires some storage.
- Using Karatsuba multiplication or fast Fourier transform techniques, one can reduce NTRU encryption and decryption times to $O(n \log n)$.

	RSA	ECC	NTRU
Encrypt/Decrypt	$O(n^3)$	$O(n^3)$	$O(n^2)$
Key size (bits)	n	n	$\approx \frac{1}{2} n \log_2 n$
Key Create	—	$O(n^3)$	$O(n^2)$
Typical n	1024	168	502

TABLE 4.9.1. Operating Characteristic of Vanilla RSA, ECC, NTRU

REMARK 4.9.2. There are many variants of NTRU that have been proposed over the years. For example, one might replace $X^N - 1$ with $X^N + 1$ and take $N = 2^k$ so that $X^N + 1$ is irreducible in $\mathbb{Z}[X]$. More generally, one can replace $X^N - 1$ with an arbitrary (monic) polynomial $\phi(X) \in \mathbb{Z}[X]$, which could be irreducible and must have small coefficients. The key is the fact that if $\mathbf{a}(X)$ and $\mathbf{b}(X)$ have small coefficients, then

$$\mathbf{a}(X) \cdot \mathbf{b}(X) \mod \phi(X) \quad \text{has small'ish coefficients.}$$

The average size of the coefficients of the product depends on the size of the roots of $\phi(X)$, which is why taking $\phi(X)$ to be a cyclotomic polynomial is especially efficient. On the other hand, using $X^N - 1$ introduces symmetries into the lattice that an attacker might exploit. This hypothetical problem can be eliminated if, for example, one takes $\phi(X) = X^N - X - 1$; see Exercise 4.10.15 for additional information.

4.10. Exercises for Lecture 4.

EXERCISE 4.10.1. Let $\mathbf{f}(X) \in \mathcal{R} := \mathbb{Z}[X]/(X^N - 1)$. We define the *reversal* (or *conjugate*) of $\mathbf{f}(X)$ to be the polynomial

$$\overline{\mathbf{f}}(X) \leftarrow \mathbf{f}(X^{N-1}) \in \mathcal{R}.$$

Since $X^N = 1$, we may write this informally as $\overline{\mathbf{f}}(X) = \mathbf{f}(X^{-1})$.

(1) Let

$$\mathbf{f}(X) := \sum_{i=0}^{N-1} a_i X^i \quad \text{and} \quad \overline{\mathbf{f}}(X) := \sum_{i=0}^{N-1} \overline{a}_i X^i.$$

Find a formula for $\overline{a}_i$ in terms of $a_0, \ldots, a_{N-1}$.

(2) Prove that reversal is a ring homomorphism $\mathcal{R} \to \mathcal{R}$, i.e., prove that

$$\overline{\mathbf{f} + \mathbf{g}} = \overline{\mathbf{f}} + \overline{\mathbf{g}} \quad \text{and} \quad \overline{\mathbf{f} \cdot \mathbf{g}} = \overline{\mathbf{f}} \cdot \overline{\mathbf{g}}.$$

(3) Define the *sup norm* of a polyomial

$$\mathbf{f}(X) := \sum_{i=0}^{N-1} a_i X^i \in \mathcal{R} \quad \text{to be} \quad \|\mathbf{f}\|_\infty := \max_{0 \le i < N} |a_i|,$$

i.e., $\|\mathbf{f}\|_\infty$ is the sup-norm of the vector of coefficients of $\mathbf{f}$. Let $B > 0$ be an integer. Compute the values of

$$\sum_{\mathbf{f} \in \mathcal{R},\, \|\mathbf{f}\|_\infty \le B} \mathbf{f}(X) \quad \text{and} \quad \sum_{\mathbf{f} \in \mathcal{R},\, \|\mathbf{f}\|_\infty \le B} \mathbf{f}(X) \cdot \overline{\mathbf{f}}(X).$$

The fact that the latter average is non-zero features in various analyses of NTRU-type cryptographic constructions.

EXERCISE 4.10.2. Prove the associative, distributive, and commutative laws for $+$ and $\star$ as described in Proposition 4.4.4.

EXERCISE 4.10.3. With the identifications described in Section 4.4, prove that

$$\mathbf{c} = \mathbf{a} \star \mathbf{b} \quad \Longleftrightarrow \quad \mathbf{c}(X) = \mathbf{a}(X) \cdot \mathbf{b}(X) \text{ in } \mathcal{R}.$$

EXERCISE 4.10.4. Let N and q be primes such that q is a primitive root modulo N; i.e., assume that $q^k \equiv 1 \pmod{N}$ if and only if $N - 1 \mid k$.

(1) Prove that the splitting field of $X^N - 1$ over $\mathbb{Z}/q\mathbb{Z}$ is a field with q^{N-1} elements.
(2) Prove that $\mathbf{a}(X) \in \mathcal{R}_q^*$ if and only if $\mathbf{a}(1) \not\equiv 0 \pmod{q}$ and $\mathbf{a}(X)$ is not a scalar multiple of $X^{N-1} + \cdots + X + 1$.
(3) Find a formula for $\#\mathcal{R}_q^*/\#\mathcal{R}_q$. Show in particular that if N is large, then this ratio is close to $1 - q^{-1}$. Notice how this quantifies the meaning of "many" in the statement of Proposition 4.4.6.
(4) Suppose instead that q has exact order k in $(\mathbb{Z}/N\mathbb{Z})^*$. Modify (1), (2), and (3) to give correct statements, and prove that those statements are correct. (Hint. Factor $X^N - 1$ in $\mathbb{F}_q[X]$ and write $\mathbb{F}_q[X]/(X^N - 1)$ as a product of quotient rings.)

EXERCISE 4.10.5. Prove the following part of Propostion 4.6.3 that was left for the reader: Let

$$(\mathbf{a}, \mathbf{b}) \in L_{\mathbf{h}} \subseteq \mathbb{Z}^{2N},$$

where $L_{\mathbf{h}}$ is defined as the $\mathbb{Z}$-span of the rows of the matrix (4.6.2). Prove that

$$\mathbf{a} \star \mathbf{h} \equiv \mathbf{b} \pmod{q}.$$

EXERCISE 4.10.6. Suppose that the coefficients of the polynomials f and g in an NTRU private key are randomly selected so that d of the coefficients equal 1 and d of the coefficients equal -1 and the remaining coefficients equal 0.[28] Suppose further that the coefficients of r and m lie in $\{-1, 0, 1\}$.

(1) Prove that if

$$q > 2d(p+1)$$

then NTRU decryption always works.

(2) Prove that the number of possible f polynomials is $N!/(N-d)!(d!)^2$, and similarly for g. Thus N and d must be large enough to preclude an attacker from simply checking all possible f values.[29]

(3) For a private key chosen as above, compare the length of the target vector $[f, g]$ in the NTRU lattice L_h to the Gaussian expected shortest length of a non-zero vector in L_h. (The Gaussian heuristic is described in Section 1.6.) Express you answer in terms of N, q, d. Then take $q \approx 2d(p+1)$ as in (a), and show that the target vector is likely to be unique shortest non-zero vector in L_h.[30] More precisely, show that the ratio of private key vector length to Gaussian expected shortest length is on the order of $1/\sqrt{Np}$.

EXERCISE 4.10.7. For a polynomial $\mathbf{a}(X) = \sum_{i=0}^{N-1} a_i X^i \in \mathcal{R}$, we define various norms

$$\|\mathbf{a}\|_\infty := \max |a_i|, \qquad \|\mathbf{a}\|_1 := \sum |a_i|, \qquad \|\mathbf{a}\|_2 := \sqrt{\sum a_i^2}. \tag{4.10.8}$$

For example, the polynomials with norm $\|\mathbf{a}\|_\infty \leq 1$ are the polynomials whose coefficients are in the set $\{-1, 0, 1\}$.

(1) Let $\mathbf{a}, \mathbf{b} \in \mathcal{R}$. Prove that

$$\|\mathbf{a} \cdot \mathbf{b}\|_\infty \leq \|\mathbf{a}\|_\infty \cdot \|\mathbf{b}\|_1. \tag{4.10.9}$$

Describe the $\mathbf{a}, \mathbf{b}$ for which (4.10.9) is an equality.

(2) Let $\mathbf{a}, \mathbf{b} \in \mathcal{R}$. Prove that

$$\|\mathbf{a} \cdot \mathbf{b}\|_2 \leq N \cdot \|\mathbf{a}\|_\infty \cdot \|\mathbf{b}\|_2. \tag{4.10.10}$$

Describe the $\mathbf{a}, \mathbf{b}$ for which (4.10.10) is an equality.

[28] The astute reader will realize that this will not work, since then $f(1) = 0$, which means that $f(X)$ cannot have an inverse in $\mathcal{R}_q$. In practice, one chooses $f(X)$ to have $d+1$ coefficients equal 1 and d coefficients equal -1. This has a small numerical effect, but little practical effect, on the results of this problem.

[29] There are so-called collision algorithms that give a the attacker a square-root advantage; cf. Exercise 4.10.14.

[30] This cannot be literally true, of course, since the rotations (4.6.5) of $[f, g]$ have the same length as $[f, g]$. So the correct conclusion is that these N vectors are most likely the shortest non-zero vectors in L_h.

(3) Let $\mathbf{b} \in \mathcal{R}$. Prove that

$$\frac{1}{3^N} \sum_{\|\mathbf{a}\|_\infty \le 1} \|\mathbf{a} \cdot \mathbf{b}\|_2^2 = N \cdot \|\mathbf{b}\|_2^2.$$

This suggests that for most $\mathbf{a}$ satisfying $\|\mathbf{a}\|_\infty \le 1$, the upper bound (4.10.10) in (2) is off by a factor of $\sqrt{N}$.

EXERCISE 4.10.11. Explain how NTRU decryption can be described as Babai's algorithm applied to the short partial basis consisting of $[f, g]$ and its rotations (4.6.5).

EXERCISE 4.10.12. For NTRU, one generally takes N to be prime. This exercise illustrates why by describing a potential attack of Craig Gentry [**12**] when N is even, say $N = 2M$. In this case the polynomial $X^N - 1$ factors, and we get a homomorphism

$$\mathbb{Z}[X]/(X^{2M} - 1) \longrightarrow \mathbb{Z}[X]/(X^M - 1) \times \mathbb{Z}[X]/(X^M + 1). \tag{4.10.13}$$

(1) What are the kernel and cokernel of the map (4.10.13)?
(2) If a is a small element of $\mathbb{Z}[X]/(X^{2M} - 1)$, show that its images in

$$\mathbb{Z}[X]/(X^M - 1) \quad \text{and} \quad \mathbb{Z}[X]/(X^M + 1)$$

are small. (Give a quantitative estimate for the sup norms of the coefficients.)
(3) Set up a lattice reduction search for an NTRU private key $[f, g]$ as a pair of lattice searches in two lattices of dimension M. This cuts the dimension of the relevant lattices in half compared to the lattice in Section 4.6.

EXERCISE 4.10.14. Any cryptographic construction in which the private key is chosen from a finite set is susceptible to a *combinatorial* or *brute-force attack* in which the attacker simply checks the possible keys. And frequently the attacker can search for "collisions" to significantly speed up the process. In this exercise we describe one such attack. We set the notation

$$\mathcal{R}_q[1] \leftarrow \left\{ \sum_{0 \le i \le N-1} a_i x^i : a_i \in \{-1, 0, 1\} \right\}$$

for the set of polynomials whose coefficients are in the set $\{-1, 0, 1\}$.
(1) Prove that $\#\mathcal{R}_q[1] = 3^N$.
(2) Suppose that Alice chooses random polynomials $\mathbf{f}, \mathbf{g} \in \mathcal{R}_q[1]$ and that her public key is the polynomial $\mathbf{h} = \mathbf{g} \cdot \mathbf{f}^{-1}$ in $\mathcal{R}_q$. An attacker tries to find Alice's private key by randomly choosing polynomials $\mathbf{a} \in \mathcal{R}_q[1]$ and checking whether $\mathbf{h} \cdot \mathbf{a}$ is in $\mathcal{R}_q[1]$. Show that the expected number of attempts is $\frac{1}{2}\#\mathcal{R}_q[1]$.
(3) We define two sets

$$\mathcal{R}_q[1]_0 \leftarrow \left\{ \sum_{0 \le i \le N/2} a_i x^i : a_i \in \{-1, 0, 1\} \right\},$$

$$\mathcal{R}_q[1]_1 \leftarrow \left\{ \sum_{N/2 < i \le N-1} a_i x^i : a_i \in \{-1, 0, 1\} \right\}.$$

Prove that

$$\#\mathcal{R}_q[1]_0 \approx \#\mathcal{R}_q[1]_1 \approx \sqrt{\#\mathcal{R}_q[1]}.$$

(4) An attacker uses Alice's public key $\mathbf{h}$ to compile the following two lists of polynomials in $\mathcal{R}_q$:

$$\mathcal{S}_0 \leftarrow \{\mathbf{h} \cdot \mathbf{a} : \mathbf{a} \in \mathcal{R}_q[1]_0\} \quad \text{and} \quad \mathcal{S}_1 \leftarrow \{\mathbf{h} \cdot \mathbf{a} : \mathbf{a} \in \mathcal{R}_q[1]_1\}.$$

Prove that there exists an $\mathbf{h} \cdot \mathbf{a}_0 \in \mathcal{S}_0$ and an $\mathbf{h} \cdot \mathbf{a}_1 \in \mathcal{S}_1$ that are very close to one another; more precisely, they satisfy

$$\mathbf{h} \cdot \mathbf{a}_0 - \mathbf{h} \cdot \mathbf{a}_1 \in \mathcal{R}_q[1].$$

Prove that $\mathbf{a}_0 - \mathbf{a}_1$ is (almost certainly) equal to $x^j \cdot \mathbf{f}$ for some $0 \le j < n$, and thus that an attacker who can compute this "almost collision" between the sets $\mathcal{S}_0$ and $\mathcal{S}_1$ has broken Alice's system.

(5) Show that after sorting $\mathcal{S}_0$ in some reasonable way, e.g., lexigraphically by coordinates, it is possible to check whether an element of $\mathcal{S}_1$ has an almost collision in $\mathcal{S}_0$ in roughly $O(\log N)$ steps. Deduce that using this collision-style attack, an attacker can break Alice's system in roughly $O\big(\sqrt{\#\mathcal{R}_q[1]} \cdot \log N\big)$ steps, which represents a substantial improvement on the brute force search described in (2).

EXERCISE 4.10.15. This exercise asks you to explore Remark 4.9.2.

(1) Generate some data to illustrate Remark 4.9.2. For example, generate a lot of random polynomials $\mathbf{a}(X), \mathbf{b}(X)$ of degree at most $N - 1$ and compare the largest coefficient of the product $\mathbf{a}(X) \cdot \mathbf{b}(X)$ in the rings

$$\mathbb{Z}[X]/(X^N - 1), \qquad \mathbb{Z}[X]/(X^N + 1), \qquad \mathbb{Z}[X]/(X^N - X - 1).$$

Do this for, say, $N = 7$ and/or $N = 23$.

(2) Let $\phi(X) \in \mathbb{Z}[X]$ be a monic polynomial of degree N. Give an heuristic argument as to why the largest coefficient of a product $\mathbf{a}(X)\mathbf{b}(X)$ in $\mathbb{Z}[X]/\big(\phi(X)\big)$ should depend on the largest root of $\phi(X)$ in $\mathbb{C}$. Make your estimate as quantitative as you can.

5. Lattice-Based Digital Signatures and Rejection Sampling

5.1. Digital Signatures. We recall from Section 3.6 that a *Digital Signature Scheme* consists of a signing function and a verification function:

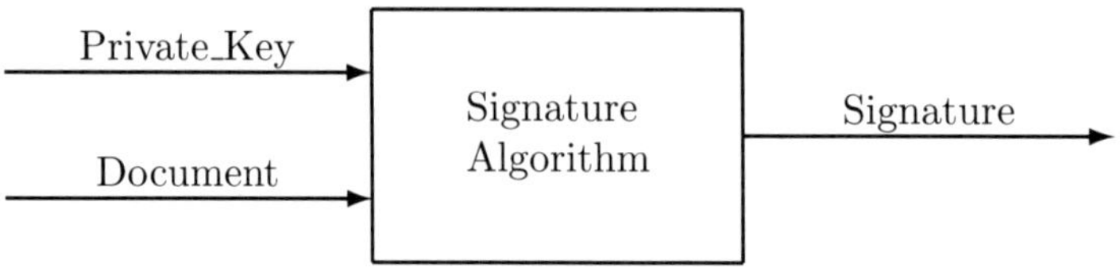

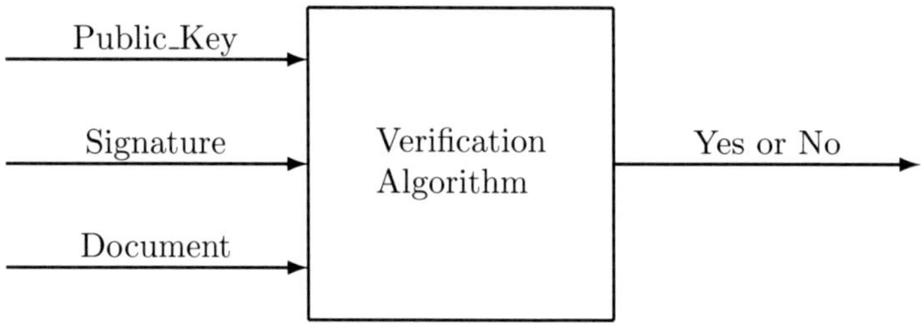

If (PUBKEY, PRIVKEY) is a valid public/private key pair and DOC is a document and SIG is a purported signature, then we want the following to be true:

$$\mathsf{Verify}(\text{PUBKEY}, \text{SIG}, \text{DOC}) = \text{Yes} \quad \Longleftrightarrow \quad \text{SIG} = \mathsf{Sign}(\text{PRIVKEY}, \text{DOC}).$$

5.2. CVP Digital Signatures — GGH. In Section 3.6.3 we briefly discussed digital signature schemes based on the integer factorization problem (IFP), the discrete logarithm problem (DLP), and the elliptic discrete logarithm problem (ECDLP). In this section we discuss lattice-based digital signature schemes whose security relies on the shortest and/or closest vector problem (SVP/CVP).

The prototypical example of a lattice-based digital signature is the *GGH digital signature scheme*, which is described in Table 5.2.2. Its public and private keys are the same as that of the GGH lattice-based cryptosystem, and its signatures are solutions to approximate CVP problems:

$$\text{Private Key} := \{\mathbf{v}_1, \ldots, \mathbf{v}_n\} := \text{a good basis } \mathcal{B}^{\text{good}} \text{ for a lattice } \mathcal{L}.$$

$$\text{Public Key} := \{\mathbf{w}_1, \ldots, \mathbf{w}_n\} := \text{a bad basis } \mathcal{B}^{\text{bad}} \text{ for the lattice } \mathcal{L}.$$

$$\text{Document} := \text{a vector } \mathbf{d} \text{ that is } \underline{\text{not}} \text{ in } \mathcal{L}.$$

REMARK 5.2.1. In practice, the "document" $\mathbf{d}$ that Alice signs is the output of a hash function applied to her actual document concatenated or otherwise combined with some random bits; cf. Section 3.7, Example 3.8.1, and Exercise 3.12.3

- Alice uses Babai's algorithm (Section 1.3) with $\mathcal{B}^{\text{good}}$ to find a lattice vector close to $\mathbf{d}$. To do this, she first writes

$$\mathbf{d} = \delta_1\mathbf{v}_1 + \cdots + \delta_n\mathbf{v}_n \quad \text{with } \delta_1, \ldots, \delta_n \in \mathbb{R}.$$

Then she rounds the coefficients to get a lattice vector

$$\mathbf{s} := \lfloor \delta_1 \rceil \mathbf{v}_1 + \cdots + \lfloor \delta_n \rceil \mathbf{v}_n \in \mathcal{L}$$

that is close to $\mathbf{d}$.
- Alice writes $\mathbf{s}$ using the bad basis $\mathcal{B}^{\text{bad}}$,

$$\mathbf{s} = s_1\mathbf{w}_1 + \cdots + s_n\mathbf{w}_n.$$

Her signature on $\mathbf{d}$ is the n-tuple $(s_1, \ldots, s_n)$.
- Bob uses the n-tuple $(s_1, \ldots, s_n)$ and the bad public basis $\mathcal{B}^{\text{bad}}$ to reconstruct $\mathbf{s}$ using the formula

$$\mathbf{s} = s_1\mathbf{w}_1 + \cdots + s_n\mathbf{w}_n.$$

The vector $\mathbf{s}$ is automatically in $\mathcal{L}$, and Bob verifies that the signature is valid by checking that $\mathbf{s}$ is sufficiently close to $\mathbf{d}$.

TABLE 5.2.2. The GGH Lattice-Based Digital Signature Scheme

5.3. Security of GGH and other CVP-based Digital Signatures. Just as with CVP-based encryption schemes, there are many security issues that must be addressed for CVP-based digital signatures:

Cominatorial Security: The spaces of keys and signature must to be large enough so that an attacker cannot check all their elements. Further, one must take into account the fact that collision search algorithms often run in time $O(\sqrt{K})$ on sets of size K; cf. Exercises 4.10.14 and 5.8.10.

Lattice Security: Just as with CVP-based public key cryptosystems, one needs to check that latttice reduction algorithms such as LLL-BKZ cannot solve the approximate CVP well enough to forge signatures.

Practicality: The GGH digital signature scheme has impractically large public keys, since a public key is a complete basis of the lattice. One might try to instead use an NTRU-type construction, although it is not entirely clear how to do that, since an NTRU lattice only provides a good basis for a half-dimensional sublattice. However, such a construction is possible; see [**21**] for details.

Transcript Security: For digital signature schemes, here is

Yet Another Thing to Worry About!

- Each signature potentially reveals some information about the private key.
- So a long transcript of signatures might compromise security.

5.4. A Transcript Attack on the GGH Digital Signature Scheme. The possibility of a transcript attack is not an idle threat. A number of people, including Gentry–Szydlo [**14**] and Nguyen–Regev [**29, 30**], developed practical transcript attacks on GGH and other early CVP-based signature schemes. We describe roughly how one might use a GGH transcript to recover the private key. A GGH signature

s on a document **d** reveals a vector in the centered fundamental domain spanned by the good basis:

$$\mathbf{s} - \mathbf{d} \in \mathcal{F}^{\text{good}} := \left\{ t_1\mathbf{v}_1 + \cdots + t_n\mathbf{v}_n : -\frac{1}{2} \le t_i \le \frac{1}{2} \right\}.$$

As illustrated in Figure 5.4.1 on page 215, a transcript consisting of only a few signatures is relatively innocuous, but a long transcript of signatures reveals a lot of information about the fundamental domain spanned by the good secret basis.

Note that Alice, Bob, and Carl each have their own lattice, and thus the good bases that they're chosen for their individual lattices define different fundamental domains. Then transcripts of their signatures fill out their individual private fundamental domains, as illustrated in Figure 5.4.2 on page 216.

5.5. Rejection Sampling to the Rescue. *Rejection sampling* is a technique from statistics in which one starts with a random process whose output has a known distribution, and by judiciously throwing away (rejecting) some of the output values, creates a new random process whose output has some other desired distribution.

Lyubashevsky [**26**, **27**] constructed a lattice-based identification scheme and digital signature using a technique that he called *aborting*, which is a version of rejection sampling.

The basic idea is that Alice generates a signature on her document and checks if it will leak information about her individual private key. If it will leak information, then she rejects that signature and generates another one. She continues to do this until finding a safe signature, which she publishes.

First Issue: How does Alice generate multiple signatures on the same document?

Remember that Alice isn't really signing her document, she's signing a hash of her document. In fact, for various reasons, it is advisable that she signs something like

$$\mathsf{Hash}\begin{pmatrix}\text{Alice's actual document concate-}\\ \text{nated with, say, 160 random bits}\end{pmatrix}.$$

So if Alice generates a signature and decides to reject it, she simply selects a new random 160 bits and tries again. Her signature would then include the 160 random bits that she uses in the signature that she doesn't reject.

How might this work for GGH? Alice's signatures give a set of points that are uniformly distributed in her fundamental domain $\mathcal{F}^{\text{good}}_{\text{Alice}}$. Similarly, Bob's signatures give points uniformly distributed in $\mathcal{F}^{\text{good}}_{\text{Bob}}$, and the same for Carl. But suppose that they fix a region that's common to all three fundamental domains; i.e., that is contained in $\mathcal{F}^{\text{good}}_{\text{Alice}}$, $\mathcal{F}^{\text{good}}_{\text{Bob}}$, and $\mathcal{F}^{\text{good}}_{\text{Carl}}$. Then they reject signatures that fall outside the common region. This is illustrated in Figure 5.5.1 on page 217, where the shaded box is common to everyone's centered fundamental domain. Hence when Alice, Bob, and Carl reveal sets of equidistributed points in the shaded box, they are yielding no information about their private bases.

1. The public parameters include a cut-off value B with the property that everyone's centered fundamental domain $\mathcal{F}^{\text{good}}$ contains the centered box
$$\left\{(x_1,\ldots,x_n) \in \mathbb{R}^n : -\frac{1}{2}B \le x_i \le \frac{1}{2}B\right\}.$$
2. Alice computes a vector $\mathbf{d}$ to sign:
$$\mathbf{d} \leftarrow \mathsf{Hash}(\text{Alice's Doc} \,\|\, \text{Random Bits}).$$
3. Alice uses her good basis $\mathcal{B}^{\text{good}}$ to find a lattice vector $\mathbf{s}$ that is close to $\mathbf{d}$.
4. If
$$\|\mathbf{s} - \mathbf{d}\|_\infty > B,$$
then Alice rejects $\mathbf{s}$, returns to Step 2, and selects new random bits.
5. Otherwise Alice accepts and publishes the signature $\mathbf{s}$ and the random bits that she used in Step 2. Note that it is only at this step that $\mathbf{s}$ becomes public, so an attacker never gets to see the rejected signatures.

TABLE 5.5.1. The GGH Digital Signature Scheme with Rejection Sampling

5.6. Transcript Security — At A Cost. Assuming that the hash function and pseudo-random number generator work as advertised, Alice's and Bob's and Carl's transcripts will have exactly the same distribution, namely they will be equidistributed over the lattice points in the common box. Thus the transcripts provide no useful information about the private key.

However, we have the added cost of rejection sampling. This leads to the following imporatant question:

> How many signatures are rejected before one of them is accepted?

For a naive implementation of GGH, the volume of the common box will be very small compared to the volume of the fundamental domain. So using rejection sampling is at best inefficient, and at worst, completely impractical.

Lyubashevsky and others use various methods to make it easier to find non-rejected signatures, while maintaining the property that the distribution of signatures is independent of the private key. For example, a signature scheme called BLISS devised by Ducas, Durmus, Lepoint, and Lyubashevsky [**11**] makes use of an NTRU-like key generation procedure and a bimodal discrete Gaussian noise distribution to produce compact equidistributed signatures. The efficiency of the scheme is quite good, especially considering the complexity of sampling discrete Gaussian distributions.

There are also lattice-based signature schemes for which the simple box rejection method is practical. These include a variant of NTRU described in [**21**], and another older scheme based on partial evaluation of polynomials called PASS [**23**]. In the next section we use PASS to illustrate how to prove that rejection sampling gives transcript security, although we will limit ourselves to describing the transcript aspects of PASS and leave a full description of the PASS signature scheme to the exercises.

5.7. An Example of a Lattice Recovery Problem and Rejection Sampling. In this section we describe how a short basis of a certain type of lattice can be recovered from a long list of CVP solutions, and how applying rejection sampling to the list of solutions prevents recovery of the short basis. In the exercises we explain how these lattices can be used to create a digital signature schemee.

We recall from Remark 4.4.5 that we defined a quotient ring

$$\mathcal{R} := \mathbb{Z}[X]/(X^N - 1),$$

and that we identified each element of $\mathcal{R}$ with its vector of coefficients in $\mathbb{Z}^N$. We also defined the *sup norm* of a polyomial to be the maximum of its coefficients, i.e.,

$$\text{the sup norm of} \quad \mathbf{f}(X) := \sum_{i=0}^{N-1} a_i X^i \in \mathcal{R} \quad \text{is} \quad \|\mathbf{f}\|_\infty := \max_{0\le i<N} |a_i|.$$

We let

$$\mathcal{R}[b] := \{\mathbf{f} \in \mathcal{R} : \|\mathbf{f}\|_\infty \le b\},$$

so for example, the polynomials in $\mathcal{R}[1]$ have coefficients in the set $\{-1, 0, 1\}$.

ALGORITHM 5.7.1 (**Prototypical Rejection Sampling Algorithm**).

1. **Parameter Selection:** *Fix a lattice dimension parameter N and a norm bound k that is significantly larger than N.*
2. **Secret Lattice Creation:** *Alice chooses a small polynomial $\mathbf{f} \in \mathcal{R}[1]$ that determines her secret lattice.*
3. **Random Polynomial Selection:** *Alice chooses a random polynomial $\mathbf{y} \in \mathcal{R}[k]$.*
4. **Hash Function:** *A hash function is applied to certain quantities associated to $\mathbf{f}$ and $\mathbf{y}$. The output from the hash function is a polynomial $\mathbf{c} \in \mathcal{R}[1]$ that depends randomly on the inputs.*[31]
5. **Signature Creation:** *Alice computes the polynomial*

$$\mathbf{s} := \mathbf{f} \star \mathbf{c} + \mathbf{y} \quad \textit{in the ring } \mathcal{R}.$$

6. **Rejection Sampling:** *If*

$$\|\mathbf{s}\|_\infty \ge k - N,$$

then Alice goes back to Step 3 and selects a new value for $\mathbf{y}$.

5. **Publication:** *Alice publishes the pair of polynomials $(\mathbf{s}, \mathbf{c})$ as her signature.*

[31]Since we are only interested in transcripts at the moment, we will not explain exactly how $\mathbf{c}$ depends on $\mathbf{f}$ and $\mathbf{y}$. It suffices to know that $\mathbf{c}$ behaves like a random variable uniformly distributed in $\mathcal{R}[1]$. See Figure 5.8.7 for further details.

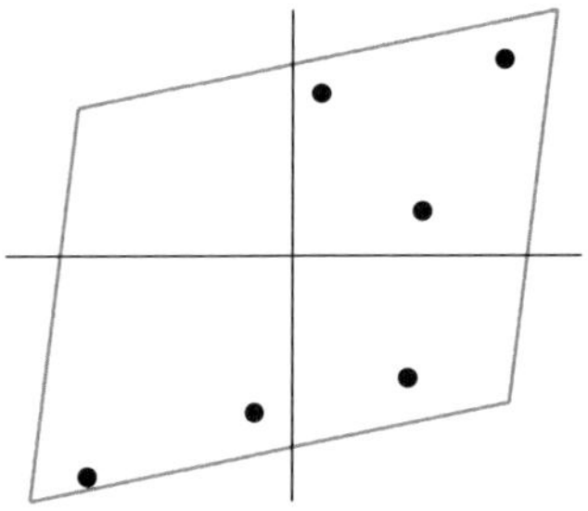

A few signatures aren't much help in reconstructing $\mathcal{F}^{\text{good}}$.

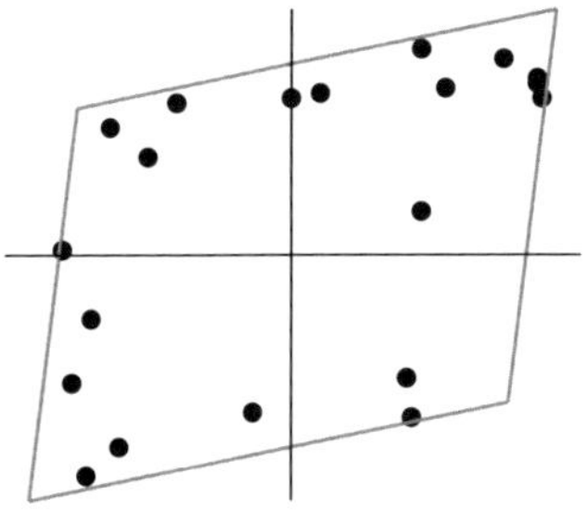

With a moderate number of signatures, the fundamental domain $\mathcal{F}^{\text{good}}$ starts to emerge.

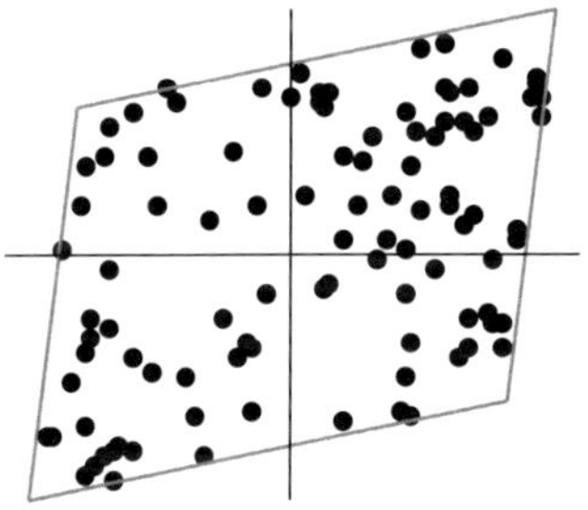

With lots of signatures, one may be able to reconstruct the fundamental domain $\mathcal{F}^{\text{good}}$.

FIGURE 5.4.1. Illustrating a Transcript Attack

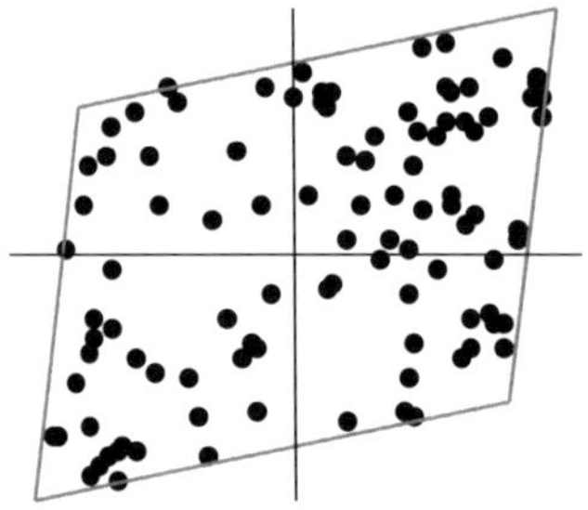

A Fundamental Domain $\mathcal{F}_{\text{Alice}}^{\text{good}}$ for Alice's Private Basis

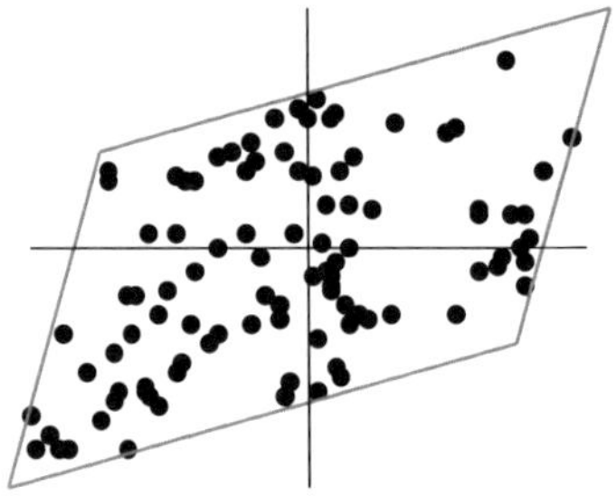

A Fundamental Domain $\mathcal{F}_{\text{Bob}}^{\text{good}}$ for Bob's Private Basis

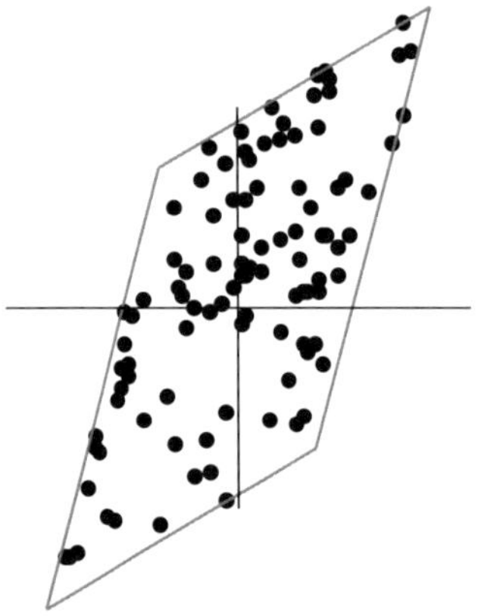

A Fundamental Domain $\mathcal{F}_{\text{Carl}}^{\text{good}}$ for Carl's Private Basis

FIGURE 5.4.2. Different Private Good Bases Have Different Transcripts

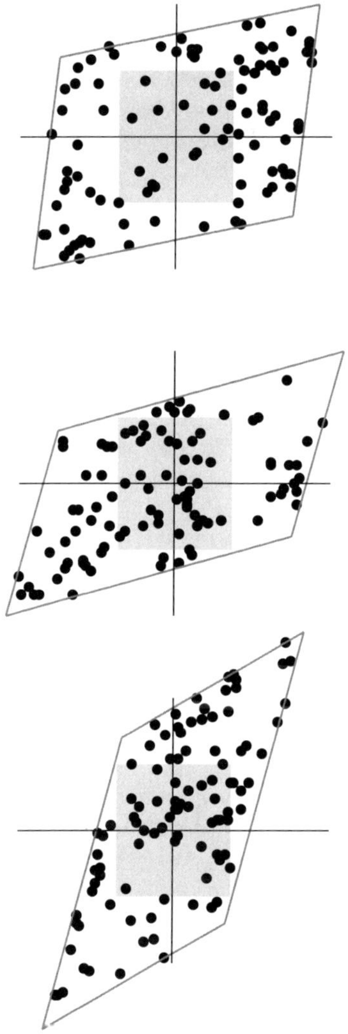

Alice's private fundamental domain $\mathcal{F}^{\text{good}}_{\text{Alice}}$ with common box

Bob's private fundamental domain $\mathcal{F}^{\text{good}}_{\text{Bob}}$ with common box

Carl's private fundamental domain $\mathcal{F}^{\text{good}}_{\text{Carl}}$ with common box

FIGURE 5.5.1. Selecting a Region Common to All Fundamental Domains

Suppose that Alice uses Algorithm 5.7.1 to create a long list of signatures

$$(\mathbf{s}_1, \mathbf{c}_1),\ (\mathbf{s}_2, \mathbf{c}_2),\ (\mathbf{s}_3, \mathbf{c}_3),\ \ldots .$$

We are going to prove that this transcript reveals no information about Alice's private key $\mathbf{f}$, but that if Alice were to skip the rejection sampling step, then information would be revealed. We also estimate the probability that a proposed signature makes it through the rejction step, since as noted in Section 5.6, rejection sampling comes a cost, and an excessively high rejection rate could make the scheme impractical. We also note that the constraint $k > N^2$ in Proposition 5.7.2 must be balanced against the underlying CVP when the system is turned into a full-fledged digital signature scheme; see Figure 5.8.7 for details.

PROPOSITION 5.7.2. *Let*

$$(\mathbf{s}_1, \mathbf{c}_1),\ (\mathbf{s}_2, \mathbf{c}_2),\ (\mathbf{s}_3, \mathbf{c}_3),\ \ldots . \tag{5.7.3}$$

be a transcript of signatures created using Algorithm 5.7.1.

(1) *If the* $\mathbf{c}_i$ *are uniformly and randomly distributed in* $\mathcal{R}[1]$,[32] *then the transcript* (5.7.3) *reveals no information about the private key* $\mathbf{f}$.
(2) *If instead the transcript is created skipping the rejection sampling step in Algorithm* 5.7.1, *then the transcript will contain information allowing recovery of the private key* $\mathbf{f}$.
(3) *If* $k > N^2$, *then the probability that a signature passes the rejection sampling step in Algorithm* 5.7.1 (*roughly*) *satisfies*

$$\operatorname{Prob}\Bigl(\|\mathbf{s}\|_\infty < k - N\Bigr) \gtrapprox \left(1 - \frac{N}{2k+1}\right)^N \approx e^{-N^2/(2k+1)}.$$

PROOF. (1) What does it mean to say that the transcript reveals no information about $\mathbf{f}$? The $\mathbf{c}_i$ polynomials are assumed to be randomly distributed, so the question is to what extent the $\mathbf{s}_i$ values contain information about $\mathbf{f}_i$. We claim that the probability that any particular valid $\mathbf{s}$ shows up in a signature does not depend on the polynomial $\mathbf{f}$. We can formulate this as a statement in conditional probability as follows:

CLAIM. For all $\mathbf{f}_0 \in \mathcal{R}[1]$ and all $\mathbf{s}_0 \in \mathcal{R}[k-N]$,

$$\operatorname{Prob}\left(\begin{matrix}\text{a signature } (\mathbf{s},\mathbf{c}) \text{ created}\\ \text{using } \mathbf{f} \text{ satisfies } \mathbf{s} = \mathbf{s}_0\end{matrix}\ \middle|\ \mathbf{f} = \mathbf{f}_0\right) \quad \text{does not depend on } \mathbf{f}_0.$$

The probability is computed over the space of randomly chosen polynomials $\mathbf{c} \in \mathcal{R}[1]$ and $\mathbf{y} \in \mathcal{R}[k]$, so we compute

$$\begin{aligned}\operatorname*{Prob}_{\substack{\mathbf{c}\in\mathcal{R}[1]\\ \mathbf{y}\in\mathcal{R}[k]}}\bigl(\mathbf{s} = \mathbf{s}_0 \mid \mathbf{f} = \mathbf{f}_0\bigr) &= \frac{1}{\#\mathcal{R}[1]\cdot\#\mathcal{R}[k]}\cdot\#\Bigl\{(\mathbf{c},\mathbf{y}) \in \mathcal{R}[1]\times\mathcal{R}[k]\ \Bigm|\ \mathbf{s}_0 = \mathbf{c}\cdot\mathbf{f}_0 + \mathbf{y}\Bigr\}\\ &= \frac{1}{\#\mathcal{R}[1]\cdot\#\mathcal{R}[k]}\cdot\#\Bigl\{\mathbf{c} \in \mathcal{R}[1]\ \Bigm|\ \mathbf{s}_0 - \mathbf{c}\cdot\mathbf{f}_0 \in \mathcal{R}[k]\Bigr\}. \qquad (5.7.4)\end{aligned}$$

We next use the fact that $\mathbf{c}$ and $\mathbf{f}_0$ are in $\mathcal{R}[1]$ to bound the coefficients of their product. More generally, for $\mathbf{a}, \mathbf{b} \in \mathcal{R}$, we can use the triangle inequality to prove

[32]We are being somewhat informal here. In a formal security proof we would assume that the algorithm that creates the $\mathbf{c}_i$ may be modeled as a random oracle whose output is completely unpredictable.

the bound[33]

$$\|\mathbf{a}\star\mathbf{b}\|_\infty = \max_{0\le n<N}\left|\sum_{i+j\equiv n \pmod N} a_i b_j\right| \le N\|\mathbf{a}\|_\infty\cdot\|\mathbf{b}\|_\infty.$$

Hence for all $\mathbf{s}_0 \in \mathcal{R}[k-N]$ and all $\mathbf{c},\mathbf{f}_0 \in \mathcal{R}[1]$ we have

$$\begin{aligned}\|\mathbf{s}_0 - \mathbf{c}\cdot\mathbf{f}_0\|_\infty &\le \|\mathbf{s}_0\|_\infty + \|\mathbf{c}\cdot\mathbf{f}_0\|_\infty\\ &\le \|\mathbf{s}_0\|_\infty + N\|\mathbf{c}\|_\infty\cdot\|\mathbf{f}_0\|_\infty\\ &\le (k-N)+N\\ &= k.\end{aligned}$$

Thus the condition $\mathbf{s}_0 - \mathbf{c}\star\mathbf{f}_0 \in \mathcal{R}[k]$ in the conditional probability (5.7.4) is vacuous. This yields

$$\operatorname*{Prob}_{\substack{\mathbf{c}\in\mathcal{R}[1]\\ \mathbf{y}\in\mathcal{R}[k]}}\big(\mathbf{s}=\mathbf{s}_0 \mid \mathbf{f}=\mathbf{f}_0\big) = \frac{1}{\#\mathcal{R}[1]\cdot\#\mathcal{R}[k]}\cdot\#\big\{\mathbf{c}\in\mathcal{R}[1]\big\} = \frac{1}{\#\mathcal{R}[k]}.$$

In particular, the probability does not depend on the choice of $\mathbf{f}_0 \in \mathcal{R}[1]$, and indeed, the probability is also independent of $\mathbf{s}_0 \in \mathcal{R}[k-N]$.
(2) Exercise 4.10.1 in Lecture 4 defines the *reversal* (or *conjugate*) of a polynomial $\mathbf{a}(X) \in \mathcal{R}$ to be the polynomial

$$\overline{\mathbf{a}}(X) \leftarrow \mathbf{a}(X^{N-1}) \in \mathcal{R}.$$

In that exercise you verified that the map

$$\mathcal{R}\longrightarrow\mathcal{R},\quad \mathbf{a}\longmapsto\overline{\mathbf{a}},$$

is a ring automorphism, and you computed the average values

$$(5.7.5)\qquad \frac{1}{\#\mathcal{R}[B]}\sum_{\mathbf{a}\in\mathcal{R}[B]}\mathbf{a}=\mathbf{0}\qquad\text{and}\qquad \frac{1}{\#\mathcal{R}[B]}\sum_{\mathbf{a}\in\mathcal{R}[B]}\mathbf{a}\star\overline{\mathbf{a}}=\frac{N}{3}(B^2+B).$$

We note in particular that the latter average is a non-zero constant polynomial in $\mathcal{R}$.

Suppose that we have a transcript that contains T different signatures. For each signature $(\mathbf{s}_i,\mathbf{c}_i)$ in the transcript, we multiply $\mathbf{s}_i$ by $\overline{\mathbf{c}}_i$, and then we take the

[33] We remark that this estimate also follows from Exercise 4.10.7 combined with the trivial triangle inequality bound $\|\mathbf{b}\|_1 \le N\|\mathbf{b}\|_\infty$.

average over the signatures in the transcript. This yields

$$\begin{aligned}
\frac{1}{T}\sum_{i=1}^{T}\mathbf{s}_i\star\overline{\mathbf{c}}_i &= \frac{1}{T}\sum_{i=1}^{T}(\mathbf{f}\star\mathbf{c}_i+\mathbf{y}_i)\star\overline{\mathbf{c}}_i \\
&= \mathbf{f}\star\left(\frac{1}{T}\sum_{i=1}^{T}\mathbf{c}_i\star\overline{\mathbf{c}}_i\right)+\frac{1}{T}\sum_{i=1}^{T}\mathbf{y}_i\star\mathbf{c}_i \\
&\approx \mathbf{f}\star\left(\frac{1}{\#\mathcal{R}[1]}\sum_{\mathbf{c}\in\mathcal{R}[1]}\mathbf{c}\star\overline{\mathbf{c}}\right)+\left(\frac{1}{\#\mathcal{R}[k]}\sum_{\mathbf{y}\in\mathcal{R}[k]}\mathbf{y}\right)\star\left(\frac{1}{\#\mathcal{R}[1]}\sum_{\mathbf{c}\in\mathcal{R}[1]}\overline{\mathbf{c}}\right)
\end{aligned}$$

since the $\mathbf{c}_i$ are uniformly distributed in $\mathcal{R}[1]$, the $\mathbf{y}_i$ are uniformly distributed in $\mathcal{R}[k]$, and the $\mathbf{c}_i$ and $\mathbf{y}_i$ are independent random variables,

$$= \mathbf{f}\star\frac{2N}{3}+\mathbf{0} \quad \text{applyiing (5.7.5) with } (\mathbf{a},B)=(\mathbf{c},1) \text{ and } (\mathbf{y},k).$$

Hence if T is sufficiently large, then rounding the coefficients of

$$\frac{3}{2NT}\sum_{i=1}^{T}\mathbf{s}_i\star\overline{\mathbf{c}}_i$$

to the nearest integer is likely to yield the secret polynomial $\mathbf{f}$.

(3) The idea is that the coefficients of $\mathbf{c}\star\mathbf{f}$ are more-or-less a random sum of N terms, where each term is a product of two independent random variables that are uniformly distributed in $\{-1,0,1\}$. Hence each coefficient of $\mathbf{c}\star\mathbf{f}$ is an N-step random walk, where each step has a $\frac{5}{9}$ probability of being 0 and a $\frac{2}{9}$ probability of being either $+1$ or -1. Hence it is unlikely that there will be a coefficient whose magnitude is much larger than a small multiple of $\sqrt{N}$. Since we are assuming that k is considerably larger than N, the probability that $\|\mathbf{f}\star\mathbf{c}+\mathbf{y}\|_\infty\le k-N$ is roughly the same as the probability that $\mathbf{y}$ itself satisfies $\|\mathbf{y}\|_\infty\le k-N$; see Remark 5.7.6. This allows us to estimate

$$\begin{aligned}
&\operatorname*{Prob}_{\substack{\mathbf{c}\leftarrow\mathcal{R}[1]\\ \mathbf{y}\leftarrow\mathcal{R}[k]}}\Bigl(\|\mathbf{c}\star\mathbf{f}+\mathbf{y}\|_\infty\le k-N\Bigr) \\
&\qquad\approx \operatorname*{Prob}_{\mathbf{y}\leftarrow\mathcal{R}[k]}\Bigl(\|\mathbf{y}\|_\infty\le k-N\Bigr) \\
&\qquad\approx \prod_{i=0}^{N-1}\operatorname*{Prob}_{\mathbf{y}\leftarrow\mathcal{R}[k]}\Bigl(|i\text{th coefficient of }\mathbf{y}|\le k-N\Bigr) \\
&\qquad\qquad \text{since the coefficients are more-or-less independent,} \\
&\qquad= \prod_{i=0}^{N-1}\frac{2(k-N)+1}{2k+1} \\
&\qquad= \left(1-\frac{N}{2k+1}\right)^N.
\end{aligned}$$

We have proven the first estimate, and for the second we use $(1-t)^N\approx e^{-tN}$, which is valid provided that tN is small. □

REMARK 5.7.6. In the proof of Proposition 5.7.2(3) we ignored the $\mathbf{c}\star\mathbf{f}$ term as being negligible. For those who object to this cavalier attitude, we explain in more detail what happens if we include the $\mathbf{c}\star\mathbf{f}$ term. As noted in the proof, the coefficients of $\mathbf{c} \star \mathbf{f}$ may be modeled by a $\frac{4}{9}N$-step random walk in which each step is equally likely to be $+1$ and -1. If we write W for the distance from the origin at the end of the walk, then a well-known calculation says that the expected values of the powers of W are

$$\mathbb{E}(W) = 0, \quad \mathbb{E}(W^2) = \frac{4}{9}N, \quad \mathbb{E}(W^3) = 0, \quad \mathbb{E}(W^4) \approx \frac{16}{27}N^2;$$

see Exercise 5.8.1. So roughly speaking, we expect that the absolute value of a random coefficient of $\mathbf{c}\star\mathbf{f}$ to be around $0.67\sqrt{N}$, and the standard deviation of this value to be around $0.63N$. The probability that a coefficient has magnitude that is (say) 8 standard deviations from the mean is vanishingly small. So we make the reasonable assumption that every coefficient of $\mathbf{c} \star \mathbf{f}$ is at most $5\sqrt{N}$. Then a small modification of our computation yields

$$\operatorname*{Prob}_{\substack{\mathbf{c}\leftarrow\mathcal{R}[1]\\ \mathbf{y}\leftarrow\mathcal{R}[k]}}\left(\|\mathbf{c}\star\mathbf{f}+\mathbf{y}\|_\infty \le k - N\right) \gtrapprox= \left(1 - \frac{N+5\sqrt{N}}{2k+1}\right)^N.$$

In practice, the dimension is taken to satisfy at least $N \ge 1000$, and we can see how little difference the the $\mathbf{c} \star \mathbf{f}$ terms makes by computing

$N = 1000$ and $K = N^2$

$$\implies \quad \left(1 - \frac{N}{2k+1}\right)^N \approx 0.606 \quad \text{and} \quad \left(1 - \frac{N+5\sqrt{N}}{2k+1}\right)^N \approx 0.560.$$

As a practical matter, it makes little difference whether Alice has a 60% success rate or a 56% success rate in her signature generation routine.

5.8. Exercises for Lecture 5.

EXERCISE 5.8.1. Let $X_1, X_2, X_3, \dots, X_n$ be independent random variables with the property that

$$\operatorname{Prob}(X_i = 1) = \frac{1}{2} \quad \text{and} \quad \operatorname{Prob}(X_i = -1) = \frac{1}{2}.$$

Let

$$W := X_1 + X_2 + \cdots + X_n.$$

Compute the expected values of W, W^2, W^3, and W^4 as functions of n.

EXERCISE 5.8.2. This exercise illustrates an elementary example of rejection sampling. Suppose that a pair of dice is rolled many times, that we write R_k for the value of the kth roll, and that we create a value A_k by the rule

$$A_k := \begin{cases} H & \text{if } R_k \le 6, \\ T & \text{if } R_k \ge 8, \\ \text{Reject} & \text{if } R_k = 7. \end{cases}$$

(1) Prove that the sequence $A_1, A_2, A_3, \dots$ is uniformly distributed on the set $\{H, T\}$. We have thus converted a non-uniformly distributed random variable R into a uniformly distributed random variable A.
(2) About how many times would we need to sample R in order to expect to get n samples of A? In other words, about how many dice rolls would we need in order to simulate n coin flips?

(3) Generalize as much as you can!

Series of Exercises for the PASS Digital Signature Scheme

Exercises 5.8.4–5.8.8 describes a transcript secure digital signature scheme called PASS[34] that is based on partial evaluation of polynomials that have small coefficients. The transcript security was proved in Section 5.7, so we concentrate on describing how PASS works and why there is an underlying lattice problem. For those interested in the history of such schemes, the original idea of using partial evaluation of small polynomials appears in [**18**], and simplifications and methods to improve efficiency were described in [**19, 22**]. Then, after the introduction of rejection sampling methods to achieve transcript security in lattice-based signature schemes, it was noted in [**23**] that such methods are particularly easy to implement for PASS.

We assume for Exercises 5.8.4–5.8.8 that the following (public) parameters have been fixed:

$$N := \text{a moderate size prime, say } 500 < N < 5000.$$
$$q := \text{a prime satisfying } q \equiv 1 \pmod{N}.$$
$$\zeta := \text{a primitive } N\text{th root of unity in } \mathbb{F}_q^*.$$

We recall from Definition 4.4.1 that we defined the convolution product of two vectors by the formula (4.4.2), and that Proposition 4.4.4 says that coordinate-wise addition and convolution product give a ring structure to $\mathbb{F}_q^N$, i.e.,

$$(\mathbb{F}_q^N, +, \star) \text{ is a ring.}$$

Alternatively, Exercise 4.10.3 says that if we identify the elements of $\mathbb{F}_q^N$ with the coefficients of polynomials of degree $N-1$, then the convolution product is given by polynomial multiplication in the quotient ring

$$\mathcal{R}_q := \mathbb{F}_q[X]/(X^N - 1).$$

We observe that there is another, easier, way to make $\mathbb{F}_q^N$ into a ring, namely via coordinate-wise multiplication. We define

$$\mathbf{a} \odot \mathbf{b} := (a_0b_0, a_1b_1, \ldots, a_{N-1}b_{N-1}),$$

and then

$$(\mathbb{F}_q^N, +, \odot) \text{ is a ring.} \tag{5.8.3}$$

We use the generator $\zeta \in \mathbb{F}_q^*$ to define the *discrete Fourier transform* (DFT) map

$$\mathcal{F} : R_q \longrightarrow \mathbb{F}_q^N, \quad \mathcal{F}\big(\mathbf{f}(X)\big) := \big(\mathbf{f}(\zeta^i)\big)_{0 \le i < N}.$$

EXERCISE 5.8.4. Prove that $\mathcal{F}$ is a ring isomorphism, where $\mathbb{F}_q^N$ is given a ring structure via (5.8.3).

The idea underlying PASS is that the secret key is a polynomial $\mathbf{f}$ having small coefficients, and the public key is a subset of the coordinates of the DFT of $\mathbf{f}$. There will be many elements of $\mathcal{R}_q$ having those DFT coordinates, but very few of them will themselves have small coefficients. This leads to the following problem

[34]For those who are curious, PASS is an acronymn for Polynomial Authentication and Signature Scheme.

DEFINITION 5.8.5. Fix a subset $\Omega \subset \{0,1,\ldots,N-1\}$. The *small partial DFT inversion problem* is as follows: For a given set of values $(b_i)_{i\in\Omega}$, find a polynomial $\mathbf{f}(X) \in \mathcal{R}_q$ satisfying

$$\mathbf{f} \text{ has small coefficients} \quad \text{and} \quad \mathbf{f}(\zeta^i) = b_i \quad \text{for all } i \in \Omega,$$

or prove that no such $\mathbf{f}(X)$ exists. To ease notation, we let

$$t := \#\Omega, \quad \text{and we write} \quad \mathcal{F}_\Omega : \mathcal{R}_q \longrightarrow \mathbb{F}_q^t, \quad \mathcal{F}_\Omega(\mathbf{f}) := \bigl(\mathbf{f}(\zeta^i)\bigr)_{i\in\Omega},$$

for the *partial DFT associated to the set* Ω.

EXERCISE 5.8.6.

(1) Prove that

$$\mathcal{F}_\Omega : \mathcal{R}_q = \mathbb{F}_q^N \longrightarrow \mathbb{F}_q^t$$

is a surjective $\mathbb{F}_q$-linear transformation

(2) We compose $\mathcal{F}_\Omega$ with the reduction modulo q map and consider the function

$$\overline{\mathcal{F}}_\Omega : \mathbb{Z}^N \xrightarrow[\text{modulo } q]{\text{reduction}} \mathbb{F}_q^N = \mathcal{R}_q \xrightarrow{\mathcal{F}_\Omega} \mathbb{F}_q^t.$$

Prove that

$$\operatorname{Ker}(\overline{\mathcal{F}}_\Omega) := \bigl\{\mathbf{f} \in \mathbb{Z}^N : \overline{\mathcal{F}}_\Omega(\mathbf{f}) = \mathbf{0}\bigr\}$$

is an N-dimensional lattice in $\mathbb{Z}^N$. (*Hint.* Note that $\operatorname{Ker}(\overline{\mathcal{F}}_\Omega)$ contains the lattice $(q\mathbb{Z})^N$.)

(3) Prove that

$$\operatorname{Det}\bigl(\operatorname{Ker}(\overline{\mathcal{F}}_\Omega)\bigr) = q^t.$$

(4) Use (2) and (3) and the Gaussian heuristic (Section 1.6) to estimate the length of the shortest non-zero vector in $\operatorname{Ker}(\overline{\mathcal{F}}_\Omega)$

EXERCISE 5.8.8. The PASS Digital Signature Scheme is described in Figure 5.8.7. Prove that if Alice uses the signing procedure in Figure 5.8.7, then Bob will accept her signature as valid.

EXERCISE 5.8.9. This exercise explains how to formulate the small partial DFT inversion problem (Definition 5.8.5) as a closest vector problem in a certain lattice of dimension N and discriminant q^t. Assume that we are given a vector $\mathbf{F} \in \mathbb{F}_q^t$, and that we are told that there is a vector $\mathbf{f} \in \mathbb{Z}^N$ satisfying

$$\|\mathbf{f}\|_\infty \le 1 \quad \text{and} \quad \overline{\mathcal{F}}_\Omega(\mathbf{f}) = \mathbf{F},$$

where $\overline{\mathcal{F}}_\Omega$ is the map described in Exercise 5.8.6. Our goal is to find $\mathbf{f}$.

(1) Explain why it is easy to find a vector $\mathbf{f}_0 \in \mathbb{Z}^N$ satisfying $\overline{\mathcal{F}}_\Omega(\mathbf{f}_0) = \mathbf{F}$, provided that we do not require that $\mathbf{f}_0$ be small.

(2) Prove that

$$\bigl\{\mathbf{g} \in \mathbb{Z}^N : \overline{\mathcal{F}}_\Omega(\mathbf{g}) = \mathbf{F}\bigr\} = \mathbf{f}_0 + \operatorname{Ker}(\overline{\mathcal{F}}_\Omega).$$

(3) Suppose that the length of the shortest non-zero vector in $\operatorname{Ker}(\overline{\mathcal{F}}_\Omega)$ is given by the Gaussian heuristic; see Exercise 5.8.6. Prove that $\mathbf{f}_0 - \mathbf{f}$ solves the CVP for the lattice $\operatorname{Ker}(\overline{\mathcal{F}}_\Omega)$ and the target vector $\mathbf{f}_0$.

Public Parameters and Functions: N, q, Ω, $\mathcal{F}_\Omega$, k, Hash.

Private Key Creation: Alice's private key is a small polynomial $\mathbf{f} \in \mathcal{R}_q$, say satisfying $\|\mathbf{f}\|_\infty \le 1$.

Public Key Creation: Alice's public key is $\mathbf{F} := \mathcal{F}_\Omega(\mathbf{f}) \in \mathbb{F}_q^t$, the partial DFT of $\mathbf{f}$.

Signing: Alice wants to sign the document Doc for her public key $\mathbf{F}$.

(1) (Commitment) Alice chooses a random small polynomial $\mathbf{y} \in \mathcal{R}_q$, say satisfying $\|\mathbf{y}\|_\infty \le k$.

(2) (Challenge) Alice uses a hash function to compute a small polynomial

$$\mathbf{c} := \mathsf{Hash}\big(\mathsf{Doc}, \mathcal{F}_\Omega(\mathbf{y}), \mathbf{F}\big),$$

where say $\|\mathbf{c}\|_\infty \le 1$. (Note how $\mathbf{c}$ serves to tie Alice's signature to her document, her public key, and her random commitment.)

(3) (Response) Alice computes

$$\mathbf{s} := \mathbf{c} \star \mathbf{f} + \mathbf{y}.$$

(4) (Rejection Sampling) If $\|s\|_\infty$ is large, say $\|s\|_\infty \ge k - N$, then Alice goes back to Step (1) and chooses a new random commitment polynomial.

(5) Alice publishes her signature

$$(\mathbf{s}, \mathbf{c}) \quad \text{on the document } \mathsf{Doc} \text{ for the public key } \mathbf{F}.$$

Verification: Bob wants to check the validity of the signature $(\mathbf{s}, \mathbf{c})$ on the document Doc for Alice's public key $\mathbf{F}$.

(1) Bob computes

$$\mathbf{Y} := \mathcal{F}_\Omega(\mathbf{s}) - \mathcal{F}_\Omega(\mathbf{c}) \odot \mathbf{F}.$$

(2) Bob accepts the signature as valid if the following two statements are true:

- $\mathbf{s}$ is small, say $\|\mathbf{s}\|_\infty < k - N$.
- $\mathbf{c}$ is equal to $\mathsf{Hash}(\mathsf{Doc}, \mathbf{Y}, \mathbf{F})$.

FIGURE 5.8.7. The PASS Digital Signature Scheme

EXERCISE 5.8.10. This exercise describes a collision algorithm to solve the small partial DFT inversion problem; cf. Exericse 4.10.14. We suppose that there is an unknown polynomial $\mathbf{f}(X) \in \mathcal{R}[1]$, i.e., having coefficients in $\{-1, 0, 1\}$, and we further suppose that we are given the value of $\mathcal{F}_\Omega(\mathbf{f})$. We define two sets

$$\mathcal{R}[1]_0 := \left\{ \sum_{0 \le i \le N/2} a_i x^i : a_i \in \{-1, 0, 1\} \right\},$$

$$\mathcal{R}[1]_1 := \left\{ \sum_{N/2 < i \le N-1} a_i x^i : a_i \in \{-1, 0, 1\} \right\},$$

whose size is roughly $\sqrt{\#\mathcal{R}[1]}$, and we use them to create two lists

$$\mathcal{S}_0 := \big\{\mathcal{F}_\Omega(\mathbf{a}) : \mathbf{a} \in \mathcal{R}[1]_0\big\} \quad \text{and} \quad \mathcal{S}_1 := \big\{\mathcal{F}_\Omega(\mathbf{f}) - \mathcal{F}_\Omega(\mathbf{a}) : \mathbf{a} \in \mathcal{R}[1]_1\big\}.$$

(1) Prove that $\mathcal{S}_0 \cap \mathcal{S}_1 \ne \emptyset$.

(2) Prove that if

$$\mathcal{F}_\Omega(\mathbf{a}_0) \in \mathcal{S}_0 \quad \text{and} \quad \mathcal{F}_\Omega(\mathbf{f}) - \mathcal{F}_\Omega(\mathbf{a}_1) \in \mathcal{S}_1$$

satisfy

$$\mathcal{F}_\Omega(\mathbf{a}_0) = \mathcal{F}_\Omega(\mathbf{f}) - \mathcal{F}_\Omega(\mathbf{a}_1),$$

then (almost certainly) $\mathbf{f} = \mathbf{a}_0 + \mathbf{a}_1$.

(3) For a given element of $\mathcal{S}_1$, explain how to check whether it is in the set $\mathcal{S}_0$ in roughly $O(\log N)$ steps. Deduce that one can recover $\mathbf{f}$ in roughly $O\big(\sqrt{\#\mathcal{R}[1]}\log N\big)$ steps.

References

[1] M. Ajtai and C. Dwork, *A public-key cryptosystem with worst-case/average-case equivalence*, STOC '97 (El Paso, TX), ACM, New York, 1999, pp. 284–293. MR1715640 ↑199

[2] M. R. Albrecht, L. Ducas, G. Herold, E. Kirshanova, E. W. Postlethwaite, and M. Stevens, *The general sieve kernel and new records in lattice reduction*, Advances in cryptology—EUROCRYPT 2019. Part II, Lecture Notes in Comput. Sci., vol. 11477, Springer, Cham, 2019, pp. 717–746, DOI 10.1007/978-3-030-17656-3_2. MR3964651 ↑180

[3] M. R. Albrecht, F. Göpfert, F. Virdia, and T. Wunderer, *Revisiting the expected cost of solving uSVP and applications to LWE*, Advances in cryptology—ASIACRYPT 2017. Part I, Lecture Notes in Comput. Sci., vol. 10624, Springer, Cham, 2017, pp. 297–322, DOI 10.1007/978-3-319-70694-8_1. MR3747701 ↑181

[4] E. Alkim, L. Ducas, T. Pöppelmann, and P. Schwab, *Post-quantum key exchange: A new hope*, 25th USENIX Conference on Security Symposium, SEC16, USENIX Association, 2016, pp. 327343. ↑181

[5] L. Babai, *On Lovász' lattice reduction and the nearest lattice point problem*, Combinatorica **6** (1986), no. 1, 1–13, DOI 10.1007/BF02579403. MR856638 ↑185

[6] H. F. Blichfeldt, *The minimum value of quadratic forms, and the closest packing of spheres*, Math. Ann. **101** (1929), no. 1, 605–608, DOI 10.1007/BF01454863. MR1512555 ↑167

[7] J. W. S. Cassels, *An introduction to the geometry of numbers*, Classics in Mathematics, Springer-Verlag, Berlin, 1997. Corrected reprint of the 1971 edition. MR1434478 ↑167

[8] Y. Chen and P. Q. Nguyen, *BKZ 2.0: better lattice security estimates*, Advances in cryptology—ASIACRYPT 2011, Lecture Notes in Comput. Sci., vol. 7073, Springer, Heidelberg, 2011, pp. 1–20, DOI 10.1007/978-3-642-25385-0_1. MR2934994 ↑180, 181

[9] J. H. Conway and N. J. A. Sloane, *Sphere packings, lattices and groups*, 3rd ed., Grundlehren der mathematischen Wissenschaften [Fundamental Principles of Mathematical Sciences], vol. 290, Springer-Verlag, New York, 1999. With additional contributions by E. Bannai, R. E. Borcherds, J. Leech, S. P. Norton, A. M. Odlyzko, R. A. Parker, L. Queen and B. B. Venkov, DOI 10.1007/978-1-4757-6568-7. MR1662447 ↑167

[10] W. Diffie and M. E. Hellman, *New directions in cryptography*, IEEE Trans. Inform. Theory **IT-22** (1976), no. 6, 644–654, DOI 10.1109/tit.1976.1055638. MR437208 ↑180

[11] L. Ducas, A. Durmus, T. Lepoint, and V. Lyubashevsky, *Lattice signatures and bimodal Gaussians*, Advances in cryptology—CRYPTO 2013. Part I, Lecture Notes in Comput. Sci., vol. 8042, Springer, Heidelberg, 2013, pp. 40–56, DOI 10.1007/978-3-642-40041-4_3. MR3126416 ↑213

[12] C. Gentry, *Key recovery and message attacks on NTRU-composite*, Advances in cryptology—EUROCRYPT 2001 (Innsbruck), Lecture Notes in Comput. Sci., vol. 2045, Springer, Berlin, 2001, pp. 182–194, DOI 10.1007/3-540-44987-6_12. MR1895433 ↑208

[13] C. Gentry, *Fully homomorphic encryption using ideal lattices*, STOC'09—Proceedings of the 2009 ACM International Symposium on Theory of Computing, ACM, New York, 2009, pp. 169–178. MR2780062 ↑199

[14] C. Gentry and M. Szydlo, *Cryptanalysis of the revised NTRU signature scheme*, Advances in cryptology—EUROCRYPT 2002 (Amsterdam), Lecture Notes in Comput. Sci., vol. 2332, Springer, Berlin, 2002, pp. 299–320, DOI 10.1007/3-540-46035-7_20. MR1975541 ↑211

[15] O. Goldreich, S. Goldwasser, and S. Halevi, *Public-key cryptosystems from lattice reduction problems*, Advances in cryptology—CRYPTO '97 (Santa Barbara, CA, 1997), Lecture Notes in Comput. Sci., vol. 1294, Springer, Berlin, 1997, pp. 112–131, DOI 10.1007/BFb0052231. MR1630399 ↑199

[16] F. Göpfert, *Securely Instantiating Cryptographic Schemes Based on the Learning with Errors Assumption*, 2016. Thesis (Ph.D.)–Technische Universität Darmstadt. ↑181

[17] P. M. Gruber and C. G. Lekkerkerker, *Geometry of numbers*, 2nd ed., North-Holland Mathematical Library, vol. 37, North-Holland Publishing Co., Amsterdam, 1987. MR893813 ↑167

[18] J. Hoffstein, B. S Kaliski Jr, D. B. Lieman, M. J. B. Robshaw, and Y. L. Yin, *Secure user identification based on constrained polynomials*, Filed October 1997, Issued June 2000. US Patent 6,076,163. ↑222

[19] J. Hoffstein, D. Lieman, and J. H. Silverman, *Polynomial rings and efficient public key authentication*, International Workshop on Cryptographic Techniques and E-Commerce (CrypTEC '89) (Hong Kong, 1989). ↑222

[20] J. Hoffstein, J. Pipher, and J. H. Silverman, *NTRU: a ring-based public key cryptosystem*, Algorithmic number theory (Portland, OR, 1998), Lecture Notes in Comput. Sci., vol. 1423, Springer, Berlin, 1998, pp. 267–288, DOI 10.1007/BFb0054868. MR1726077 ↑199

[21] J. Hoffstein, J. Pipher, J. M. Schanck, J. H. Silverman, and W. Whyte, *Transcript secure signatures based on modular lattices*, Post-quantum cryptography, Lecture Notes in Comput. Sci., vol. 8772, Springer, Cham, 2014, pp. 142–159, DOI 10.1007/978-3-319-11659-4_9. MR3278400 ↑211, 213

[22] J. Hoffstein and J. H. Silverman, *Polynomial rings and efficient public key authentication. II*, Cryptography and computational number theory (Singapore, 1999), Progr. Comput. Sci. Appl. Logic, vol. 20, Birkhäuser, Basel, 2001, pp. 269–286. MR1944732 ↑222

[23] J. Hoffstein, J. Pipher, J. M. Schanck, J. H. Silverman, and W. Whyte, *Practical signatures from the partial Fourier recovery problem*, Applied cryptography and network security, Lecture Notes in Comput. Sci., vol. 8479, Springer, Cham, 2014, pp. 476–493, DOI 10.1007/978-3-319-07536-5_28. MR3219007 ↑213, 222

[24] A. K. Lenstra, H. W. Lenstra Jr., and L. Lovász, *Factoring polynomials with rational coefficients*, Math. Ann. **261** (1982), no. 4, 515–534, DOI 10.1007/BF01457454. MR682664 ↑167, 178

[25] A. K. Lenstra, J. P. Hughes, M. Augier, J. W. Bos, T. Kleinjung, and C. Wachter, *Public keys*, Advances in cryptology—CRYPTO 2012, Lecture Notes in Comput. Sci., vol. 7417, Springer, Heidelberg, 2012, pp. 626–642, DOI 10.1007/978-3-642-32009-5_37. MR3005994 ↑193

[26] V. Lyubashevsky, *Lattice-based identification schemes secure under active attacks*, Public key cryptography—PKC 2008, Lecture Notes in Comput. Sci., vol. 4939, Springer, Berlin, 2008, pp. 162–179, DOI 10.1007/978-3-540-78440-1_10. MR2570228 ↑212

[27] V. Lyubashevsky, *Fiat-Shamir with aborts: applications to lattice and factoring-based signatures*, Advances in cryptology—ASIACRYPT 2009, Lecture Notes in Comput. Sci., vol. 5912, Springer, Berlin, 2009, pp. 598–616, DOI 10.1007/978-3-642-10366-7_35. MR2593089 ↑212

[28] D. Micciancio, *The shortest vector in a lattice is hard to approximate to within some constant*, SIAM J. Comput. **30** (2001), no. 6, 2008–2035, DOI 10.1137/S0097539700373039. MR1856567 ↑167

[29] P. Q. Nguyen and O. Regev, *Learning a parallelepiped: cryptanalysis of GGH and NTRU signatures*, J. Cryptology **22** (2009), no. 2, 139–160, DOI 10.1007/s00145-008-9031-0. MR2496387 ↑211

[30] P. Q. Nguyen and O. Regev, *Learning a parallelepiped: cryptanalysis of GGH and NTRU signatures*, Advances in cryptology—EUROCRYPT 2006, Lecture Notes in Comput. Sci., vol. 4004, Springer, Berlin, 2006, pp. 271–288, DOI 10.1007/11761679_17. MR2423548 ↑211

[31] P. Q Nguyen, *Hermite's constant and lattice algorithms*, The LLL algorithm (2010), 19–69. MR2722178 ↑180

[32] P. Q. Nguyen and B. Vallée (eds.), *The LLL algorithm*, Information Security and Cryptography, Springer-Verlag, Berlin, 2010. Survey and applications, DOI 10.1007/978-3-642-02295-1. MR2722178 ↑

[33] W. M. Schmidt, *Diophantine approximation*, Lecture Notes in Mathematics, vol. 785, Springer, Berlin, 1980. MR568710 ↑167

[34] C. P. Schnorr, *Lattice reduction by random sampling and birthday methods*, STACS 2003, Lecture Notes in Comput. Sci., vol. 2607, Springer, Berlin, 2003, pp. 145–156, DOI 10.1007/3-540-36494-3_14. MR2066588 ↑180

[35] C.-P. Schnorr and M. Euchner, *Lattice basis reduction: improved practical algorithms and solving subset sum problems*, Math. Programming **66** (1994), no. 2, Ser. A, 181–199, DOI 10.1007/BF01581144. MR1297061 ↑180

[36] C.-P. Schnorr and M. Euchner, *Lattice basis reduction: improved practical algorithms and solving subset sum problems*, Fundamentals of computation theory (Gosen, 1991), Lecture Notes in Comput. Sci., vol. 529, Springer, Berlin, 1991, pp. 68–85, DOI 10.1007/3-540-54458-5_51. MR1136071 ↑180

[37] C. L. Siegel, *Lectures on the geometry of numbers*, Springer-Verlag, Berlin, 1989. Notes by B. Friedman; Rewritten by Komaravolu Chandrasekharan with the assistance of Rudolf Suter; With a preface by Chandrasekharan, DOI 10.1007/978-3-662-08287-4. MR1020761 ↑167

[38] Y. Yu and L. Ducas, *Second order statistical behavior of LLL and BKZ*, Selected areas in cryptography—SAC 2017, Lecture Notes in Comput. Sci., vol. 10719, Springer, Cham, 2018, pp. 3–22, DOI 10.1007/978-3-319-72565-9_1. MR3775576 ↑180

Department of Mathematics, Box 1917, Brown University, Providence, Rhode Island 02912 USA

Email address: `joseph_silverman@brown.edu`

IAS/Park City Mathematics Series
Volume **29**, 2025
https://doi.org/10.1090/pcms/029/00894

Rational Points on Varieties and the Brauer–Manin Obstruction

Bianca Viray

Prologue

A central object of interest in arithmetic geometry is the set of k-rational points on a smooth projective geometrically integral k-variety X. In these notes, we focus on the fundamental problem of determining whether this set, denoted $X(k)$, is nonempty.

If you are handed a variety, e.g., $X := V(x^3 + 2y^3 + 10z^3) \subset \mathbb{P}^2$, and asked to determine if it has rational points, a natural first step is to try to find some solutions where the coordinates are small (in other words, solutions of *small height*). If you are lucky, you can find a point of small height (try to do so with the example above), and you will thereby have proved that the set $X(\mathbb{Q}) \neq \emptyset$.

In fact, if $X(k) \neq \emptyset$, then an approach like this will give a proof of nonemptyness. Any point $P \in X(k)$ has a (finite) *height* H, and searching in a box of bounded height $\{x \in \mathbb{P}^n(k) : H(x) \leq B\}$ is a finite procedure. Therefore, increasing B by a constant amount and searching in these larger boxes will (eventually) result in you finding the point P.

However, if $X(k) = \emptyset$, then this approach will never terminate. Indeed, if your search keeps failing, you cannot tell whether you need to search in an even larger box, or whether actually there is no rational point. Searching *cannot* certify an absence of k-points; for that, we need another method. These notes focus on one such method that is known as the *Brauer–Manin obstruction.*

The goal of these notes. There are many excellent references on the Brauer–Manin obstruction (e.g., [**11**, **37**, **41**] just to name a few). We do not endeavor to improve on (or replicate) those references here. Rather, these notes should be thought of as a guidebook to the field, giving an overview of the current landscape, with pointers on where to go if you, the reader, wish to explore more. Just as travel guidebooks have particular biases (affordability, nature, good for kids, etc.), so do these notes. They give *my* perspective on how the feedback loop between computation and theory currently manifests in the study of rational points and the Brauer–Manin obstruction.

Acknowledgments. These notes were prepared to accompany a lecture series at the 2022 Park City Mathematics Institute 2022 Graduate Summer School. My perspective on this subject, particularly that presented in Sections 3 and 4, has been greatly influenced by my collaborations with Brendan Creutz, which I have been honored to take part in.

I thank the organizers of the PCMI 2022 Graduate Summer School, Jennifer Balakrishnan, Bjorn Poonen, and Akshay Venkatesh, for the opportunity to speak, and I thank the PCMI Director, Rafe Mazzeo, and the PCMI Program Manager, Dena Vigil, for their assistance and for running a wonderful program. These notes were greatly improved by comments and questions from the graduate student participants at the PCMI summer school; I am grateful for their interest, engagement, and patience with an early draft of these notes. Thomas Carr and Carlos Rivera skillfully served as TA's for my lecture series; I thank them for finding many typos and improving the exposition. I also thank Alex Galarraga, Ting Gong, Leo Mayer, Haoming Ning, Caelan Ritter, Alex Wang, Olivier Wittenberg, and the referee for their careful readings and comments, which have improved the final version.

Notation and conventions. Throughout, we will use F to denote an arbitrary field. We write F^{sep} for a fixed separable closure, and $\overline{F}$ for a fixed algebraic closure containing F^{sep}. We use G_F to denote the absolute Galois group $\mathrm{Gal}(F^{\mathrm{sep}}/F)$.

We reserve k to denote a number field. We write Ω_k for the set of places of k, and for any $v \in \Omega_k$ we let k_v denote the completion. If $v \in \Omega_k$ is nonarchimedean, then we write $\mathcal{O}_v \subset k_v$ for the valuation ring, $\mathfrak{m}_v$ for the maximal ideal, $\mathbb{F}_v := \mathcal{O}_v/\mathfrak{m}_v$ for the residue field, and k_v^{nr} for the maximal unramified extension of k_v. We write $\mathbb{A}_k := \prod'(k_v, \mathcal{O}_v)$ for the adèle ring of k, i.e., the restricted product $\{(x_v) \in \prod_v k_v : \#\{v : x_v \notin \mathcal{O}_v\} < \infty\}$.

We will use X to denote a smooth projective geometrically integral variety over k. For any extension F/k we write X_F for the base change of X to $\operatorname{Spec} F$. We also will use the conventions $\overline{X} := X_{\overline{k}}$ and $X^{\mathrm{sep}} := X_{k^{\mathrm{sep}}}$.

1. Obstructions to the existence of rational points

Given a smooth projective geometrically integral variety X over a number field k, there is no known general method to determine if the set $X(k)$ is nonempty.[1] Instead, we find a set S that 1) contains $X(k)$ and 2) seems to be computable (at least in some cases). Then, if we can prove that $S = \emptyset$, this will imply that $X(k) = \emptyset$. The set S is typically called an *obstruction set* since the emptyness of S *obstructs* the existence of k-points.

1.1. Local obstructions.

1.1. Local obstructions. The simplest obstruction set comes from the embeddings of k into one of its completions k_v. Precisely, if v is a place of k, then the embedding $k \hookrightarrow k_v$ induces an inclusion $X(k) \subset X(k_v)$. So if there are no k_v-points on X, then there are also no rational points. In addition, if v is nonarchimedean, then Hensel's lemma can be used to show that there is some n such that $X(k_v) \neq \emptyset$ if and only if $X(\mathcal{O}_v/\mathfrak{m}_v^n) \neq \emptyset$ [**33**]. In particular, it is a finite computation to determine if $X(k_v) \neq \emptyset$ for nonarchimedean places v (and the same is true for archimedean places, see [**37**, Remark 2.6.4]). (Interestingly, analogous effectivity results are not known for local fields of positive characteristic)

[1] In fact, it may be an undecidable problem! This question of decidability is known as Hilbert's tenth problem; see [**36**] for more details.

EXERCISE 1.1.1. Show that

$$X := V(x^2 + y^2 + 7z^2) \subset \mathbb{P}^2$$

has no $\mathbb{Q}_7$-points. Conclude that $X(\mathbb{Q}) = \emptyset$.

We may package these obstructions together using the adèles $\mathbb{A}_k$ of k. Just as above, the embedding $k \hookrightarrow \mathbb{A}_k$ allows us to view $X(k)$ as a subset of $X(\mathbb{A}_k)$, and one can prove (see [**37**, Section 2.6.3]) that

$$X(\mathbb{A}_k) = \prod_{v \in \Omega_k}{}'(X(k_v), X(\mathcal{O}_v)).$$

Since, by assumption, X is projective, the valuative criterion for properness implies that the inclusion $X(\mathcal{O}_v) \subset X(k_v)$ is an equality. Hence, $X(\mathbb{A}_k) = \prod X(k_v)$.

It turns out that even though $X(\mathbb{A}_k)$ encapsulates solubility conditions at all of the (infinitely many) completions at once, it is still a finite computation to determine whether $X(\mathbb{A}_k) \neq \emptyset$.

PROPOSITION 1.1.2 (c.f. [**37**, Theorem 7.7.2]). *Let X be a smooth projective geometrically integral variety over a number field k. Then the set of places where X fails to have k_v-points is finite and is contained in an effectively computable set.*

This proposition implies that given X one can effectively compute a finite set $S \subset \Omega_k$ such that $X(k_v) \neq \emptyset$ for all $v \notin S$. Then for each $v \in S$, it is a finite computation to determine whether $X(k_v) \neq \emptyset$ (see the first paragraph of this section). Since S is finite, this proves our assertion that determining whether $X(\mathbb{A}_k) \neq \emptyset$ is a finite computation.

PROOF OF PROPOSITION 1.1.2. Since X is smooth and geometrically integral, then by slicing with sufficiently general hyperplanes, we may obtain a smooth, geometrically integral curve $C \subset X$. By generic smoothness, C necessarily has good reduction away from a finite set of places (see [**37**, Section 3.2] for more details). Furthermore, the Hasse-Weil bounds [**37**, Corollary 7.2.1] imply that, for any place of good reduction with $\#\mathbb{F}_v$ sufficiently large, the reduction $C \bmod v$ is guaranteed to have a smooth $\mathbb{F}_v$-point. Hensel's lemma [**37**, Theorem 3.5.63] then implies that C, and hence X, has $\mathbb{Q}_v$-points for such v. Thus we have proved that $X(k_v) \neq \emptyset$ for all v outside the following subset

$$\{v|\infty\} \cup \{C \bmod v \text{ singular}\} \cup \{\#\mathbb{F}_v < N\},$$

where N is some positive integer given by the Hasse-Weil bounds. Since this set is finite and effectively computable, we have proved the desired result. □

EXERCISE 1.1.3. For this problem, the following specific consequence of Hensel's Lemma will be useful: If p is a prime and $u \in \mathbb{Z}_p^\times$, then

$$u \in \mathbb{Z}_p^{\times 2} \Leftrightarrow \begin{cases} u \bmod p \in \mathbb{F}_p^{\times 2} & \text{if } p \neq 2, \\ u \equiv 1 \pmod 8 & \text{if } p = 2. \end{cases}$$

(1) Let p be an odd prime and let $a, b, c \in \mathbb{Z} \smallsetminus p\mathbb{Z}$. Show that $\{ax^2 : x \in \mathbb{F}_p\}$ and $\{c - by^2 : y \in \mathbb{F}_p\}$ both have cardinality $\frac{p+1}{2}$ and that, therefore, the sets contain a common value. Use this result to show that

$$X := V(ax^2 + by^2 + cz^2) \subset \mathbb{P}^2$$

has a $\mathbb{Q}_p$-point.

(2) Determine whether $X := V(5x^2 + 7y^2 - 3z^2) \subset \mathbb{P}^2$ has $\mathbb{A}_{\mathbb{Q}}$-points.
(3) Let $a, b, c \in \mathbb{Z}$ be squarefree, pairwise relatively prime integers. Prove that $X := V(ax^2 + by^2 + cz^2)$ has $\mathbb{A}_{\mathbb{Q}}$-points if and only if a, b, c are not all the same sign and $ax^2 + by^2 + cz^2$ has solutions in $\mathbb{Z}/8abc\mathbb{Z}$ with the property that for every $p|8abc$, at least two of the coordinates are nonzero modulo p.

1.1.1. *The local-to-global principle.*

DEFINITION 1.1.4. A class of varieties $\mathcal{C}$ satisfies the *local-to-global principle* if, for all $X \in \mathcal{C}$,

$$X(\mathbb{A}_k) \neq \emptyset \Leftrightarrow X(k) \neq \emptyset.$$

EXAMPLES 1.1.5.

(1) Quadrics, in any number of variables, satisfy the local-to-global principle by the Hasse-Minkowski theorem [**26**, Chapter VI, Section 3].
(2) *Severi-Brauer varieties*, i.e., varieties that are geometrically *isomorphic* to projective space, satisfy the local-to-global principle by the Albert-Brauer-Hasse-Noether theorem [**34**, Part III, Theorem 5.8] (see [**18**, Chapter 5] for the relationship between Severi-Brauer varieties and the Brauer group).

Note that every variety in one of the two classes in the above example is geometrically rational, i.e., over $\overline{k}$ there is a birational map $\overline{X} \dashrightarrow \mathbb{P}^n_{\overline{k}}$. Classes of varieties with more complicated geometry often have members that fail the local-to-global principle. The first example of this failure dates back to the 1940's and is due to Lind and Reichardt, independently.

THEOREM 1.1.6 ([**28**, **38**]). *Let $C \subset \mathbb{P}^3$ be the smooth projective genus 1 curve defined by the two quadrics $2y^2 = w^2 - 17z^2$ and $wz = x^2$. Then*

$$C(\mathbb{A}_{\mathbb{Q}}) \neq \emptyset \quad \text{and} \quad C(\mathbb{Q}) = \emptyset.$$

To prove that C has no rational points despite the presence of the adelic points, we need a *refined* obstruction set, that is, an intermediate set S that sits between the set of rational points and the set of adelic points.

1.2. An introduction to the Brauer–Manin obstruction. The goal of this section is to define the Brauer–Manin set $X(\mathbb{A}_k)^{\mathrm{Br}}$, and show that it is a refined obstruction set, i.e., that we have the following containments:

$$X(k) \subset X(\mathbb{A}_k)^{\mathrm{Br}} \subset X(\mathbb{A}_k).$$

To do so, we must first introduce the Brauer group.

1.2.1. *The Brauer group of a field.* Let F be a field.

DEFINITION 1.2.1.

(1) A *central simple algebra* over F is a finite dimensional F-algebra $\mathcal{A}$ whose center is exactly F and that has no nontrivial two-sided ideals.
(2) Two central simple algebras $\mathcal{A}, \mathcal{B}$ over F are said to be *Brauer equivalent* if there exist positive integers n, m such that $\mathcal{A} \otimes_F \mathrm{M}_m(F) \simeq \mathcal{B} \otimes_F \mathrm{M}_n(F)$.

THEOREM 1.2.2 ([**18**, Proposition 2.4.8 and Theorem 4.4.7]). *Let F be a field. The set of Brauer equivalence classes of central simple algebras over F forms a torsion abelian group under tensor product, where the identity element is the class of $\mathrm{M}_n(F)$. Moreover, this group is isomorphic to the Galois cohomology group $\mathrm{H}^2(G_F, (F^{\mathrm{sep}})^\times)$.*

DEFINITION 1.2.3. The group from Theorem 1.2.2 is known as the *Brauer group of F* and is denoted $\operatorname{Br} F$.

EXAMPLE 1.2.4. Assume that F has characteristic different from 2 and let $a, b \in F^\times$. Then the *quaternion algebra*

$$\mathcal{A}_{a,b} := F \otimes F \cdot i \otimes F \cdot j \otimes F \cdot ij, \quad i^2 = a, j^2 = b, ji = -ij$$

is a central simple F-algebra. It has order dividing 2 in $\operatorname{Br} F$ [**18**, Corollary 1.5.3].

EXERCISE 1.2.5. Let F be a field of characteristic different from 2 and let $a, b \in F^\times$.

Prove that $\mathcal{A}_{a,b} \simeq \mathrm{M}_2(F)$ if and only if there is some $x, y, z \in F$, not all zero, such that $ax^2 + by^2 = z^2$.

Conclude that $(a, b) := [\mathcal{A}_{a,b}] \in \operatorname{Br} F$ is trivial if and only if the conic $C_{a,b}$: $ax^2 + by^2 = z^2$ has an F-rational point, and that $C_{a,b}(F) \neq \emptyset$ if and only if $a \in \mathrm{N}(k(\sqrt{b})^\times)$. (By symmetry this is equivalent to $b \in \mathrm{N}(k(\sqrt{a})^\times)$).

REMARK 1.2.5.1. One can also show that $\mathcal{A}_{a,b} \otimes_k \mathcal{A}_{a,c} \simeq \mathrm{M}_2(\mathcal{A}_{a,bc})$ (see [**18**, Lemma 1.5.2]), which implies that in $\operatorname{Br} F$, we have $(a, b)(a, c) = (a, bc)$.

The above example with quaternion algebras and conics is a special case of a general correspondence. To any central simple algebra over F (e.g., a quaternion algebra), one can associate a Severi-Brauer variety (e.g., a conic; see Example 1.1.5(2)), and vice versa [**18**, Section 5.2].

Our interest in the Brauer group stems from the fact that, over a number field k, the Brauer group can be viewed as encoding many generalizations of quadratic reciprocity; this is encapsulated by the fundamental exact sequence of global class field theory.

THEOREM 1.2.6 ([**37**, Theorems 1.5.34 and 1.5.36]). *Let k be a number field. For each place v, there is an injective homomorphism*

$$\mathrm{inv}_v \colon \operatorname{Br} k_v \to \mathbb{Q}/\mathbb{Z},$$

that is an isomorphism for nonarchimedean v. For archimedean v, the image is $\frac{1}{2}\mathbb{Z}/\mathbb{Z}$ if $k_v = \mathbb{R}$ and 0 if $k_v = \mathbb{C}$. Furthermore, these isomorphisms fit together in the following short exact sequence

$$0 \to \operatorname{Br} k \to \oplus_v \operatorname{Br} k_v \xrightarrow{\sum_v \mathrm{inv}_v} \mathbb{Q}/\mathbb{Z} \to 0. \tag{1.2.7}$$

REMARK 1.2.7.1. When working with explicit computations, one has to take care that the maps inv_v are defined in a globally compatible way. An arbitrary collection of isomorphisms $\phi_v \colon \operatorname{Br} k_v \to \mathbb{Q}/\mathbb{Z}$ for nonarchimedean v will not necessarily give an exact sequence as in (1.2.7). See [**11**, Definition 13.1.7] for a definition of inv_v. The explicit examples in these notes will restrict to 2-torsion elements to avoid this subtlety.

EXAMPLE 1.2.8. Let $a, b \in k^\times$ and let v be a place of k. Using Example 1.2.4, one can check that $\mathcal{A}_{a,b} \in \operatorname{Br} k_v$ is nontrivial if and only if the Hilbert symbol $(a, b)_v = -1$. (Indeed, the definition of the Hilbert symbol is that $(a, b)_v = -1$ if $C_{a,b}(k_v) = \emptyset$ and $(a, b)_v = 1$ otherwise.) Thus, the fact $\sum_v \mathrm{inv}_v(\mathcal{A}_{a,b}) = 0$ is exactly the product formula

$$\prod_v (a, b)_v = 1,$$

which, over $\mathbb{Q}$, is also equivalent to quadratic reciprocity.

EXERCISE 1.2.9.

(1) Let p be an odd prime and let a be an integer that is relatively prime to p. Prove that the Hilbert symbol $(a,p)_p$ equals the Legendre symbol $\left(\frac{a}{p}\right)$.

(2) Let p, q be distinct odd primes. Prove $(p,q)_2 = (-1)^{\frac{(p-1)(q-1)}{4}}$.

(3) Let p, q be distinct odd primes. Using Exercise 1.1.3, prove that

$$\prod_v (p,q)_v = (p,q)_p(p,q)_q(p,q)_2 = \left(\frac{q}{p}\right)\left(\frac{p}{q}\right)(-1)^{\frac{(p-1)(q-1)}{4}}.$$

(4) Let p be an odd prime. Following the same strategy as above, prove that

$$\prod_v (2,p)_v = \left(\frac{2}{p}\right)(-1)^{\frac{(p^2-1)}{8}}.$$

(5) Let p be an odd prime. Following the same strategy as above, prove that

$$\prod_v (-1,p)_v = \left(\frac{-1}{p}\right)(-1)^{\frac{(p-1)}{2}}$$

(6) Prove that for all $a, b \in \mathbb{Q}^\times$, $\prod_v (a,b)_v = 1$ if and only if for all pairs of distinct odd primes p, q, we have

$$\prod_v (p,q)_v = 1, \quad \prod_v (2,p)_v = 1, \quad \prod_v (-1,p)_v = 1.$$

(7) Combine the previous parts to observe that the product formula over $\mathbb{Q}$ is equivalent to quadratic reciprocity.

For more extensive introductions on the Brauer group of a field, see [**11**, Chap. 1; **18**; **30**, Chap. 4].

1.2.2. *The Brauer group of a variety.* The notion of the Brauer group of a field can be generalized to the Brauer group of a scheme using étale cohomology.

DEFINITION 1.2.10. The Brauer group of a scheme X is $\operatorname{Br} X := \mathrm{H}^2_{\text{et}}(X, \mathbb{G}_m)$.[2]

For our purposes, the following properties will be particularly useful.

THEOREM 1.2.11.

(1) *For any field F we have* $\operatorname{Br} F = \operatorname{Br} \operatorname{Spec} F$.

(2) *The Brauer group is functorial, i.e., if there is a morphism of varieties $f\colon X \to Y$, then we have a homomorphism $f^*\colon \operatorname{Br} Y \to \operatorname{Br} X$.*

(3) *If X is a smooth geometrically integral variety over a field F of characteristic 0, then the inclusion of the generic point $\iota\colon \operatorname{Spec} \mathbf{k}(X) \to X$ gives an injective homomorphism $\iota^*\colon \operatorname{Br} X \hookrightarrow \operatorname{Br} \mathbf{k}(X)$.*

(4) *The Brauer group is a birational invariant of smooth projective varieties, i.e., if $f\colon X \dashrightarrow Y$ is a birational map between two smooth projective varieties, then $f^*\colon \operatorname{Br} \mathbf{k}(Y) \to \operatorname{Br} \mathbf{k}(X)$ sends $\operatorname{Br} Y$ isomorphically to $\operatorname{Br} X$.*

[2]There are several ways to extend the definition of the Brauer group from a field to a scheme. The definition we use here is sometimes called the *cohomological Brauer group*, since, in complete generality, it is not necessarily equal to other possible definitions of the Brauer group of a scheme. However, under the standing assumptions of these notes (i.e., smooth projective varieties over a field of characteristic 0), all of the generalizations coincide, so we will simply refer to this as the Brauer group.

(5) *If X is a smooth projective variety and $U \subset X$ is an open subset such that* $\operatorname{codim}(X \smallsetminus U) \geq 2$, *then* $\operatorname{Br} X = \operatorname{Br} U$.

PROOF. The first two properties follow from the definition together with properties of étale cohomology. For the last three properties, see [**11**, Theorem 3.5.5], [**11**, Corollary 6.2.11], and [**11**, Theorem 3.7.6], respectively. □

See [**11**; **37**, Section 6.6] for more details on the Brauer group of a scheme.

1.2.3. *An example of a Brauer–Manin obstruction using quadratic reciprocity.* Before introducing the full Brauer–Manin obstruction in the next section, we first give an extended treatment of a particular type of Brauer–Manin obstruction using conics.

Let X be a smooth projective variety over $\mathbb{Q}$ and assume that we have a conic bundle $\pi\colon \mathcal{C} \to X$ over X where the morphism is smooth. Then for every extension $F/\mathbb{Q}$, we have a set map

$$X(F) \to \{\text{conics over } F\}, \quad x \mapsto \mathcal{C}_x := \pi^{-1}(x).$$

In particular, by applying this to $F = \mathbb{Q}$ and $F = \mathbb{Q}_v$ for all places v, we obtain the following commutative diagram.

$$\begin{array}{ccc} X(\mathbb{Q}) & \hookrightarrow & X(\mathbb{A}_{\mathbb{Q}}) = \prod_v X(\mathbb{Q}_v) \\ \downarrow & & \downarrow \\ \{\text{conics over } \mathbb{Q}\} & \hookrightarrow & \prod_v \{\text{conics over } \mathbb{Q}_v\}, \end{array} \tag{1.2.12}$$

where the bottom inclusion is given by diagonal basechange.

To see how we can use the above diagram to define an obstruction set we must first explore the following rephrasing of quadratic reciprocity that we alluded to in Example 1.2.8.

THEOREM 1.2.13 (Rephrasing of quadratic reciprocity I). *Let $C : V(ax^2 + by^2 = z^2) \subset \mathbb{P}^2_{\mathbb{Q}}$ be a smooth conic over $\mathbb{Q}$. Then*

$$\{v \in \Omega_{\mathbb{Q}} : C(\mathbb{Q}_v) = \emptyset\}$$

is a finite set with even *cardinality.*

The finiteness is a special case of Proposition 1.1.2; the surprising fact is that the cardinality is always even! Why is this surprising? Let's rephrase the theorem again.

THEOREM 1.2.14 (Rephrasing of quadratic reciprocity II). *Let $C : V(ax^2 + by^2 = z^2) \subset \mathbb{P}^2_{\mathbb{Q}}$ be a smooth conic over $\mathbb{Q}$. Then*

$$\#\{p \text{ prime} : C(\mathbb{Q}_p) = \emptyset\} \equiv 1 \bmod 2 \iff a < 0 \text{ and } b < 0.$$

EXERCISE 1.2.15. Verify that Theorems 1.2.13 and 1.2.14 are indeed equivalent.

Let's unravel the statement of Theorem 1.2.14. It says that if you know whether C is soluble over $\mathbb{Q}_p$ for *all* primes p, that tells you something about the positivity of a and b. This is a strong and surprising statement! There is *no* notion of positivity in the p-adics. Positivity is a concept from the real numbers. But quadratic reciprocity implies that p-adic information for *all* the primes p can detect some positivity information. `:exploding_head:`

Now we return to the commutative diagram (1.2.12) and apply Theorem 1.2.13. Then we obtain the following.

$$\begin{array}{ccccc} X(\mathbb{Q}) & & \hookrightarrow & & \prod_v X(\mathbb{Q}_v) \\ \downarrow & & & & \downarrow \\ \{\text{conics}/\mathbb{Q}\} & \hookrightarrow & \{(C_v) : \#\{v : C_v(\mathbb{Q}_v) = \emptyset\} \equiv 0 \bmod 2\} & \hookrightarrow & \prod_v \{\text{conics}/\mathbb{Q}_v\} \end{array}$$

The intermediate set on the bottom row provides a construction of an intermediate set on the top row.

$$\begin{array}{ccccc} X(\mathbb{Q}) & \hookrightarrow & \{(x_v) : \#\{v : \mathcal{C}_{x_v}(\mathbb{Q}_v) = \emptyset\} \equiv 0 \bmod 2\} & \hookrightarrow & \prod_v X(\mathbb{Q}_v) \\ \downarrow & & \downarrow & & \downarrow \\ \{\text{conics}/\mathbb{Q}\} & \hookrightarrow & \{(C_v) : \#\{v : C_v(\mathbb{Q}_v) = \emptyset\} \equiv 0 \bmod 2\} & \hookrightarrow & \prod_v \{\text{conics}/\mathbb{Q}_v\} \end{array}$$

And thus we obtain a refined obstruction set!

To conclude this section, let us see how this can be phrased in terms of the Brauer group. By Example 1.2.4 and Exercise 1.2.5, we have seen that we can associate a Brauer class to a conic. Further, we saw in Example 1.2.8 that quadratic reciprocity is encoded by the 2-torsion in the fundamental exact sequence of global class field theory. Thus, we may express the top row of the above diagram in the following, equivalent, way:

$$X(\mathbb{Q}) \hookrightarrow \{(x_v) : \textstyle\sum_v \operatorname{inv}_v[\mathcal{C}_{x_v}] = 0 \in \mathbb{Q}/\mathbb{Z}\} \hookrightarrow X(\mathbb{A}_{\mathbb{Q}}) = \textstyle\prod_v X(\mathbb{Q}_v).$$

In fact, we can work with the Brauer group from the beginning. Given a conic bundle $\pi\colon \mathcal{C} \to X$ where the morphism is smooth, one can associate a Brauer class $\mathcal{A} := [\mathcal{C}] \in \operatorname{Br} X$. Further, given an extension $F/\mathbb{Q}$ and a map $x\colon \operatorname{Spec} F \to X$ (equivalently a point $x \in X(F)$), the Brauer class $x^*\mathcal{A}$ coming from Theorem 1.2.11(2) is exactly the class $[\mathcal{C}_x]$ given by the fiber of π above x. More generally, one can replace $\mathbb{Q}$ with any number field k.

1.2.4. *The Brauer–Manin set.* In the previous section, we saw how a conic bundle $\pi\colon \mathcal{C} \to X$ carves out a refined obstruction set and how this can be phrased in terms of the Brauer group. In this section, we will show how an arbitrary element of the Brauer group of a variety X (i.e., not just those Brauer classes corresponding to conic bundles) carves out a subset of $X(\mathbb{A}_k)$ that contains the k-rational points. This subset was introduced by Manin [**29**] and is known as the *Brauer–Manin set.*

For any extension F/k, the functoriality of the Brauer group (Theorem 1.2.11(2)) gives a pairing

$$\operatorname{Br} X \times X(F) \to \operatorname{Br} F, (\alpha, x) \mapsto \langle \alpha, x\rangle := x^*\alpha,$$

where we view the point $x \in X(F)$ as a map $x\colon \operatorname{Spec} F \to X$.

If X is projective, then $X(\mathbb{A}_k) = \prod_v X(k_v)$ and so we may apply these pairings componentwise to obtain a pairing

$$\operatorname{Br} X \times X(\mathbb{A}_k) \to \prod_v \operatorname{Br} k_v.$$

Using integral models and properties of Brauer groups of local fields, one can show that the image of this pairing actually lands in $\bigoplus_v \operatorname{Br} k_v$, i.e., that for every $\alpha \in \operatorname{Br} X$ there exists a finite set $S_\alpha \subset \Omega_k$ such that $\langle \alpha, -\rangle\colon X(k_v) \to \operatorname{Br} k_v$ is identically 0 for all $v \notin S_\alpha$ [**37**, Proposition 8.2.1]. Therefore, we have the following commutative diagram.

$$\begin{array}{ccc} \operatorname{Br} X \times X(k) & \hookrightarrow & \operatorname{Br} X \times X(\mathbb{A}_k) \\ \downarrow & & \downarrow \\ \operatorname{Br} k & \longrightarrow & \oplus_v \operatorname{Br} k_v \end{array}$$

Recall from Theorem 1.2.6, the bottom vertical arrow fits into an exact sequence, so we may extend our commutative diagram as follows:

$$\begin{array}{ccccccccc} & & \operatorname{Br} X \times X(k) & \hookrightarrow & \operatorname{Br} X \times X(\mathbb{A}_k) & \searrow & & & \\ & & \downarrow & & \downarrow & & & & \\ 0 & \longrightarrow & \operatorname{Br} k & \longrightarrow & \oplus_v \operatorname{Br} k_v & \xrightarrow{\sum \operatorname{inv}_v} & \mathbb{Q}/\mathbb{Z} & \longrightarrow & 0, \end{array}$$

where the diagonal map is determined by the commutativity. Observe that since the bottom row is a complex, then the induced map $\operatorname{Br} X \times X(k) \to \mathbb{Q}/\mathbb{Z}$ is identically 0. In particular, for all $\alpha \in \operatorname{Br} X$, the set $X(k)$ is contained in the set of *adelic points orthogonal to* α, i.e.,

$$X(k) \subset X(\mathbb{A}_k)^\alpha := \{(x_v) \in X(\mathbb{A}_k) : \langle \alpha, (x_v) \rangle = 0\} \subset X(\mathbb{A}_k).$$

(We will see later that $X(\mathbb{A}_k)^\alpha$ can be a proper subset of $X(\mathbb{A}_k)$, e.g., Lemma 2.2.1.1). Taking the intersection $X(\mathbb{A}_k)^\alpha$ for all α, we obtain the *Brauer–Manin set*

(1.2.16)
$$X(\mathbb{A}_k)^{\operatorname{Br}} := \bigcap_{\alpha \in \operatorname{Br} X} X(\mathbb{A}_k)^\alpha = \{(x_v) \in X(\mathbb{A}_k) : \langle \alpha, (x_v) \rangle = 0 \, \forall \alpha \in \operatorname{Br} X\}.$$

EXERCISE 1.2.17. Let X be a smooth projective geometrically integral variety over a number field k and let π denote the structure morphism $\pi\colon X \to \operatorname{Spec} k$.

(1) Let $\alpha_0 \in \operatorname{Br} k$. Show that $X(\mathbb{A}_k)^{\pi^* \alpha_0} = X(\mathbb{A}_k)$.

(2) Let $\alpha, \beta \in \operatorname{Br} X$. Show that

$$X(\mathbb{A}_k)^\alpha \cap X(\mathbb{A}_k)^\beta = \bigcap_{\gamma \in \{\alpha^i \beta^j : i, j \in \mathbb{Z}\}} X(\mathbb{A}_k)^\gamma$$

These exercises show that the Brauer–Manin set depends only on the quotient $\operatorname{Br} X / \operatorname{im}(\pi^*\colon \operatorname{Br} k \to \operatorname{Br} X)$. We write $\operatorname{Br}_0 X := \operatorname{im}(\pi^*\colon \operatorname{Br} k \to \operatorname{Br} X)$ and refer to elements in $\operatorname{Br}_0 X$ as *constant Brauer classes.*

2. Computing the Brauer–Manin obstruction

We have successfully defined a refined obstruction set, the Brauer–Manin set

$$X(k) \subset X(\mathbb{A}_k)^{\operatorname{Br}} \subset X(\mathbb{A}_k),$$

but we have yet to see whether this is *useful.* Utility has a theoretical component and a practical or computational component.

(1) [Theory] Do there exist varieties where $X(\mathbb{A}_k)^{\operatorname{Br}} = \emptyset$ yet $X(\mathbb{A}_k) \neq \emptyset$?
(2) [Practice/Computation] Can we compute the Brauer–Manin set?

The answer to the first question is a resounding *YES!* In fact, there are many such examples. When Manin introduced this obstruction, he showed that several of the known failures of the local-to-global principle could be explained by the Brauer–Manin obstruction. Since then, several more examples have been constructed, which together lead to the expectation that if a class of varieties can have nontrivial Brauer group, it is likely that there exists a variety in that class with a Brauer–Manin obstruction to the existence of rational points.[3]

The answer to the second question is more mixed. On the one hand, the Brauer–Manin set *has* been computed in several examples, and for some classes of varieties, it is more or less standard to do so (e.g., conic bundles over $\mathbb{P}^1$ [**11**, Proposition 11.3.4]). On the other hand, there is no general effectivity result for the Brauer–Manin set, nor is there an approach that is known to work in full generality. The best effectivity results to date are due to Kresch and Tschinkel [**23, 24**]. In the first paper, Kresch and Tschinkel study the obstruction coming from the *algebraic Brauer group*

$$\operatorname{Br}_1 X := \ker\left(\operatorname{Br} X \to \operatorname{Br} \overline{X}\right);$$

the *algebraic Brauer–Manin set* $X(\mathbb{A}_k)^{\operatorname{Br}_1}$ is the set of adelic points orthogonal to every Brauer class in $\operatorname{Br}_1 X$, i.e., $X(\mathbb{A}_k)^{\operatorname{Br}_1} := \cap_{\alpha\in \operatorname{Br}_1 X} X(\mathbb{A}_k)^{\alpha}$. The algebraic Brauer–Manin set can be (and often is) larger than the Brauer–Manin set. Kresch and Tscinkel prove that if $\operatorname{Pic}\overline{X}$ is torsion free, then $X(\mathbb{A}_k)^{\operatorname{Br}_1}$ is effectively computable. In the case of surfaces (and still under the assumption that $\operatorname{Pic}\overline{X}$ is torsion-free), they can improve their results to show that for any positive integer n, there is an effectively computable set X_n such that

$$X(\mathbb{A}_k)^{\operatorname{Br}} \subset X_n \subset X(\mathbb{A}_k)^{\operatorname{Br} X[n]}.$$

In particular, if X is a surface with torsion-free geometric Picard group and a bound on the exponent of $\operatorname{Br} X/\operatorname{Br}_0 X$ is effectively computable (e.g., if X is a geometrically rational surface[4]), then the Brauer–Manin set is effectively computable.

Despite the lack of general algorithms for computing the Brauer–Manin obstruction, there is a general framework that is often helpful for at least computing the structure of $\operatorname{Br} X/\operatorname{Br}_0 X$ as an abstract torsion abelian group.

2.1. The Hochschild-Serre spectral sequence and a filtration of the Brauer group. We leverage the Hochschild-Serre spectral sequence in étale cohomology (applied to the Galois cover $\overline{X} \to X$ and the sheaf $\mathbb{G}_m$):

$$\mathrm{H}^p\left(G_k, H^q_{\text{et}}(\overline{X}, \mathbb{G}_m)\right) \Longrightarrow H^{p+q}_{\text{et}}(X, \mathbb{G}_m). \tag{2.1.1}$$

to give a filtration of the Brauer group. The exact sequence of low degree terms (see [**37**, Proposition 6.7.1 and Corollary 6.7.8]) is

$$0 \to \operatorname{Pic} X \to (\operatorname{Pic}\overline{X})^{G_k} \to \operatorname{Br} k \to \ker\left(\operatorname{Br} X \to \operatorname{Br}\overline{X}\right) \to \mathrm{H}^1(G_k, \operatorname{Pic}\overline{X}) \to 0;$$

(here we use the assumption that k is a number field and so $\mathrm{H}^3(G_k, \overline{k}^{\times}) = 0$ [**37**, Remark 6.7.10] to obtain the rightmost 0). Using the definition of $\operatorname{Br}_0 X$ (see the

[3] Unless there is an "obvious" reason why not, e.g., some classes of varieties always have a rational point, like in the case of del Pezzo surfaces of degree 1.

[4] Any geometrically rational surface is k-birational to a conic bundle over a conic or a del Pezzo surface [**37**, Corollary 9.3.21]. If X is a conic bundle over a conic then $\operatorname{Br} X/\operatorname{Br}_0 X$ has exponent at most 2 [**11**, Proposition 11.3.4 and its proof], and if X is a del Pezzo surface, then $\operatorname{Br} X/\operatorname{Br}_0 X$ has exponent dividing 60 [**13**, Theorem 4.1].

paragraph after Exercise 1.2.17) and $\operatorname{Br}_1 X$ (see the paragraphs above), we obtain a short exact sequence

$$0 \to \operatorname{Br}_0 X \to \operatorname{Br}_1 X \to \mathrm{H}^1(G_k, \operatorname{Pic} \overline{X}) \to 0. \tag{2.1.2}$$

The quotient $\operatorname{Br} X / \operatorname{Br}_1 X$ is called the *transcendental Brauer group* of X.

The transcendental Brauer group can also be studied using the Hochschild-Serre spectral sequence. Indeed, the higher degree terms yield the following exact sequence (again using the assumption that k is a number field and so $\mathrm{H}^3(G_k, \overline{k}^{\times}) = 0$) [**11**, 5.24]

$$0 \to \frac{\operatorname{Br} X}{\operatorname{Br}_1 X} \to (\operatorname{Br} \overline{X})^{G_k} \to \mathrm{H}^2(G_k, \operatorname{Pic} \overline{X}). \tag{2.1.3}$$

Thus, if we have a good enough understanding of $\operatorname{Pic} \overline{X}$ and $\operatorname{Br} \overline{X}$ as Galois modules, we can leverage (2.1.2) and (2.1.3) to compute the Brauer group modulo constants. See [**11**, Section 5.4] for more details on these approaches.

2.2. Properties of local evaluation maps. If we have successfully computed the Brauer group (or the quotient $\operatorname{Br} X / \operatorname{Br}_0 X$), then the task that remains is describing the set $X(\mathbb{A}_k)^{\alpha}$ for α a set of generators of $\operatorname{Br} X / \operatorname{Br}_0 X$.

Given an $\alpha \in \operatorname{Br} X$, there are some theoretic results that simplify calculation of the map $\langle \alpha_v, - \rangle \colon X(k_v) \to \mathbb{Q}/\mathbb{Z}$.

PROPOSITION 2.2.1. *Let X be a smooth projective variety over a number field k, let v be a place of k and let $\alpha \in \operatorname{Br} X$.*

1. *The map $\langle \alpha_v, - \rangle \colon X(k_v) \to \mathbb{Q}/\mathbb{Z}$ is locally constant.* [**37**, Proposition 8.2.9(a)]
2. *If $\alpha_v \in \operatorname{Br}_0 X_v$, then the map $\langle \alpha_v, - \rangle \colon X(k_v) \to \mathbb{Q}/\mathbb{Z}$ is constant.*
3. *If v is a nonarchimedean place of good reduction (i.e., there exists a smooth proper model $\mathcal{X} \to \mathcal{O}_v$) and $\alpha_{k_v^{nr}} = 0$, then the map $\langle \alpha_v, - \rangle \colon X(k_v) \to \mathbb{Q}/\mathbb{Z}$ is constant.* [**10**, Lemma 2.2(ii)]

The vast majority of examples in the literature with $X(\mathbb{A}_k)^{\{\alpha_i\}} = \emptyset$ actually have $\langle (\alpha_i)_v, - \rangle \colon X(k_v) \to \mathbb{Q}/\mathbb{Z}$ constant for all i and for all places v (e.g., Iskovskikh's Châtelet surface [**37**, Proposition 8.2.15] and Birch and Swinnerton-Dyer's quartic del Pezzo surface [**3**, Theorem 3] which is discussed in Exercise 2.3.1). Based on this, one might hope that constant evaluation everywhere is common, which would greatly simplify computation of the Brauer–Manin set. However, these examples are misleading! The examples in the literature were typically constructed to ensure this happens exactly so that the authors could prove that $X(\mathbb{A}_k)^{\{\alpha_i\}} = \emptyset$. In fact, experimentation shows that the expectation should be the opposite, namely:

Whenever possible, there should be a place v where the image of $\langle \alpha_v, - \rangle \colon X(k_v) \to \mathbb{Q}/\mathbb{Z}$ is large.

Note that large image of $\langle \alpha_v, - \rangle$ has implications for the Brauer–Manin set.

LEMMA 2.2.1.1. *Let X be a smooth projective variety over a number field k that has $X(\mathbb{A}_k) \neq \emptyset$. Let $\alpha \in \operatorname{Br} X$ be a nontrivial element.*

1. *If there is a place v where the image of $\langle \alpha_v, - \rangle \colon X(k_v) \to \mathbb{Q}/\mathbb{Z}$ has at least 2 elements, then $X(\mathbb{A}_k)^{\alpha} \subsetneq X(\mathbb{A}_k)$.*
2. *If there is a place v where the image of $\langle \alpha_v, - \rangle \colon X(k_v) \to \mathbb{Q}/\mathbb{Z}$ is equal to $\frac{1}{\operatorname{ord}(\alpha)}\mathbb{Z}/\mathbb{Z}$, then $X(\mathbb{A}_k)^{\alpha} \neq \emptyset$.*

REMARK 2.2.1.2. Functoriality of the Brauer group gives, for any $P_v \in X(k_v)$, a *group homomorphism* $P_v^*\colon \operatorname{Br} X \to \operatorname{Br} k_v$. Thus, the image of $\langle \alpha_v, - \rangle\colon X(k_v) \to \mathbb{Q}/\mathbb{Z}$ is always contained in $\frac{1}{\operatorname{ord}(\alpha)}\mathbb{Z}/\mathbb{Z}$, so the condition in part (2) is saying that the image of $\langle \alpha_v, - \rangle$ is as large as possible.

PROOF. Let v_0 be a place satisfying the assumptions of part (1) or part (2) and let $(P_v) \in X(\mathbb{A}_k)$.

If v_0 satisfies the assumptions of part (1), then there exists a point $Q_{v_0} \in X(k_{v_0})$ such that $\langle \alpha_{v_0}, P_{v_0}\rangle \neq \langle \alpha_{v_0}, Q_{v_0}\rangle$. Define the adelic point (Q_v) to be equal to (P_v) at all places $v \neq v_0$ and to be equal to Q_{v_0} at $v = v_0$. Since inv_v is an isomorphism, the difference

$$\sum_v \operatorname{inv}_v\langle \alpha, P_v\rangle - \sum_v \operatorname{inv}_v\langle \alpha, Q_v\rangle = \operatorname{inv}_v\left(\langle \alpha_{v_0}, P_{v_0}\rangle - \langle \alpha_{v_0}, Q_{v_0}\rangle\right)$$

must be nonzero, so at least one of (P_v), (Q_v) must not be orthogonal to α. Hence, $X(\mathbb{A}_k)^\alpha \subsetneq X(\mathbb{A}_k)$.

Now assume that v_0 satisfies the assumptions of part (2). Then, there exists a point $Q_{v_0} \in X(k_{v_0})$ such that $\operatorname{inv}_{v_0}\langle \alpha_{v_0}, Q_{v_0}\rangle = \sum_{v\neq v_0} \operatorname{inv}_v\langle \alpha_v, P_v\rangle$. Define the adelic point (Q_v), as above, i.e., so that Q_v is equal to (P_v) at all places $v \neq v_0$ and is equal to Q_{v_0} at $v = v_0$. Then, by construction, $(Q_v) \in X(\mathbb{A}_k)^\alpha$. □

The experimental observation articulated above is corroborated by several theoretic results, the earliest of which is due to Harari in 1994.

PROPOSITION 2.2.2 ([**20**, Proposition 6.1.1]). *Let $\pi\colon \mathcal{V} \to \mathbb{P}^1_k$ be a projective morphism of smooth varieties, and let V denote the generic fiber of π. Assume that $\mathcal{V}(\mathbb{A}_k) \neq \emptyset$ and that π has a geometric section.*

If $\operatorname{Br}\mathcal{V} = \operatorname{Br}_0\mathcal{V}$ and $\operatorname{Br} V \neq \operatorname{Br}_0 V$, then there are infinitely many rational points $\theta \in \mathbb{P}^1(k)$ such that $\mathcal{V}_\theta(\mathbb{A}_k)^{\operatorname{Br}} \subsetneq \mathcal{V}_\theta(\mathbb{A}_k)$.

The idea of the proof is to essentially show that for $\alpha \in \operatorname{Br} V \smallsetminus \operatorname{Br}\mathcal{V}$, there must be enough variation of the images of the local pairing with α to obtain the result.

Later work of Bright [**4**, Theorem 5.16] shows that if the residue field of k_v is sufficiently large, and the special fiber of a model $\mathcal{X}$ over $\mathcal{O}_v$ of X/k_v has rich enough geometry, then there exists a Brauer class $\alpha \in \operatorname{Br} X_v$ whose image is as large as possible. Moreover, with additional assumptions on the geometry of the special fiber, one can arrange for multiple classes $\alpha_i \in \operatorname{Br} X_v$ whose images are as large and as independent as possible [**4**, Theorem 5.24].

More recently, Pagano [**35**], following work of Bright and Newton [**5**], constructed an example with surjective evaluation map at a place of good reduction, thereby showing that any generalization of Proposition 2.2.1(3) must retain some assumption on α.

2.3. Extended exercise computing a Brauer–Manin obstruction.

EXERCISE 2.3.1. Let $X \subset \mathbb{P}^4$ be given by the vanishing of the following two quadrics

$$st - x^2 + 5y^2, \quad (s+t)(s+2t) - x^2 + 5z^2.$$

This variety was first studied by Birch and Swinnerton-Dyer [**3**].

(1) Use the Hasse-Weil bounds to show that any smooth genus 1 curve over a finite field $\mathbb{F}$ has an $\mathbb{F}$-point.

(2) Show that an intersection of quadrics in $\mathbb{P}^3$ is a genus 1 curve, and prove that $X \cap V(z)$ is smooth modulo p for all $p \neq 2, 5$. Use this to deduce that $X(\mathbb{Q}_p) \neq \emptyset$ for all $p \neq 2, 5$.

(3) Use the special case of Hensel's Lemma given in Exercise 1.1.3 to show that $X(\mathbb{Q}_5) \neq \emptyset$ and that $X(\mathbb{Q}_2) \neq \emptyset$. Then prove that $X(\mathbb{A}_{\mathbb{Q}}) \neq \emptyset$.

(4) Using Exercise 1.2.5, show that $(5, \frac{s}{t})$ and $(5, \frac{s+t}{s+2t})$ are trivial in $\operatorname{Br} \mathbf{k}(X)$.

(5) Using the previous part and Remark 1.2.5.1 show that, in $\operatorname{Br} \mathbf{k}(X)$

$$\alpha := \left(5, \frac{s+t}{s}\right) = \left(5, \frac{s+2t}{s}\right) = \left(5, \frac{s+t}{t}\right) = \left(5, \frac{s+2t}{t}\right).$$

Additionally show that for every point $P \in X \smallsetminus V(s,t)$, there is an open set $P \in U_P \subset X \smallsetminus V(s,t)$ such that at least one of $\frac{s+t}{s}, \frac{s+t}{t}, \frac{s+2t}{s}, \frac{s+2t}{t}$ is regular and invertible on U. This proves that $\alpha \in \operatorname{Br} U$. Now use Theorem 1.2.11(5) to prove that $\alpha \in \operatorname{Br} X$.

(6) Show that $X(\mathbb{A}_{\mathbb{Q}})^{\alpha} = \emptyset$ and hence $X(\mathbb{A}_{\mathbb{Q}})^{\mathrm{Br}} = \emptyset$. (*Sketch*: First show that for all $p \neq 5$ and $P_p \in X(\mathbb{Q}_p)$, at least one of $\frac{s+t}{s}, \frac{s+t}{t}, \frac{s+2t}{s}, \frac{s+2t}{t}$ is a p-adic unit at P_p. Then, noting that $\mathbb{Q}_p(\sqrt{5})/\mathbb{Q}_p$ is unramified for $p \neq 5$, use Exercise 1.2.5 to deduce that $\langle \alpha, P_p \rangle = 0 \in \operatorname{Br} \mathbb{Q}_p$. Lastly, show that $\langle \alpha, P_5 \rangle \neq 0 \in \operatorname{Br} \mathbb{Q}_5$ for all $P_5 \in X(\mathbb{Q}_5)$.)

3. Capturing the Brauer–Manin obstruction to the existence of points

We saw in Section 2.2 that "unless forced otherwise" we should typically expect $X(\mathbb{A}_k)^{\alpha} \subsetneq X(\mathbb{A}_k)$ and in fact, that Brauer elements should not only cause obstructions, but cause *independent* obstructions as much as they are able. This shows that we cannot hope to meaningfully decrease the work involved to compute $X(\mathbb{A}_k)^{\mathrm{Br}}$, that to compute the Brauer–Manin set, we really need to determine all the local pairings for a complete set of generators of $\operatorname{Br} X / \operatorname{Br}_0 X$.

However, if we are merely interested in the existence of rational points (rather than finer information about them), we can make do with answering the possibly weaker decision problem: determining whether $X(\mathbb{A}_k)^{\mathrm{Br}} \neq \emptyset$.

Why may this problem be easier? One can show that a consequence of the topological properties of the Brauer–Manin pairing (Proposition 2.2.1) and the compactness of $X(\mathbb{A}_k)$[5] is that if $X(\mathbb{A}_k)^{\mathrm{Br}} = \emptyset$ then this is witnessed by a finite collection of Brauer classes.

EXERCISE 3.0.1.

(1) Using Proposition 2.2.1(3), show that $\mathrm{ev}_{\alpha} \colon X(\mathbb{A}_k) \to \mathbb{Q}/\mathbb{Z}$ is locally constant.
(2) Let $\alpha \in \operatorname{Br} X$. Show that $X(\mathbb{A}_k)^{\alpha}$ is open and closed in the adelic topology. Conclude that $X(\mathbb{A}_k)^{\mathrm{Br}}$ is closed.
(3) Assume that $X(\mathbb{A}_k)^{\mathrm{Br}} = \emptyset$. Show that there exists a finite set $B \subset \operatorname{Br} X$ such that $X(\mathbb{A}_k)^{B} = \emptyset$.

[5]Compactness of $X(\mathbb{A}_k)$ follows from the compactness of $X(k_v)$ for local fields [**37**, Proposition 2.6.1], the definition of the topology on $X(\mathbb{A}_k)$ [**37**, Section 2.6.3], and Tychonoff's theorem.

In this section we consider the question of whether any properties of this finite subset of Brauer classes can be determined *a priori*, i.e., without computing $X(\mathbb{A}_k)^{\mathrm{Br}}$. With this goal in mind, we make the following definitions.

DEFINITION 3.0.2. Let X be a smooth projective variety over a number field k and let $\mathcal{B} \subset \operatorname{Br} X$.

(1) The set $\mathcal{B}$ *captures the Brauer–Manin obstruction* (to the existence of rational points) if $X(\mathbb{A}_k)^{\mathrm{Br}} = \emptyset \Rightarrow X(\mathbb{A}_k)^{\mathcal{B}} = \emptyset$.
(2) The set $\mathcal{B}$ *completely captures the Brauer–Manin obstruction* (to the existence of rational points) if, for all $\mathcal{B}' \subset \operatorname{Br} X$, we have[6]

$$X(\mathbb{A}_k)^{\mathcal{B}'} = \emptyset \Rightarrow X(\mathbb{A}_k)^{\mathcal{B}\cap\mathcal{B}'} = \emptyset.$$

Although these definitions are relatively recent, prior results can be phrased in this language.

THEOREM 3.0.3.

(1) *If C is a smooth projective degree d genus 1 curve, then* $\operatorname{Br} C[d^\infty]$ *completely captures the Brauer–Manin obstruction. (Follows from* [**29**]*.)*
(2) *If X is a smooth projective cubic surface (equivalently, a degree 3 del Pezzo) then* $\operatorname{Br} X[3]$ *completely captures the Brauer–Manin obstruction* [**44**, Corollary 1].
(3) *If X is a smooth projective del Pezzo surface of degree 3 or 4, then there exists an $\alpha \in \operatorname{Br} X$ such that $\{\alpha\}$ captures the Brauer–Manin obstruction* [**7**, Lemma 3.4 and the following Remark 2].

Given a class of varieties, one could hope to find a $\mathcal{B} \subset \operatorname{Br} X$ that captures or completely captures the Brauer–Manin obstruction where $X(\mathbb{A}_k)^{\mathcal{B}}$ is (more) easily computable. Recent work of Skorobogatov–Zarhin [**42**], Creutz and myself [**15**], Nakahara [**31**], and Creutz [**14**] indicates this is possible for varieties closely related to abelian varieties.

THEOREM 3.0.4.

(1) [**15**, Theorem 1.2] *Let V/k be a torsor under an abelian variety. Then* $\operatorname{Br} V[\operatorname{per}(V)^\infty]$ *completely captures the Brauer–Manin obstruction.*
(2) [**15**, Theorem A.1] *(building on* [**15**, **42**]*) Let Y/k be a Kummer variety. Then* $\operatorname{Br} V[2^\infty]$ *completely captures the Brauer–Manin obstruction.*
(3) [**31**, Theorem 1.5] *Let $\pi\colon X \to \mathbb{P}^n$ be a fibration of torsors under abelian varieties and let P denote the period of the generic fiber X_η (considered as a torsor under an abelian variety over $\mathbf{k}(\eta)$). Then* $\operatorname{Br} X[P^\infty]$ *completely captures the Brauer–Manin obstruction.*
(4) [**14**, Theorem 1] *Let V/k be a torsor under an abelian variety. Then the subgroup of locally constant Brauer classes completely captures the Brauer–Manin obstruction.*

However, there is a limit to how much can be determined about a capturing or completely capturing subgroup *a priori*. In joint work with Creutz and Voloch, we explored whether there exist easily describable proper subgroups $\mathcal{B}$ that capture or completely capture the Brauer–Manin obstruction for curves [**17**]. While our

[6]This terminology came out of discussions with Brendan Creutz. We use 'capture' in the sense of "record or express accurately".

results do not rule out the possibility of a capturing or completely capturing proper subgroup, they do suggest that any such subgroup likely does not depend only on the genus and the degree.

Furthermore, in joint work with Berg, Pagano, Poonen, Stoll, Triantafillou, and Vogt, we showed that completely capturing subgroups can be arbitrarily large. We constructed conic bundles that had empty Brauer–Manin set, but that required arbitrarily many Brauer classes to obstruct all adelic points [**1**, Theorem 1.1].

3.1. Open questions.

3.1.1. *del Pezzo surfaces.* The older results summarized in Theorem 3.0.3 show that for cubic and quartic del Pezzo surfaces, there is always a single Brauer class that captures the Brauer–Manin obstruction. For degree 2 del Pezzo surfaces (the only other case that can have empty Brauer set and nonempty adelic points), Nakahara's result from Theorem 3.0.4(3) implies that the 2-primary subgroup of the Brauer group completely captures the Brauer–Manin obstruction. However, for degree 2 del Pezzo surfaces it is not known how many elements are needed to detect a Brauer–Manin obstruction. (In general, there is no bound on how many Brauer classes are needed to detect an obstruction [**1**, Theorem 1.1].) Corn constructed an example of a degree 2 del Pezzo surfaces where two Brauer classes are required to detect the obstruction [**13**, Example 9.4]. However, the $\mathbb{F}_2$-dimension of $\operatorname{Br} X/(\operatorname{Br}_0 X + 2\operatorname{Br} X)$ can be as large as 6 [**13**, Theorem 4.1]. Are there degree 2 del Pezzo surfaces that require 3 Brauer classes to detect the obstruction? 4 Brauer classes? 6 Brauer classes?

3.1.2. *K3 surfaces.* As mentioned in Theorem 3.0.4(2), Skorobogatov proved that the 2-primary Brauer–Manin obstruction completely captures the Brauer–Manin obstruction on Kummer surfaces. Since Kummer surfaces are a particular class of K3 surfaces, it is natural to ask whether the 2-primary subgroup captures the Brauer–Manin obstruction for all K3 surfaces. Unfortunately, this is false, as work of Gvirtz, Loughran, and Nakahara shows [**19**, Theorem 1.5]. In addition, Berg and Várilly-Alvarado construct a K3 surface of Picard rank 1 where the 2-primary subgroup fails to completely capture the Brauer–Manin obstruction [**2**, Theorem 1.1]. In both of these papers, the authors show that the obstruction can be detected by 3-torsion. Thus, one can ask whether the 6-primary subgroup captures (or completely captures) the Brauer–Manin obstruction on K3 surfaces or whether classes of order prime to 6 can also obstruct the existence of rational points.

4. Behavior of the Brauer–Manin obstruction over extensions

Thus far, we have considered the question of determining whether $X(k) \neq \emptyset$ or whether $X(\mathbb{A}_k)^{\operatorname{Br}} \neq \emptyset$ over a *fixed* ground field k. However, the arithmetic content of X is much richer than that. In some sense, the arithmetic of X is the set of all algebraic points $X(\overline{k})$ together with its action of the absolute Galois group $G_k := \operatorname{Gal}(\overline{k}/k)$.

Thus, we can ask: what can we say about the sets $X(\mathbb{A}_L)^{\operatorname{Br}}$ as L ranges over extensions of k? Is knowledge of $X(\mathbb{A}_k)^{\operatorname{Br}}$ enough to give us some information about what happens when we range over algebraic extensions?

As we saw in Section 2.2, asking to completely describe the Brauer sets $X(\mathbb{A}_L)^{\operatorname{Br}}$ is likely too fine of a question since "unless there is a reason otherwise" we expect Brauer classes to carve out a proper subset of adelic points. This is true even when X stays fixed and only the field varies, as the following results of Liang and Nakahara–Roven demonstrate.

THEOREM 4.0.1 ([**27**, Theorem 3.1]). *For all number fields k, there exists a Châtelet surface V/k and a finite extension L/k such that*

$$V(\mathbb{A}_k)^{\mathrm{Br}} = V(\mathbb{A}_k) \neq \emptyset, \quad \text{and} \quad V(\mathbb{A}_L)^{\mathrm{Br}} \subsetneq V(\mathbb{A}_L).$$

THEOREM 4.0.2 ([**32**, Theorem 1.1]). *Let k be a number field, let $a \in k^\times$ and let $p(x)$ be a degree 4 separable polynomial over k. Let X be the Châtelet surface given by $y^2 - az^2 = p(x)$ and let L be an extension where $p(x)$ splits completely and $X(\mathbb{A}_L) \neq \emptyset$. Then $X(\mathbb{A}_L)^{\mathrm{Br}} \subsetneq X(\mathbb{A}_L)$ if a is not totally positive or if there exists a finite place v such that $p(x) \bmod v$ is not separable and $a \notin k_v^{\times 2}$.*

In light of these results, we must modify our question of interest. Instead of asking to describe the Brauer–Manin sets over all extensions, we can instead ask whether we can say anything about the set of field extensions

$$\{L/k \text{ finite} : X(\mathbb{A}_L)^{\mathrm{Br}} = \emptyset\}$$

or its complement

$$\{L/k \text{ finite} : X(\mathbb{A}_L)^{\mathrm{Br}} \neq \emptyset\}$$

based on information about X over k.

This is a relatively new research direction, and, as we will see, the results thus far are mainly limited to geometrically rational surfaces. More exploration is needed to determine whether the phenomena we will see manifests in greater generality.

4.1. Compatibility of the Brauer–Manin pairing with corestriction. The functoriality of the Brauer group and compatibility of the Brauer–Manin pairing with corestriction leads to the following containments of Brauer sets under extensions.

LEMMA 4.1.0.1 ([**16**, Lemma 3.1]). *Let K/k be an extension of number fields, let Y/k be a smooth projective geometrically integral variety over k, and let $B \subset \mathrm{Br}\, Y_K$. Then $Y(\mathbb{A}_k)^{\mathrm{Cor}_{Y_K/Y_k}(B)} \subset Y(\mathbb{A}_K)^B$. In particular,*

1. *if $Y(\mathbb{A}_k)^{\mathrm{Br}} \neq \emptyset$ then $Y(\mathbb{A}_K)^{\mathrm{Br}} \neq \emptyset$, and*
2. *for any $d|[K:k]$, we have $Y(\mathbb{A}_k) \subset Y(\mathbb{A}_K)^{(\mathrm{Br}\, Y)[d]}$.*

REMARK 4.1.0.2. In an early draft of [**16**], Creutz and I proved parts (1) and (2) in our case of interest, quartic del Pezzo surfaces. Upon review of that draft, Wittenberg observed that the lemma held more generally and outlined a proof of the version given here.

From Lemma 4.1.0.1, we can immediately deduce that if $X(\mathbb{A}_k)^{\mathrm{Br}} \neq \emptyset$, then

$$\{L/k \text{ finite} : X(\mathbb{A}_L)^{\mathrm{Br}} = \emptyset\} = \emptyset \quad \text{and} \quad \{L/k \text{ finite} : X(\mathbb{A}_L)^{\mathrm{Br}} \neq \emptyset\} = \{L/k \text{ finite}\}.$$

Thus, the main case of interest for this question is when $X(\mathbb{A}_k)^{\mathrm{Br}} = \emptyset$.

4.2. Vanishing of the Brauer–Manin obstruction. In this section we retain the assumption that $X(\mathbb{A}_k)^{\mathrm{Br}} = \emptyset$ and consider sufficient conditions for the Brauer–Manin obstruction to *vanish* over the extension L/k, i.e., such that $X(\mathbb{A}_L)^{\mathrm{Br}} \neq \emptyset$.

From Lemma 4.1.0.1, we can deduce that if $X(\mathbb{A}_k) \neq \emptyset$ *and* the $[L:k]$-torsion subgroup of $\mathrm{Br}\, X_L$ captures the Brauer–Manin obstruction, then the Brauer–Manin obstruction vanishes over L. Here are some example cases where these properties are known.

PROPOSITION 4.2.1 (Corollary of Theorem 3.0.3 and Lemma 4.1.0.1).

(1) *Let* $\pi\colon X \to \mathbb{P}^1_k$ *be an everywhere locally soluble conic bundle. Then there exists a finite extension* K/k *such that for all even degree extensions* L/k *with* L *linearly disjoint from* K*, we have* $X(\mathbb{A}_L)^{\mathrm{Br}} \neq \emptyset$.
(2) *Let* $X \subset \mathbb{P}^3_k$ *be an everywhere locally soluble cubic surface. Then there exists a finite extension* K/k *such that for all extensions* L/k *with* L *linearly disjoint from* K *and* $[L : k] \equiv 0 \bmod 3$*, we have* $X(\mathbb{A}_L)^{\mathrm{Br}} \neq \emptyset$.
(3) *Let* $X \subset \mathbb{P}^4_k$ *be an everywhere locally soluble quartic del Pezzo surface. Then there exists a finite extension* K/k *such that for all even degree extensions* L/k *with* L *linearly disjoint from* K*, we have* $X(\mathbb{A}_L)^{\mathrm{Br}} \neq \emptyset$.

More generally, we can obtain a result of this flavor whenever X is a variety whose geometric Picard group is torsion free and whose geometric Brauer group is trivial (see also [**22**]).

For the varieties in Proposition 4.2.1, the Brauer–Manin obstruction is conjecturally the only obstruction to the existence of rational points. Thus, Proposition 4.2.1 suggests the existence of points over numerous field extensions. What can we prove unconditionally?

PROPOSITION 4.2.2. *Let* $\pi\colon X \to \mathbb{P}^1$ *be an everywhere locally soluble conic bundle. Then, for any finite set of places* $S \subset \Omega_k$ *and any collection of quadratic étale algebras* $(F_v)_{v\in S}$*, there exists a quadratic extension* L/k *such that* $L\otimes_k k_v \simeq F_v$ *and* $X(L) \neq \emptyset$.

PROOF. By the implicit function theorem, for any $x_v \in X(k_v)$ there exists an open set $U_v \subset \mathbb{P}^1(k_v)$ such that $\pi(x_v) \subset U_v \subset \pi(X(k_v))$. By weak approximation on $\mathbb{P}^1$, there exists a $t \in \mathbb{P}^1(k)$ such that $t \in U_v$ for all $v \in S$. In particular, the fiber X_t is a conic that has k_v-points for all $v \in S$. Let $T := \{v \in \Omega_k : X_t(k_v) = \emptyset\}$; note that T is finite (see Proposition 1.1.2) and disjoint from S. By [**34**, Part II, Definition 1.3 and Theorem 5.6], $\mathrm{inv}_F \circ \mathrm{Res}_{F/k_v} = [F : k_v] \circ \mathrm{inv}_{k_v}$ for any extension of local fields F/k_v. Therefore, for any quadratic extension K/k, $X_t(\mathbb{A}_K) \neq \emptyset$ if and only if v does not split in K for all $v \in T$. Since S and T are disjoint and finite, there exists a quadratic extension L/k that has the desired splitting behavior at all $v \in S \cup T$. Since conics satisfy the local-to-global principle and $X_t(\mathbb{A}_L) \neq \emptyset$, this proves that $X_t(L) \neq \emptyset$, as desired. □

For locally soluble quartic del Pezzo surfaces, we can unconditionally prove the existence of a quadratic point [**16**, Proposition 4.7 and Remark 4.8]; however, it is a nonexplicit construction and doesn't easily give control of local behavior. Thus, our unconditional results over number fields fall far short of what is predicted by Proposition 4.2.1.

If $X(\mathbb{A}_k) = \emptyset$, then Lemma 4.1.0.1 no longer can be leveraged to show that the Brauer–Manin obstruction vanishes over an extension. Indeed, the lack of adelic points over k limits the construction of adelic points over the extension, and so it is intrinsically more difficult to prove that the obstruction vanishes. Despite this added difficulty, Roven [**40**] and Creutz and myself [**16**] can obtain some positive results in this direction.

THEOREM 4.2.3 ([**40**]). *Let* $\pi\colon X \to \mathbb{P}^1_k$ *be a conic bundle. Then there exists a finite extension* K/k *such that for all quadratic* L/k *linearly disjoint from* K*, we have*

$$X(\mathbb{A}_L) \neq \emptyset \Leftrightarrow X(\mathbb{A}_L)^{\mathrm{Br}} \neq \emptyset.$$

THEOREM 4.2.4 ([**40**]). *Let* $\pi\colon X \to \mathbb{P}^1_k$ *be a Châtelet surface. Then there exists a finite extension* K/k *such that for all even degree* L/k *linearly disjoint from* K*, we have*

$$X(\mathbb{A}_L) \neq \emptyset \Leftrightarrow X(\mathbb{A}_L)^{\mathrm{Br}} \neq \emptyset.$$

REMARK 4.2.4.1. Colliot-Thélène, Sansuc, and Swinnerton-Dyer [**8, 9**] have proved that for Châtelet surfaces over number fields, the set of k-rational points are dense in the Brauer–Manin set. Using this one can deduce from Theorem 4.2.4 that for all even degree L/k linearly disjoint from K, X satisfies the local-to-global principle over L.

THEOREM 4.2.5 (Special case of [**16**, Theorem 1.2]). *Let* $X \subset \mathbb{P}^4$ *be a quartic del Pezzo surface over a number field* k*. Then for all places* v*, there exists a quadratic extension* L_w/k_v *such that* $X(L_w) \neq \emptyset$*. Furthermore, if all rank* 4 *quadrics containing* X *are defined over* k*, then there exists a quadratic extension* L/k *such that* $X(\mathbb{A}_L)^{\mathrm{Br}} \neq \emptyset$*.*

4.3. Persistence of a Brauer–Manin obstruction. We conclude in this section by reviewing the complementary question of conditions on L/k that guarantee that a Brauer–Manin obstruction persists, i.e., that $X(\mathbb{A}_k)^{\mathrm{Br}} = \emptyset$ implies that $X(\mathbb{A}_L)^{\mathrm{Br}} = \emptyset$. We should think of this as a harder condition to achieve, because it typically becomes easier to obtain points over extensions.

However, there are some simple varieties where we can prove persistence.

THEOREM 4.3.1 (Corollary of Springer's theorem [**43**]). *Let* X *be a smooth quadric over a number field* k*. Then a Brauer–Manin obstruction persists over any odd degree extension, i.e.,* $X(\mathbb{A}_k)^{\mathrm{Br}} = \emptyset$ *implies that* $X(\mathbb{A}_L)^{\mathrm{Br}} = \emptyset$ *for all odd degree extensions* L/k*.*

THEOREM 4.3.2 (Corollary of [**6**, Theorem C] and [**7**, Remark 3]). *Let* X *be a quartic del Pezzo surface over a number field* k*. Then a Brauer–Manin obstruction persists over any odd degree extension, i.e.,* $X(\mathbb{A}_k)^{\mathrm{Br}} = \emptyset$ *implies that* $X(\mathbb{A}_L)^{\mathrm{Br}} = \emptyset$ *for all odd degree extensions* L/k*.*

THEOREM 4.3.3 ([**39**, Theorem 1.1]). *Let* X *be a smooth cubic surface over a number field* k*. Then a Brauer–Manin obstruction persists over any extension with degree coprime to three, i.e.,* $X(\mathbb{A}_k)^{\mathrm{Br}} = \emptyset$ *implies that* $X(\mathbb{A}_L)^{\mathrm{Br}} = \emptyset$ *for all* L/k *with* $3 \nmid [L:k]$*.*

References

[1] J. Berg, C. Pagano, B. Poonen, M. Stoll, N. Triantafillou, B. Viray, and I. Vogt, *Brauer-Manin obstructions requiring arbitrarily many Brauer classes*, Bull. Lond. Math. Soc. **56** (2024), no. 5, 1587–1604, DOI 10.1112/blms.12999. MR4743882 ↑243

[2] J. Berg and A. Várilly-Alvarado, *Odd order obstructions to the Hasse principle on general K3 surfaces*, Math. Comp. **89** (2020), no. 323, 1395–1416, DOI 10.1090/mcom/3485. MR4063322 ↑243

[3] B. J. Birch and H. P. F. Swinnerton-Dyer, *The Hasse problem for rational surfaces*, J. Reine Angew. Math. **274(275)** (1975), 164–174, DOI 10.1515/crll.1975.274-275.164. MR429913 ↑239, 240

[4] M. Bright, *Bad reduction of the Brauer-Manin obstruction*, J. Lond. Math. Soc. (2) **91** (2015), no. 3, 643–666, DOI 10.1112/jlms/jdv005. MR3355119 ↑240

[5] M. Bright and R. Newton, *Evaluating the wild Brauer group*, Invent. Math. **234** (2023), no. 2, 819–891, DOI 10.1007/s00222-023-01210-8. MR4651011 ↑240

[6] J.-L. Colliot-Thélène and D. Coray, *L'équivalence rationnelle sur les points fermés des surfaces rationnelles fibrées en coniques* (French), Compositio Math. **39** (1979), no. 3, 301–332. MR550646 ↑246

[7] J.-L. Colliot-Thélène and B. Poonen, *Algebraic families of nonzero elements of Shafarevich-Tate groups*, J. Amer. Math. Soc. **13** (2000), no. 1, 83–99, DOI 10.1090/S0894-0347-99-00315-X. MR1697093 ↑242, 246

[8] J.-L. Colliot-Thélène, J.-J. Sansuc, and P. Swinnerton-Dyer, *Intersections of two quadrics and Châtelet surfaces. II*, J. Reine Angew. Math. **374** (1987), 72–168, DOI 10.1515/crll.1987.374.72. MR876222 ↑246

[9] J.-L. Colliot-Thélène, J.-J. Sansuc, and P. Swinnerton-Dyer, *Intersections of two quadrics and Châtelet surfaces. I*, J. Reine Angew. Math. **373** (1987), 37–107, DOI 10.1515/crll.1987.373.37. MR870307 ↑246

[10] J.-L. Colliot-Thélène and A. N. Skorobogatov, *Good reduction of the Brauer-Manin obstruction*, Trans. Amer. Math. Soc. **365** (2013), no. 2, 579–590, DOI 10.1090/S0002-9947-2012-05556-5. MR2995366 ↑239

[11] J.-L. Colliot-Thélène and A. N. Skorobogatov, *The Brauer-Grothendieck group*, Ergebnisse der Mathematik und ihrer Grenzgebiete. 3. Folge. A Series of Modern Surveys in Mathematics [Results in Mathematics and Related Areas. 3rd Series. A Series of Modern Surveys in Mathematics], vol. 71, Springer, Cham, 2021, DOI 10.1007/978-3-030-74248-5. MR4304038 ↑229, 233, 234, 235, 238, 239

[12] J.-L. Colliot-Thélène and P. Swinnerton-Dyer, *Hasse principle and weak approximation for pencils of Severi-Brauer and similar varieties*, J. Reine Angew. Math. **453** (1994), 49–112, DOI 10.1515/crll.1994.453.49. MR1285781 ↑

[13] P. Corn, *The Brauer-Manin obstruction on del Pezzo surfaces of degree 2*, Proc. Lond. Math. Soc. (3) **95** (2007), no. 3, 735–777, DOI 10.1112/plms/pdm015. MR2368282 ↑238, 243

[14] B. Creutz, *There are no transcendental Brauer-Manin obstructions on abelian varieties*, Int. Math. Res. Not. IMRN **9** (2020), 2684–2697, DOI 10.1093/imrn/rny098. MR4095422 ↑242

[15] B. Creutz and B. Viray, *Degree and the Brauer-Manin obstruction*, Algebra Number Theory **12** (2018), no. 10, 2445–2470, DOI 10.2140/ant.2018.12.2445. With an appendix by Alexei N. Skorobogatov. MR3911136 ↑242

[16] B. Creutz and B. Viray, *Quadratic points on intersections of two quadrics*, Algebra Number Theory **17** (2023), no. 8, 1411–1452, DOI 10.2140/ant.2023.17.1411. MR4633628 ↑244, 245, 246

[17] B. Creutz, B. Viray, and J. F. Voloch, *The d-primary Brauer-Manin obstruction for curves*, Res. Number Theory **4** (2018), no. 2, Paper No. 26, 16, DOI 10.1007/s40993-018-0120-3. MR3807414 ↑242

[18] P. Gille and T. Szamuely, *Central simple algebras and Galois cohomology*, Cambridge Studies in Advanced Mathematics, vol. 101, Cambridge University Press, Cambridge, 2006, DOI 10.1017/CBO9780511607219. MR2266528 ↑232, 233, 234

[19] D. Gvirtz, D. Loughran, and M. Nakahara, *Quantitative arithmetic of diagonal degree 2 K3 surfaces*, Math. Ann. **384** (2022), no. 1-2, 135–209, DOI 10.1007/s00208-021-02280-w. MR4476231 ↑243

[20] D. Harari, *Méthode des fibrations et obstruction de Manin* (French), Duke Math. J. **75** (1994), no. 1, 221–260, DOI 10.1215/S0012-7094-94-07507-8. MR1284820 ↑240

[21] V. A. Iskovskih, *A counterexample to the Hasse principle for systems of two quadratic forms in five variables* (Russian), Mat. Zametki **10** (1971), 253–257. MR286743 ↑

[22] D. Kanevsky, *Application of the conjecture on the Manin obstruction to various Diophantine problems*, Astérisque **147-148** (1987), 307–314, 345. Journées arithmétiques de Besançon (Besançon, 1985). MR891437 ↑245

[23] A. Kresch and Y. Tschinkel, *Effectivity of Brauer-Manin obstructions on surfaces*, Adv. Math. **226** (2011), no. 5, 4131–4144, DOI 10.1016/j.aim.2010.11.012. MR2770443 ↑238

[24] A. Kresch and Y. Tschinkel, *Effectivity of Brauer-Manin obstructions*, Adv. Math. **218** (2008), no. 1, 1–27, DOI 10.1016/j.aim.2007.11.017. MR2409407 ↑238

[25] A. Kresch and Y. Tschinkel, *On the arithmetic of del Pezzo surfaces of degree 2*, Proc. London Math. Soc. (3) **89** (2004), no. 3, 545–569, DOI 10.1112/S002461150401490X. ↑

[26] T. Y. Lam, *Introduction to quadratic forms over fields*, Graduate Studies in Mathematics, vol. 67, American Mathematical Society, Providence, RI, 2005, DOI 10.1090/gsm/067. MR2104929 ↑232

[27] Y. Liang, *Noninvariance of weak approximation properties under extension of the ground field*, Michigan Math. J. **73** (2023), no. 4, 675–692, DOI 10.1307/mmj/20205984. MR4634976 ↑244

[28] C.-E. Lind, *Untersuchungen über die rationalen Punkte der ebenen kubischen Kurven vom Geschlecht Eins* (German), University of Uppsala, Uppsala, 1940. Thesis. MR22563 ↑232

[29] Y. I. Manin, *Le groupe de Brauer-Grothendieck en géométrie diophantienne*, Actes du Congrès International des Mathématiciens (Nice, 1970), Gauthier-Villars Éditeur, Paris, 1971, pp. 401–411. MR427322 ↑236, 242

[30] J. S. Milne, *Étale cohomology*, Princeton Mathematical Series, No. 33, Princeton University Press, Princeton, NJ, 1980. MR559531 ↑234

[31] M. Nakahara, *Index of fibrations and Brauer classes that never obstruct the Hasse principle*, Adv. Math. **348** (2019), 512–522, DOI 10.1016/j.aim.2019.03.012. MR3928654 ↑242

[32] M. Nakahara and S. Roven, *Weak approximation on Châtelet surfaces*, Manuscripta Math. **174** (2024), no. 3-4, 1133–1149, DOI 10.1007/s00229-024-01548-0. MR4756234 ↑244

[33] A. Nerode, *A decision method for p-adic integral zeros of diophantine equations*, Bull. Amer. Math. Soc. **69** (1963), 513–517, DOI 10.1090/S0002-9904-1963-10979-9. MR168460 ↑230

[34] J. Neukirch, *Class field theory*, Springer, Heidelberg, 2013. The Bonn lectures, edited and with a foreword by Alexander Schmidt; Translated from the 1967 German original by F. Lemmermeyer and W. Snyder; Language editor: A. Rosenschon, DOI 10.1007/978-3-642-35437-3. MR3058613 ↑232, 245

[35] M. Pagano, *An example of a Brauer-Manin obstruction to weak approximation at a prime with good reduction*, Res. Number Theory **8** (2022), no. 3, Paper No. 63, 15, DOI 10.1007/s40993-022-00353-6. MR4476267 ↑240

[36] B. Poonen, *Undecidability in number theory*, Notices Amer. Math. Soc. **55** (2008), no. 3, 344–350. MR2382821 ↑230

[37] B. Poonen, *Rational points on varieties*, Graduate Studies in Mathematics, vol. 186, American Mathematical Society, Providence, RI, 2017, DOI 10.1090/gsm/186. MR3729254 ↑229, 230, 231, 233, 235, 237, 238, 239, 241

[38] H. Reichardt, *Einige im Kleinen überall lösbare, im Grossen unlösbare diophantische Gleichungen* (German), J. Reine Angew. Math. **184** (1942), 12–18, DOI 10.1515/crll.1942.184.12. MR9381 ↑232

[39] C. Rivera and B. Viray, *Persistence of the Brauer-Manin obstruction on cubic surfaces*, Math. Res. Lett. **29** (2022), no. 6, 1881–1889, DOI 10.4310/mrl.2022.v29.n6.a11. MR4589381 ↑246

[40] S. Roven and A. Wang, *On the Hasse Principle for conic bundles over even degree extensions*, Preprint, `arXiv:2508.07564`. ↑245, 246

[41] A. Skorobogatov, *Torsors and rational points*, Cambridge Tracts in Mathematics, vol. 144, Cambridge University Press, Cambridge, 2001, DOI 10.1017/CBO9780511549588. MR1845760 ↑229

[42] A. N. Skorobogatov and Y. G. Zarhin, *Kummer varieties and their Brauer groups*, Pure Appl. Math. Q. **13** (2017), no. 2, 337–368, DOI 10.4310/PAMQ.2017.v13.n2.a5. MR3858012 ↑242

[43] T. A. Springer, *Quadratic forms over fields with a discrete valuation. II. Norms*, Indag. Math. **18** (1956), 238–246. Nederl. Akad. Wetensch. Proc. Ser. A **59**. MR76794 ↑246

[44] P. Swinnerton-Dyer, *The Brauer group of cubic surfaces*, Math. Proc. Cambridge Philos. Soc. **113** (1993), no. 3, 449–460, DOI 10.1017/S0305004100076106. MR1207510 ↑242

University of Washington, Department of Mathematics, Box 354350, Seattle, Washington 98195, USA

Email address: `bviray@uw.edu`

IAS/Park City Mathematics Series
Volume **29**, 2025
https://doi.org/10.1090/pcms/029/00895

Rigid Cocycles and Singular Moduli for Real Quadratic Fields

J. B. Vonk

Abstract. These lectures aim to give an introduction to the theory of rigid cocycles, and to discuss their role in the analytic construction of singular moduli for real quadratic fields. Special emphasis will lie on the computational techniques and experiments that informed the development of the subject.

Contents

1. Singular moduli for imaginary quadratic fields

The theme of this PCMI is "number theory informed by computation". One mathematical story that may be described by this phrase is the theory of complex multiplication (CM). Several of its most celebrated results have been discovered or directly informed by explicit experimentation and computation. Notably, this includes the work of Gross–Zagier on heights of Heegner points.

The goal of these notes is to illustrate the role of computations in the development of CM theory, and to discuss how this tradition continues in recent investigations of a nascent RM theory. The focus lies on the theory of rigid cocycles.

Outline. After a brief introduction on CM theory, we recall some reduction theory of binary quadratic forms in §2. We define *rational cocycles* for $\mathrm{SL}_2(\mathbf{Z})$ in §3, and in §4 we consider their p-adic limits, defining *rigid cocycles* for $\mathrm{SL}_2(\mathbf{Z}[1/p])$. Special values of rigid cocycles at RM points define invariants which behave like RM counterparts of the differences of singular moduli of Gross and Zagier.

2020 *Mathematics Subject Classification.* Primary 11Fxx; Secondary 11Gxx.
Key words and phrases. Number theory, Modular forms.

1.1. Singular moduli. The story begins with classical results from the late 19th century German school, on the theory of complex multiplication. Several comprehensive treatments exist: inspiring historical sources the reader may wish to consult are Klein–Fricke [**47,48**] and Fricke–Klein [**32,33**] as well as the insightful account of the works of Eisenstein and Kronecker by Weil [**67**].

Our primary interest is Klein's modular j-invariant

$$\begin{aligned} j(q) &= \left(1+240\sum_{n\geq 1}\frac{n^3q^n}{1-q^n}\right)^3 \div \left(q\prod_{n\geq 1}(1-q^n)^{24}\right) \\ &= q^{-1}+744+196884q+21493760q^2+\ldots \end{aligned}$$

with $q=\exp(2\pi i\tau)$. It is a holomorphic function (though with a simple pole at the cusp ∞), invariant under the action of $\mathrm{SL}_2(\mathbf{Z})$ by linear fractional transformations on the argument τ, defined on the Poincaré or Lobachevsky hyperbolic plane

$$\begin{aligned} \mathfrak{H}_\infty &:= \{\tau\in\mathbf{C} \;:\; \mathrm{Im}(\tau)>0\} \\ &= \{q\in\mathbf{C} \;:\; |q|<1\}. \end{aligned}$$

We will study *singular moduli*, which are the values of the j-function at CM points τ, i.e. elements $\tau\in\mathfrak{H}_\infty$ that satisfy a quadratic equation over $\mathbf{Q}$, necessarily of negative discriminant. Singular moduli are always algebraic integers. Let us naively tabulate some singular moduli at purely imaginary CM points:

τ	$j(\tau)$	Trace	Norm
$\sqrt{-1}$	12^3	$2^6\cdot 3^3$	$2^6\cdot 3^3$
$\sqrt{-2}$	20^3	$2^6\cdot 5^3$	$2^6\cdot 5^3$
$\sqrt{-3}$	54000	$2^4\cdot 3^3\cdot 5^3$	$2^4\cdot 3^3\cdot 5^3$
$\sqrt{-4}$	66^3	$2^3\cdot 3^3\cdot 11^3$	$2^3\cdot 3^3\cdot 11^3$
$\sqrt{-5}$	$(26\sqrt{5}+20)^3$	$2^7\cdot 5^3\cdot 79$	$2^{12}\cdot 5^3\cdot 11^3$
$\sqrt{-6}$	$1707264\sqrt{2}+2417472$	$2^7\cdot 3^3\cdot 1399$	$2^{12}\cdot 3^6\cdot 17^3$
$\sqrt{-7}$	255^3	$3^3\cdot 5^3\cdot 17^3$	$3^3\cdot 5^3\cdot 17^3$
$\sqrt{-8}$	$(130\sqrt{2}+190)^3$	$2^4\cdot 5^6\cdot 11\cdot 19$	$2^6\cdot 5^6\cdot 23^3$
$\sqrt{-9}$	$44330496\sqrt{3}+76771008$	$2^7\cdot 3^2\cdot 133283$	$2^{12}\cdot 3^3\cdot 11^3\cdot 23^3$
$\sqrt{-10}$	$(162\sqrt{5}+390)^3$	$2^7\cdot 3^3\cdot 5^2\cdot 13\cdot 379$	$2^{12}\cdot 3^6\cdot 5^3\cdot 29^3$

The next entry of the table would be $j(\sqrt{-11})$, whose minimal polynomial

$$f(x)=x^3-1122662608x^2+270413882112x-653249011576832$$

has a non-abelian splitting field over $\mathbf{Q}$ with Galois group S_3. It is the Hilbert class field of $\mathbf{Q}(\sqrt{-11})$. We note for future reference that the norm of $j(\sqrt{-11})$, which is the absolute value of the constant coefficient, has prime factorisation

$$653249011576832=2^{12}\cdot 11^3\cdot 17^3\cdot 29^3.$$

These examples invite several observations and results, as noticed by early practitioners of the theory of complex multiplication; we single out two:

(1) **Field of definition.** A singular modulus generates the *ring class field* of the quadratic order of its argument τ, of discriminant $\Delta<0$. These fields

are important, for instance, in the classical question of the representability of primes by quadratic forms, e.g. characterising primes of the form

$$p = x^2 + ny^2.$$

Class field theory reduces this question to the explicit knowledge of a set of generators of the ring class field. Specifically: excluding divisors of $4n$, it is precisely (see Cox [**10**]) the set of primes p that *split completely* in the ring class field of discriminant $-4n$, namely

$$\mathbf{Q}\left(\sqrt{-n}, j(\sqrt{-n})\right).$$

Historically, a set of generators was first described in situations where the ring class field is abelian over $\mathbf{Q}$. This is the subject of *genus theory.* The most classical version is due to Gauß [**34**], and it covers all ten entries in the above table. To give an example, the fact that the ring class field of conductor -20 is $\mathbf{Q}(\sqrt{-5}, \sqrt{5})$, where the set of split primes p is characterised by the simultaneous solubility of x^2+5 and x^2-5 over $\mathbf{F}_p$, allows one to deduce a famous conjecture of Euler, stating that

$$p = x^2 + 5y^2 \iff p = 5 \quad \text{or} \quad p \equiv 1, 9 \pmod{20}.$$

In general, a characterisation by a simple congruence condition on the prime p does not exist, owing to the fact that the ring class field is not generally abelian, i.e. not contained in a cyclotomic field. We observed for instance that the ring class field of discriminant -44 is an S_3-extension. Two other singular moduli of historical notability are

$$\begin{aligned} j(\sqrt{-14}) &= 2^3\left(323 + 228\sqrt{2} + (231 + 161\sqrt{2})\sqrt{2\sqrt{2}-1}\right)^3, \\ j(\sqrt{-27}) &= 2^4 \cdot 3 \cdot 5^3 \left(5285131824\sqrt[3]{2} + 6658848836\sqrt[3]{2} + 8389623817\right). \end{aligned}$$

The former was computed by Weber [**66**, Section 144]. It is an integer in a cyclic extension of degree 4 of $\mathbf{Q}(\sqrt{-14})$; the ring class field of discriminant -56 is a degree 8 dihedral extension of $\mathbf{Q}$. The latter is a generator of the number field $\mathbf{Q}(\sqrt[3]{2})$, proving a criterion conjectured by Euler:

$$p = x^2 + 27y^2 \iff \begin{cases} p \equiv 1 \pmod{3} \\ x^3 - 2 \equiv 0 \pmod{p} \text{ has a solution.} \end{cases}$$

(2) **Arithmetic factorisations.** Arguably the most powerful applications of singular moduli came from a detailed study of their prime factorisations, which was first undertaken entirely experimentally. An early collection of tables may be found in the survey by Greenhill [**36**], published in 1889. The rich arithmetic properties that reside in these prime factorisations are clearly acknowledged in the experimental work of Berwick [**1**], who tabulated the prime factorisations of the two quantities

$$j(\tau) \quad \text{and} \quad j(\tau) - 1728$$

whenever the singular modulus $j(\tau)$ is of degree at most three. As Berwick notes, several striking patterns arise from these tables. For instance:

10. An inspection of the results of §§ 4, 5, 8, 9 suggests the possibility of several general theorems, unattempted here, concerning the factors of j and $j-1728$.

(i) $$\Delta \equiv 3 \pmod 8, \quad j \equiv 0 \pmod{2^{15}};$$
$$\tfrac{1}{4}\Delta \equiv 3 \pmod 8, \quad j \equiv 0 \pmod{2^{4}};$$
$$\Delta \text{ or } \tfrac{1}{4}\Delta \text{ or } \tfrac{1}{16}\Delta \equiv 7 \pmod 8, \quad j \not\equiv 0 \pmod 2;$$

FIGURE 1.1. An observation of Berwick [**1**].

Berwick makes several observations, and we will pick out a few that fit into the themes that reappear later in our discussion of RM theory.

First, the question of *which primes* can arise in the factorisation. Inspecting the ten entries of our own table above, we might for instance conjecture that for primes q we have

$$q \mid \operatorname{Nm} j(\sqrt{-n}) \quad \Rightarrow \quad \begin{cases} q \equiv 0, 2 \pmod 3 \\ q < 3n. \end{cases}$$

This was also noticed by Berwick. Furthermore, he notes that such primes satisfy a congruence condition modulo the discriminant $\Delta = -4n$ summarised by a single condition on the Kronecker symbol:

$$\left(\frac{\Delta}{q}\right) \neq 1.$$

Second, for a given prime q that arises in the factorisation, the q-adic valuation of that norm can be studied. Berwick makes several conjectures for small values of q, predicting a lower bound on this valuation. The lower bound for the case $q = 2$ appears in the excerpt Figure 1.1.

1.2. The work of Gross–Zagier. Following the observations of Berwick, the study of singular moduli took a spectacular turn several decades later, in the modern era. This time around, in the context of the Birch–Swinnerton-Dyer conjecture, with the landmark results of Gross and Zagier [**39, 40**].

Documented in the 1983 letter of Zagier to Gross [**68**], we find an elaboration of the computational experiments of Berwick. We can infer from the letter that Gross and Zagier had already proved several observations of Berwick about

$$\operatorname{Nm} j(\tau) \qquad \text{and} \qquad \operatorname{Nm}(j(\tau) - 1728). \tag{1.2}$$

In his letter, Zagier discovers a more general phenomenon. Observe that both quantities in (1.2) are the norms of *differences* of singular moduli, since

$$j\left(\frac{1+\sqrt{-3}}{2}\right) = 0, \qquad j(\sqrt{-1}) = 1728.$$

One may wonder whether the factorisations of general differences of singular moduli behave similarly. Consider a pair of CM points τ_1 and τ_2 of coprime fundamental discriminants $\Delta_1, \Delta_2 < 0$, and define the integer

$$J_\infty(\tau_1, \tau_2) := \operatorname{Nm}_{\mathbf{Q}}\big(j(\tau_1) - j(\tau_2)\big) \qquad \in \mathbf{Z}.$$

One the first page of Zagier's letter, we find the following table of some of these integers, associated to CM points whose discriminant has class number one.

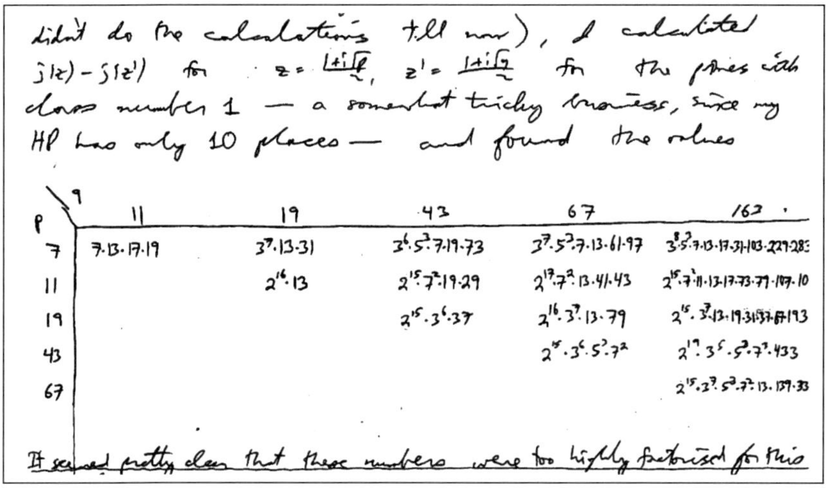

didn't do the calculations till now), I calculated $j(z) - j(z')$ for $z = \frac{1+i\sqrt{p}}{2}$, $z' = \frac{1+i\sqrt{q}}{2}$ for the primes with class number 1 — a somewhat tricky business, since my HP has only 10 places — and found the values

$p \backslash q$	11	19	43	67	163
7	$7\cdot13\cdot17\cdot19$	$3^7\cdot13\cdot31$	$3^6\cdot5^3\cdot7\cdot19\cdot73$	$3^7\cdot5^3\cdot7\cdot13\cdot61\cdot97$	$3^8\cdot5^3\cdot7\cdot13\cdot17\cdot31\cdot103\cdot229\cdot283$
11		$2^{16}\cdot13$	$2^{15}\cdot7^2\cdot19\cdot29$	$2^{17}\cdot7^2\cdot13\cdot41\cdot43$	$2^{15}\cdot7^2\cdot11\cdot13\cdot17\cdot73\cdot79\cdot107\cdot10$
19			$2^{15}\cdot3^6\cdot37$	$2^{16}\cdot3^7\cdot13\cdot79$	$2^{15}\cdot3^7\cdot13\cdot19\cdot31\cdot37\cdot67\cdot193$
43				$2^{15}\cdot3^6\cdot5^3\cdot7^2$	$2^{17}\cdot3^5\cdot5^3\cdot7^3\cdot433$
67					$2^{15}\cdot3^7\cdot5^3\cdot7^2\cdot13\cdot139\cdot33$

It seemed pretty clear that these numbers were too highly factorised for this

FIGURE 1.3. Excerpt from the letter of Zagier [**68**]. Reprinted here with permission.

A crucial observation, scribbled at the bottom, is that the prime divisors q of these integers are small. Zagier also observes congruences modulo $\Delta_1\Delta_2$ satisfied by those primes q, and proceeds to completely determine the q-adic valuation

$$\mathrm{ord}_q\, J_\infty(\tau_1, \tau_2).$$

In his reply, dated 18 February 1983, Gross gives a different determination of this q-adic valuation, using properties of CM elliptic curves. This leads to two independent proofs of the same result, which have the following features:

- The proof in Zagier's latter is analytic, and studies the Fourier coefficients of a family of Hilbert modular forms introduced by Hecke [**43**]. Denote $L = \mathbf{Q}(\sqrt{\Delta_1}, \sqrt{\Delta_2})$ and F for its real quadratic subfield. Write χ for the genus character of F corresponding to L/F. Hecke defines a non-holomorphic Eisenstein family in the variables z_1, z_2 on $\mathfrak{H}_\infty$ by

$$E_s(z_1, z_2) := \sum_{[\mathfrak{a}] \in \mathrm{Cl}_F^+} \chi(\mathfrak{a})\mathrm{N}\mathfrak{m}(\mathfrak{a})^{1+2s} E_s^{\mathfrak{a}}(z_1, z_2),$$

where the non-holomorphic series $E_s^{\mathfrak{a}}(z_1, z_2)$ is defined as follows. Set

$$y_1 := \mathrm{Im}(z_1), \qquad y_2 := \mathrm{Im}(z_2)$$

and denote the algebraic conjugation of $F/\mathbf{Q}$ by $a \mapsto a'$. Define

$$E_s^{\mathfrak{a}}(z_1, z_2) \quad := \sum_{(m,n) \in \mathfrak{a}^2/\mathcal{O}_F^\times} \frac{y_1^s y_2^s}{(mz_1 + n)(m'z_2 + n')|mz_1 + n|^{2s}|m'z_2 + n'|^{2s}}$$

where the sum extends over non-zero pairs (m, n) of elements in $\mathfrak{a}$, modulo the diagonal action of totally positive units in F. This sum converges absolutely for real values of $s > 0$, and transforms under $\mathrm{SL}_2(\mathcal{O}_F)$ like a modular form of parallel weight one. The argument relies on the computation of the Fourier expansion of

(1) its diagonal restriction $E_s(z,z)$ (vanishes at $s=0$)
(2) its analytic first order derivative with respect to s
(3) its holomorphic projection, contained in the space of holomorphic forms of weight two for the full modular group, which is trivial:
$$M_2(\mathrm{SL}_2(\mathbf{Z})) = \{0\}.$$

Through an ingenious direct computation, the argument shows that its first Fourier coefficient can be expressed in the form

$$\log J_\infty(\tau_1,\tau_2) \;-\; \sum_q \mathrm{Int}_q \cdot \log(q)$$

for an explicit character sum Int_q. Since this Fourier coefficient is zero, this gives an explicit description of the integer $J_\infty(\tau_1,\tau_2)$. It is striking that this proof does not use CM theory, and *neither uses nor implies* that the quantities $j(\tau_1) - j(\tau_2)$ (without the norm) are algebraic.

- The proof in the letter of Gross offers a different set of insights, and relies on a study of the pair of CM elliptic curves (E_1, E_2) associated to (τ_1,τ_2). Note that both E_1 and E_2 have potentially good reduction, and since the reduction map induces an injection on endomorphism rings, they can only reduce to the same curve $\overline{E}$ in finite characteristic q if that curve is supersingular, i.e. we get injections
$$\mathcal{O}_{K_1}, \mathcal{O}_{K_2} \;\hookrightarrow\; R := \mathrm{End}(\overline{E}) \subset B_{\infty q}$$
where R is a maximal order in the definite quaternion algebra $B_{\infty q}$ ramified at $\{q,\infty\}$. Gross inverts this procedure and reduces the problem of computing the q-adic valuation
$$\mathrm{ord}_q\, J_\infty(\tau_1,\tau_2)$$
to a counting problem of conjugacy classes of such pairs of embeddings. When unfolded and carefully counted, this yields precisely the same expressions as the ideal sums Int_q appearing in the analytic proof.

These two proofs involve very different ideas. When combined, they amount to the computation of the Néron height pairing $\langle P_1, P_2\rangle$ of the Heegner divisors

$$P_1 = \sum_{\sigma\in \mathrm{Pic}(\mathcal{O}_{K_1})} (\sigma\tau_1) - (\infty), \qquad P_2 = \sum_{\sigma\in \mathrm{Pic}(\mathcal{O}_{K_2})} (\sigma\tau_2) - (\infty)$$

on the modular curve $X(1)$. The contribution at finite primes q is given by the intersection multiplicity Int_q whereas the archimedean contribution is $\log J_\infty(\tau_1,\tau_2)$. The series in $M_2(\mathrm{SL}_2(\mathbf{Z}))$ appearing in Zagier's letter is therefore

$$\mathrm{Proj}_{\mathrm{hol}}\left[\frac{\partial}{\partial s} E_s(z,z)\right]_{s=0} \;=\; \sum_{n\geq 1} \langle P_1, T_n P_2\rangle q^n.$$

Since the modular curve $X_0(1)$ has genus zero the global height is trivial, and this series vanishes. In later work, Gross–Zagier [**40**] and Gross–Kohnen–Zagier [**41**] consider instead the height pairing of Heegner divisors on $X_0(N)$, which does not necessarily vanish, and relate the series to derivatives of L-functions.

1.3. Real quadratic fields. It is natural to wonder what can be said on the results discussed so far for pairs $\Delta_1, \Delta_2 > 0$ of *positive* discriminants. The goal of these notes is to discuss a recent conjectural approach to singular moduli for *real quadratic fields*, based on the notion of rigid meromorphic cocycles.

The question of generalisation to real quadratic, or arbitrary number fields, has been around since Kronecker, and is the objective of Hilbert's 12th problem. To discuss related progress, it is important to clarify exactly what properties of singular moduli one is trying to generalise to other number fields.

If our goal is to find explicit generators for abelian extensions of number fields, an important approach came from the conjectures of Stark [**63**] which predict that leading terms of L-series of Artin representations produce generators via a refinement of the Dirichlet class number formula. Little is known about Stark's conjecture; the case of real quadratic fields remains open. A refinement of the p-adic version of Gross [**37**] was recently proved by Dasgupta–Kakde [**25, 26**], giving an analytic formula for p-units in abelian extensions of totally real fields, and thus a satisfactory answer to the problem of finding *generators* for those fields.

If, on the other hand, we want the rich arithmetic factorisations which gave rise to the results of Gross–Zagier, we need a different approach. The p-units provided by the refinements of Gross–Stark based on L-functions have trivial factorisations at primes $q \neq p$. Purely archimedean approaches have been attempted, and cycle integrals of the j-function were shown to mirror in several key aspects the analytic properties of generating series of singular moduli as the appear in Kudla–Rapoport–Yang [**51**] by Kaneko [**46**] and Duke–Imamoḡlu–Tóth [**28, 29**]. So far, cycle integrals of the j-function have not yielded algebraic numbers that reflect the arithmetic of singular moduli. We mention also a conjectural programme based on non-commutative geometry and C^*-algebras due to Manin [**55**].

1.4. Computational tools. It is remarkable how the study of singular moduli has been directly informed by computation, using tools of ever increasing technological sophistication. The 1889 article of Greenhill [**36**] gives an overview of the fruits of the manual labour of the likes of Abel, Jacobi, Kronecker, Weber, and others. Powerful methods to obtain closed expressions in radicals for singular moduli are discussed at length in Weber's 1908 *Lehrbuch der Algebra* [**66**].

Two decades later, in the 1928 work of Berwick [**1**] we read:

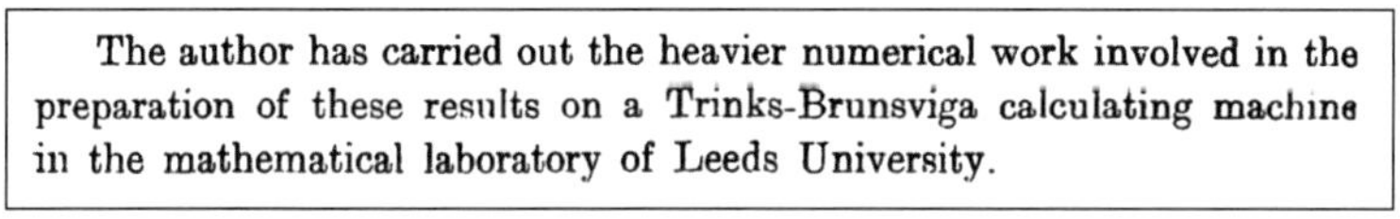
The author has carried out the heavier numerical work involved in the preparation of these results on a Trinks-Brunsviga calculating machine in the mathematical laboratory of Leeds University.

FIGURE 1.4. Berwick's acknowledgement [**1**].

Over half a century later, Zagier acknowledges in his letter an HP with 10 decimal places, see Figure 1.3. Not only does this take advantage of a computing power much superior to that of the Trinks–Brunsviga calculating machine used by Berwick, but also of its greater portability, since Zagier was writing his letter while he was travelling in Japan in early 1983.

Since those days, the degree of mechanisation has transformed beyond recognition. This greatly enhances our ability to continue, as well as expand, the tradition of explicit experimentation established by many generations before us. Perhaps this ability should even be viewed as a duty. We possess, after all, an immense privilege of which early practitioners in CM theory could only dream.

2. Binary quadratic forms

In this section, we recall some classical aspects of the theory of quadratic forms, including reduction theory, with an emphasis on indefinite forms. Most results mentioned here are due to Gauß [**34**].

2.1. Binary quadratic forms. A (binary integral) quadratic form is an element

$$\langle a,b,c\rangle := aX^2 + bXY + cY^2 \quad \in \mathbf{Z}[X,Y].$$

It is called *primitive* if $\gcd(a,b,c) = 1$. There is a right $\mathrm{SL}_2(\mathbf{Z})$-action on $\mathbf{Z}[X,Y]$ by ring automorphisms, defined on generators by

$$\gamma : \begin{cases} X \longmapsto & pX + qY \\ Y \longmapsto & rX + sY \end{cases} \qquad \forall \gamma = \begin{pmatrix} p & q \\ r & s \end{pmatrix} \in \mathrm{SL}_2(\mathbf{Z}).$$

This action preserves the set of quadratic forms, respects primitivity, and preserves the *discriminant* $\Delta = b^2 - 4ac$ of a quadratic form $\langle a,b,c\rangle$. A form of discriminant Δ is called *definite* if $\Delta < 0$ and *indefinite* if $\Delta > 0$. The set of all primitive quadratic forms with fixed discriminant Δ, satisfying $a > 0$ when $\Delta < 0$, is denoted by $\mathcal{F}_\Delta$. To a quadratic form $F = \langle a,b,c\rangle$ in $\mathcal{F}_\Delta$ we associate

$$A_F := \begin{pmatrix} -b & -2c \\ 2a & b \end{pmatrix}, \qquad A_F^2 = \Delta.$$

When $\Delta \neq 0$ the matrix A_F has two eigenlines, with eigenvalues[1] $\sqrt{\Delta}$ and $-\sqrt{\Delta}$ respectively. They define canonical elements r_F and r'_F in $\mathbf{P}^1(\mathbf{C})$, respectively, which we call the *first* and *second* roots of F. When Δ is not a square, we have

$$r_F = \frac{-b + \sqrt{\Delta}}{2a}, \qquad r'_F = \frac{-b - \sqrt{\Delta}}{2a} \qquad \in \mathbf{C}.$$

The group $\mathrm{SL}_2(\mathbf{Z})$ acts on $\mathcal{F}_\Delta$, where the orbits are typically infinite. Any orbit has a convenient visualisation via the *Conway topograph* [**9**, **42**]. It is a tree in $\mathcal{H}_\infty$ together with a labelling of the connected components of its complement. The tree is the inverse image of the closed interval $[0, 1728]$ for the j-function, i.e.

$$\mathcal{T}\mathrm{op} \;:=\; j^{-1}\left([0,1728]\right),$$

[1]Throughout these notes, we fix embeddings of $\overline{\mathbf{Q}}$ into both $\mathbf{C}$ and $\mathbf{C}_p$.

It is the planar embedding of a 3-regular tree, and has a simply transitive action of $\mathrm{SL}_2(\mathbf{Z})$ on the set of *oriented* edges, see Serre [**61**]. It is depicted here.

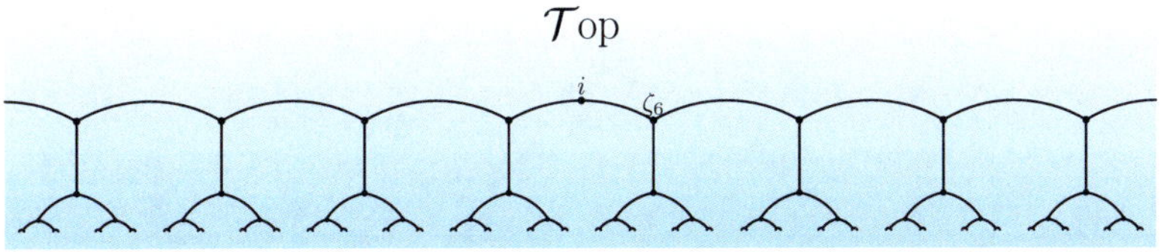

Each connected component C of $\mathcal{H}_\infty \backslash \mathcal{T}\mathrm{op}$ is adjacent to a unique cusp $(u : v) \in \mathbf{P}^1(\mathbf{Q})$. Normalise u, v to be coprime integers, and label the component C by the integer $F(u,v) = au^2 + buv + cv^2$. In particular, the regions adjacent to $\infty = (1:0)$ and $(0:1)$ are labelled by the integers a and c respectively.

EXAMPLE 2.1. Consider the indefinite form $F = \langle 2, -3, -2\rangle$ of discriminant $\Delta = 25 = 5^2$. Since this discriminant is square (such forms are sometimes called *isotropic*), its first and second roots are rational, in this case given by the cusps

$$r_F = (-1:2), \qquad r'_F = (2:1) \in \mathbf{P}^1(\mathbf{Q})$$

The form takes both negative and positive values at the cusps, and the topograph has a finite set of edges (labelled in red in the picture below) separating the corresponding regions according to sign, and connecting the two components labelled with zero, corresponding to the two roots of F. We depict the topograph here.

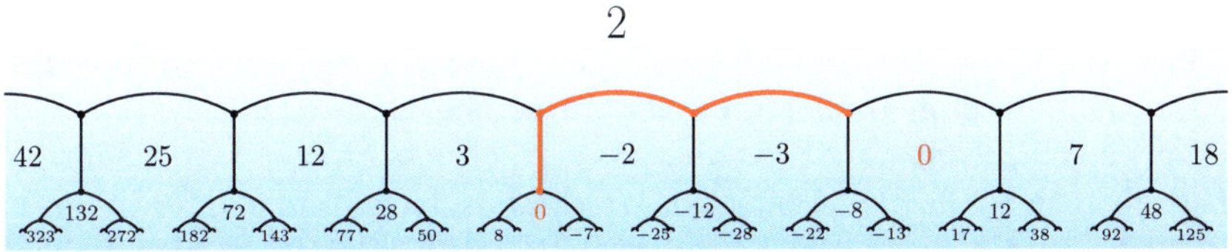

EXAMPLE 2.2. Consider the indefinite form $F = \langle 1, 0, -3\rangle$ of discriminant $\Delta = 12$. Its first and second roots are irrational, in this case given by

$$r_F = \sqrt{3}, \qquad r'_F = -\sqrt{3}$$

The form takes both negative and positive values at the cusps, and the topograph has an infinite set of edges (labelled in red in the picture below) separating the regions according to sign. This infinite path of edges in the topograph of an indefinite form with non-square discriminant is called the *river*.

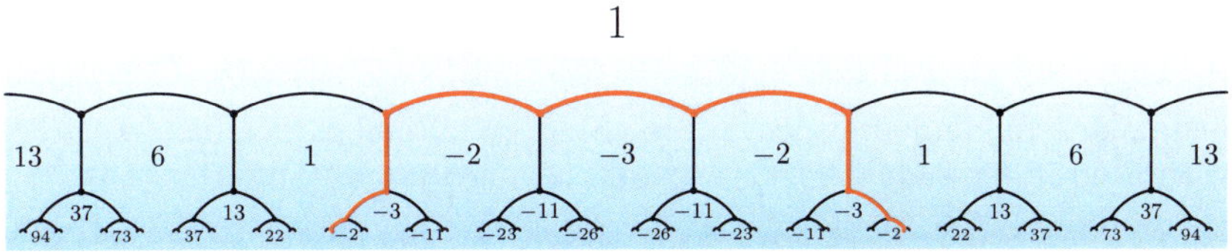

REMARK 2.3. From three pairwise adjacent numbered regions, one easily reconstructs the entire topograph, and the associated quadratic forms, using a very simple rule [**9**, p.9]. Around the oriented edge corresponding to $\langle a, b, c\rangle$ (which one can always transform to the standard edge) the topograph looks like

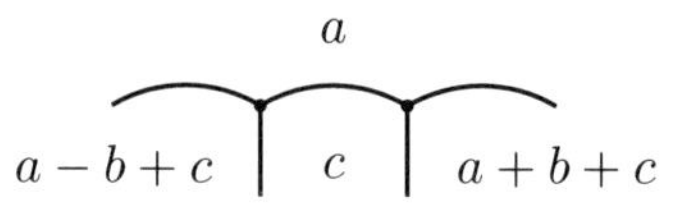

2.2. Class groups. For a fixed discriminant $\Delta \neq 0$, the set of orbits $\mathcal{F}_\Delta/\operatorname{SL}_2(\mathbf{Z})$ is finite and endowed with the structure of an abelian group. More specifically:

- When $\Delta = n^2$ with $n \geq 1$, every $\operatorname{SL}_2(\mathbf{Z})$-orbit contains a unique element of the form $\langle a, n, 0\rangle$ with $0 \leq a < n$ (see exercises) and the map that sends an orbit to the class of a modulo n gives a bijection

$$\begin{array}{ccc} \mathcal{F}_\Delta/\operatorname{SL}_2(\mathbf{Z}) & \xrightarrow{\sim} & (\mathbf{Z}/n\mathbf{Z})^\times \\ \langle a, n, 0\rangle & \longmapsto & [a] \end{array}$$

- When Δ is not a square, there is a bijection between $\mathcal{F}_\Delta/\operatorname{SL}_2(\mathbf{Z})$ and the narrow class group of the quadratic order of discriminant Δ, given by

$$\begin{array}{ccc} \mathcal{F}_\Delta/\operatorname{SL}_2(\mathbf{Z}) & \xrightarrow{\sim} & \operatorname{Pic}^+\left(\mathbf{Z}\left[\frac{\Delta+\sqrt{\Delta}}{2}\right]\right) \\ \langle a, b, c\rangle & \longmapsto & \left[\left(a, \frac{-b+\sqrt{\Delta}}{2}\right)\right] \end{array}$$

In both cases, the set of $\operatorname{SL}_2(\mathbf{Z})$-orbits of primitive forms inherits the structure of a finite abelian group. It should be noted that the composition law on quadratic forms historically predates the introduction of ideal class groups, and when endowed with this operation, the above bijections become isomorphisms.

2.3. Reduction theory. The aim of reduction theory is to identify distinguished elements inside $\operatorname{SL}_2(\mathbf{Z})$-orbits of primitive quadratic forms. The notion is qualitatively different for definite and indefinite forms.

DEFINITION 2.4. Let $F = \langle a, b, c\rangle$ be a quadratic form of discriminant $\Delta \neq 0$.

- When $\Delta < 0$, we say F is *reduced* if

$$|b| \leq a \leq c,$$
with $b \geq 0$ if either equality holds.

- When $\Delta > 0$, we say that F is

nearly reduced if $ac < 0$,
reduced if $ac < 0$ and $b > |a + c|$.

When $\Delta < 0$, a form F is reduced if and only if its first root r_F is contained in the standard fundamental domain for $\operatorname{SL}_2(\mathbf{Z})$. When $\Delta > 0$ non-square, F is nearly reduced if and only if $r_F r'_F < 0$, and reduced if and only if

$$r_F r'_F < 0 \quad \text{and} \quad |r_F| < 1 < |r'_F|.$$

Reducedness for $\Delta > 0$ non-square may therefore be rephrased in terms of the topograph as the property that the *river* passes through i, and flows along one of the vertical edges with $\mathrm{Re}(z) = 1/2$ or $-1/2$, but not both. As an example, we depict here the case of the reduced form $F = \langle 1, -1, -1 \rangle$ of discriminant $\Delta = 5$.

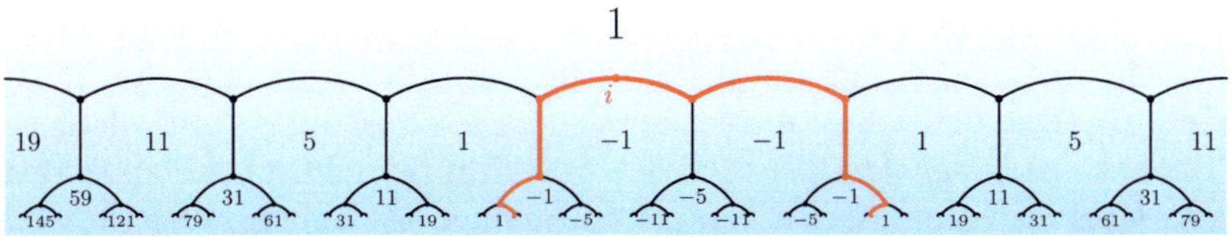

The following classical result is due to to Gauß, and its proof is well-known.

PROPOSITION 2.5. *Suppose Δ is a non-square discriminant. There are finitely many reduced forms of discriminant Δ, and any $\mathrm{SL}_2(\mathbf{Z})$-orbit of $\mathcal{F}_\Delta$ contains at least one.*

The finiteness is easily seen. When $\Delta > 0$ there are clearly finitely many forms with $ac < 0$, and when $\Delta < 0$ a reduced form satisfies $|b| \le a$ and $3a^2 \le |\Delta|$ and therefore there are finitely many options for each of the coefficients.

The proof that every $\mathrm{SL}_2(\mathbf{Z})$-orbit contains at least one reduced form is constructive, and uses an explicit *reduction algorithm*. It proceeds as follows.

(1) Apply the unique power of the translation matrix

$$T = \begin{pmatrix} 1 & 1 \\ 0 & 1 \end{pmatrix} \;:\; \langle a, b, c \rangle \longmapsto \langle a, b + 2a, a + b + c \rangle$$

that yields a quadratic form $\langle a, b, c \rangle$ satisfying

$$\begin{cases} -|a| < b \le |a| & \text{if } \Delta > 0 \text{ and } |a| \ge \sqrt{\Delta}, \text{ or } \;\Delta < 0 \\ \sqrt{\Delta} - 2|a| < b \le \sqrt{\Delta} & \text{if } \Delta > 0 \text{ and } |a| < \sqrt{\Delta}. \end{cases}$$

(2) If the form is reduced, stop. Otherwise, repeat the previous step after applying the matrix

$$S = \begin{pmatrix} 0 & -1 \\ 1 & 0 \end{pmatrix} \;:\; \langle a, b, c \rangle \longmapsto \langle c, -b, a \rangle.$$

Assume now that F is an element of $\mathcal{F}_\Delta$ where $\Delta > 0$ is a non-square discriminant. One can show that this algorithm produces a reduced form after at most

$$\frac{1}{2} \log_2 \left(\frac{|a|}{\sqrt{\Delta}} \right) + 2$$

steps. When continued, it produces all the reduced forms in the $\mathrm{SL}_2(\mathbf{Z})$-orbit. From the powers of the translation T that were applied at each step, one also enumerates all *nearly* reduced forms in the orbit (see exercises). Likewise, from these transformations one computes a generator

$$\gamma_F = \begin{pmatrix} r & s \\ t & u \end{pmatrix} \in \mathrm{SL}_2(\mathbf{Z})$$

of the stabiliser of F in $\mathrm{SL}_2(\mathbf{Z})$. The quantity $\varepsilon := t \cdot r_F + u$ is a fundamental unit of norm 1 in the quadratic order of discriminant Δ. The generator γ_F for which

$\varepsilon > 1$, is called the *automorph* of F. Indefinite reduction theory is discussed in detail in Gauß [**34**], Buell [**6**, Chapter 3], and Buchmann–Vollmer [**5**, Chapter 6].

3. Rational cocycles

To prepare us for the rigid cocycles and RM singular moduli discussed in § 4, we first explore all the structural steps for the simpler *rational cocycles*, which are 1-cocycles for the modular group $\mathrm{SL}_2(\mathbf{Z})$ valued in rational functions on $\mathbf{P}^1(\mathbf{C})$.

There are three fundamental steps in the process of producing arithmetic invariants. First, we consider *additive* cocycles, valued in the additive group of rational functions $\mathbf{C}(z)$. The main actor is a cocycle defined by Knopp [**49, 50**]. We then discuss *multiplicative lifts* of these additive cocycles with respect to the logarithmic derivative

$$\mathrm{dlog} : \mathbf{C}(z)^{\times} \longrightarrow \mathbf{C}(z).$$

Finally, we explain how to *evaluate* the resulting multiplicative cocycle, yielding a systematic supply of well-defined arithmetic invariants.

General notation. Let G be a group and M a left G-module. Denote the action by $\star$. The group of 1-cocycles $Z^1(G, M)$ is the set of maps $\varphi : G \to M$ that satisfy

$$\varphi(\gamma_1\gamma_2) = \varphi(\gamma_1) + \gamma_1 \star \varphi(\gamma_2) \qquad \forall\, \gamma_1, \gamma_2 \in G. \tag{3.1}$$

The subgroup of 1-coboundaries $B^1(G, M)$ consists of all maps $\varphi : G \to M$ for which there exists $g_\varphi \in M$ satisfying

$$\varphi(\gamma) = (1 - \gamma) \star g_\varphi \qquad \forall\, \gamma \in G.$$

The first cohomology group is defined by $\mathrm{H}^1(G, M) := Z^1(G, M)/B^1(G, M)$.

In this chapter we will consider $G = \mathrm{SL}_2(\mathbf{Z})$ acting on $M = \mathbf{C}(z)$ or $\mathbf{C}(z)^{\times}$. This eliminates some technicalities that we encounter for $G = \mathrm{SL}_2\,\mathbf{Z}[1/p]$ and its action on meromorphic functions $M = \mathrm{Mer}_p$ or $\mathrm{Mer}_p^{\times}$ on $\mathcal{H}_p$ considered in § 4.

3.1. Additive cocycles.

Consider $\mathbf{C}(z)$, the additive group of rational functions on the Riemann sphere $\mathbf{P}^1(\mathbf{C})$ with coordinate z. It is a left $\mathrm{GL}_2(\mathbf{C})$-module endowed with the weight two action, defined by

$$\begin{pmatrix} a & b \\ c & d \end{pmatrix} \star f(z) \;:=\; \frac{ad - bc}{(-cz + a)^2} \cdot f\left(\frac{dz - b}{-cz + a}\right) \tag{3.2}$$

Definition 3.3. *Rational cocycles* are elements of $Z^1\big(\mathrm{SL}_2(\mathbf{Z}), \mathbf{C}(z)\big)$. The subgroup of *parabolic* cocycles $Z^1_{\mathrm{par}}\left(\mathrm{SL}_2(\mathbf{Z}), \mathbf{C}(z)\right)$ consists of those 1-cocycles φ that are trivial on the subgroup of translations, i.e. that satisfy

$$\varphi(T) = 0, \qquad \text{where } T = \begin{pmatrix} 1 & 1 \\ 0 & 1 \end{pmatrix}.$$

Finally, we define *rational coboundaries* to be the elements of $B^1\big(\mathrm{SL}_2(\mathbf{Z}), \mathbf{C}(z)\big)$.

We first investigate a toy example of a rational cocycle, which is in fact a cocycle for the larger group $\mathrm{GL}_2(\mathbf{C})$. Its arithmetic interest is very limited, but it gives us opportunity to introduce some useful tools for the richer cocycles to follow.

Toy cocycle

Choose a base point $b = (u : v) \in \mathbf{P}^1(\mathbf{C})$, and define

$$p_b : \begin{array}{ccc} \mathrm{GL}_2(\mathbf{C}) & \longrightarrow & \mathbf{C}(z), \\ \gamma & \longmapsto & L(b) - L(\gamma b), \end{array} \qquad \text{where } L((r:s)) := \frac{s}{sz - r}.$$

We will prove the following claims.

- The map p_b defines a cocycle.
- This rational cocycle only depends on $b \in \mathbf{P}^1(\mathbf{C})$ up to a coboundary.

Whereas both may be checked by a direct calculation, this is neither particularly pleasant, nor enlightening. We use instead the following elementary lemma.

LEMMA 3.4. *Let G be a group endowed with*

- *a left action on a non-empty set X,*
- *a left action on an abelian group M.*

Let $m : X \times X \longrightarrow M$ be a G-equivariant map such that for all $x, y, z \in X$

$$(3.5) \qquad \begin{cases} m(x,x) = 0 & \text{“antisymmetry”} \\ m(x,z) = m(x,y) + m(y,z) & \text{“additivity”} \end{cases}$$

Then for any $x \in X$, we obtain a 1-cocycle $\varphi_x \in Z^1(G, M)$ defined by

$$\varphi_x : \gamma \longmapsto m(x, \gamma x),$$

whose cohomology class $[\varphi_x] \in \mathrm{H}^1(G, M)$ is independent of the choice of $x \in X$.

PROOF. Clearly φ_x is a 1-cocycle, since for every $\gamma_1, \gamma_2 \in G$ we have

$$\begin{aligned} \varphi_x(\gamma_1\gamma_2) &= m(x, \gamma_1\gamma_2 x) \\ &= m(x, \gamma_1 x) + m(\gamma_1 x, \gamma_1\gamma_2 x) \\ &= m(x, \gamma_1 x) + \gamma_1 \star m(x, \gamma_2 x) \\ &= \varphi_x(\gamma_1) + \gamma_1 \star \varphi_x(\gamma_2). \end{aligned}$$

The independence of the cohomology class follows from

$$\varphi_x(\gamma) - \varphi_y(\gamma) = (1 - \gamma) \star m(x, y).$$

□

REMARK 3.6. Note that the proof is essentially the standard passage between homogeneous and non-homogeneous cocycles from homological algebra. The lemma extends formally to higher cocycles $Z^n(G, M)$ from G-equivariant maps $X^n \to M$ satisfying an appropriate homogeneous cocycle condition.

To establish the required claims about the toy example p_b, it suffices to apply the lemma to the function

$$m : \begin{array}{ccc} \mathbf{P}^1(\mathbf{C}) \times \mathbf{P}^1(\mathbf{C}) & \longrightarrow & \mathbf{C}(z) \\ (r, s) & \longmapsto & L(r) - L(s). \end{array}$$

where $G = \mathrm{GL}_2(\mathbf{C})$ acts on $\mathbf{P}^1(\mathbf{C})$ by Möbius transformations. The antisymmetry and additivity (3.5) are clear, so it remains to verify the G-equivariance of m.

LEMMA 3.7. *For any pair of base points $r, s \in \mathbf{P}^1(\mathbf{C})$ we have the relation*

$$\gamma \star (L(r) - L(s)) = L(\gamma r) - L(\gamma s), \qquad \forall\, \gamma \in \mathrm{GL}_2(\mathbf{C}).$$

PROOF. Whereas this lemma may be proved by a direct calculation, it is more pleasantly proved using the space of meromorphic differentials Ω^1 on $\mathbf{P}^1_{\mathbf{C}}$, viewed as a $\mathrm{GL}_2(\mathbf{C})$-module for the action defined by

$$\gamma \star f(z)dz := f(\gamma^\dagger z)d(\gamma^\dagger z), \qquad \text{where } \gamma^\dagger := \det(\gamma)\gamma^{-1}.$$

It is easily seen from the definition that this makes the following map into an isomorphism of $\mathrm{GL}_2(\mathbf{C})$-modules:

$$\mathrm{diff} : \mathbf{C}(z) \longrightarrow \Omega^1 \;;\; f(z) \longmapsto f(z)dz.$$

Note that for any $r, s \in \mathbf{P}^1(\mathbf{C})$ the differential

$$\omega_{r,s} := \mathrm{diff}(L(r) - L(s))$$

is of the third kind (meaning it has only simple poles). Taking residue divisors $\mathrm{div} : \Omega^1 \to \mathrm{Div}^0\, \mathbf{P}^1(\mathbf{C})$, we obtain for any $\gamma \in \mathrm{GL}_2(\mathbf{C})$ that

$$\begin{aligned} \mathrm{div}(\gamma \star \omega_{r,s}) &= (\gamma r) - (\gamma s) \\ \mathrm{div}(\omega_{\gamma r, \gamma s}) &= (\gamma r) - (\gamma s) \end{aligned}$$

Since both $\gamma \star \omega_{r,s}$ and $\omega_{\gamma r, \gamma s}$ are differentials of the third kind, and there are no non-zero holomorphic differentials on $\mathbf{P}^1(\mathbf{C})$, they must be equal, as required. □

More interesting examples of rational cocycles were constructed by Knopp [**49**, **50**]. They are associated to quadratic forms $F \in \mathcal{F}_\Delta$ with $\Delta > 0$. We first make some definitions on geodesics. For any indefinite form Q, we write $\mathrm{geo}(Q)$ to denote the oriented geodesic in the extended Poincaré upper half plane

$$\mathcal{H}^*_\infty := \mathcal{H}_\infty \cup \mathbf{P}^1(\mathbf{Q}),$$

which is an oriented semicircle running from the second root r'_Q to the first root r_Q. Fix a form $F \in \mathcal{F}_\Delta$ with $\Delta > 0$, and define the set

$$X_F := \mathcal{H}^*_\infty \setminus \{\mathrm{geo}(Q) : Q \sim F\},$$

i.e. the complement in the extended Poincaré upper half plane of the (measure zero) $\mathrm{SL}_2(\mathbf{Z})$-orbit of $geo(F)$. The set X_F is uncountably infinite, and contains the rational cusps $\mathbf{P}^1(\mathbf{Q})$ if and only if the discriminant Δ of F is non-square.

To define the Knopp cocycle, we need a notion of intersection numbers of geodesics. Choose $x, y \in X_F$, and define

$$\mathrm{sgn}_Q(x, y) \in \{0, 1, -1\}$$

to be the intersection number of $\mathrm{geo}(Q)$ and the hyperbolic geodesic $\mathrm{geo}(x, y)$ in $\mathcal{H}^*_\infty$ from x to y, taken with respect to a fixed orientation of the plane. The following figure illustrates this for the right handed orientation.

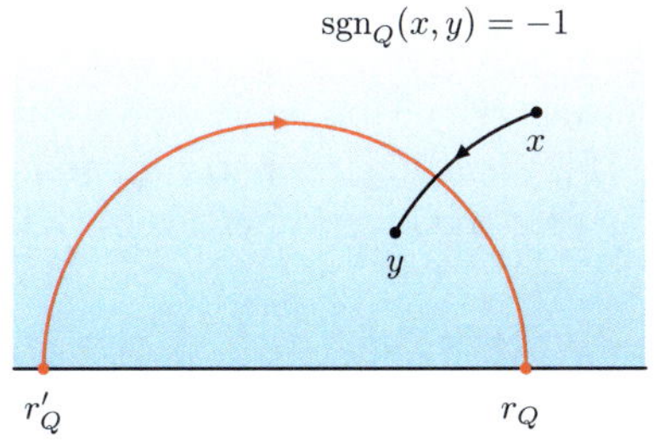

The $\mathrm{SL}_2(\mathbf{Z})$-orbit of F is infinite, but for a fixed $x, y \in X_F$ it contains only finitely many forms Q whose geodesic $\mathrm{geo}(Q)$ intersects $\mathrm{geo}(x, y)$ non-trivially.

LEMMA 3.8. *Let $F \in \mathcal{F}_\Delta$ with $\Delta > 0$, and $x, y \in X_F$. Then*

$$\left|\{Q \sim F \; : \; \mathrm{sgn}_Q(x, y) \neq 0\}\right| < \infty.$$

PROOF. The images of *geo*(F) and $\mathrm{geo}(x, y)$ on the orbifold modular curve

$$X(1) := \mathrm{SL}_2(\mathbf{Z})\backslash\mathcal{H}_\infty^*$$

are compact geodesics, and thus have a finite set S of intersection points. The quotient map from the set of intersection points

$$\{\mathrm{geo}(Q) \cap \mathrm{geo}(x, y) : Q \sim F\}$$

to S is finite to one, since the stabiliser in $\mathrm{SL}_2(\mathbf{Z})$ of any point in $\mathcal{H}_\infty$ is finite. $\square$

The crucial property, on which the entire theory hinges, is the *additivity* of the intersection numbers. Let $x, y, z \in X_F$, then we have (for any $Q \sim F$) that

$$\mathrm{sgn}_Q(x, y) + \mathrm{sgn}_Q(y, z) = \mathrm{sgn}_Q(x, z) \tag{3.9}$$

since the composition of $\mathrm{geo}(x, y)$ and *geo*(y, z) is homotopic to *geo*(x, z).

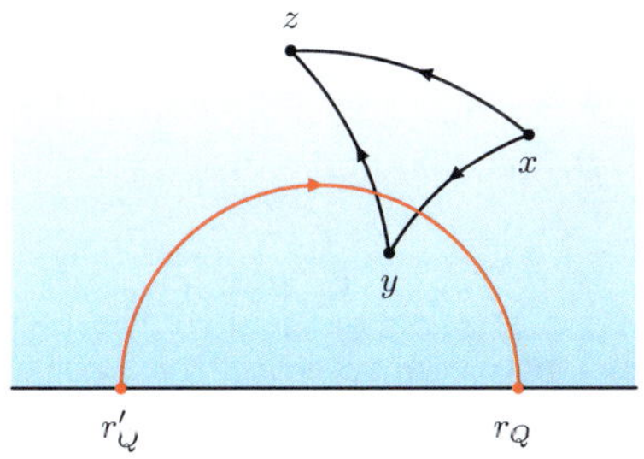

The additivity is crucial, and allows us by Lemma 3.4 to define two cocycles by averaging over the infinite orbit $\{Q \sim F\} = F \cdot \mathrm{SL}_2(\mathbf{Z})$.

- The function

$$\begin{array}{ccc} X_F \times X_F & \longrightarrow & \mathbf{Z} \\ (x, y) & \longmapsto & \displaystyle\sum_{Q \sim F} \mathrm{sgn}_Q(x, y) \end{array}$$

  is clearly antisymmetric and $\mathrm{SL}_2(\mathbf{Z})$-equivariant (where $\mathbf{Z}$ has the trivial action). By (3.9) it is also additive. As a consequence, we obtain for any $b \in X_F$ a homomorphism $\mathrm{SL}_2(\mathbf{Z}) \longrightarrow \mathbf{Z}$ defined by

$$\gamma \longmapsto \sum_{Q \sim F} \mathrm{sgn}_Q(b, \gamma b) = 0 \tag{3.10}$$

which must necessarily be identically zero, since the abelianisation of $\mathrm{SL}_2(\mathbf{Z})$ is finite. This vanishing property for sums of intersection numbers underlies the later convergence properties of rigid cocycles.

- The function

$$\begin{array}{ccc} X_F \times X_F & \longrightarrow & \mathbf{C}(z) \\ (x, y) & \longmapsto & \displaystyle\sum_{Q \sim F} \mathrm{sgn}_Q(x, y) \cdot L(r_Q). \end{array}$$

is clearly antisymmetric. It is $\mathrm{SL}_2(\mathbf{Z})$-invariant by Lemma 3.7, and it is additive by (3.9). We now formally obtain the *Knopp cocycle* from this function by virtue of Lemma 3.4.

Knopp cocycle

Let $F \in \mathcal{F}_\Delta$ with $\Delta > 0$. Choose a base point $b \in X_F$ and define

$$\begin{array}{cccc} \mathrm{kn}_{b,F} : & \mathrm{SL}_2(\mathbf{Z}) & \longrightarrow & \mathbf{C}(z) \\ & \gamma & \longmapsto & \displaystyle\sum_{Q \sim F} \mathrm{sgn}_Q(b, \gamma b) \cdot L(r_Q) \end{array}$$

The Knopp cocycle $\mathrm{kn}_{b,F}$ is a rational cocycle, whose cohomology class

$$[\mathrm{kn}_{b,F}] \in \mathrm{H}^1(\mathrm{SL}_2(\mathbf{Z}), \mathbf{C}(z))$$

is independent of the base point b. It is frequently convenient to choose the base point $b = \infty := (1:0) \in \mathbf{P}^1(\mathbf{C})$, and we know precisely when this is possible by

$$\infty \in X_F \iff \Delta \text{ non-square}.$$

Whenever Δ is not a square, we may therefore choose $b = \infty$. In this case, we will drop the base point from our notation and simply write kn_F for the Knopp cocycle $kn_{\infty,F}$. Note that this cocycle is parabolic, i.e.

$$\mathrm{kn}_F \in Z^1_{\mathrm{par}}\big(\mathrm{SL}_2(\mathbf{Z}), \mathbf{C}(z)\big).$$

The cocycle kn_F can be efficiently computed, in terms of the reduction theory of the quadratic form F discussed in § 2. More precisely, the group $\mathrm{SL}_2(\mathbf{Z})$ is generated by the matrices S and T, where the cocycle has the values

$$\begin{array}{llll} \mathrm{kn}_F(T) & = & 0 & \text{where } T = \left(\begin{smallmatrix} 1 & 1 \\ 0 & 1 \end{smallmatrix}\right) \\ \mathrm{kn}_F(S) & = & \displaystyle\sum_{Q = \langle a,b,c \rangle \in \Sigma_F} \frac{\mathrm{sgn}\,(a)}{z - r_Q}, & \text{where } S = \left(\begin{smallmatrix} 0 & -1 \\ 1 & 0 \end{smallmatrix}\right) \end{array}$$

where Σ_F is the set of *nearly reduced* forms in the orbit of F:

$$\Sigma_F := \big\{ \langle a, b, c \rangle \sim F \ : \ ac < 0 \big\}.$$

The reduction algorithm of § 2.3 computes the set Σ_F efficiently. It is instructive to compute a few non-trivial examples of Knopp cocycles, and to verify the resulting functional equations, as in the following (simplest) example.

EXAMPLE 3.11. Consider the form $F = \langle 1, 1, -1\rangle$ of discriminant $\Delta = 5$. We have

$$\Sigma_F = \{\langle -1, 1, 1\rangle, \langle -1, -1, 1\rangle, \langle 1, -1, -1\rangle, \langle 1, 1, -1\rangle\}$$

so that the value of the Knopp cocycle at S is given by

$$\mathrm{kn}_F(S) = \frac{\sqrt{5}}{z^2 - z - 1} + \frac{\sqrt{5}}{z^2 + z - 1}.$$

The fact that this rational function is the special value at S of a parabolic rational cocycle translates into the following identities

$$\begin{aligned} (1 + S) \star \mathrm{kn}_F(S) &= 0, \\ (1 + (ST) + (ST)^2) \star \mathrm{kn}_F(S) &= 0. \end{aligned}$$

These functional equations impose very restrictive conditions on $\mathrm{kn}_F(S)$. In fact, any rational function satisfying them arises as the value at S of a parabolic rational cocycle, providing a means to classify all parabolic cocycles [**7**] (see exercises).

REMARK 3.12. Let φ be a rational cocycle. A *modular integral* for φ is a holomorphic function $\mathcal{G}$ on $\mathcal{H}_\infty^*$ that satisfies

$$\varphi(\gamma) = \mathcal{G}(z) \parallel (1 - \gamma^{-1}), \qquad \forall \gamma \in \mathrm{SL}_2(\mathbf{Z})$$

where we use the (right) weight two slash action. Note that the space of invariant forms $M_2(\mathrm{SL}_2(\mathbf{Z})) = 0$ is trivial, so modular integrals are unique if they exist.

The modular integral of the toy cocycle p is the Eisenstein series of weight two. Recall that this is the holomorphic function $G_2(z)$ on $\mathcal{H}_\infty^*$ defined by

$$\begin{aligned} G_2(z) &:= \sum_{a \in Z} \sum_{b \in \mathbf{Z}}{}' \frac{1}{(az + b)^2} \\ &= \frac{\pi^2}{3}\Big(1 - 24 \sum_n \sigma_1(n) q^n\Big), \qquad \sigma_1(n) = \sum_{d|n} d, \end{aligned}$$

where $q = \exp(2\pi i z)$. Its transformation law is given by

$$G_2(z) \parallel (1 - \gamma^{-1}) = 2\pi i \cdot p(\gamma)$$

where $p = p_\infty$ is the toy cocycle corresponding to the base point $b = \infty \in \mathbf{P}^1(\mathbf{C})$. This transformation law of $G_2(z)$ is equivalent to Legendre's period relation, and occupies a central place in the arithmetic theory of elliptic curves [**67**].

The modular integral of the Knopp cocycle lies much deeper, and its arithmetic properties were investigated by Duke–Imamoḡlu–Tóth [**28, 30**]. They construct a modular integral $\mathcal{G}_F$ with q-expansion

$$\mathcal{G}_F = \sum_{n \geq 1} \left(\int_{z_0}^{\gamma_F z_0} j_n(z) \frac{dz}{F(z)} \right) q^n, \qquad j_n(q) = q^{-n} + O(q) \in \mathbf{C}(j)$$

where γ_F is the automorph of F. In other words, it is a generating series for the cycle integrals of the j-function, mentioned in the introduction. We have

$$\mathcal{G}_F \parallel (1 - \gamma^{-1}) = \mathrm{kn}_F(\gamma) - kn_{-F}(\gamma).$$

3.2. Multiplicative cocycles. We now consider the *multiplicative* group $\mathbf{C}(z)^\times$ of non-zero rational functions on $\mathbf{P}^1(\mathbf{C})$. It is a left $\mathrm{GL}_2(\mathbf{C})$-module for the weight zero action, defined by

$$\begin{pmatrix} a & b \\ c & d \end{pmatrix} \cdot f(z) \quad := \quad f\left(\frac{dz-b}{-cz+a}\right)$$

The multiplicative module $\mathbf{C}(z)^\times$ is related to the additive module $\mathbf{C}(z)$ by the logarithmic derivative, which is a morphism of $\mathrm{GL}_2(\mathbf{C})$-modules

$$\mathrm{dlog} \ : \ \mathbf{C}(z)^\times \longrightarrow \mathbf{C}(z) \ ; \ f(z) \longmapsto \left(\frac{d}{dz} f(z)\right) \cdot f(z)^{-1}$$

whose kernel is the subgroup of constant functions $\mathbf{C}^\times \subset \mathbf{C}(z)^\times$.

We will now investigate whether the additive cocycles we constructed in § 3.1 can be lifted to multiplicative cocycles under the dlog map. Note that the rational function $L(c)$, for any $c = (u : v) \in \mathbf{P}^1(\mathbf{C})$ is the image of

$$[z - c] := vz - u \in \mathbf{C}(z)^\times/\mathbf{C}^\times$$

under dlog. Therefore both the toy cocycle p_b and the Knopp cocycle $\mathrm{kn}_{b,F}$ are valued in the image of the logarithmic derivative

$$\mathrm{dlog} : \mathbf{C}(z)^\times/\mathbf{C}^\times \hookrightarrow \mathbf{C}(z),$$

and as a consequence they lift formally to multiplicative cocycles modulo scalars. Such a multiplicative lift is easily found explicitly:

$$\left.\begin{array}{lccl} \text{Toy:} & \gamma & \longmapsto & [z - \gamma b] \\ \text{Knopp:} & \gamma & \longmapsto & \displaystyle\prod_{Q \sim F} [z - r_Q]^{\mathrm{sgn}_Q(b, \gamma b)} \end{array}\right\} \in Z^1_{\mathrm{par}}\left(\mathrm{SL}_2(\mathbf{Z}), \frac{\mathbf{C}(z)^\times}{\mathbf{C}^\times}\right)$$

An important question is whether we can resolve the scalar ambiguity, and lift to multiplicative cocycles valued in $\mathbf{C}(z)^\times$ rather than merely $\mathbf{C}(z)^\times/\mathbf{C}^\times$.

The toy cocycle. The multiplicative lift of the toy cocycle p_b can be written down explicitly. We will consider the quotient map

$$\begin{array}{rccl} \pi \ : & \mathbf{A}^2(\mathbf{C}) \setminus \{0\} & \longrightarrow & \mathbf{P}^1(\mathbf{C}) \\ & (u, v) & \longmapsto & (u : v) \end{array}$$

which is a morphism of left $\mathrm{GL}_2(\mathbf{C})$-modules for the natural left action on $\mathbf{A}^2(\mathbf{C})$. By lifting our earlier definitions for the morphism π, we will be able to lift p_b to a multiplicative cocycle for the logarithmic derivative dlog.

LEMMA 3.13. *Consider $X = \mathbf{A}^2(\mathbf{C}) \setminus \{0\}$ and define the map*

$$\begin{array}{rccl} m : & X \times X & \longrightarrow & \mathbf{C}(z)^\times \\ & (r, s), (u, v) & \longmapsto & (r - sz)/(u - vz) \end{array}$$

Then m is antisymmetric, additive, and $\mathrm{GL}_2(\mathbf{C})$-equivariant.

PROOF. It is clear that m is antisymmetric and additive. Standard properties of the determinant imply that m is G-equivariant, since

$$\gamma \star m(x,y) = \frac{\det\left[\begin{pmatrix} r \\ s \end{pmatrix}, \gamma^{-1}\begin{pmatrix} z \\ 1 \end{pmatrix}\right]}{\det\left[\begin{pmatrix} u \\ v \end{pmatrix}, \gamma^{-1}\begin{pmatrix} z \\ 1 \end{pmatrix}\right]} = \frac{\det\left[\gamma\begin{pmatrix} r \\ s \end{pmatrix}, \begin{pmatrix} z \\ 1 \end{pmatrix}\right]}{\det\left[\gamma\begin{pmatrix} u \\ v \end{pmatrix}, \begin{pmatrix} z \\ 1 \end{pmatrix}\right]} = m(\gamma x, \gamma y)$$

□

For any lift $\widetilde{b} \in \mathbf{A}^2 \setminus \{0\}$ of the base point $b \in \mathbf{P}^1(\mathbf{C})$, we may apply Lemma 3.4 to the function m in order to obtain a completely explicit associated cocycle

$$P_b^\times : \mathrm{GL}_2(\mathbf{C}) \longrightarrow \mathbf{C}(z)^\times \tag{3.14}$$

The cocycle $P_b^\times$ is independent of the chosen lift $\widetilde{b}$ of b, and we easily see that

$$\mathrm{dlog}(P_b^\times) = p_b.$$

REMARK 3.15. Note that Lemma 3.13 implies Lemma 3.7 by applying the logarithmic derivative. Arguably, the proof we give here of the (stronger) Lemma 3.13 is also simpler and more elementary than the proof of Lemma 3.7 above.

The Knopp cocycle. Constructing an explicit multiplicative lift of the Knopp cocycle is possible but less straightforward, see [**20**]. We instead take the abstract but more general route, merely proving its existence. Define

$$Z_f^1\left(\mathrm{SL}_2(\mathbf{Z}), \mathbf{C}(z)^\times\right) \subset Z^1\left(\mathrm{SL}_2(\mathbf{Z}), \mathbf{C}(z)^\times\right)$$

to be the subgroup of cocycles which are parabolic *modulo scalars*, i.e. become parabolic after they are composed with the natural projection map

$$\mathbf{C}(z)^\times \longrightarrow \mathbf{C}(z)^\times/\mathbf{C}^\times. \tag{3.16}$$

LEMMA 3.17. *The natural projection map* (3.16) *induces an isomorphism:*

$$12Z_f^1\left(\mathrm{SL}_2(\mathbf{Z}), \mathbf{C}(z)^\times\right) \xrightarrow{\sim} 12Z_{\mathrm{par}}^1\left(\mathrm{SL}_2(\mathbf{Z}), \mathbf{C}(z)^\times/\mathbf{C}^\times\right).$$

PROOF. We first prove a corresponding cohomological statement. The Mayer–Vietoris sequence [**61**] applied to the amalgamated product

$$\mathrm{SL}_2(\mathbf{Z}) = (\mathbf{Z}/4\mathbf{Z}) *_{(\mathbf{Z}/2\mathbf{Z})} (\mathbf{Z}/6\mathbf{Z})$$

gives us, by reduction to the cohomology of finite cyclic groups, that

$$\begin{aligned} \mathrm{H}^1(\mathrm{SL}_2(\mathbf{Z}), \mathbf{C}^\times) &= \mathbf{Z}/12\mathbf{Z} \\ \mathrm{H}^2(\mathrm{SL}_2(\mathbf{Z}), \mathbf{C}^\times) &= 0. \end{aligned}$$

Now consider the short exact sequence of $\mathrm{SL}_2(\mathbf{Z})$-modules

$$1 \longrightarrow \mathbf{C}^\times \longrightarrow \mathbf{C}(z)^\times \longrightarrow \mathbf{C}(z)^\times/\mathbf{C}^\times \longrightarrow 1.$$

From the associated long exact sequence in cohomology, we extract

$$\mathbf{Z}/12\mathbf{Z} \longrightarrow \mathrm{H}^1\left(\mathrm{SL}_2(\mathbf{Z}), \mathbf{C}(z)^\times\right) \longrightarrow \mathrm{H}^1\left(\mathrm{SL}_2(\mathbf{Z}), \mathbf{C}(z)^\times/\mathbf{C}^\times\right) \longrightarrow 0.$$

Note that the following two groups of parabolic coboundaries are trivial:

$$B_f^1\left(\mathrm{SL}_2(\mathbf{Z}), \mathbf{C}(z)^\times\right) = B_{\mathrm{par}}^1\left(\mathrm{SL}_2(\mathbf{Z}), \mathbf{C}(z)^\times/\mathbf{C}^\times\right) = 0$$

To see this, note that a rational function (modulo scalars) whose divisor is invariant under translation must be constant, so that the associated coboundary is trivial.

This means that any parabolic cocycle valued in $\mathbf{C}(z)^\times/\mathbf{C}^\times$ may be lifted to a cocycle in $\mathbf{C}(z)^\times$ whose 12-th power is unique. The statement follows. □

By virtue of Lemma 3.17 we may uniquely lift any parabolic cocycle modulo scalars, after we take the 12-th power (and even without the 12th power, at the cost of uniqueness). In the case of our two running examples, we denote

$$p^\times, \mathrm{kn}_F^\times \in Z_f^1\left(\mathrm{SL}_2(\mathbf{Z}), \mathbf{C}(z)^\times\right)$$

for the unique lifts of 12th power of p and kn_F, respectively. Note that we already constructed $p^\times$ explicitly, it is the 12th power of $P_{(1:0)}^\times$ in (3.14).

3.3. Values of cocycles. Now that we have constructed an infinite supply of multiplicative rational cocycles, it remains to explain how they may be used to produce well-defined invariants. This will be done by assigning *values* to a cocycle, which we associate to indefinite quadratic forms of non-square discriminant.

Suppose we are given a pair (φ, G) where

- $\varphi \in Z^1(\mathrm{SL}_2(\mathbf{Z}), \mathbf{C}(z)^\times)$ a multiplicative rational cocycle,
- $G \in \mathcal{F}_\Delta$ a quadratic form with $\Delta > 0$ non-square,

then we define the *value* of φ at G to be the number

$$\varphi[G] := \varphi(\gamma_G)(r_G) \quad \in \mathbf{P}^1(\mathbf{C})$$

where γ_G is the automorph of the quadratic form G, which is the distinguished free generator for the stabiliser of G in $\mathrm{SL}_2(\mathbf{Z})$ we defined in § 2.3.

4. Singular moduli for real quadratic fields

We have defined an infinite collection of rational cocycles, and discussed a three-step procedure to produce meaningful invariants. To wit;

(1) we started by defining *additive* cocycles in § 3.1,
(2) then, we lifted them to *multiplicative* cocycles in § 3.2,
(3) finally, we *evaluated* the resulting cocycle at a quadratic form in § 3.3.

The invariants we are interested in are p-adic limits of these values. The p-adic limit allows the values to converge to global elements outside the biquadratic field. We perform numerical experiments, and make observations that resemble closely the properties of CM singular moduli discussed in the introduction.

4.1. Rigid cocycles. We begin by explaining how to construct p-adic limits of the Knopp cocycles [**20, 21**]. The definitions are largely the same as those in § 3, and proceed by repeating the three-step procedure while replacing

$$\begin{array}{rrll} & \mathrm{SL}_2(\mathbf{Z}) & \text{acting on} & \mathbf{C}(z)^\times \\ \text{by} & \Gamma := \mathrm{SL}_2(\mathbf{Z}[1/p]) & \text{acting on} & \mathrm{Mer}^\times \end{array}$$

where $\mathrm{Mer}^\times$:=non-zero meromorphic functions on the p-adic half plane $\mathcal{H}_p$. At this point assume some more background from our readers. An introduction to the geometry of the p-adic half plane can be found in [**27**].

Consider the projective line $\mathbf{P}^1$ over $\mathbf{Q}_p$ with homogeneous variables $(z_1 : z_2)$, and for any integer $n \geq 0$ define the affinoid open subset

$$\mathcal{H}_p^{\leq n} := \left\{ (z_1, z_2) \text{ primitive:} \begin{array}{c} |sz_1 + rz_2|_p \geq p^{-n} \\ \forall (r,s) \in \mathbf{Z}^2 \text{ primitive} \end{array} \right\} \subset \mathbf{P}^1(\mathbf{C}_p)$$

FIGURE 4.1. The Bruhat–Tits tree for $p = 5$.

where a pair of numbers in $\mathbf{C}_p$ is called *primitive* if they are integral and at least one of them is a p-adic unit. The affinoid $\mathcal{H}_p^{\leq n}$ is obtained from $\mathbf{P}^1(\mathbf{C}_p)$ by removing open disks of radius p^{-n} around rational points. The increasing union of these affinoids forms an admissible open covering of the p-adic half plane

$$\mathcal{H}_p := \lim_{n\to\infty} \mathcal{H}_p^{\leq n},$$

a rigid analytic space over $\mathbf{Q}_p$. The set of its $\mathbf{C}_p$-points is

$$\mathcal{H}_p(\mathbf{C}_p) = \mathbf{P}^1(\mathbf{C}_p)\backslash\mathbf{P}^1(\mathbf{Q}_p).$$

It is often referred to as "upper" half plane, though $\mathcal{H}_p$ is connected, unlike its archimedean counterpart $\mathbf{P}^1(\mathbf{C})\backslash\mathbf{P}^1(\mathbf{R})$ which is the disjoint union of "upper" and "lower" half planes. The p-adic half plane has a natural reduction map to the *Bruhat–Tits* tree. It is a $(p+1)$-regular tree. The reduction map sends the affinoid $\mathcal{H}_p^{\leq 0}$ to a distinguished vertex v_0, and the affinoid $\mathcal{H}_p^{\leq n}$ to the finite subtree spanned by the vertices of distance at most n to v_0, see [**27**, § 1.3].

DEFINITION 4.2. A meromorphic function on $\mathcal{H}_p$ is the uniform limit, with respect to the supremum norm, of rational functions on each affinoid $\mathcal{H}_p^{\leq n} \subset \mathbf{P}^1(\mathbf{C}_p)$ for $n \geq 0$. The space of meromorphic functions defined over a field extension $L \supset \mathbf{Q}_p$ is denoted by Mer_L. When $L = \mathbf{C}_p$ we simply write $\mathrm{Mer} := \mathrm{Mer}_{\mathbf{C}_p}$.

We endow the additive group $(\mathrm{Mer}, +)$ of meromorphic functions with the left weight two action of $\mathrm{GL}_2(\mathbf{C}_p)$, defined by

$$\begin{pmatrix} a & b \\ c & d \end{pmatrix} \star f(z) \;:=\; \frac{ad-bc}{(-cz+a)^2} \cdot f\left(\frac{dz-b}{-cz+a}\right)$$

whereas the multiplicative group $\mathrm{Mer}^\times$ of non-zero meromorphic functions is a left $\mathrm{GL}_2(\mathbf{C}_p)$-module, with respect to the weight zero action

$$\begin{pmatrix} a & b \\ c & d \end{pmatrix} \cdot f(z) \;:=\; f\left(\frac{dz-b}{-cz+a}\right)$$

so that the logarithmic derivative provides a morphism of left $\mathrm{GL}_2(\mathbf{C}_p)$-modules

$$\begin{array}{rccc} \mathrm{dlog}: & \mathrm{Mer}^\times & \longrightarrow & \mathrm{Mer} \\ & f & \longmapsto & f'/f \end{array}$$

REMARK 4.3. Additive cocycles are only considered for the weight two action here. We refer to Negrini [**56**] for a systematic treatment of higher weight additive cocycles, which admit an analogue of the Shimura–Shintani correspondence.

Arithmetic phenomena transpire after restricting to the *Ihara group*

$$\Gamma := \mathrm{SL}_2(\mathbf{Z}[1/p]) \leq \mathrm{GL}_2(\mathbf{C}_p).$$

It is a discrete subgroup of $\mathrm{SL}_2(\mathbf{R}) \times \mathrm{SL}_2(\mathbf{Q}_p)$ which is dense in each factor. Analogies of Γ with a Hilbert modular group were studied by Ihara [**44, 45**]. As before, we consider cocycles modulo scalars, and attempt to lift the scalar ambiguity. This is more involved than it was for the group $\mathrm{SL}_2(\mathbf{Z})$. Define the groups of

$$\begin{array}{ccc} \text{rigid cocycles} & := & Z^1(\Gamma, \mathrm{Mer}^\times) \\ \cap & & \cap \\ \text{theta cocycles} & := & Z^1(\Gamma, \mathrm{Mer}^\times/\mathbf{C}_p^\times). \end{array}$$

For $\mathrm{SL}_2(\mathbf{Z})$ the space of lifting obstructions $\mathrm{H}^2(\mathrm{SL}_2(\mathbf{Z}), \mathbf{C}^\times) = 0$ is trivial. The Ihara group Γ is cohomologically richer; in general a theta cocycle does not lift to a rigid cocycle. Its action on the Bruhat–Tits tree gives [**61**, II.1.4]

$$\Gamma \simeq \mathrm{SL}_2(\mathbf{Z}) *_{\Gamma_0(p)} \mathrm{SL}_2(\mathbf{Z})$$

so that we obtain from Mayer–Vietoris an exact sequence

$$(\mathbf{Z}/12\mathbf{Z})^2 \longrightarrow \mathrm{H}^1(\Gamma_0(p), \mathbf{C}_p^\times) \longrightarrow \mathrm{H}^2(\Gamma, \mathbf{C}_p^\times) \longrightarrow 0$$

Since $\mathrm{H}^1(\Gamma_0(p), \mathbf{Z})$ has rank $2g+1$, where g is the genus of $X_0(p)$, we see that the lifting obstruction lies in a $\mathbf{C}_p^\times$-torus of the same rank. This presents issues in defining the RM values of a theta cocycle, which we circumvent as follows.

Let G be a quadratic form of positive discriminant for which p is inert in the associated quadratic order. Its stabiliser in the Ihara group Γ is of rank one

$$\mathrm{Stab}_\Gamma(G) = \{\pm 1\} \times \gamma_G^{\mathbf{Z}} \;\leq\; \mathrm{SL}_2(\mathbf{Z}) \;\leq\; \Gamma$$

Let φ be any theta cocycle, then we may lift its *restriction to* $\mathrm{SL}_2(\mathbf{Z})$ to a cocycle $\widetilde{\varphi} \in Z^1\left(\mathrm{SL}_2(\mathbf{Z}), \mathrm{Mer}^\times\right)$. The RM value of φ at G is the well-defined quantity

$$\varphi[G] := \widetilde{\varphi}^{12}(\gamma_G)(r_G) \in \mathbf{P}^1(\mathbf{C}_p).$$

REMARK 4.4. We note that RM values may be defined more generally at any indefinite form G discriminant $\Delta > 0$ such that p is non-split, see [**22**]. When we numerically compute RM values, as a rule we omit the 12th power appearing in the definition to assure well-definedness. The reason is that algebraic recognition routines based on LLL [**54**] are more likely to succeed when the height of the algebraic number is small, making the 12th power undesirable.

4.2. The p-adic Knopp cocycle. We already know one example of a rigid cocycle: the construction of the toy cocycle goes through without any changes. Its RM values are algebraic, but of limited interest (see exercises). To construct more interesting theta cocycles, we adapt the Knopp cocycle in § 3.

Fix a quadratic form $F \in \mathcal{F}_\Delta$ with $\Delta > 0$, and a base point $b \in X_F$.

CASE 1. Assume first that p is **nonsplit** in the quadratic algebra of F, i.e. that the Kronecker symbol (Δ/p) is not one. Define the *additive* cocycle Θ_F by

$$\Theta_F : \gamma \longmapsto \sum_{Q \in F \cdot \Gamma} \mathrm{sgn}_Q(b, \gamma b) L(r_Q), \qquad \Theta_F \in Z^1(\Gamma, \mathrm{Mer}).$$

Note that the only difference with the definition of the Knopp cocycle is in the index set; this time Q runs over all quadratic forms in the Γ-orbit of F. As before, we then lift the additive cocycle to a *multiplicative* theta cocycle

$$\Theta_F^\times : \gamma \longmapsto \prod_{Q \in F \cdot \Gamma} [z - r_Q]^{\mathrm{sgn}_Q(b,\gamma b)}, \qquad \Theta_F^\times \in Z^1(\Gamma, \mathrm{Mer}^\times/\mathbf{C}_p^\times). \tag{4.5}$$

The cohomology class is independent of the choice of base point, and since p is nonsplit, the cusp ∞ is contained in X_F and the cohomology class is parabolic. We may represent it uniquely by a parabolic cocycle, valued in $\mathrm{Mer}^\times/\mathbf{C}_p^\times$.

REMARK 4.6. When $\mathrm{SL}_2(\mathbf{Z})$-equivalence is replaced by Γ-equivalence, the statement of Lemma 3.8 is false, and the sums and products occurring in the above definitions are *infinite*. The concomitant convergence issues, which we have kept from the reader, are slightly subtle, see [**20**]. We content ourselves here by pointing out the two main ingredients for this convergence:

- The divisor of Θ_F and $\Theta_F^\times$ is discrete, Lemma 3.8 can be used to show that the index set is naturally filtered by *finite* subsets

$$\mathcal{S}_n := \left\{Q \in F \cdot \Gamma \ : \ \mathrm{sgn}_Q(b, \gamma b) \neq 0, \ \ r_Q \in \mathcal{H}_p^{\leq n}\right\}. \tag{4.7}$$

- When restricted to the affinoids $\mathcal{H}_p^{\leq n}$, the divisors of Θ_F and $\Theta_F^\times$ are of degree zero, since the vanishing of (3.10) can be used to show that

$$\sum_{Q \in \mathcal{S}_n} \mathrm{sgn}_Q(b, \gamma b) = 0. \tag{4.8}$$

Convergence is easier when one makes a symmetrisation $\Theta_{F,\mathrm{sym}}$ of the cocycle, as we do when p splits. It is similar to the symmetrised (rational) cocycles of Duke–Imamoḡlu–Tóth [**30**], where the corresponding symmetrisation of the knot assures that it is null-homologous. We mention the work of Simon [**62**] and Rickards [**60**], who determine the linking numbers without symmetrisation.

CASE 2. When p is **split** in the quadratic algebra of F, i.e. when $(\Delta/p) = 1$, we define only the *symmetrised* additive cocycle $\Theta_{F,\mathrm{sym}} \in Z^1(\Gamma, \mathrm{Mer})$ by

$$\Theta_{F,\mathrm{sym}} : \gamma \longmapsto \sum_{Q \in F \cdot \Gamma} \mathrm{sgn}_Q(b, \gamma b) \left(L(r_Q) - L(r'_Q)\right),$$

The convergence of this sum is manifest, since the terms uniformly converge to zero on the affinoids $\mathcal{H}_p^{\leq n}$. We lift the symmetrised additive cocycle to a symmetrised multiplicative theta cocycle $\Theta_{F,\mathrm{sym}}^\times \in Z^1(\Gamma, \mathrm{Mer}^\times/\mathbf{C}_p^\times)$ by

$$\Theta_{F,\mathrm{sym}}^\times : \gamma \longmapsto \prod_{Q \in F \cdot \Gamma} \left(\frac{[z - r_Q]}{[z - r'_Q]}\right)^{\mathrm{sgn}_Q(b,\gamma b)}.$$

The cohomology class is independent of the choice of base point, but it is not necessarily parabolic. When Δ is not a square, the cusp ∞ is contained in X_F and there is a unique parabolic representative. When Δ is a square this is false, e.g. for the *winding cocycle* appearing below, corresponding to the case $\Delta = 1$.

The RM values of the p-adic Knopp cocycles associated to a p-adically non-split form F (Case 1) are of primary interest. Henceforth we write

$$\Theta_p^{\times}[F,G] := \Theta_F^{\times}[G] \qquad \in \mathbf{P}^1(\mathbf{Q}_{p^2}),$$

where $\mathbf{Q}_{p^2}$ is the unramified quadratic extension of $\mathbf{Q}_p$, which contains all the roots of the forms in the Γ-orbits of F and G. These RM values satisfy:

- The invariant is (multiplicatively) anti-symmetric, in the sense that
$$\Theta_p^{\times}[F,G] = \Theta_p^{\times}[G,F]^{-1}.$$
- The invariant only depends on the $\mathrm{SL}_2(\mathbf{Z})$-orbits of the forms F and G, giving us for a fixed pair of discriminants $\Delta_1, \Delta_2 > 0$ with
$$\left(\frac{\Delta_1}{p}\right) = \left(\frac{\Delta_2}{p}\right) = -1$$
a finite collection
$$\left\{ \Theta_p^{\times}[F,G] \;:\; \begin{array}{rcl} \mathrm{disc}(F) & = & \Delta_1 \\ \mathrm{disc}(G) & = & \Delta_2 \end{array} \right\} \subset \mathbf{P}^1(\mathbf{C}_p)$$
canonically indexed by $\mathrm{Cl}_1^+ \times \mathrm{Cl}_2^+$, the product of (narrow) class groups of the quadratic orders of discriminants Δ_1 and Δ_2 respectively.
- We may define an involution
$$w_\infty : \langle a,b,c\rangle \longmapsto \langle -a,b,-c\rangle$$
on the set of quadratic forms of any fixed discriminants. This involution changes the first root r_Q to minus the second root $-r_Q'$. It reflects the geodesic $\mathrm{geo}(Q)$ along the y-axis and negates intersection numbers. Therefore, the effect of this involution on the RM values is
$$\Theta_p^{\times}[w_\infty F, w_\infty G] = \Theta_p^{\times}[F,G]^{-1}$$

4.3. Computational observations. The explicit nature of the theta cocycles constructed in § 4.2 makes it possible to experiment with examples. In early 2017, we computed a first example of an RM value of a p-adic Knopp cocycle. We chose the prime $p = 3$ and the indefinite quadratic forms

$$\begin{array}{rcll} F & = & \langle 1,1,-1\rangle & \text{discriminant } \Delta_1 = 5, \\ G & = & \langle 1,8,-4\rangle & \text{discriminant } \Delta_2 = 80. \end{array}$$

The RM value of the 3-adic rigid cocycle associated to F at G, was found to satisfy

$$\Theta_3^{\times}[F,G] \equiv \frac{24\sqrt{-1}-7}{25} \pmod{3^{200}}.$$

The appearance of an algebraic number was encouraging. What seemed striking is that it generates the narrow ring class field of conductor 4 of $\mathbf{Q}(\sqrt{5})$, and its factorisation involves primes above $5 \neq 3$. This motivated a systematic computational exploration, to understand the arithmetic of these invariants.

The enormous benefit of historical hindsight on experimentation with CM singular moduli (discussed in the introduction) facilitates informed guesses for phenomena that we might expect. The reader is encouraged to use the algorithms [**19**] to make their own observations from experiments.

REMARK 4.9. The infinite product $\Theta_F^\times$ may be computed directly up to any desired precision $O(p^n)$. However, the size of the set $\mathcal{S}_n$ in (4.7) grows exponentially in n. An algorithm running polynomially in n is described in [**20**], which computes the product (4.5) restricted to $\mathcal{S}_n$ iteratively in n, through a recursion formula. The price we pay for this polynomial running time, is that the output is only the restriction of $\Theta_F^\times$ to the standard affinoid $\mathcal{H}_p^{\leq 0}$, and that its RM values are only correct modulo powers of the fundamental unit ε_F in the quadratic order of F.

EXAMPLE 4.10. Let $p = 2$ and choose the quadratic forms

$$\begin{array}{rcll} F & = & \langle 1,1,-1\rangle & \text{discriminant } \Delta_1 = 5, \\ G & = & \langle 1,3,-3\rangle \text{ and } \langle -1,3,3\rangle & \text{discriminant } \Delta_2 = 21. \end{array}$$

The class group of quadratic forms of discriminant 21 is isomorphic to $\mathbf{Z}/2\mathbf{Z}$, and the two choices of G are representatives of the two classes. Using the polynomial time algorithm for computing RM invariants [**20**] we compute the quantity

$$\Theta_2^\times[F,G] \pmod{2^{300}}.$$

We suspect this is an algebraic number, and use the LLL agorithm [**54**] to attempt to find an algebraic relation between small powers of this number, see [**8**, § 2.7] for more on this method. The algorithm returns the quartic polynomial

$$91x^4 + 112x^3 + 123x^2 + 112x + 91. \tag{4.11}$$

The two choices for G yield different 2-adic numbers $\Theta_2^\times[F,G]$ computed to precision $O(2^{300})$, but the algebraic recognition nonetheless yields the same polynomial (4.11). The polynomial (4.11) has 4 roots, and further experimentation reveals that the two "missing" roots are accounted for by the values

$$\Theta_2^\times[w_2F,G] = \Theta_2^\times[F,w_2G], \qquad \text{where } \ w_p : \langle a,b,c\rangle \mapsto \langle a,pb,p^2c\rangle.$$

These invariants can also be computed using the algorithms of [**20**], and appear to give a full set of Galois conjugates over $\mathbf{Q}$. The splitting field of (4.11) is

$$\mathbf{Q}(\sqrt{-3},\sqrt{-35}).$$

Its roots are generators for H_1H_2 over K_1K_2; the compositum of the narrow Hilbert class fields of $K_1 = \mathbf{Q}(\sqrt{5})$ and $K_2 = \mathbf{Q}(\sqrt{21})$, which are given by, respectively,

$$H_1 = \mathbf{Q}(\sqrt{5}), \qquad H_2 = \mathbf{Q}(\sqrt{-3},\sqrt{-7}).$$

EXAMPLE 4.12. We go a little bit deeper into the same theme as the previous example, and explore the arithmetic of the algebraic numbers we find on a slightly larger (and richer) example. This time, we let $p = 3$ and choose quadratic forms

$$\begin{array}{ll} F = \langle 1,-1,-1\rangle & \text{discriminant } \Delta_1 = 5, \\ G = \langle 1,6,-2\rangle & \text{discriminant } \Delta_2 = 44. \end{array}$$

Using LLL, we find that $\Theta_3^\times[F,G] \pmod{3^{200}}$ is the root of the polynomial

$$48841x^8 + 115280x^6 + 164562x^4 + 115280x^2 + 48841.$$

Once again, $\Theta_3^\times[F,G]$ is found to be a generator of H_1H_2 over K_1K_2, the splitting field of this polynomial is the triquadratic field

$$K = \mathbf{Q}(\sqrt{5}, \sqrt{11}, \sqrt{-1}).$$

Now let us look at the arithmetic properties of this generator. Its prime factorisation is concentrated at primes dividing the constant term

$$48841 = 13^2 \cdot 17^2.$$

We note that both these primes are inert in both K_1 and K_2. This observation is strikingly similar to what was noticed by Berwick, Gross, and Zagier, and encourages us to tabulate the positive integers of the form

$$N_x := \frac{\Delta_1\Delta_2 - x^2}{4}$$

for which we find

x	N_x	x	N_x	x	N_x
0	$5 \cdot 11$	6	$2 \cdot 23$	12	19
2	$2 \cdot 3^3$	8	$3 \cdot \mathbf{13}$	14	$2 \cdot 3$
4	$3 \cdot \mathbf{17}$	10	$2 \cdot 3 \cdot 5$		

This reflects what Gross and Zagier observed in their experiments. We may further sharpen it by noticing the special role of the prime $p = 3$. Indeed, it seems that in fact $pq \mid N_x$ for all the primes q that divide our invariant, i.e. the factorisation is concentrated above primes dividing a positive integer of the form

$$\frac{\Delta_1\Delta_2 - x^2}{4p}. \tag{4.13}$$

REMARK 4.14. The reader may have wondered why the RM invariants we find experimentally are never algebraic *integers*. After all, this was true for the quantities $j(\tau_1) - j(\tau_2)$ studied by Gross–Zagier. One explanation is that complex conjugation plays a different role in RM theory, acting on these invariants via

$$\overline{\Theta_p^\times[F,G]} = \Theta_p^\times[w_\infty F, w_\infty G] = \Theta_p^\times[F,G]^{-1}.$$

This explains why all the polynomials we find appear to be *palindromic.*

EXAMPLE 4.15. The previous example shows that the prime p is reflected in the arithmetic of our RM invariants. The setup of Gross–Zagier corresponds to the choice $p = \infty$, and the fact that we are able to vary the prime p begs the question what the corresponding variation in the RM invariants looks like. To investigate this, we compute $\Theta_p^\times[F,G]$ in a few instances, where we fix the quadratic form

$$F = \langle 1, 3, -1\rangle \qquad \text{discriminant} \quad \Delta_1 = 13.$$

We now take forms G of discriminant $\Delta_2 > 0$ and compute each time the p-adic invariant for two different choices of p, where p is chosen to be inert with respect to both discriminants Δ_1 and Δ_2.

For G of discriminant $\Delta_2 = 12$ we find the following values of $\Theta_p^\times[F,G]$:

$p=5$	$p=7$
$\frac{1\pm 4\sqrt{-3}}{7}$	$\frac{3\pm 4\sqrt{-1}}{5}$

For G of discriminant $\Delta_2 = 45$ we find the following values of $\Theta_p^{\times}[F,G]$:

$p=2$	$p=7$
$\frac{150824917\pm 100674475\sqrt{-3}}{2\cdot 7^2\cdot 13\cdot 37\cdot 67\cdot 73}$	$\frac{1\pm\sqrt{-15}}{2^2}$

For G of discriminant $\Delta_2 = 108$ we find the following values of $\Theta_p^{\times}[F,G]$:

$p=5$	$p=7$
$\frac{1237487\pm 857860\sqrt{-3}}{7^2\cdot 19\cdot 31\cdot 67}$	$\frac{128\pm 2046\sqrt{-1}}{2\cdot 5^2\cdot 41}$

A few observations are in order here:

- It appears that the factorisations are more rich for smaller values of p. This is in line with our previous observations; for a fixed Δ_1 and Δ_2, there are fewer positive integers of the form (4.13) when p is larger.
- When a prime q divides an integer of the form $(\Delta_1\Delta_2 - x^2)/4p$, then the same is obviously true with the roles of p and q reversed. Inspecting the above examples, we see that there even appears to be a relation between

$$\mathrm{ord}_{\mathfrak{p}}\,\Theta_q^{\times}[F,G] \;\leftrightarrow\; \mathrm{ord}_{\mathfrak{q}}\,\Theta_p^{\times}[F,G].$$

To better understand the prime factorisations of the RM invariants $\Theta_p^{\times}[F,G]$, we compute a larger example that involves rich factorisations with large exponents. This example will also provide algebraic invariants that do not live in genus fields, i.e. whose splitting fields are non-abelian over $\mathbf{Q}$.

EXAMPLE 4.16. We now investigate the set of RM invariants for

$$(\Delta_1, \Delta_2) = (13, 621).$$

We expect rich factorisations by inspecting the positive integers of the form

$$N_x = \frac{\Delta_1\Delta_2 - x^2}{4}$$

which we tabulate here for future reference:

x	N_x	x	N_x	x	N_x	x	N_x	x	N_x
1	$2 \cdot 1009$	19	$2^3 \cdot 241$	37	$2^2 \cdot 419$	55	$2 \cdot 631$	73	$2 \cdot 7^3$
3	$2^5 \cdot 3^2 \cdot 7$	21	$2^2 \cdot 3^2 \cdot 53$	39	$2 \cdot 3^2 \cdot 7 \cdot 13$	57	$2 \cdot 3^2 \cdot 67$	75	$2^2 \cdot 3^2 \cdot 17$
5	$2^2 \cdot 503$	23	$2 \cdot 23 \cdot 41$	41	$2 \cdot 17 \cdot 47$	59	$2^2 \cdot 7 \cdot 41$	77	$2^3 \cdot 67$
7	$2 \cdot 17 \cdot 59$	25	$2 \cdot 7^2 \cdot 19$	43	$2^2 \cdot 389$	61	$2^6 \cdot 17$	79	$2 \cdot 229$
9	$2 \cdot 3^3 \cdot 37$	27	$2^2 \cdot 3^3 \cdot 17$	45	$2^3 \cdot 3^3 \cdot 7$	63	$2 \cdot 3^3 \cdot 19$	81	$2 \cdot 3^3 \cdot 7$
11	$2^2 \cdot 7 \cdot 71$	29	$2^4 \cdot 113$	47	$2 \cdot 733$	65	$2 \cdot 13 \cdot 37$	83	$2^3 \cdot 37$
13	$2^3 \cdot 13 \cdot 19$	31	$2 \cdot 7 \cdot 127$	49	$2 \cdot 709$	67	$2^7 \cdot 7$	85	$2^2 \cdot 53$
15	$2 \cdot 3^2 \cdot 109$	33	$2 \cdot 3^2 \cdot 97$	51	$2^3 \cdot 3^2 \cdot 19$	69	$2^2 \cdot 3^2 \cdot 23$	87	$2 \cdot 3^2 \cdot 7$
17	$2 \cdot 7 \cdot 139$	35	$2^4 \cdot 107$	53	$2^2 \cdot 7 \cdot 47$	71	$2 \cdot 379$	89	$2 \cdot 19$

Note that the narrow class group $\mathrm{Cl}_1^+ = 1$ of discriminant $\Delta_1 = 13$ is trivial, whereas the narrow class group $\mathrm{Cl}_2^+ \simeq \mathbf{Z}/6\mathbf{Z}$ of discriminant $\Delta_2 = 621$ is cyclic of order 6. We therefore produce, for every choice of prime p that is non-split for both discriminants, a total of twelve p-adic RM invariants

$$\left\{\Theta_p^\times[F,G]\right\} \cup \left\{\Theta_p^\times[w_pF,G]\right\}.$$

We wish to recognise them as algebraic numbers. This is a challenging task; our previous observations and the entries in the above table cause us to expect algebraic numbers of very large height, particularly when $p = 2$. We noticed that the heights are more manageable for the symmetrised invariants

$$\frac{\Theta_p^\times[w_pF,G]}{\Theta_p^\times[F,G]}.$$

We will now compute these six symmetrised invariants for a variety of small primes p that are inert for both discriminants $(\Delta_1, \Delta_2) = (13, 621)$. The prime $p = 2$ presents special difficulties, since it produces algebraic numbers of immense height. Computing to precision $O(2^{1000})$, we recognised the 2-adic symmetrised invariants to be roots of the following degree six polynomial:

$$\begin{aligned}
&532662811974216268987046368230622959690070361192975999934916\ x^6\\
-&278367526244451072555505377961835322613068104302177742390746\ x^5\\
-&292977016274297008338188853638915462702409980987593334148135\ x^4\\
+&879582695503881002603098558912072457112885628056565606298805\ x^3\\
-&292977016274297008338188853638915462702409980987593334148135\ x^2\\
-&278367526244451072555505377961835322613068104302177742390746\ x\\
+&532662811974216268987046368230622959690070361192975999934916
\end{aligned}$$

The discriminant of the number field defined by this polynomial is $3^7 \cdot 23^2$, and its Galois closure is the narrow ring class field H_2 of discriminant 621; it is a dihedral extension of $\mathbf{Q}$ of degree 12. The factorisation of the constant term is

$$2^2 \cdot 7^7 \cdot 19^4 \cdot 37^5 \cdot 47^2 \cdot 59^2 \cdot 67^3 \cdot 97^2 \cdot 109^2 \cdot 229 \cdot 241 \cdot 379 \cdot 631 \cdot 709 \cdot 733 \cdot 1009.$$

For different choices of primes p, the reader may prefer to try to predict some of its factorisations based on the above table of integers N_x and then compare with the LLL recognition, which yield the following algebraic polynomials:

p	Minimal polynomial	Factorisation
7	$4378144x^6 + 5762700x^5 + 9490680x^4 + 11616641x^3$ $+9490680x^2 + 5762700x + 4378144$	$2^5 \cdot 41 \cdot 47 \cdot 71$
19	$64x^6 + 72x^5 + 207x^4 + 142x^3 + 207x^2 + 72x + 64$	2^6
41	$7x^6 + 6x^5 + 6x^4 + 10x^3 + 6x^2 + 6x + 7$	7
47	$28x^6 + 54x^5 + 39x^4 + 14x^3 + 39x^2 + 54x + 28$	$2^2 \cdot 7$
59	$4x^6 + 3x^4 + 2x^3 + 3x^2 + 4$	2^2
71	$7x^6 + 6x^5 + 6x^4 + 10x^3 + 6x^2 + 6x + 7$	7

We invite the reader to reflect on these numerical examples, and to compare to previously made observations. Several things may be noticed that confirm our earlier observations, and several more can be made. The next section contains an overview of the main conjectures in [**20**] that were informed by computations of this sort. The amount of rich arithmetic appearing in this data does not exclude the possibility that an industrious reader could expand these observations to include some that eluded the authors of [**20**]. The theme of this PCMI Summer School is "number theory informed by computation", and is perhaps best enjoyed by actively taking part in the act of being informed by computation; be it by the data included here, or by new experiments using the algorithms [**19**].

4.4. Main conjectures. We now state the main conjectures of [**20**], which were informed by multitudes of examples of the sort we discussed above. We do not attempt to be maximally general in the statements, in favour of simplicity. Let $\Delta_1, \Delta_2 > 0$ be coprime discriminants and p a prime such that

$$\left(\frac{\Delta_1}{p}\right) = \left(\frac{\Delta_2}{p}\right) = -1.$$

Consider the real quadratic fields

$$K_1 = \mathbf{Q}(\sqrt{\Delta_1}), \qquad K_2 = \mathbf{Q}(\sqrt{\Delta_2}),$$

whose compositum L is a real biquadratic field. Let H_1 and H_2 be the narrow ring class fields of discriminants Δ_1 and Δ_2 respectively. The Galois groups $G_i := \mathrm{Gal}(H_i/\mathbf{Q})$ are generalised dihedral, canonically split by Frobenius at p, and

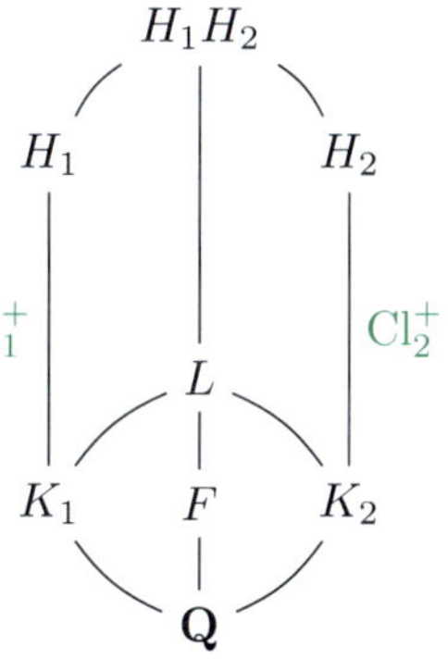

$$G_1 \simeq \mathrm{Cl}_1^+ \rtimes \langle \mathrm{Frob}_p \rangle$$
$$G_2 \simeq \mathrm{Cl}_2^+ \rtimes \langle \mathrm{Frob}_p \rangle$$

where the pair of isomorphisms $\mathrm{Cl}_i^+ \simeq \mathrm{Gal}(H_i/K_i)$ for $i = 1, 2$ is provided by the global Artin map from class field theory. We make the following assumption:

Assumption: $(p-1)$ divides 12.

Equivalently, we assume that $X_0(p)$ has genus 0, or explicitly, that

$$p \in \{2, 3, 5, 7, 13\}.$$

This assumption is made to get the most straightforward statements of the main conjectures. We see why it is natural: when $X_0(p)$ has genus zero, the group of lifting obstructions is essentially trivial. Any concerned reader who feels the urge to object to such strong restrictions on p should keep in mind that:

- the work of Gross–Zagier restricts to $p \in \{\infty\}$,
- we address later in § 4.5 what happens for general primes p.

With the above notation and assumptions, we conjecture the following.

CONJECTURE 4.17. *For any quadratic forms* $(F, G) \in \mathcal{F}_{\Delta_1} \times \mathcal{F}_{\Delta_2}$ *we have*

$$\Theta_p^\times[F, G] \in H_1 H_2.$$

Furthermore, the set of these invariants is finite, and permuted simply transitively by the action of the Galois group $\mathrm{Gal}(H_1H_2/L) \simeq \mathrm{Cl}_1^+ \times \mathrm{Cl}_2^+$.

For a more precise version of the reciprocity law, see [**20**, Conjecture 3.14]. It is analogous to the well-known Shimura reciprocity law for CM singular moduli.

The factorisations of these algebraic numbers relate to intersection numbers of geodesics on *Shimura curves*. Let q be a prime, and $R \subset B_{pq}$ a maximal order in the indefinite quaternion algebra over $\mathbf{Q}$ ramified at p and q. The Shimura curve

$$X_{pq} := R_1^\times \backslash \mathcal{H}_\infty$$

is a compact algebraic curve defined over $\mathbf{Q}$, where $R_1^\times$ is the group of units of norm one, viewed as a subgroup of $\mathrm{SL}_2(\mathbf{R})$ via some chosen embedding.

Consider a quadratic order $\mathcal{O}$. An *optimal embedding* is an injective ring homomorphism $\mathcal{O} \hookrightarrow R$ that does not extend to a larger order in $\mathrm{Frac}(O)$. Choose a pair of optimal embeddings of the orders $\mathcal{O}_1, \mathcal{O}_2$ of discriminants Δ_1, Δ_2

$$\alpha_1 : \mathcal{O}_1 \hookrightarrow R, \qquad \alpha_2 : \mathcal{O}_2 \hookrightarrow R$$

Note that for the set of such embeddings to be non-empty, the primes p and q must both be nonsplit with respect to both Δ_1 and Δ_2.

- Choose an embedding $R \subset M_2(\mathbf{R})$, then there are associated oriented geodesics $geo(\alpha_1)$ and $geo(\alpha_2)$ in the upper half plane $\mathcal{H}_\infty$, connecting the fixed points of the images of the order. Define the *intersection*

  $$\mathrm{sgn}(\alpha_1, \alpha_2) \in \{0, 1, -1\}$$

  to be the intersection of the oriented geodesics $geo(\alpha_1)$ and $geo(\alpha_2)$.
- The (q-adic) *multiplicity* of the embeddings

  $$m_q(\alpha_1, \alpha_2) \in \mathbf{Z}_{\geq 1}$$

  is defined to be the largest integer t such that the images of the embeddings α_1 and α_2 coincide in the ring $R/q^{t-1}R$.

Finally, we define the q-weighted intersection number by

$$\mathrm{Int}_q(\alpha_1,\alpha_2) \quad := \quad \sum_{b\in\Gamma_1\backslash R_1^\times/\Gamma_2} \mathrm{sgn}(\alpha_1, b\alpha_2 b^{-1})\cdot m_q(\alpha_1, b\alpha_2 b^{-1})$$

where Γ_1 and Γ_2 are the (infinite) stabilisers of the images of the two embeddings in the group $R_1^\times$. If we were to omit the q-adic multiplicity m_q, we would recover the usual intersection pairing of the homology classes generated by the images of $geo(\alpha_1)$ and $geo(\alpha_2)$ in $\mathrm{H}_1(X_{pq},\mathbf{Z})$ on the Shimura curve X_{pq}.

CONJECTURE 4.18. *With the notation and assumptions as above, for every prime* $\mathfrak{q}$ *above* q *in* $H_1H_2/\mathbf{Q}$*, there is a pair of optimal embeddings* (α_1,α_2) *such that*

$$\begin{aligned}\mathrm{ord}_{\mathfrak{q}}\left(\Theta_p^\times[F,G]\right) &= \mathrm{Int}_q(\alpha_1,\alpha_2)\\ \mathrm{ord}_{\mathfrak{q}}\left(\Theta_p^\times[w_pF,G]\right) &= \mathrm{Int}_q(w_p\alpha_1,\alpha_2)\end{aligned}$$

where $w_p\alpha_1$ *is the embedding* α_1 *conjugated by the Atkin–Lehner involution at* p.

The factorisation conjecture [**20**, Conjecture 3.27] is more precise, and stipulates how the embeddings (α_1,α_2) are put in correspondence with the different choices of primes $\mathfrak{q}$ above q. The conjecture can be phrased as a statement about G-sets, where $G=\mathrm{Gal}(H_1H_2/\mathbf{Q})$ acts on the set of optimal embeddings, as described in the work of Eichler [**31**], see Gross [**38**] and Voight [**64**, Chapter 30].

REMARK 4.19. Besides extensive amounts of explicit examples, this conjecture was inspired by the algebraic proof of the factorisation of differences of CM singular moduli [**39**]. The expression for $\mathrm{ord}_q(j(\tau_1)-j(\tau_2))$ is related to ours by interchanging ∞ and p, i.e. by counting embeddings of *imaginary* quadratic orders

$$\begin{aligned}\alpha_1 &: \mathcal{O}_1 \hookrightarrow R_1\subset B_{\infty q}\\ \alpha_2 &: \mathcal{O}_2 \hookrightarrow R_2\subset B_{\infty q}\end{aligned}$$

in maximal orders of a *definite* quaternion algebra, of which there are finitely many, up to conjugation. The *intersection* $\mathrm{sgn}(\alpha_1,\alpha_2)$ measures whether the embeddings land in the same maximal order up to conjugacy. For example, consider

$$\begin{aligned} j\left(\frac{1+\sqrt{-67}}{2}\right)-j\left(\frac{1+\sqrt{-163}}{2}\right) &= -2^{15}\cdot 3^3\cdot 5^3\cdot 11^3+2^{18}\cdot 3^3\cdot 5^3\cdot 23^3\cdot 29^3\\ &= 2^{15}\cdot 3^7\cdot 5^3\cdot 7^2\cdot 13\cdot 139\cdot 331.\end{aligned}$$

The prime $q=3$ appears in this factorisation, and the exponent may be predicted as follows: The quaternion algebra $B_{\infty 3}$ has a unique maximal order up to conjugation, making the intersection $\mathrm{sgn}(-,-)$ is identically 1. For the 3-weighted intersection number, we find

$$\begin{aligned}\mathrm{Int}_3(\alpha_1,\alpha_2) &= \sum_{b\in\Gamma_1\backslash R_1^\times/\Gamma_2} \mathrm{sgn}(\alpha_1, b\alpha_2 b^{-1})\cdot m_3(\alpha_1, b\alpha_2 b^{-1})\\ &= 1+1+1+1+1+2=7.\end{aligned}$$

Here $R_1^\times$ is of order 12 and we have $\Gamma_1=\Gamma_2=\{\pm1\}$. The unique term with multiplicity 2 comes from the embeddings (with congruent images modulo 3)

$$\begin{cases}\alpha_1 & : \frac{-1+\sqrt{-67}}{2} \mapsto -\frac{1+x+8y}{2}\\ b\alpha_2 b^{-1} & : \frac{-1+\sqrt{-163}}{2} \mapsto -\frac{1+7x-4y}{2}\end{cases}$$

where we use generators $B_{3\infty} = \langle 1, x, y, z\rangle$ with $x^2 = z^2 = -3, y^2 = -1$.

EXAMPLE 4.20. Efficient methods for computing intersection numbers of geodesics on Shimura curves were developed by Rickards [**58, 59**], complete with a PARI/GP implementation that yielded hundreds of pages of explicit data to work with. We revisit Example 4.16 for $p = 2$, where

$$(\Delta_1, \Delta_2) = (13, 621).$$

Due to practical challenges in recognising algebraic numbers $\Theta_p^\times[F, G]$ quite this gargantuan, we found it more feasible in practice to instead recognise both

$$\Theta_p^\times[F, G] \div \Theta_p^\times[w_p F, G] \qquad \text{and} \qquad \Theta_p^\times[F, G] \times \Theta_p^\times[w_p F, G].$$

The polynomial for the former was written out (for better or worse) in Example 4.16. We spare the reader the polynomial for the latter. The conjecture concerns the slopes of their Newton polygons, depicted here for the prime $q = 7$.

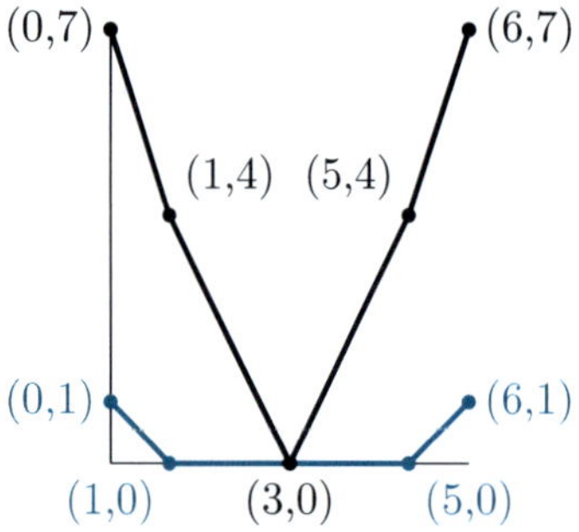

On the genus one Shimura curve

$$X_{14} \;:\; y^2 = -x^4 + 13x^2 - 128$$

the 7-weighted intersection numbers between geodesics of discriminants 13 and 621 are

$$\begin{pmatrix} 1 & 1 & 1 & -1 & -1 & -1 \\ -1 & -2 & -1 & 1 & 2 & 1 \end{pmatrix}$$

where the rows are interchanged by w_p.

Observe that the Newton slopes of the two polynomials coincide with the difference and sum of the two rows of intersection numbers, respectively. This agrees with the prediction made by Conjecture 4.18.

4.5. General primes. Without the hypothesis that $X_0(p)$ has genus zero, we do not expect $\Theta_p^\times[F, G]$ to be algebraic. The work of Gross–Zagier on singular moduli [**39**] is in essence the local computation of the height pairing of Heegner divisors

$$\langle P_{\Delta_1}, P_{\Delta_2}\rangle = 0 \qquad \text{on } X_0(1).$$

For us, the underlying geometric object is $X_0(p)$, which yields the closest analogues to [**39**] in situations where this curve is also of genus $g = 0$. For general primes p, this suggests the quantity $\Theta_p^\times[F, G]$ should perhaps be thought of as the analytic contribution to a certain, yet to be defined, height pairing. This viewpoint is further developed in the works [**23, 24**].

In the absence of algebraic cycles, it is difficult to foresee how this will pan out, but in the present context we might view the problems as being caused by the fact that the group of *lifting obstructions* $\mathrm{H}^2(\Gamma, \mathbf{C}_p^\times)$ is a torus of rank $2g+1$, and stands in the way of meaningfully lifting the theta cocycle $\Theta_F^\times$ to a rigid cocycle. When $g = 0$, it was shown in [**20**] that this may be done up to powers of a fundamental unit. If we take this more pedestrian view of our difficulties to obtain algebraic numbers for general p, there are two options to address them:

(1) **Kill the lifting obstruction.** For instance, let $p = 11$ and choose

$$\begin{array}{rclll} F & = & \langle 1,3,-3\rangle, & \text{discriminant} & \Delta_1 = 13, \\ G & = & \langle 1,4,-4\rangle, & \text{discriminant} & \Delta_2 = 32. \end{array}$$

If we attempt to find a minimal polynomial for $\Theta_{11}^{\times}[F,G]$ using LLL, the results are unconvincing, and no clear algebraic numbers have so far been recognised. Note that the space of weight two cusp forms on $\Gamma_0(11)$ is killed by the Hecke operator $(w_p - 1)$. We find that the invariant

$$\Theta_{11}^{\times}[w_{11}F,G] \div \Theta_{11}^{\times}[F,G]$$

computed modulo 11^{50} satisfies

$$13x^4 + 12x^3 + 14x^2 + 12x + 13,$$

with splitting field $\mathbf{Q}(\sqrt{-1},\sqrt{-3})$. The splitting field over K_1K_2 is H_1H_2.

In general, when T is a Hecke operator that kills the space of weight two cusp forms on $\Gamma_0(p)$, the quantity $\Theta_p^{\times}[TF,G]$ defined multiplicatively in the obvious way, is expected to be an algebraic number in H_1H_2.

(2) **Cherish the lifting obstruction.** A more gentle approach is preferable. The lifting obstructions of *analytic* theta cocycles, i.e. cocycles for Γ valued in $\mathcal{A}^{\times}/\mathbf{C}_p^{\times}$ where $\mathcal{A}^{\times}$ is the multiplicative group of invertible analytic functions on $\mathcal{H}_p$, form a multiplicative lattice Λ in $\mathrm{H}^2(\Gamma,\mathbf{C}_p^{\times})$. The quotient can be identified, up to an elementary factor coming from the Eisenstein line, with two copies of the Jacobian of $X_0(p)$.

$$\begin{array}{ccc} \mathrm{H}^1(\Gamma,\mathcal{A}^{\times}/\mathbf{C}_p^{\times}) & \longrightarrow & \Lambda \\ \cap & & \cap \\ \mathrm{H}^1(\Gamma,\mathcal{M}^{\times}/\mathbf{C}_p^{\times}) & \longrightarrow & \mathrm{H}^2(\Gamma,\mathbf{C}_p^{\times}) \\ & & \downarrow \\ & & (\mathbf{C}_p^{\times}/\langle p\rangle) \times J_0(p)^2 \end{array}$$

When $p = 11$, it is an elliptic curve, with minimal Weierstraß equation

$$J_0(11) : y^2 + y = x^3 - x^2 - 10x - 20.$$

In [**22**] we consider the image of the lifting obstruction of $\Theta_F^{\times}$. For instance, when $F = \langle 1,3,-3\rangle$ it equals $(\alpha, P, 0) \bmod 11^{100}$, where

$$\alpha = \frac{-103+24\sqrt{-7}}{121}, \qquad P = \left(\frac{-3-\sqrt{-7}}{2}, \frac{-3-\sqrt{-7}}{2}\right) \in J_0(11).$$

We see that the lifting obstruction is of independent interest, and appears to encode global points on modular Jacobians. This alternative viewpoint on Stark–Heegner points [**15**] is explored further in [**17, 22**].

4.6. Recent analytic approaches. Whereas the *geometric* nature of RM singular moduli remains elusive, we mention recent works that have explored the more accessible (p-adic) *analytic* structures analogous to those in CM theory.

Borcherds products. The work of Zagier [**70**] on *traces* of CM singular moduli has unveiled that their generating series may be understood in terms of the theory of *Borcherds products*, which provide a morphism

$$\Psi_{\mathrm{Bor}} \ : \ M_{1/2}^{+,!}(\Gamma_0(4)) \quad \longrightarrow \quad \mathrm{H}^0(\mathrm{SL}_2(\mathbf{Z}), \mathrm{Mer}_{\infty}^{\times}).$$

valued in meromorphic functions on $\mathcal{H}_\infty$ with poles at CM divisors. It is tempting to view rigid cocycles, whose poles occur at RM divisors on the p-adic upper half plane, as analogues of these Borcherds products. This analogy was explored in [**22**], where a similar morphism is constructed:

$$\Psi_{\mathrm{Rig}} \ : \ M_{1/2}^{+,!}(\Gamma_0(4p)) \ \longrightarrow \quad \mathrm{H}^1(\Gamma, \mathrm{Mer}^{\times}).$$

The natural generalisation suggested by this analogy with Borcherds products is to consider rigid meromorphc cocycles for orthogonal groups $O(r,s)$ of general signatures, realised in Darmon–Gehrmann–Lipnowski [**16**]. This suggests a natural place for rigid cocycles in an emerging p-adic Kudla programme.

Analytic families of modular forms. The recent papers [**17, 18**] have explored analogues of the analytic arguments of Gross–Zagier [**39**] in the degenerate case where $F = \langle 0,1,0 \rangle$ is the split form of discriminant 1. We refer to the corresponding theta cocycle as the *winding cocycle*, denoted

$$\Theta_{\mathrm{wind}}^{\times} := \Theta_F^{\times} \in Z^1(\Gamma, \mathrm{Mer}_p^{\times} / \mathbf{C}_p^{\times})$$

The main conjectures for its RM values reduce to the well-known properties of *p-adic Gross–Stark units.* Since the form F is split, the factorisation conjecture predicts that it is a unit at all primes $q \neq p$. At primes above p, it predicts (with the convention that $B_{pp} = M_2(\mathbf{Q})$) that the factorisations are given by classical partial L-values, after applying Meyer's theorem, showing that the algebraic numbers we predict are precisely the p-adic Gross–Stark units in this degenerate case. Of course, we already have the more powerful and more general analytic formula for p-adic Gross–Stark units due to Dasgupta–Kakde [**25, 26**]. Nonetheless, this proof yields some theoretical evidence for the non-degenerate case of RM singular moduli that have been the focus of these lectures.

The strategy of [**17, 18**] resembles the arguments on Hecke's Eisenstein family discussed in the introduction. Consider a quadratic form G of discriminant $\Delta > 0$ for which p is inert. Define the real quadratic field $K = \mathbf{Q}(\sqrt{\Delta})$ and its narrow Hilbert class field H. Choose an odd character

$$\psi : \mathrm{Cl}_K^+ \longrightarrow \bar{\mathbf{Q}}^{\times}.$$

There is a holomorphic Hilbert Eisenstein series of weight $(1,1)$ and level $\Gamma_0(p)$ over K, whose q-expansion at the class of the different is

$$E(\psi) := L_p(\psi, 0) + 4 \sum_{\nu \in \mathfrak{d}_+^{-1}} \left(\sum_{p \nmid I \mid (\nu)\mathfrak{d}} \psi(I) \right) e^{2\pi i(\nu_1 z_1 + \nu_2 z_2)}.$$

Since p is inert in K, the p-adic L-function $L_p(\psi, s)$ has an expectional zero at $s = 0$, and the series $E(\psi)$ is p-adically cuspidal. The approach is to p-adically deform this series. All p-adic deformations as *Hilbert eigenforms* are described by the local geometry of the eigenvariety $\mathscr{E} \longrightarrow \mathscr{W}$ at the parallel weight one point corresponding to $E(\psi)$. The local geometry is described by Betina–Dimitrov–Shih [**3**]; the cuspidal part is étale over weight space $\mathscr{W}$, and two lower-dimensional Eisenstein families, which exist only in parallel weight, intersect it transversely.

The papers [**17, 18**] each consider a different p-adic family specialising to $E(\psi)$. The p-adic Eisenstein family in parallel weight $(1+s, 1+s)$ is used in [**17**]. The advantage is that its Fourier expansion is completely explicit, but it yields more crude RM invariants. In contrast, [**18**] considers the cuspidal p-adic family of anti-parallel weight $(1+s, 1-s)$. This family lies deeper, and its Fourier expansion is determined from the deformation theory of the Galois representation

$$\rho = 1 \oplus \psi.$$

This representation is p-irregular, causing technical complications [**2, 3, 57**]. Galois cohomological arguments relate the deformation theory via global class field theory to the p-adic logarithm of the Gross–Stark unit $u_\psi \in \mathcal{O}_H[1/p]^\times \otimes \mathbf{Q}$.

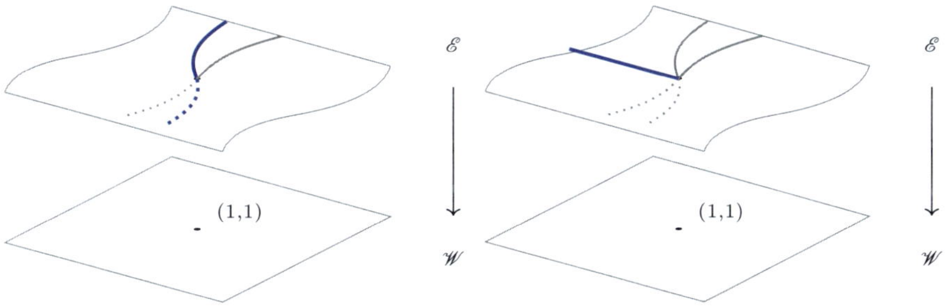

For both choices of p-adic family appearing in [**17, 18**] one considers

(1) its diagonal restriction (vanishes at $s = 0$)
(2) its analytic first order derivative with respect to s
(3) its *ordinary projection* $\lim U_p^{n!}$, contained in the space of weight two holomorphic modular forms

$$M_2(\Gamma_0(p)).$$

This reflects the structure of the analytic proof of Gross–Zagier [**39**], discussed in the introduction. These operations are applied to the two p-adic families, and the first Fourier coefficient of the resulting form is explicitly computed:

- for the elementary *parallel* weight Eisenstein family [**17**], it is

$$\log_p \left(\mathrm{Nm}_{\mathbf{Q}_{p^2}} \Theta^\times_{\mathrm{wind}}[G] \right),$$

- for (a small modification of) the *anti-parallel* weight family [**18**], it is

$$\log_p \left(\Theta^\times_{\mathrm{wind}}[G] \right).$$

The resulting weight two form is contained in $M_2(\Gamma_0(p))$. Projecting onto the Eisenstein line (has no effect when p is a genus zero prime), the relation between its zeroth and first Fourier coefficients gives the (degenerate) main conjectures.

REMARK 4.21. The proof of the degenerate RM conjectures in [**17**, **18**] raises the natural question of whether the original setup of CM points on modular curves may be attacked p-adically using arguments like the above. This question is answered affirmatively in the forthcoming PhD thesis of Mike Daas [**11**, **12**].

More precisely, Daas considers CM points on Shimura curves X_D associated to indefinite quaternion algebras of discriminant D. When the curve X_D is of genus zero, one may choose a generator j_D of its function field, and compute the (well-defined) cross ratio of its values at CM divisors. Giampietro and Darmon [**35**] study these quantities experimentally, and formulate a conjecture for their factorisations, which resemble those of Gross–Zagier in many respects. For instance, they compute for (P_1, P_2) of discriminants $(-43, -163)$ that

$$\mathrm{Nm}_{\mathbf{Q}}\left[\frac{j_6(P_1) - j_6(P_2))(j_6(P_1') - j_6(P_2'))}{j_6(P_1) - j_6(P_2'))(j_6(P_1') - j_6(P_2))}\right] = \left(\frac{2 \cdot 29 \cdot 257 \cdot 277}{73 \cdot 137 \cdot 241}\right)^2$$

on the quaternionic Shimura curve X_6 of genus zero. Daas considers p-adic deformations, where p is a divisor of D, of the same Eisenstein series considered by Hecke and Gross–Zagier, with associated Galois representation $\rho = 1 \oplus \chi$, and computes the Fourier expansion of the form

$$\mathrm{Proj}_{\mathrm{ord}}\left[\frac{\partial}{\partial s} E(1,\chi)_s^D(z,z)\right]_{s=0} \qquad \in S_2(\Gamma_0(D))$$

to equal the p-adic logarithm of the cross-ratio invariant, together with contributions for primes q that are an explicit multiple of $\log_p(q)$. Daas proves the conjectures of Giampietro and Darmon [**35**] about the factorisations of these quaternionic singular moduli by showing that the above series must vanish. Remarkably, this proof does not use CM theory, and does not make reference to the QM abelian surfaces for which the Shimura curve X_D is a moduli space.

REMARK 4.22. The proof of the degenerate RM conjectures in [**18**] also yields an *algorithm* for computing p-adic Gross–Stark units, using an idea of Hecke and Klingen–Siegel. This idea was used to compute p-adic L-functions [**53**], and using the anti-parallel family this was described in [**4**] and is upgraded to an algorithm to *simultaneously* compute p-adic Gross–Stark units and Stark–Heegner points by Håvard Damm–Johnsen in his fortcoming DPhil thesis [**13**, **14**].

For example, let $K = \mathbf{Q}(\sqrt{136})$ which has narrow class group $\mathrm{Cl}^+(K) \simeq \mathbf{Z}/4\mathbf{Z}$. Choose the prime $p = 19$, which is inert in K. Applying the three operations to the anti-parallel weight family (the ordinary projection is computed using the algorithms of Lauder [**52**, **65**]), Damm–Johnsen computes all the Fourier coefficients, *except for the constant term*, of the generating series of the RM values of the winding cocycle constructed in [**18**]

$$\log_p(u_G) + \sum_{n \geq 1} \log_p\left(\Theta_{\mathrm{wind}}^{\times}[T_n G]\right) q^n$$

for all choices of G of discriminant 136, up to precision 19^{50}. The computation takes less than three seconds. Projecting the result onto the Eisenstein line in the space $M_2(\Gamma_0(p))$, which can be done using only the higher Fourier coefficients, one *recovers* a numerical value for the constant term $\log_p(u_G)$. The resulting numerical value of the Gross–Stark unit u_G was recognised, using LLL routines, to be a root of the polynomial

$$361x^4 + 508x^3 + 310x^2 + 508x + 361.$$

It generates the narrow Hilbert class field over K.

It is clear that the limits of analytic arguments of the above sort should be explored further. Though significant new ideas are needed, one may hope that they could be applicable beyond the degenerate setups considered here. As far as we are from fully understanding RM singular moduli, it is clear that explicit experimentation has an important role to play in informing future progress.

4.7. Acknowledgments. The author wishes to thank Henri Darmon, Alice Pozzi, Isabella Negrini, James Rickards, Hendrik Lenstra, Håvard Damm–Johnsen, Mike Daas, as well as the anonymous referee, for helpful conversations and for suggesting several useful corrections to an early version of the manuscript.

5. Exercises

(1) Prove that the set of reduced forms of a fixed discriminant $D \neq 0$ is finite.
Let $F = \langle a, b, c\rangle$ be of non-square discriminant D. Show that

- if F is *definite*, then F is reduced if and only if

$$r_F \in \mathcal{D},$$

where $\mathcal{D}$ is the standard fundamental domain for the action of $\mathrm{SL}_2(\mathbf{Z})$ in the Poincaré upper half plane $\mathfrak{H}_\infty := \{z \in \mathbf{C} : \mathrm{Im}(z) > 0\}$.

- if F is *indefinite*, then F is reduced if and only if $\left|\sqrt{D} - 2|a|\right| < b < \sqrt{D}$. Show that this is furthermore equivalent to the following condition on the roots:

$$r'_F r_F < 0 \quad \text{and} \quad |r_F| < 1 < |r'_F|.$$

(2) Compute the set Σ_F of nearly reduced forms in the $\mathrm{SL}_2(\mathbf{Z})$-orbit of the quadratic form F, which we defined by

$$\Sigma_F := \big\{\langle a, b, c\rangle \sim F \ : \ ac < 0\big\},$$

for the forms $F = \langle -1, 4, 4\rangle, \langle 5, 11, -5\rangle$, and $\langle -4825, -15989, -13246\rangle$.
Characterise all forms F of non-square discriminant $D > 0$ for which the sets Σ_F are symmetric under the involutions

$$\begin{array}{rclcl} s_1 & : & \langle a,b,c\rangle & \longmapsto & \langle -a,-b,-c\rangle \\ s_2 & : & \langle a,b,c\rangle & \longmapsto & \langle a,-b,c\rangle \end{array}$$

in terms of the associated class in the (narrow) Picard group of $\mathbf{Z}\left[\frac{D+\sqrt{D}}{2}\right]$.

(3) Consider a multiplicative cocycle

$$\Theta \in Z^1\left(\mathrm{SL}_2(\mathbf{Z}), \mathbf{C}(z)^\times\right).$$

Its *value* at a form $G \in \mathcal{F}_D$ with $D > 0$ non-square is defined by

$$\Theta[G] := \Theta(\gamma_G)(r_G)$$

where γ_G is the automorph of G. Show that for a fixed Θ, the value $\Theta[G]$ only depends on the $\mathrm{SL}_2(\mathbf{Z})$-orbit of G.

(4) (Warm-up cocycle I) For any $c = (r, s) \in \mathbf{P}^1(\mathbf{Q})$ define the function

$$L(c) := \frac{s}{sz - r}$$

and consider the map

$$\begin{array}{rccc} p_c: & \mathrm{GL}_2(\mathbf{Q}) & \longrightarrow & \mathbf{C}(z), \\ & \gamma & \longmapsto & L(c) - L(\gamma c). \end{array}$$

Show that
- this is a 1-cocycle,
- its cohomology class

$$[p_c] \in \mathrm{H}^1(\mathrm{GL}_2(\mathbf{Q}), \mathbf{C}(z))$$

is independent of the choice of cusp c,
- if $F \in \mathcal{F}_D$ for $D > 0$ non-square, the multiplicative lift $p^\times$ of its restriction to the subgroup $\mathrm{SL}_2(\mathbf{Z})$ has value at F equal to

$$p^\times[F] = \varepsilon_D^{12}$$

where $\varepsilon_D > 1$ is the fundamental unit of norm 1 in the quadratic order $\mathbf{Z}\left[\frac{D+\sqrt{D}}{2}\right]$ of discriminant D.

(5) (Warm-up cocycle II) For any $c = (r, s) \in \mathbf{P}^1(\mathbf{Q})$ define the function

$$N(c) := \frac{1}{(sz - r)^2},$$

where choose $\gcd(r, s) = 1$, and consider the map

$$\begin{array}{rccc} q_c: & \mathrm{SL}_2(\mathbf{Q}) & \longrightarrow & \mathbf{C}(z), \\ & \gamma & \longmapsto & N(c) - N(\gamma c). \end{array}$$

Show that
- this is a 1-cocycle,
- its cohomology class

$$[q_c] \in \mathrm{H}^1(\mathrm{SL}_2(\mathbf{Z}), \mathbf{C}(z))$$

is independent of the choice of cusp c.

REMARK. Note that this cocycle is not in the image of dlog.

(6) (Knopp cocycle) Choose a cusp $c \in \mathbf{P}^1(\mathbf{Q})$. Define the map

$$kn_{c,F} : \mathrm{SL}_2(\mathbf{Z}) \longrightarrow \mathbf{C}(z)$$

by setting

$$kn_{c,F}(\gamma) = \sum_{Q \sim F} \frac{\mathrm{sgn}_{c,Q}(\gamma)}{z - r(Q)}.$$

where the numerator is defined by

$$\mathrm{sgn}_{c,Q}(\gamma) := \begin{cases} 1 & \text{if } Q(c) > 0 > Q(\gamma c), \\ -1 & \text{if } Q(c) < 0 < Q(\gamma c), \\ 0 & \text{else.} \end{cases}$$

- Show that $kn_{c,F}$ is a 1-cocycle.
- Show that its cohomology class

  $$[kn_{c,F}] \in \mathrm{H}^1(\mathrm{SL}_2(\mathbf{Z}), \mathbf{C}(z))$$

  is independent of the choice of cusp c.
- Show that the multiplicative lift

  $$kn_F^{\times} \in Z^1(\mathrm{SL}_2(\mathbf{Z}), \mathbf{C}(z)^{\times})$$

  satisfies

  $$kn_F^{\times}(T) = \varepsilon_D^{12},$$

  where $\varepsilon_D > 1$ is the fundamental unit of norm 1 in the quadratic order $\mathbf{Z}\left[\frac{D+\sqrt{D}}{2}\right]$ of discriminant D.

(7) Let $f(z) \in \mathbf{C}(z)$. Show that there exists a $\varphi \in Z^1_{\mathrm{par}}\big(\mathrm{SL}_2(\mathbf{Z}), \mathbf{C}(z)\big)$ with the property that $\varphi(S) = f(z)$ if and only if $f(z)$ satisfies the identities

$$\begin{cases} 0 & = (1+S)\star f \\ 0 & = (1+ST+(ST)^2)\star f \end{cases}$$

where S and T are the generators of $\mathrm{SL}_2(\mathbf{Z})$ defined in the main text.

(8) Define the map $t : \mathbf{R}_{\geq 0} \longrightarrow \mathbf{R}_{\geq 0}$ by

$$t(x) := \begin{cases} x-1 & \text{if} x \geq 1 \\ x/(1-x) & \text{if} 0 \leq x < 1 \end{cases}$$

Show that the periodic orbits for iteration of t are the sets $\{0\}$ and

$$\mathcal{S}_F := \{r(Q) \ : \ Q \in \Sigma_F\},$$

for $F \in \mathcal{F}_D$, with $D > 0$ non-square.

(9) Use the previous exercise to show that the cocycles p, q, kn_F introduced above generate the group of parabolic (additive) rational cocycles

$$Z^1_{\mathrm{par}}\left(\mathrm{SL}_2(\mathbf{Z}), \mathbf{C}(z)\right).$$

References

[1] W. E. H. Berwick, *Modular Invariants Expressible in Terms of Quadratic and Cubic Irrationalities*, Proc. London Math. Soc. (2) **28** (1928), no. 1, 53–69, DOI 10.1112/plms/s2-28.1.53. MR1575872 ↑251, 252, 255

[2] A. Betina, M. Dimitrov, and A. Pozzi, *On the failure of Gorensteinness at weight 1 Eisenstein points of the eigencurve*, Amer. J. Math. **144** (2022), no. 1, 227–265, DOI 10.1353/ajm.2022.0004. MR4367418 ↑283

[3] A. Betina and M. Dimitrov and S.C. Shih, Eisenstein points on the Hilbert cuspidal eigenvariety, Preprint, `arXiv:2311.08361`, 2023. ↑283

[4] S. Blackwell and A. Davies and R. Tanburn and J. Vonk, *On spaces of Maaß forms and overconvergent forms*, Preprint (2021). ↑284

[5] J. Buchmann and U. Vollmer, *Binary quadratic forms: An algorithmic approach*, Algorithms and Computation in Mathematics, vol. 20, Springer, Berlin, 2007. MR2300780 ↑260

[6] D. A. Buell, *Binary quadratic forms*, Springer-Verlag, New York, 1989. Classical theory and modern computations, DOI 10.1007/978-1-4612-4542-1. MR1012948 ↑260

[7] Yj. Choie and D. Zagier, *Rational period functions for* PSL(2, **Z**), A tribute to Emil Grosswald: number theory and related analysis, Contemp. Math., vol. 143, Amer. Math. Soc., Providence, RI, 1993, pp. 89–108, DOI 10.1090/conm/143/00992. MR1210514 ↑265

[8] H. Cohen, *A course in computational algebraic number theory*, Graduate Texts in Mathematics, vol. 138, Springer-Verlag, Berlin, 1993, DOI 10.1007/978-3-662-02945-9. MR1228206 ↑273

[9] J. H. Conway, *The sensual (quadratic) form*, Carus Mathematical Monographs, vol. 26, Mathematical Association of America, Washington, DC, 1997. MR1478672 ↑256, 258

[10] D. A. Cox, *Primes of the form x^2+ny^2: Fermat, class field theory and complex multiplication*, A Wiley-Interscience Publication, John Wiley & Sons, Inc., New York, 1989. MR1028322 ↑251

[11] M. A. Daas, *CM-values of p-adic Θ-functions*, Res. Math. Sci. **12** (2025), no. 3, Paper No. 53, 39, DOI 10.1007/s40687-025-00521-x. MR4942016 ↑284

[12] M. Daas, *PhD Thesis*, Leiden University, 2024. ↑284

[13] H. Damm-Johnsen, *Modular algorithms for Gross-Stark units and Stark-Heegner points*, LuCaNT: LMFDB, computation, and number theory, Contemp. Math., vol. 796, Amer. Math. Soc., [Providence], RI, [2024] ©2024, pp. 261–284, DOI 10.1090/conm/796/16005. MR4732691 ↑284

[14] H. Damm-Johnsen, *DPhil Thesis*, University of Oxford, 2024. ↑284

[15] H. Darmon, *Integration on $\mathcal{H}_p \times \mathcal{H}$ and arithmetic applications*, Ann. of Math. (2) **154** (2001), no. 3, 589–639, DOI 10.2307/3062142. MR1884617 ↑281

[16] H. Darmon and L. Gehrmann and M. Lipnowski, *Rigid meromorphic cocycles for orthogonal groups*, Preprint (2023). ↑282

[17] H. Darmon and A. Pozzi and J. Vonk, *Diagonal restrictions of p-adic Eisenstein families*, Math. Ann. **379** (2021), no. 1, 503–548, DOI doi.org/10.1007/s00208-020-02086-2. ↑281, 282, 283, 284

[18] H. Darmon, A. Pozzi, and J. Vonk, *The values of the Dedekind-Rademacher cocycle at real multiplication points*, J. Eur. Math. Soc. (JEMS) **26** (2024), no. 10, 3987–4032, DOI 10.4171/jems/1344. MR4768413 ↑282, 283, 284

[19] H. Darmon and J. Vonk, `https://pub.math.leidenuniv.nl/~vonkjb/code/rmmc/rmmc.html`. ↑273, 277

[20] H. Darmon and J. Vonk, *Singular moduli for real quadratic fields: a rigid analytic approach*, Duke Math. J. **170** (2021), no. 1, 23–93, DOI 10.1215/00127094-2020-0035. MR4194897 ↑267, 268, 271, 273, 277, 278, 279, 280

[21] H. Darmon and J. Vonk, *Arithmetic intersections of modular geodesics*, J. Number Theory **230** (2022), 89–111, DOI 10.1016/j.jnt.2020.12.012. MR4327950 ↑268

[22] H. Darmon and J. Vonk, *Real quadratic Borcherds products*, Pure Appl. Math. Q. (Special Issue in honor of Dick Gross) (2022). ↑270, 281, 282

[23] H. Darmon and J. Vonk, *p-adic Green's functions for real quadratic geodesics*, Res. Number Theory **11** (2025), no. 1, Paper No. 40, 26, DOI 10.1007/s40993-025-00621-1. MR4868105 ↑280

[24] H. Darmon and J. Vonk, *Heights of RM divisors and real quadratic singular moduli*, Preprint (2023). ↑280

[25] S. Dasgupta and M. Kakde, *On the Brumer-Stark conjecture*, Ann. of Math. (2) **197** (2023), no. 1, 289–388, DOI 10.4007/annals.2023.197.1.5. MR4513146 ↑255, 282

[26] S. Dasgupta and M. Kakde, *Brumer–Stark units and Hilbert's 12th problem*, Duke Math. J. **173** (2024), no. 8, 1477–1555, DOI 10.1215/00127094-2023-0039. ↑255, 282

[27] S. Dasgupta and J. Teitelbaum, *The p-adic upper half plane*, p-adic geometry, Univ. Lecture Ser., vol. 45, Amer. Math. Soc., Providence, RI, 2008, pp. 65–121, DOI 10.1090/ulect/045/03. MR2482346 ↑268, 269

[28] W. Duke, Ö. Imamoḡlu, and Á. Tóth, *Cycle integrals of the j-function and mock modular forms*, Ann. of Math. (2) **173** (2011), no. 2, 947–981, DOI 10.4007/annals.2011.173.2.8. MR2776366 ↑255, 265

[29] W. Duke, Ö. Imamoḡlu, and Á. Tóth, *Real quadratic analogs of traces of singular moduli*, Int. Math. Res. Not. IMRN **13** (2011), 3082–3094, DOI 10.1093/imrn/rnq159. MR2817687 ↑255

[30] W. Duke, Ö. Imamoḡlu, and Á. Tóth, *Modular cocycles and linking numbers*, Duke Math. J. **166** (2017), no. 6, 1179–1210, DOI 10.1215/00127094-3793032. MR3635902 ↑265, 271

[31] M. Eichler, *Zur Zahlentheorie der Quaternionen-Algebren* (German), J. Reine Angew. Math. **195** (1955), 127–151 (1956), DOI 10.1515/crll.1955.195.127. MR80767 ↑279
[32] R. Fricke and F. Klein, *Lectures on the theory of automorphic functions. Vol. 1*, CTM. Classical Topics in Mathematics, vol. 3, Higher Education Press, Beijing, 2017. Translated from the German original [MR0183872] by Arthur M. DuPre. MR3838411 ↑250
[33] R. Fricke and F. Klein, *Lectures on the theory of automorphic functions. Vol. 2*, CTM. Classical Topics in Mathematics, vol. 4, Higher Education Press, Beijing, 2017. Translated from the German original [MR0183872] by Arthur M. DuPre. MR3838412 ↑250
[34] C. F. Gauss, *Disquisitiones Arithmeticae*, Lipsiae : Apud G. Fleischer, 1801. ↑251, 256, 260
[35] S. Giampietro and H. Darmon, *A p-adic approach to singular moduli on Shimura curves*, Involve **15** (2022), no. 2, 345–365, DOI 10.2140/involve.2022.15.345. MR4462162 ↑284
[36] A. G. Greenhill, *Table of Complex Multiplication Moduli*, Proc. Lond. Math. Soc. **21** (1889/90), 403–422, DOI 10.1112/plms/s1-21.1.403. MR1575441 ↑251, 255
[37] B. Gross, *p-Adic L-series at $s = 0$*, J. Fac. Sci. Univ. Tokyo **28** (1981), 979–994. ↑255
[38] B. H. Gross, *Heights and the special values of L-series*, Number theory (Montreal, Que., 1985), CMS Conf. Proc., vol. 7, Amer. Math. Soc., Providence, RI, 1987, pp. 115–187. MR894322 ↑279
[39] B. H. Gross and D. B. Zagier, *On singular moduli*, J. Reine Angew. Math. **355** (1985), 191–220, DOI 10.1515/crll.1985.355.191. MR772491 ↑252, 279, 280, 282, 283
[40] B. H. Gross and D. B. Zagier, *Heegner points and derivatives of L-series*, Invent. Math. **84** (1986), no. 2, 225–320, DOI 10.1007/BF01388809. MR833192 ↑252, 254
[41] B. Gross, W. Kohnen, and D. Zagier, *Heegner points and derivatives of L-series. II*, Math. Ann. **278** (1987), no. 1-4, 497–562, DOI 10.1007/BF01458081. MR909238 ↑254
[42] A. Hatcher, *The topology of numbers* (2017). ↑256
[43] E. Hecke, *Analytische funktionen und algebraische zahlen* (German), Abh. Math. Sem. Univ. Hamburg **3** (1924), no. 1, 213–236, DOI 10.1007/BF02954625. MR3069428 ↑253
[44] Y. Ihara, *On congruence monodromy problems. Vol. 1*, University of Tokyo, Department of Mathematics, Tokyo, 1968. Lecture Notes, No. 1. MR0289518 ↑270
[45] Y. Ihara, *On congruence monodromy problems. Vol. 2*, University of Tokyo, Department of Mathematics, Tokyo, 1969. Lecture Notes, No. 2. MR289519 ↑270
[46] M. Kaneko, *Observations on the 'values' of the elliptic modular function $j(\tau)$ at real quadratics*, Kyushu J. Math. **63** (2009), no. 2, 353–364, DOI 10.2206/kyushujm.63.353. MR2568779 ↑255
[47] F. Klein and R. Fricke, *Lectures on the theory of elliptic modular functions. Vol. 1*, CTM. Classical Topics in Mathematics, vol. 1, Higher Education Press, Beijing, 2017. Translated from the German original [MR0247996] by Arthur M. DuPre. MR3838340 ↑250
[48] F. Klein and R. Fricke, *Lectures on the theory of elliptic modular functions. Vol. 2*, CTM. Classical Topics in Mathematics, vol. 2, Higher Education Press, Beijing, 2017. Translated from the German original [MR0247997] by Arthur M. DuPre. MR3838339 ↑250
[49] M. I. Knopp, *Rational period functions of the modular group*, Duke Math. J. **45** (1978), no. 1, 47–62. With an appendix by Georges Grinstein. MR485700 ↑260, 262
[50] M. I. Knopp, *Rational period functions of the modular group. II*, Glasgow Math. J. **22** (1981), no. 2, 185–197, DOI 10.1017/S0017089500004663. MR623004 ↑260, 262
[51] S. S. Kudla, M. Rapoport, and T. Yang, *On the derivative of an Eisenstein series of weight one*, Internat. Math. Res. Notices **7** (1999), 347–385, DOI 10.1155/S1073792899000185. MR1683308 ↑255
[52] A. G. B. Lauder, *Computations with classical and p-adic modular forms*, LMS J. Comput. Math. **14** (2011), 214–231, DOI 10.1112/S1461157011000155. MR2831231 ↑284
[53] A. Lauder and J. Vonk, *Computing p-adic L-functions of totally real fields*, Math. Comp. **91** (2022), no. 334, 921–942, DOI 10.1090/mcom/3678. MR4379982 ↑284
[54] A. K. Lenstra, H. W. Lenstra Jr., and L. Lovász, *Factoring polynomials with rational coefficients*, Math. Ann. **261** (1982), no. 4, 515–534, DOI 10.1007/BF01457454. MR682664 ↑270, 273
[55] Yu. I. Manin, *Real multiplication and noncommutative geometry (ein Alterstraum)*, The legacy of Niels Henrik Abel, Springer, Berlin, 2004, pp. 685–727. MR2077591 ↑255
[56] I. Negrini, *A Shimura-Shintani correspondence for rigid analytic cocycles of higher weight*, Forum Math. **35** (2023), no. 2, 549–571, DOI 10.1515/forum-2022-0235. MR4554488 ↑269

[57] A. Pozzi, *The Eigencurve at Weight One Eisenstein Points*, ProQuest LLC, Ann Arbor, MI, 2019. Thesis (Ph.D.)–McGill University (Canada). MR4197785 ↑283

[58] J. Rickards, *Computing intersections of closed geodesics on the modular curve*, J. Number Theory **225** (2021), 374–408, DOI 10.1016/j.jnt.2020.11.024. MR4259279 ↑280

[59] J. Rickards, *Counting intersection numbers of closed geodesics on Shimura curves*, Res. Number Theory **9** (2023), no. 2, Paper No. 20, 45, DOI 10.1007/s40993-023-00428-y. MR4565144 ↑280

[60] J. Rickards, Linking number of modular knots, Preprint, arXiv:2301.01334, 2023. ↑271

[61] J.-P. Serre, *Trees*, Springer-Verlag, Berlin-New York, 1980. Translated from the French by John Stillwell. MR607504 ↑257, 267, 270

[62] C.-L. Simon, *Linking numbers of modular knots*, Geom. Topol. **29** (2025), no. 6, 3241–3270, DOI 10.2140/gt.2025.29.3241. MR4965187 ↑271

[63] H. M. Stark, *L-functions at $s = 1$. IV. First derivatives at $s = 0$*, Adv. in Math. **35** (1980), no. 3, 197–235, DOI 10.1016/0001-8708(80)90049-3. MR563924 ↑255

[64] J. Voight, *Quaternion algebras*, Graduate Texts in Mathematics, vol. 288, Springer, Cham, [2021] ©2021, DOI 10.1007/978-3-030-56694-4. MR4279905 ↑279

[65] J. Vonk, *Computing overconvergent forms for small primes*, LMS J. Comput. Math. **18** (2015), no. 1, 250–257, DOI 10.1112/S1461157015000042. MR3349318 ↑284

[66] H. Weber, *Lehrbuch der Algebra III.*, Chelsea, New-York, 1908. ↑251, 255

[67] A. Weil, *Elliptic functions according to Eisenstein and Kronecker*, Classics in Mathematics, Springer-Verlag, Berlin, 1999. Reprint of the 1976 original. MR1723749 ↑250, 265

[68] D. Zagier, *Letter to Gross*, 1983. Available at https://www.mpim-bonn.mpg.de/webfm_send/144. ↑252, 253

[69] D. Zagier, *Modular points, modular curves, modular surfaces and modular forms*, Workshop Bonn 1984 (Bonn, 1984), Lecture Notes in Math., vol. 1111, Springer, Berlin, 1985, pp. 225–248, DOI 10.1007/BFb0084592. MR797423 ↑

[70] D. Zagier, *Traces of singular moduli*, Motives, polylogarithms and Hodge theory, Part I (Irvine, CA, 1998), Int. Press Lect. Ser., vol. 3, Int. Press, Somerville, MA, 2002, pp. 211–244. MR1977587 ↑282

Mathematical Institute, Snellius Building, Niels Bohrweg 1, 2333 CA Leiden, Netherlands

Email address: j.b.vonk@math.leidenuniv.nl

IAS/Park City Mathematics Series
Volume **29**, 2025
https://doi.org/10.1090/pcms/029/00896

Around the Inverse Galois Problem

Olivier Wittenberg

ABSTRACT. The inverse Galois problem asks whether any finite group can be realised as the Galois group of a Galois extension of the rationals. This problem and its refinements have stimulated a large amount of research in number theory and algebraic geometry in the past century, ranging from Noether's problem (letting X denote the quotient of the affine space by a finite group acting linearly, when is X rational?) to the rigidity method (if X is not rational, does it at least contain interesting rational curves?) and to the arithmetic of unirational varieties (if all else fails, does X at least contain interesting rational points?). The goal of the present notes, which formed the basis for three lectures given at the Park City Mathematics Institute in August 2022, is to provide an introduction to these topics.

The inverse Galois problem is a simple-looking but fundamental open question of number theory on which tools coming from diverse areas of mathematics can be brought to bear. These lectures aim to explain the problem as well as a few of the many methods that have been developed to attack it, emphasising a geometric point of view whenever possible.

The first lecture introduces the problem together with a refinement of it first considered by Grunwald, and presents the strategy of Hilbert and Noether—a strategy based on Hilbert's irreducibility theorem and laid out more than a hundred years ago. This leads us to the notion of versal torsor, and to questions of rationality, stable rationality, retract rationality for quotient varieties. The second lecture is devoted to the regular inverse Galois problem, which is about the construction of Galois covers of curves. We present the rigidity method in detail, via Hurwitz moduli spaces. The third and final lecture explains how Grunwald's problem is expected to be controlled by the Brauer–Manin obstruction to weak approximation on the quotient varieties appearing in the Hilbert–Noether strategy, and discusses the descent method and its applications to the inverse Galois problem and to its variants.

Several important topics could not fit into these three lectures and had to be left aside, such as the connection between the inverse Galois problem and the construction of Galois representations (see [**88**], [**104**] for some examples), the directions in which the rigidity method has been developed beyond its base case (see e.g. [**24**], [**95**], [**58**, Chapter III] among many others), or the study of embedding problems.

2020 *Mathematics Subject Classification.* Primary 12F12; Secondary 14E08, 14G05.
Key words and phrases. Inverse Galois problem, Noether's problem, rigidity method, Brauer-Manin obstruction.

Additional material on the inverse Galois problem can be found in [**19**, **47**, **58**, **60**, **87**, **93**, **94**].

Acknowledgments

These lectures were delivered in August 2022 in Park City on the occasion of the PCMI 2022 graduate summer school, whose organisers are gratefully acknowledged. The author thanks Diego Izquierdo for leading the problem sessions, Enric Florit and Boris Kunyavskiĭ for correcting oversights in the first version of the notes, as well as Jean-Louis Colliot-Thélène, Joachim König, Jean-Pierre Serre, Tamás Szamuely and the referee for numerous useful comments that helped improve the exposition.

1. From Galois to Hilbert and Noether

1.1. Introduction. Galois theory turns the collection of all number fields into a profinite group, the absolute Galois group $\mathrm{Gal}(\bar{\mathbf{Q}}/\mathbf{Q})$ of $\mathbf{Q}$. The study of this group and of its representations has been a cornerstone of number theory for more than a century. Yet, even such a basic question as the following one remains wide-open to this day: do all finite groups appear as quotients of $\mathrm{Gal}(\bar{\mathbf{Q}}/\mathbf{Q})$? This is the so-called "inverse Galois problem".

The same question can be asked about the absolute Galois group of an arbitrary field k: given a finite group G and a field k, does there exist a Galois field extension K of k such that $\mathrm{Gal}(K/k) \simeq G$? Obviously, the answer is in the negative for some fields k that have a small absolute Galois group (e.g. the fields $\mathbf{C}$ and $\mathbf{R}$, trivially; or the field $\mathbf{Q}_p$, as its absolute Galois group is prosolvable). When k is a number field, a positive answer is known when G is solvable (Shafarevich, see [**72**, Chapter IX, §6] and the references therein), when G is a symmetric or an alternating group (Hilbert [**43**]), when G is a sporadic group other than M_{23} (Shih, Fried, Belyi, Matzat, Thompson, Hoyden–Siedersleben, Zeh–Marschke, Hunt, Pahlings, see [**58**]), when G belongs to various infinite families of non-abelian simple groups of Lie type (e.g. the groups $\mathrm{PSL}_2(\mathbf{F}_p)$, according to Shih, Malle, Clark, Zywina; see [**104**]); but the problem remains open over $\mathbf{Q}$ even for such a small group as $\mathrm{SL}_2(\mathbf{F}_{13})$, of order 2184, or as the simple group $\mathrm{PSL}_3(\mathbf{F}_8)$ (for the latter, see [**103**], to be complemented with [**25**]).

Several variants or generalisations of the inverse Galois problem have been considered in the literature. Here is one of them. Given a number field k, we denote the set of its places by Ω and the completion of k at $v \in \Omega$ by k_v.

Problem 1.1.1 (Grunwald). *Let k be a number field and $S \subset \Omega$ be a finite subset. Let G be a finite group. For each $v \in S$, let K_v be a Galois field extension of k_v such that the group $\mathrm{Gal}(K_v/k_v)$ can be embedded into G. Does there exist a Galois field extension K of k such that $\mathrm{Gal}(K/k) \simeq G$ and such that for all $v \in S$, the completion of K/k at any place of K lying above v is isomorphic to K_v/k_v?*

The Grunwald–Wang theorem, which was proved by Wang [**100**] following the work of Grunwald [**36**] and which has an interesting history (see [**2**, Chapter X, footnote on p. 73] and [**67**, Chapter VIII, §2, p. 234, Notes]), gives a complete answer to Problem 1.1.1 when G is abelian, via class field theory. In particular, the answer to Grunwald's problem is negative for $G = \mathbf{Z}/8\mathbf{Z}$ and $k = \mathbf{Q}$ (see Proposition 1.6.2 below), but it is positive, for any number field k and any finite abelian group G, as soon as S does not contain any place dividing 2. For an arbitrary finite group G, the Grunwald problem is expected to have a positive answer whenever G does not contain any place dividing the order of G. This is the "tame" Grunwald problem, a terminology coined in [**22**].

Other variants include embedding problems (given a Galois field extension ℓ/k, a finite group G and a surjection $\varphi : G \twoheadrightarrow \mathrm{Gal}(\ell/k)$, can one embed ℓ/k into a Galois field extension K/k such that $G \simeq \mathrm{Gal}(K/k)$, the map φ being identified with the restriction map $\mathrm{Gal}(K/k) \twoheadrightarrow \mathrm{Gal}(\ell/k)$?) or the question of resolving the inverse Galois problem with additional constraints, such as the constraint that a given finite collection of elements of k be norms from K (a problem raised in [**30**]).

1.2. Torsors and Galois extensions. Let us start by reformulating the inverse Galois problem in terms of torsors. Hereafter, a *variety* over a field k is a separated scheme of finite type over k (which may be disconnected or otherwise reducible) and $\bar{k}$ denotes an algebraic closure of k.

DEFINITION 1.2.1. Let $\pi : Y \to X$ be a finite morphism between varieties over a field k. Let G be a finite group acting on Y, in such a way that π is G-equivariant (for the trivial action of G on X). We say that π is a *G-torsor*, or that Y is a G-torsor over X, if π is étale and G acts simply transitively on the fibres of the map $Y(\bar{k}) \to X(\bar{k})$ induced by π.

When G is a finite group acting on a variety Y, we denote by Y/G the quotient variety, characterised by the universal property of quotients, when it exists. Let us recall that the quotient Y/G exists if Y is quasi-projective; the projection $\pi : Y \to Y/G$ is then finite and surjective; it is étale if the action of G on Y is free (by which we mean that G acts freely on the set $Y(\bar{k})$); and in the affine case, if $Y = \mathrm{Spec}(A)$, then $Y/G = \mathrm{Spec}(A^G)$ (see [**69**, Chapter II, §7 and Chapter III, §12]).

It is easy to see that a finite G-equivariant morphism $\pi : Y \to X$ is a G-torsor if and only if G acts freely on Y and π induces an isomorphism $Y/G \xrightarrow{\sim} X$. Thus, in particular, a Galois field extension K/k with Galois group G is the same thing as an irreducible G-torsor over k (that is, over $\mathrm{Spec}(k)$); and this, in turn, is the same thing as an irreducible variety of dimension 0, over k, endowed with a simply transitive action of G.

This rewording leads to a slight change in perspective, first emphasised by Hilbert and Noether: in order to solve the inverse Galois problem for G, we can now start with any irreducible quasi-projective variety Y endowed with a free action of G; setting $X = Y/G$, we obtain a G-torsor $\pi : Y \to X$; it is then enough to look for rational points $x \in X(k)$ such that the fibre $\pi^{-1}(x)$ is irreducible. Indeed, this fibre is in any case a G-torsor over k.

REMARK 1.2.2. Given a subgroup $H \subseteq G$, any H-torsor $Y \to X$ gives rise to a G-torsor $Y' \to X$. Namely, if H acts on the left on Y, we let it act on the right on $G \times Y$ by $(g, y) \cdot h = (gh, h^{-1}y)$ and observe that $Y' = (G \times Y)/H$ inherits a free left action of G, turning the projection $Y' \to X$ into a G-torsor. The variety Y' is (canonically) a disjoint union indexed by G/H of varieties each of which is (non-canonically, in general) isomorphic, over X, to Y. Conversely, if $Y' \to X$ is a G-torsor and X is connected, then any connected component Y of Y' is an H-torsor over X for some subgroup H (namely, the stabiliser of Y), and Y' coincides with $(G \times Y)/H$. All in all, when X is connected, the data of a G-torsor

$Y' \to X$ together with the choice of a connected component of Y' is equivalent to the data of a subgroup $H \subseteq G$ and of a connected H-torsor $Y \to X$. In particular, we see that if G is a finite group and k is a field, the data of a G-torsor over k together with the choice of a connected component is equivalent to the data of a Galois field extension K/k endowed with an embedding $\mathrm{Gal}(K/k) \hookrightarrow G$.

1.3. Hilbert's irreducibility theorem. When the base of the G-torsor $\pi : Y \to X$ is an open subset of $\mathbf{P}^1_k$, with k a number field, and its total space Y is irreducible, the existence of rational points $x \in X(k)$ such that the fibre $\pi^{-1}(x)$ is irreducible is guaranteed by Hilbert's irreducibility theorem, which we state next.

THEOREM 1.3.1 (Hilbert). *Let k be a number field. Let $X \subseteq \mathbf{P}^1_k$ be a dense open subset. Let $\pi : Y \to X$ be an irreducible étale covering (i.e. a finite étale morphism from an irreducible variety). There exists $x \in X(k)$ such that $\pi^{-1}(x)$ is irreducible.*

Theorem 1.3.1 is classically formulated in the following equivalent way: given an irreducible two-variable polynomial $f(s,t)$ with coefficients in a number field k, there exist infinitely many $t_0 \in k$ such that $f(s,t_0)$ is an irreducible one-variable polynomial with coefficients in k. In fact, the set of such t_0 is not just infinite: asymptotically, it contains 100% of the elements of k, when they are ordered by height (see [**86**, §13.1, Theorem 3]).

A proof of Theorem 1.3.1 can be found in [**86**, §9.2, §9.6], where the next corollary is also established.

COROLLARY 1.3.2. *Same statement, with X now a dense open subset of $\mathbf{P}^n_k$ for some $n \geq 1$.*

Combining Corollary 1.3.2 with the remarks of §1.2 leads to an observation, originating from Hilbert's work, which is extremely effective for the inverse Galois problem. Before stating it in Corollary 1.3.3 below, let us recall that a variety X over a field k is said to be *rational* if it is birationally equivalent to an affine space, i.e. if it contains a dense open subset isomorphic to a dense open subset of an affine space; when X is irreducible and reduced, this means that its function field $k(X)$ is a purely transcendental extension of k.

COROLLARY 1.3.3. *Let k be a number field. Let G be a finite group. If there exist an irreducible quasi-projective variety Y over k and a faithful action of G on Y such that the quotient Y/G is rational, then the inverse Galois problem admits a positive solution for G over k.*

PROOF. As G acts faithfully on Y, it acts freely on a dense open subset of Y, say V. Indeed, for $g \in G$, the locus in Y fixed by g is a closed subset of Y of codimension ≥ 1; one can take for V the complement of the (finite) union of these fixed loci. After shrinking V, we may assume that V/G is isomorphic to an open subset of $\mathbf{P}^n_k$. Corollary 1.3.2 can now be applied to the projection $V \to V/G$. □

EXAMPLE 1.3.4. The order 3 automorphism of $\mathbf{P}^1_k$ given, in homogeneous coordinates, by $[x:y] \mapsto [y : y-x]$ induces a faithful action of $G = \mathbf{Z}/3\mathbf{Z}$ on $\mathbf{P}^1_k$. The quotient $\mathbf{P}^1_k/G$ is rational since it is a unirational curve (Lüroth's theorem). Thus, any number field admits a cyclic extension of degree 3.

Ekedahl proved the following useful generalisation of Hilbert's irreducibility theorem. We recall that Ω denotes the set of places of a number field k.

THEOREM 1.3.5 ([**26**]). *Let $\pi : Y \to X$ be a finite étale morphism between geometrically irreducible varieties over a number field k. Let $S \subset \Omega$ be a finite subset. If $X(k) \neq \varnothing$, then there exists a nonempty open subset $\mathfrak{U} \subset \prod_{v \in \Omega \setminus S} X(k_v)$ such that for all $x \in X(k) \cap \mathfrak{U}$, the fibre $\pi^{-1}(x)$ is irreducible.*

In the above statement, we view $X(k)$ as diagonally embedded into the product $\prod_{v\in\Omega\setminus S} X(k_v)$, which we endow with the product of the v-adic topologies. When X is rational, the set $X(k)\cap\mathfrak{U}$ is automatically nonempty, by the weak approximation property for affine spaces. In general, though, this set can be empty. A proof of Theorem 1.3.5, at least in the Galois case,[1] which is the only one that we shall use (we use it in the proof of Proposition 1.5.5 below), can be found in [**26**, Theorem 1.3].

1.4. Noether's problem: statement. The following problem, raised by Emmy Noether, takes on particular importance in view of Corollary 1.3.3.

PROBLEM 1.4.1 (Noether). Let G be a finite group and k be a field. Choose an embedding $G \hookrightarrow S_n$ for some $n \geq 1$. Let G act on $\mathbf{A}^n_k$ through this embedding by permuting the coordinates. Is the quotient $\mathbf{A}^n_k/G$ rational over k?

By Corollary 1.3.3, when k is a number field, a positive answer to Noether's problem for G over k implies a positive answer to the inverse Galois problem for G over k. Beyond this implication, Noether's problem has become a central problem in the study of rationality, and has been the focus of much research for its own sake.

EXAMPLE 1.4.2. Noether's problem has a positive answer, over any field, for the symmetric group $G = S_n$. Indeed, for $G = S_n$, the quotient $\mathbf{A}^n_k/G$ is in fact isomorphic to $\mathbf{A}^n_k$, as the ring $k[x_1, \ldots, x_n]^{S_n}$ of symmetric polynomials coincides with the polynomial ring in the elementary symmetric polynomials. Thus, in particular, every number field admits a Galois field extension with group S_n, for every $n \geq 1$.

EXAMPLE 1.4.3. Noether's problem has a positive answer, for any $n \geq 1$ and any embedding $G \hookrightarrow S_n$, when G is an abelian group of exponent e and k is a field that contains the eth roots of unity and whose characteristic does not divide e. In particular, it has a positive answer for all abelian groups over $\mathbf{C}$. This is a theorem of Fischer [**28**].

EXAMPLE 1.4.4. Noether's problem has a positive answer, over any field, for the alternating group $G = A_5$. This is a theorem of Maeda [**57**]. On the other hand, as soon as $n \geq 6$, Noether's problem for $G = A_n$ is still open, over any field.

Noether knew that her problem has a positive answer for small groups (namely, for all subgroups of S_4). In general, however, its answer is often negative, as we discuss in some detail in §1.6 below.

[1] It can be checked that Theorem 1.3.5, in the Galois case, still holds when Y is only assumed to be irreducible, instead of geometrically irreducible. Under this weaker assumption on Y, the Galois case does imply the general case; hence the validity of Theorem 1.3.5 as stated (and even slightly more generally than stated, since this weaker assumption on Y also suffices in the non-Galois case).

1.5. Versal torsors. For some G-torsors $\pi : Y \to X$, the existence of rational points $x \in X(k)$ such that $\pi^{-1}(x)$ is irreducible is not only a sufficient condition for a positive answer to the inverse Galois problem for G over k, but it is also necessary. These are the *versal* torsors.

DEFINITION 1.5.1. Let G be a finite group, let k be a field and let X be a variety over k. A G-torsor $\pi : Y \to X$ is *weakly versal* if for any field extension k'/k with k' infinite, every G-torsor over k' can be realised as the fibre of π above a k'-point of X. It is *versal* if for any dense open subset $U \subseteq X$, the induced G-torsor $\pi^{-1}(U) \to U$ is weakly versal.

EXAMPLE 1.5.2. Choose an embedding $G \hookrightarrow S_n$ for some $n \geq 1$ and let G act on $\mathbf{A}^n_k$ through this embedding by permuting the coordinates. Let Y be the open subset of $\mathbf{A}^n_k$ consisting of the points whose coordinates are all pairwise distinct. Then G acts freely on Y and it can be checked, as a consequence of Hilbert's[2] Theorem 90, according to which the Galois cohomology set $H^1(k', \mathrm{GL}_n)$ is a singleton for any field k', that the resulting torsor $\pi : Y \to X = Y/G$ is versal (see [**33**, Example 5.5]).

EXAMPLE 1.5.3. Choose an embedding $G \hookrightarrow \mathrm{SL}_n(k)$ for some $n \geq 1$ and let G act on the algebraic group SL_n over k through this embedding by right multiplication. This action is free and it can again be checked, as a consequence of Hilbert's Theorem 90, that the resulting torsor $\pi : \mathrm{SL}_n \to \mathrm{SL}_n/G$ is versal. The argument for this is the same as in [**33**, Example 5.5] once one knows that the Galois cohomology set $H^1(k', \mathrm{SL}_n)$ is a singleton for any field k'; the latter fact easily follows from Hilbert's Theorem 90.

REMARK 1.5.4. Two varieties V and W over k are called *stably birationally equivalent* if $V \times \mathbf{A}^r_k$ and $W \times \mathbf{A}^s_k$ are birationally equivalent for some r, s. It can be shown that for any finite group G, the varieties $\mathbf{A}^m_k/G$ and SL_n/G appearing in Examples 1.5.2 and 1.5.3, for all values of m, n and all possible choices of embeddings $G \hookrightarrow S_m$ and $G \hookrightarrow \mathrm{SL}_n(k)$, all fall into the same stable birational equivalence class of varieties over k. This is the so-called "no-name lemma", see [**17**, Corollary 3.9].

The notion of versality, in the context of these notes,[3] is motivated by the following observation, which is an improved version of Corollary 1.3.3:

PROPOSITION 1.5.5. *Let k be a number field. Let $S_0 \subset \Omega$ be a finite subset. Let G be a finite group. Suppose that there exist an irreducible smooth quasi-projective variety Y over k and a free action of G on Y satisfying the following two conditions:*

(i) *the variety $X = Y/G$ satisfies weak approximation off S_0, i.e. the diagonal embedding $X(k) \hookrightarrow \prod_{v \in \Omega \setminus S_0} X(k_v)$ has dense image;*
(ii) *the G-torsor $\pi : Y \to X$ is weakly versal.*

[2]Despite its name Hilbert's Theorem 90, this theorem, for arbitrary n, is due to Speiser [**90**].

[3]Outside of the context discussed here, the notion of versality notably also gives rise to the definition of the "essential dimension" of a finite group G over a field k—this is the minimal dimension of a versal G-torsor defined over k—which is interesting in its own right and has been the focus of many works (see [**7, 10, 61, 62, 79, 80**]). Even determining the essential dimension of $\mathbf{Z}/8\mathbf{Z}$ over $\mathbf{Q}$ is a highly nontrivial task (see [**29**]).

Then Grunwald's problem admits a positive answer for G over k, for any finite subset $S \subset \Omega$ disjoint from S_0.

PROOF. We shall need the following classical lemma, whose proof can be found in [**76**, Proposition 3.5.74] and whose statement holds for any finite étale morphism π.

LEMMA 1.5.6 (Krasner). *For $v \in \Omega$, the isomorphism class of the variety $\pi^{-1}(x_v)$ over k_v is a locally constant function of $x_v \in X(k_v)$ with respect to the v-adic topology.*

Fix Galois field extensions K_v/k_v for $v \in S$ as in Problem 1.1.1 and choose embeddings $\mathrm{Gal}(K_v/k_v) \hookrightarrow G$ for $v \in S$. By Remark 1.2.2, these choices give rise to G-torsors over k_v for $v \in S$. By weak versality, the latter come from k_v-points $x_v \in X(k_v)$. Lemma 1.5.6 provides, for every $v \in S$, a neighbourhood $\mathcal{U}_v \subset X(k_v)$ of x_v such that $\pi^{-1}(x'_v)$ and $\pi^{-1}(x_v)$ are isomorphic, as varieties over k_v, for all $x'_v \in \mathcal{U}_v$. In particular, by Remark 1.2.2 again, for all $v \in S$ and all $x'_v \in \mathcal{U}_v$, the fibre $\pi^{-1}(x'_v)$ is isomorphic, over k_v, to a disjoint union of copies of $\mathrm{Spec}(K_v)$.

The weak versality of π also implies that $X(k) \neq \varnothing$. Theorem 1.3.5 therefore provides a nonempty open subset $\mathcal{U}^0 \subset \prod_{v \in \Omega \setminus (S \cup S_0)} X(k_v)$ such that $\pi^{-1}(x)$ is irreducible for all $x \in X(k) \cap \mathcal{U}^0$. Let $\mathcal{U} = \left(\prod_{v \in S} \mathcal{U}_v\right) \times \mathcal{U}^0$. As the variety X satisfies weak approximation off S_0, the set $X(k) \cap \mathcal{U}$ is nonempty. We fix $x \in X(k) \cap \mathcal{U}$. The fibre $\pi^{-1}(x)$ is now an irreducible G-torsor (i.e. $\mathrm{Spec}(K)$ for some Galois field extension K/k with Galois group G) whose scalar extension from k to k_v, for each $v \in S$, is a disjoint union of copies of $\mathrm{Spec}(K_v)$. This proves the proposition. □

As smooth rational varieties satisfy weak approximation, one can apply Proposition 1.5.5 with $S_0 = \varnothing$ whenever the variety Y/G is rational and the torsor $Y \to Y/G$ is weakly versal. In view of Example 1.5.2, we deduce:

COROLLARY 1.5.7. *Given a finite group G and a number field k, a positive answer to Noether's problem for G and k implies a positive answer to Grunwald's problem for G and k, for any $S \subset \Omega$.*

Corollary 1.5.7 was first established by Saltman (see [**82**, Theorem 5.1, Theorem 5.9]). As an example of an application, Corollary 1.5.7 implies that over any number field k, Grunwald's problem has a positive answer for S_n and for A_5 over k, without the need to exclude any place from $S \subset \Omega$ (see Example 1.4.2 and Example 1.4.4).

1.6. Noether's problem: some counterexamples. The hope that a positive solution to the inverse Galois problem might in general come from a positive solution to Noether's problem turned out, however, to be too optimistic. Indeed, Noether's problem seems to have a negative solution more often than not, as we briefly discuss below.

1.6.1. *Counterexamples among abelian groups.* Noether's problem has a negative answer even for cyclic groups over $\mathbf{Q}$. Swan and Voskresenskiĭ discovered, at the end of the 1960's, the counterexample $\mathbf{Z}/47\mathbf{Z}$ over $\mathbf{Q}$ (see [**91**], [**96**]). An even smaller counterexample, the group $\mathbf{Z}/8\mathbf{Z}$ over $\mathbf{Q}$, was then exhibited in [**27**], [**54**], [**98**]. As Saltman [**82**] later observed, Corollary 1.5.7 provides a direct proof that Noether's problem admits a negative answer for $\mathbf{Z}/8\mathbf{Z}$ over $\mathbf{Q}$. Indeed, by this corollary, it suffices to show that Grunwald's problem has a negative answer for $G = \mathbf{Z}/8\mathbf{Z}$, $k = \mathbf{Q}$ and $S = \varnothing$, and this is exactly what Wang had done in the 1940's:

PROPOSITION 1.6.2 (Wang). *In a cyclic field extension $K/\mathbf{Q}$ of degree 8, the prime 2 cannot be inert. In other words, the completion of a cyclic field extension $K/\mathbf{Q}$ of degree 8 at a place dividing 2 cannot be the unramified extension of $\mathbf{Q}_2$ of degree 8.*

An elementary proof can be found in [**92**, p. 29, end of §5].

Further work on Noether's problem for abelian groups, by Endo, Miyata, Voskresenskiĭ and Lenstra, led to a complete characterisation, by Lenstra [**54**], of the stable rationality of the quotient $\mathbf{A}^n_k/G$ appearing in Problem 1.4.1 (and even of its rationality, in the case where G acts through its regular representation), when G is a finite abelian group and k is an arbitrary field. This characterisation is in terms of the arithmetic of cyclotomic number fields. For cyclic groups over $\mathbf{Q}$, it reads as follows (see [**55**, §3]):

THEOREM 1.6.3 (Lenstra). *Let $n \geq 1$ be an integer. Let $G = \mathbf{Z}/n\mathbf{Z}$ faithfully act on $\mathbf{A}^n_{\mathbf{Q}}$ by cyclically permuting the coordinates. The following conditions are equivalent:*

(1) *The variety $\mathbf{A}^n_{\mathbf{Q}}/G$ is rational.*
(2) *The variety $\mathbf{A}^n_{\mathbf{Q}}/G$ is stably rational.*
(3) *The integer n is not divisible by 8, and for every prime factor p of n, if s denotes the p-adic valuation of n, the cyclotomic ring $\mathbf{Z}\big[\zeta_{(p-1)p^{s-1}}\big]$ contains an element whose norm is equal to p or to $-p$.*

We recall that a variety is said to be *stably rational* if its product with an affine space of large enough dimension is rational. Stable rationality is known to be strictly weaker than rationality in general, even over $\mathbf{C}$, see [**6**].

Theorem 1.6.3 when n is a prime number is due to Voskresenskiĭ [**97**]. Even when n is prime, determining whether condition (3) of Theorem 1.6.3 does or does not hold for a given n is in general a hard problem; for instance, it is only recently that this condition was shown to fail for $n = 59$ (see [**44**, Added remark 3.2]). Even more recently, based on Theorem 1.6.3, on a height estimate due to Amoroso and Dvornicich [**1**] and on extensive computer calculations run by Hoshi [**44**], among other tools, Plans [**75**] was able to give a complete answer to Noether's problem for cyclic groups over $\mathbf{Q}$:

THEOREM 1.6.4 (Plans). *The conditions of Theorem 1.6.3 are equivalent to the following:*

(4) *The integer n divides $2^2{\cdot}3^m{\cdot}5^2{\cdot}7^2{\cdot}11{\cdot}13{\cdot}17{\cdot}19{\cdot}23{\cdot}29{\cdot}31{\cdot}37{\cdot}41{\cdot}43{\cdot}61{\cdot}67{\cdot}71$ for some integer $m \geq 0$.*

In particular, Noether's problem has a negative answer over $\mathbf{Q}$ for $G = \mathbf{Z}/p\mathbf{Z}$ for all but finitely many prime numbers p.

1.6.5. *Counterexamples over* $\mathbf{C}$. For non-abelian groups, Noether's problem has a negative answer even over $\mathbf{C}$. Saltman [**83**] gave the first counterexamples over $\mathbf{C}$. His results were then generalised by Bogomolov [**8**], who established the following theorem (see [**17**, §7] and [**34**, §6.6] for accounts of its proof):

THEOREM 1.6.6 (Bogomolov's formula). *Let $n \geq 1$ and $G \subset \mathrm{SL}_n(\mathbf{C})$ be a finite subgroup. The unramified Brauer group of the complex variety SL_n/G is isomorphic to*

$$\mathrm{Ker}\Big(H^2(G,\mathbf{Q}/\mathbf{Z}) \to \prod H^2(H,\mathbf{Q}/\mathbf{Z})\Big), \tag{1.6.7}$$

where the product ranges over all bicyclic *subgroups $H \subseteq G$ (i.e. abelian subgroups of G that are generated by at most two elements).*

We recall that the Brauer group, defined by Grothendieck as $H^2_{\text{ét}}(-,\mathbf{G}_{\mathrm{m}})$, is a stable birational invariant among smooth proper varieties over a field of characteristic 0, and that the *unramified Brauer group* of a variety over a field of characteristic 0 is by definition the Brauer group of any smooth proper variety birationally equivalent to it; for instance, the unramified Brauer group of $\mathbf{A}^n_{\mathbf{C}}$ is trivial. Thus, if the unramified Brauer group of a variety over $\mathbf{C}$ does not vanish, then this variety is not stably rational, a fortiori it is not rational. The unramified Brauer group was first considered and used as a tool for rationality questions by Saltman [**83**,**84**]. For smooth proper unirational varieties over $\mathbf{C}$, it coincides with the invariant that had earlier been employed by Artin and Mumford [**3**] to give "elementary" examples of complex unirational threefolds failing to be rational. For a thorough treatment of the Brauer group, we refer the reader to [**18**].

In view of Remark 1.5.4, Bogomolov's formula gives an easy recipe for computing the unramified Brauer group of the variety $\mathbf{A}^n_{\mathbf{C}}/G$ that appears in Noether's problem over $\mathbf{C}$. The kernel (1.6.7) can be computed to be nonzero for some p-groups G, thus yielding counterexamples to Noether's problem over $\mathbf{C}$ (see [**17**, Example 7.5], [**34**, §6.7]).

Other counterexamples over $\mathbf{C}$ were later produced by Peyre [**74**] based on a further stable birational invariant called *unramified cohomology of degree* 3 and introduced by Colliot-Thélène and Ojanguren [**14**]. The unramified Brauer group coincides with unramified cohomology of degree 2.

Many more results about Noether's problem can be found in the survey [**45**].

1.7. Retract rationality. Saltman introduced a useful weakening of the notion of stable rationality: a variety X over a field k is said to be *retract rational* if there exist an integer $n \geq 1$, a dense open subset $U \subseteq \mathbf{A}^n_k$ and a morphism $U \to X$ that admits a rational section. Retract rationality is a stable birational invariant.

In the situation of Noether's problem, it can happen that the variety $\mathbf{A}^n_k/G$ fails to be rational and even to be stably rational, but is nevertheless retract rational. For instance this is so when $G = \mathbf{Z}/47\mathbf{Z}$ and $k = \mathbf{Q}$:

THEOREM 1.7.1 (Saltman [**82**]). *Taking up the notation of Problem 1.4.1, assume that G is abelian, that k has characteristic 0, and, letting 2^r denote the highest power of 2 that divides the exponent of G, that the cyclotomic field extension $k(\zeta_{2^r})/k$ is cyclic. Then the quotient $\mathbf{A}^n_k/G$ is retract rational over k.*

Theorem 1.7.1 can in particular be applied to all finite abelian groups of odd order. Thus, retract rationality is weaker than rationality (compare with Theorem 1.6.4). Nevertheless, as far as the applications to the inverse Galois problem are concerned, it is just as good: indeed, as Saltman observed, smooth retract rational varieties over number fields are easily seen to satisfy weak approximation, so that Proposition 1.5.5 can be applied with $S_0 = \varnothing$ whenever the variety Y/G is retract rational and the torsor $Y \to Y/G$ is weakly versal.

Combining this observation with Theorem 1.7.1 and with Proposition 1.5.5, we deduce, in view of Example 1.5.2, that Grunwald's problem has a positive answer over any number field k, without excluding any place, for all abelian groups G satisfying the assumption of Theorem 1.7.1—a conclusion that already resulted from the Grunwald–Wang theorem, but whose proof now fits into the framework of Hilbert's and Noether's general strategy, even though, according to Theorem 1.6.4, Noether's problem itself has a negative answer for many of these groups G (perhaps even for "almost all" of them?).

Conversely, by the same token, Wang's negative answer to Grunwald's problem (see Proposition 1.6.2) implies that when $G = \mathbf{Z}/8\mathbf{Z}$ and $k = \mathbf{Q}$, the quotient $\mathbf{A}^n_k/G$ fails not only to be stably rational but also to be retract rational. Similarly, the negative answers to Noether's problem over $\mathbf{C}$ discussed in §1.6.5 are in fact counterexamples to the retract rationality of the quotients $\mathbf{A}^n_{\mathbf{C}}/G$ in question. Thus, despite the wider scope of applicability of the Hilbert–Noether method when rationality is replaced with the weaker notion of retract rationality, further ideas are necessary to address arbitrary finite groups.

2. Regular inverse Galois problem

2.1. Statement. We saw, in §1, that Noether's problem does not always admit a positive answer, i.e. the quotient variety $\mathbf{A}^n_k/G$ can fail to be rational, or stably rational, or even retract rational. A simple way out, if one still wants to apply Hilbert's irreducibility theorem, is to look for rational *subvarieties* of $\mathbf{A}^n_k/G$, in particular rational curves. To take advantage of the geometry of the situation, it is natural to focus on those rational curves whose inverse image in $\mathbf{A}^n_k$ is geometrically irreducible and meets the locus $Y \subset \mathbf{A}^n_k$ on which G acts freely. By the versality of the torsor $Y \to Y/G$ (Example 1.5.2), finding such curves is the same as solving the *regular inverse Galois problem* (when k is perfect):

PROBLEM 2.1.1 (regular inverse Galois). Let k be a field. Let G be a finite group. Do there exist a smooth, projective, geometrically irreducible curve C over k and a finite morphism $\pi : C \to \mathbf{P}^1_k$ such that the corresponding extension of function fields $k(C)/k(t)$ is Galois with $\mathrm{Gal}(k(C)/k(t)) \simeq G$?

When k is a perfect field, this is equivalent to asking for the existence of a field extension of $k(t)$ with Galois group G in which k is algebraically closed, i.e. a field extension that is regular over k. Following standard practice, we shall refer to such a field extension as a *regular Galois extension of* $k(t)$ *with group* G.

When k is a number field, a positive answer to Problem 2.1.1 for k and G implies a positive answer to the inverse Galois problem for k and G, by Hilbert's irreducibility theorem. Over an arbitrary field and for an arbitrary finite group, the inverse Galois problem and Noether's problem both have negative answers in general, as we have seen; in contrast, Problem 2.1.1 might well always have a positive answer.

REMARK 2.1.2. It follows from the Bertini theorem that if k is infinite and perfect, a positive answer to Noether's problem for k and G implies a positive answer to the regular inverse Galois problem for k and G (see [**48**, Théorème 6.3]). In fact, for such k, one can check that the retract rationality of $\mathbf{A}^n_k/G$ already implies a positive answer to the regular inverse Galois problem for k and G.

2.2. Riemann's existence theorem. A solution to the regular inverse Galois problem over k gives rise, by scalar extension, to a solution over any field extension of k. Thus, in order to find a solution over $\mathbf{Q}$, it is necessary to first solve the problem over $\mathbf{C}$ and over $\bar{\mathbf{Q}}$. The key tool for this is Riemann's existence theorem, which allows one to transform this algebraic question into a purely topological one.

THEOREM 2.2.1 (Riemann's existence theorem). *Let k be an algebraically closed subfield of $\mathbf{C}$. Let X be a variety over k. The natural functor*

$$\Big(\textit{étale coverings of } X\Big) \to \Big(\textit{finite topological coverings of } X(\mathbf{C})\Big)$$

that maps $Y \to X$ to $Y(\mathbf{C}) \to X(\mathbf{C})$ is an equivalence of categories.

An *étale covering* of X is a variety over k endowed with a finite étale morphism to X. A topological covering is *finite* if its fibres are finite. Theorem 2.2.1 in the above formulation is proved in [**35**]. To be precise, the case where $k = \mathbf{C}$ is [**35**, Exp. XII, Théorème 5.1] and builds on Grothendieck's reworking of Serre's GAGA theorems; the case of an arbitrary algebraically closed subfield of $\mathbf{C}$ then results from it by [**35**, Exp. XIII, Corollaire 3.5].

COROLLARY 2.2.2. *Let k be an algebraically closed subfield of $\mathbf{C}$. Let X be a connected variety over k. Let $x \in X(k)$. For any finite group G, isomorphism classes of G-torsors (resp. of connected G-torsors) $Y \to X$ endowed with a lift $y \in Y(k)$ of x are canonically in one-to-one correspondence with homomorphisms $\pi_1(X(\mathbf{C}), x) \to G$ (resp. with surjective homomorphisms $\pi_1(X(\mathbf{C}), x) \twoheadrightarrow G$). Changing the choice of y amounts to conjugating the homomorphism by an element of G.*

PROOF. Indeed, this follows from Theorem 2.2.1 combined with the well-known equivalence of categories between the category of topological coverings of $X(\mathbf{C})$ and the category of sets endowed with an action of $\pi_1(X(\mathbf{C}), x)$ (see [**93**, Theorem 2.3.4]). The homomorphism $\pi_1(X(\mathbf{C}), x) \to G$ corresponding to $Y \to X$ sends $\gamma \in \pi_1(X(\mathbf{C}), x)$ to the unique $g \in G$ such that $\gamma y = yg$, where we are taking the convention that the action of G on Y is a right action and that the monodromy action of $\pi_1(X(\mathbf{C}), x)$ on the fibre of $Y(\mathbf{C}) \to X(\mathbf{C})$ above x is a left action. □

REMARK 2.2.3 (reminder on monodromy groups and Galois groups). Let k and X be as in Corollary 2.2.2. Let $x \in X(\mathbf{C})$. The *monodromy group* M of an étale covering $Y \to X$ is, by definition, the largest quotient of $\pi_1(X(\mathbf{C}), x)$ through which the monodromy action of this group on the fibre of $Y(\mathbf{C}) \to X(\mathbf{C})$ above x factors. Assume that X is normal and irreducible and let $Y' \to Y \to X$ be a tower of irreducible étale coverings such that the field extension $k(Y')/k(X)$ is a Galois closure of $k(Y)/k(X)$. Let $G = \mathrm{Gal}(k(Y')/k(X))$. Then $Y' \to X$ is the normalisation of X in $k(Y')$; as such, it receives an action of G, with respect to which it is a G-torsor; in addition, the surjective homomorphism $\pi_1(X(\mathbf{C}), x) \twoheadrightarrow G$ that Corollary 2.2.2 associates with $Y' \to X$ and with the choice of a lift $y' \in Y'(k)$ of x induces an isomorphism $M \xrightarrow{\sim} G$. (Changing the choice of the lift y' amounts to composing this isomorphism with an inner automorphism.) Thus, computing the Galois group of the Galois closure of the field extension $k(Y)/k(X)$ is tantamount to computing a monodromy group in the topological setting.

2.3. Classifying Galois covers of the projective line over C or over $\bar{\mathbf{Q}}$. Let us apply Theorem 2.2.1 to the open subsets of the projective line. The fundamental group of the complement of finitely many points in $\mathbf{P}^1(\mathbf{C})$ is easy to describe:

PROPOSITION 2.3.1. *Let $X \subseteq \mathbf{P}^1_{\mathbf{C}}$ be a dense open subset. Write $\mathbf{P}^1_{\mathbf{C}} \setminus X = \{b_1, \dots, b_r\}$. Let $x \in X(\mathbf{C})$. The group $\pi_1(X(\mathbf{C}), x)$ admits a presentation with r generators $\gamma_1, \dots, \gamma_r$ and a unique relation $\gamma_1 \cdots \gamma_r = 1$, such that γ_i belongs, for every $i \in \{1, \dots, r\}$, to the conjugacy class in $\pi_1(X(\mathbf{C}), x)$ of a local counterclockwise loop around b_i.*

What the last sentence of Proposition 2.3.1 means is this: if N_i denotes a small enough open neighbourhood of b_i in $\mathbf{P}^1(\mathbf{C})$ that is biholomorphic to the unit disc, then a loop contained in $N_i \setminus \{b_i\}$ and going once around b_i in the counterclockwise direction gives rise, after choosing a path from x to a point of this loop, to an element of $\pi_1(X(\mathbf{C}), x)$ whose conjugacy class does not depend on the chosen path. The content of Proposition 2.3.1 is that these paths can be chosen in such a way that the γ_i generate $\pi_1(X(\mathbf{C}), x)$ and satisfy the relation $\gamma_1 \cdots \gamma_r = 1$. This is elementary and well-known.

Using Proposition 2.3.1, we can draw the following corollary from Riemann's existence theorem. Corollary 2.3.2 completely describes G-torsors over dense open subsets of the projective line over algebraically closed subfields of $\mathbf{C}$, and implies a positive solution to the regular inverse Galois problem over such fields. (The notation $\mathrm{ni}^*_r(G)$ appearing in its statement refers to the name Nielsen, see [**94**, §9.2], [**81**, §3.1].)

COROLLARY 2.3.2. *Let k be an algebraically closed subfield of $\mathbf{C}$. Let $X \subseteq \mathbf{P}^1_k$ be a dense open subset. Write $\mathbf{P}^1_{\mathbf{C}} \setminus X = \{b_1, \dots, b_r\}$. Let G be a finite group. Consider the set of r-tuples $(g_1, \dots, g_r) \in G^r$ such that $g_1 \cdots g_r = 1$ and that $g_1, \dots, g_r$ generate G. Let $\mathrm{ni}^*_r(G)$ denote the quotient of this set by the action of G by simultaneous conjugation. The set of isomorphism classes of irreducible G-torsors over X is in bijection with $\mathrm{ni}^*_r(G)$ (through a bijection that is canonically determined once a presentation of $\pi_1(X(\mathbf{C}), x)$ as in Proposition 2.3.1 is fixed).*

PROOF. By Corollary 2.2.2, isomorphism classes of irreducible G-torsors over X are canonically in one-to-one correspondence with conjugacy classes of surjections $\pi_1(X(\mathbf{C}), x) \twoheadrightarrow G$. Apply Proposition 2.3.1 to conclude. □

COROLLARY 2.3.3. *For any finite group G, the regular inverse Galois problem admits a positive answer over $\bar{\mathbf{Q}}$.*

PROOF. Let r be an integer, large enough that G can be generated by $r-1$ elements. Pick r points of $\mathbf{P}^1(\bar{\mathbf{Q}})$ and let $X \subset \mathbf{P}^1_{\bar{\mathbf{Q}}}$ denote their complement. As $\mathrm{ni}^*_r(G) \neq \varnothing$, Corollary 2.3.2 ensures the existence of an irreducible G-torsor $p : Y \to X$. As Y is normal and p is finite, the normalisation of $\mathbf{P}^1_{\bar{\mathbf{Q}}}$ in the function field of Y is a smooth curve C over $\bar{\mathbf{Q}}$ containing Y as a dense open subset, equipped with a finite morphism $\pi : C \to \mathbf{P}^1_{\bar{\mathbf{Q}}}$ that extends p. As p is a G-torsor, the function field extension $\bar{\mathbf{Q}}(C)/\bar{\mathbf{Q}}(t)$ is Galois with group G (see §1.2). □

2.4. Monodromy of some non-Galois covers of the projective line. Proposition 2.3.1 is also useful for computing the monodromy of ramified covers of the complex projective line that are not necessarily Galois, via the following result.

PROPOSITION 2.4.1. *Let C be a smooth, projective, irreducible curve over $\mathbf{C}$, endowed with a finite morphism $\pi : C \to \mathbf{P}^1_{\mathbf{C}}$. Let $X \subseteq \mathbf{P}^1_{\mathbf{C}}$ be a dense open subset over which π is étale. Fix $x \in X(\mathbf{C})$ and write $\mathbf{P}^1_{\mathbf{C}} \setminus X = \{b_1, \ldots, b_r\}$. Let M denote the monodromy group of π, i.e. the largest quotient of $\pi_1(X(\mathbf{C}), x)$ that still acts on $\pi^{-1}(x)$. After choosing a bijection $\pi^{-1}(x) \simeq \{1, \ldots, n\}$, we view M as a transitive subgroup of the symmetric group S_n. There exist $\mu_1, \ldots, \mu_r \in M$ satisfying the following three properties:*

(1) *the elements $\mu_1, \ldots, \mu_r$ generate the group M;*
(2) *their product $\mu_1 \cdots \mu_r$ is the identity of M;*
(3) *for each $i \in \{1, \ldots, r\}$, the element $\mu_i \in S_n$ is a product of cycles whose lengths are the ramification indices of π at the points of $\pi^{-1}(b_i)$.*

PROOF. Applying Proposition 2.3.1 and letting μ_i denote the image of γ_i in M, we obtain (1) and (2). Property (3) only depends on the conjugacy class of γ_i and is a standard calculation of the monodromy of the étale coverings of the punctured unit disc. □

EXAMPLE 2.4.2. Let C be a smooth, projective, irreducible curve over an algebraically closed subfield k of $\mathbf{C}$, endowed with a morphism $\pi : C \to \mathbf{P}^1_k$ of degree $n \geq 1$. Assume that all ramification points have ramification index 2 and that no two of them lie in the same fibre of π. Then the Galois group of a Galois closure of the function field extension $k(C)/k(t)$ is the full symmetric group S_n. Indeed, Remark 2.2.3 and Proposition 2.4.1 show that this Galois group is a transitive subgroup of S_n generated by transpositions; the only such subgroup is S_n itself.

REMARK 2.4.3. Let k be a field of characteristic 0. Let $k(t) \subseteq K \subseteq K' \subset \overline{k(t)}$ be a tower of fields, where $\overline{k(t)}$ denotes an algebraic closure of $k(t)$, where $K/k(t)$ is a finite extension and where $K'/k(t)$ is its Galois closure inside $\overline{k(t)}$. Let us assume that k is algebraically closed in K. The field k need not, in general, be algebraically closed in K'. (For example, if $k = \mathbf{Q}$ and $K = \mathbf{Q}(t^{1/n})$, then $K' = \mathbf{Q}(\zeta_n)(t^{1/n})$, where ζ_n denotes a primitive nth root of unity.) This pathology, however, cannot occur if the underlying topological monodromy group is the full symmetric group, or, more generally, if it is a self-normalising subgroup of the ambient symmetric group. Indeed, let k' denote the algebraic closure of k in K', set $G = \mathrm{Gal}(K'/k(t))$ and $G_{\mathrm{geom}} = \mathrm{Gal}(K'/k'(t))$. Letting $\bar{k}$ denote the algebraic closure of k in $\overline{k(t)}$, we remark that $K' \otimes_{k'} \bar{k}$ and $K \otimes_k \bar{k}$ are fields and that the field extension $K' \otimes_{k'} \bar{k}/\bar{k}(t)$ is a Galois closure of $K \otimes_k \bar{k}/\bar{k}(t)$, so that its Galois group G_{geom} can be viewed as the topological monodromy group associated with $K/k(t)$ (see Remark 2.2.3). Fix a primitive element $\alpha_1 \in K$ over $k(t)$. Denote by $\alpha_1, \ldots, \alpha_n \in K'$ the collection of its Galois conjugates. As G acts faithfully on the α_i's, there is a sequence of inclusions $G_{\mathrm{geom}} \subseteq G \subseteq S_n$. As k'/k is a Galois field extension, the group G_{geom} is normal in G; hence, if G_{geom} is self-normalising in S_n, then $G = G_{\mathrm{geom}}$ and k is algebraically closed in K'. Thus, for instance, if the curve C and the morphism π of Example 2.4.2 come by scalar extension from a curve and a morphism defined over $\mathbf{Q}$, and if $K/\mathbf{Q}(t)$ denotes the function field extension given by the latter morphism, then a Galois closure of $K/\mathbf{Q}(t)$ has Galois group S_n.

In conjunction with Remark 2.4.3, Example 2.4.2 leads to many concrete examples of regular Galois extensions of $\mathbf{Q}(t)$ with group S_n. Let us recall, however, that the mere existence of regular Galois extensions of $\mathbf{Q}(t)$ with group S_n already followed from the positive answer to Noether's problem for S_n over $\mathbf{Q}$ (see Example 1.4.2 and Remark 2.1.2). As Noether's problem is open for the alternating group A_n over $\mathbf{Q}$ as soon as $n \geq 6$, it is of interest to note that Proposition 2.4.1 also leads to concrete examples of regular Galois extensions of $\mathbf{Q}(t)$ with group A_n for all values of n, as we now illustrate.

EXAMPLE 2.4.4. Let C be a smooth, projective, geometrically irreducible curve over a subfield k of $\mathbf{C}$, endowed with a morphism $\pi : C \to \mathbf{P}^1_k$ of degree $n \geq 3$. Assume that π has exactly three ramification points, that these ramification points are rational points of C lying above $0, 1, \infty \in \mathbf{P}^1(k)$, with ramification indices e_0, e_1, e_∞, respectively, and that $(e_0, e_1, e_\infty) = (n, n-1, 2)$ if n is even and $(e_0, e_1, e_\infty) = (n-1, n, 2)$ if n is odd. Let $K'/k(t)$ denote a Galois closure of the function field extension $k(C)/k(t)$. We first note that K' is a regular Galois extension of $k(t)$ with group S_n. Indeed, when k is algebraically closed, Remark 2.2.3 and Proposition 2.4.1 imply that $\mathrm{Gal}(K'/k(t))$ is a transitive subgroup of S_n that contains a cycle of order $n-1$ and a transposition, but the only such subgroup is S_n itself (see [**87**, Lemma 4.4.3]); by Remark 2.4.3, the case of arbitrary k follows. Secondly, we claim that there exists $\alpha \in k^*$ such that αt is a square in K'. Setting $u = \sqrt{\alpha t}$, this will imply that K' is a regular Galois extension of $k(u)$ with group A_n (where u can now be viewed as a free variable), as desired. To verify this claim, we note that the topological monodromy of the double cover of $\mathbf{P}^1_k$ corresponding to the (unique) quadratic subextension $L/k(t)$ of $K'/k(t)$ is obtained by composing the topological monodromy of π with the signature morphism $S_n \to \mathbf{Z}/2\mathbf{Z}$. As the local monodromy of π at 1 is given by a cycle of odd length, it follows that $L/k(t)$ is unramified outside of 0 and ∞, and, hence, that $L = k(\sqrt{\alpha t})$ for some $\alpha \in k^*$ (as k is algebraically closed in L).

For explicit equations to which Example 2.4.4 can be applied, see [**87**, §4.5].

2.5. Looking for covers over non-algebraically closed ground fields. Now that we know that the regular inverse Galois problem has a positive answer over $\bar{\mathbf{Q}}$, we can try to find solutions over $\mathbf{Q}$ or at least over overfields of $\mathbf{Q}$ as small as possible. This has been achieved over the completions of $\mathbf{Q}$, thus yielding, for all finite groups, a positive answer to the regular inverse Galois problem over $\mathbf{R}$ (Krull and Neukirch [**53**]) and over the field $\mathbf{Q}_p$ of p-adic numbers for every prime p (Harbater [**39**]). Pop [**78**] generalised these results as follows[4,5]:

THEOREM 2.5.1 (Harbater and Pop). *The regular inverse Galois problem has a positive answer over any large field, for any finite group.*

By definition, a field k is *large* when every smooth curve over k that has a rational point has infinitely many of them. Examples include all fields that are complete with respect to an absolute value, such as $\mathbf{R}$ and $\mathbf{Q}_p$, as well as infinite algebraic extensions of finite fields or more generally all so-called pseudo-algebraically closed

[4]It is not immediately clear that the article [**78**] establishes Theorem 2.5.1 as we have stated it, without assuming the field to be perfect: in our definition of the regular inverse Galois problem, the sought-for field extension of $k(t)$ was required to admit a *smooth* projective model, which could fail over imperfect fields. However, in any case, Theorem 2.5.1 as we have stated it is proved in [**68**, Théorème 1.1].

[5]Theorem 2.5.1 (at least for perfect large fields, see the previous footnote) also follows from the results of Harbater [**39**], see [**40**, §4.5].

fields (fields over which every smooth geometrically connected curve has infinitely many rational points).

The proofs of Theorem 2.5.1 given by Harbater and by Pop rely, in the formal or in the rigid analytic context over a complete discretely valued ground field, on the construction, by patching, of appropriate "topological coverings", and on a variant, in the corresponding context, of Riemann's existence theorem. Over $\mathbf{C}$, the underlying patching construction is presented in [**93**, §3.5].

Theorem 2.5.1 had previously been established by Fried and Völklein [**31**] in the case of pseudo-algebraically closed fields of characteristic 0. From this special case they deduced the following result in positive characteristic:

THEOREM 2.5.2 (Fried and Völklein). *Let G be a finite group. The regular inverse Galois problem has a positive answer for G over $\mathbf{F}_p(t)$ for all but finitely many primes p.*

Colliot-Thélène later shed new light on Theorem 2.5.1 by recasting it as a theorem about the existence of suitable rational curves on the varieties $\mathbf{A}^n_k/G$ appearing in Noether's problem, and by noting that even though these varieties can fail to be rational, they are in any case *rationally connected*, which opens the door to applications of the theory of deformation of rational curves on rationally connected varieties over large fields—a theory developed, in great generality, by Kollár [**50**]. Over large fields of characteristic 0, a geometric proof of Theorem 2.5.1 that proceeds by constructing rational curves on $\mathbf{A}^n_k/G$ was thus given in [**12**]. See also [**51**], [**52**], [**68**] for generalisations.

Unfortunately, no method for the systematic construction of rational curves on rationally connected varieties over $\mathbf{Q}$ is known; more generally, the various methods on which all known proofs of Theorem 2.5.1 rely fall short of solving any case of the regular inverse Galois problem over a given number field. As of today, all known constructions of realisations of finite groups as regular Galois groups over $\mathbf{Q}$ exploit more or less ad hoc ideas. One of the most successful approaches is the rigidity method, initiated by Shih, Fried, Belyi, Matzat and Thompson in the 1970's and the 1980's, which we discuss next, in §§2.6–2.7.

REMARK 2.5.3. In practice, the rigidity method is particularly useful for realising finite simple groups. Considering the regular inverse Galois problem for various classes of non-simple groups, especially central extensions, also leads to interesting challenges, better dealt with by other methods. We refer the reader to the work of Mestre [**63–65**], who solved it for the central extensions of A_n for all n, as well as for $\mathrm{SL}_2(\mathbf{F}_7)$ and for the unique nontrivial central extension of the Mathieu group M_{12} by $\mathbf{Z}/2\mathbf{Z}$ (starting from known regular realisations of $\mathrm{PSL}_2(\mathbf{F}_7)$ and of M_{12}). To put these results into perspective, let us note that Noether's problem has negative answers, over $\mathbf{Q}$, for the unique nontrivial central extensions of A_6 and of A_7 by $\mathbf{Z}/2\mathbf{Z}$ (Serre [**33**, Theorem 33.25]).

2.6. Hurwitz spaces. Even though Hurwitz spaces are not necessary for the description and the implementation of the rigidity method, their introduction makes the theory rather transparent; in addition, they are indeed indispensable for some of its refinements. Hurwitz spaces are moduli spaces of smooth projective irreducible covers of the projective line. We shall consider them only in characteristic 0. In addition, we shall restrict attention to the moduli space of G-covers; the term

"G-cover" is another name for the regular Galois extensions of $k(t)$ with group G that we have been considering since the beginning of §2:

DEFINITION 2.6.1. Let G be a finite group. Let k be a field. A *G-cover* over k is a smooth, proper, geometrically irreducible curve C over k endowed, on the one hand, with a finite morphism $\pi : C \to \mathbf{P}^1_k$ such that the corresponding extension of function fields $k(C)/k(t)$ is Galois, and, on the other hand, with an isomorphism $G \xrightarrow{\sim} \mathrm{Gal}(k(C)/k(t))$. (In particular G acts faithfully on C and the morphism $\pi^{-1}(U) \to U$ induced by π is a G-torsor for any dense open subset $U \subset \mathbf{P}^1_k$ above which π is étale.)

The group of automorphisms of any G-cover, i.e. the group of automorphisms of C that respect not only the morphism π but also the given isomorphism $G \xrightarrow{\sim} \mathrm{Gal}(k(C)/k(t))$, is the centre of G. We shall assume, until the end of §2, that G has trivial centre. This is not too serious a restriction (as any finite group is a quotient of a finite group with trivial centre, see [**31**, Lemma 2]) and it will ensure that our moduli space is a variety rather than a stack (as the objects that we want to classify have no nontrivial automorphism).

To prepare for the statement of the next theorem, we need to introduce some notation. When k has characteristic 0, the *branch locus* of $\pi : C \to \mathbf{P}^1_k$ is by definition the smallest reduced 0-dimensional subvariety B of $\mathbf{P}^1_k$ such that π is étale over $\mathbf{P}^1_k \setminus B$. Its *degree* is the cardinality of $B(\bar{k})$, where $\bar{k}$ denotes an algebraic closure of k. For any integer $r \geq 1$, we denote by $\mathcal{U}^r \subset (\mathbf{P}^1_{\mathbf{Q}})^r$ the locus of r-tuples with pairwise distinct components, and by $\mathcal{U}_r$ the quotient of $\mathcal{U}^r$ by the natural action of the symmetric group S_r. Thus $\mathcal{U}_r$ is a smooth variety over $\mathbf{Q}$, and for any field k of characteristic 0, the set $\mathcal{U}_r(k)$ can be identified with the set of reduced 0-dimensional subvarieties of $\mathbf{P}^1_k$ of degree r, i.e. with the set of subsets of $\mathbf{P}^1(\bar{k})$ of cardinality r that are stable under $\mathrm{Gal}(\bar{k}/k)$.

THEOREM 2.6.2 (Fried and Völklein [**31**]). *Let G be a finite group with trivial centre and $r \geq 1$ be an integer. With G and r, one can canonically associate a smooth variety $\mathcal{H}_{G,r}$ over $\mathbf{Q}$ such that for any field k of characteristic 0, the set $\mathcal{H}_{G,r}(k)$ is the set of isomorphism classes of G-covers over k whose branch locus has degree r. It is equipped with a finite étale morphism $\rho : \mathcal{H}_{G,r} \to \mathcal{U}_r$ that maps the isomorphism class of a G-cover to its branch locus.*

The variety $\mathcal{H}_{G,r}$ is called a *Hurwitz space.*

A modern approach to Theorem 2.6.2 consists in defining G-covers not just over fields, as in Definition 2.6.1, but more generally over schemes; one then proves that the resulting moduli functor on the category of schemes of characteristic 0 is representable, by $\mathcal{H}_{G,r}$. This is the approach adopted by Wewers [**101**], who works more generally over $\mathbf{Z}$ (with tame covers) and without assuming that the centre of G is trivial (thus obtaining a moduli stack $\mathcal{H}_{G,r}$). See [**81**].

We note that Hurwitz spaces were first contemplated by Hurwitz [**46**], and, with a functorial point of view, by Fulton [**32**]. However, these authors only considered (non-Galois) covers with "simple" ramification, i.e. such that all ramification points have ramification index 2 and no two of them lie over the same branch point. This is insufficient for the purposes of the regular inverse Galois problem (see Example 2.4.2).

Let us come back to our motivation. It is tautological that for any finite group G with trivial centre, the regular inverse Galois problem admits a positive answer for G over $\mathbf{Q}$ if and only if there exists an integer $r \geq 1$ such that $\mathcal{H}_{G,r}(\mathbf{Q}) \neq \varnothing$. For the question of the existence of a rational point on one of the varieties $\mathcal{H}_{G,r}$ to be tractable, one needs, in turn, some understanding of their geometry.

The varieties $\mathcal{H}_{G,r}$ can be described in a very explicit combinatorial fashion, at least geometrically, thanks to the finite étale morphism $\rho : \mathcal{H}_{G,r} \to \mathcal{U}_r$ given by Theorem 2.6.2. Let us pick up the notation $\mathrm{ni}_r^*(G)$ introduced in Corollary 2.3.2 and write $\mathrm{ni}_r(G) \subseteq \mathrm{ni}_r^*(G)$ for the subset formed by the conjugacy classes of those r-tuples $(g_1, \ldots, g_r)$ such that none of the g_i's is equal to 1. It then follows from (the proof of) Corollary 2.3.2 that the fibre of ρ above any complex point of $\mathcal{U}_r$ can be identified with $\mathrm{ni}_r(G)$. In addition, the fundamental group of $\mathcal{U}_r(\mathbf{C})$ admits a down-to-earth presentation (as a quotient of the Artin braid group by one relation) and its action on $\mathrm{ni}_r(G)$ can also be made explicit (see [**31**, §1.3]). Thus, for instance, the task of describing the irreducible components of the variety $(\mathcal{H}_{G,r})_{\bar{\mathbf{Q}}}$ becomes equivalent to that of computing the orbits of a certain action of the braid group on $\mathrm{ni}_r(G)$. Unfortunately, as r increases, even this "simple" task quickly becomes computationally infeasible for modern computers (see e.g. [**37**]).

2.7. The rigidity method. This method consists in cleverly identifying irreducible components of $\mathcal{H}_{G,r}$ that contain rational points for somehow "trivial" reasons. To explain it, we need to refine the étale covering ρ that appears in Theorem 2.6.2.

2.7.1. *Algebraic local monodromy.* Let $\pi : C \to \mathbf{P}^1_k$ be a G-cover over a field k of characteristic 0. Let $b_i \in \mathbf{P}^1(k)$ be a rational branch point. Let $X \subset \mathbf{P}^1_k$ be a dense open subset over which π is étale.

Under the assumption that k is a subfield of $\mathbf{C}$, we have associated with π and b_i, in §2.3, a canonical conjugacy class of G, namely the conjugacy class of the element g_i appearing in Corollary 2.3.2. This is the "local monodromy" of π at b_i. We recall that it is the image, by a surjection $\pi_1(X(\mathbf{C}), x) \twoheadrightarrow G$ that is well-defined up to conjugation, of the conjugacy class of a local counterclockwise loop around b_i. To make this topological definition fit in with the moduli picture of Theorem 2.6.2 and in particular to understand how it behaves with respect to the action of the group of automorphisms of k, we need to make it algebraic.

We do this as follows. Let $\bar{k}$ be an algebraic closure of k. The completion $\bar{k}(t)_{b_i}$ of $\bar{k}(t)$ at the discrete valuation defined by b_i is isomorphic to the field of formal power series $\bar{k}((u))$, whose algebraic closure is the field of Puiseux series $\bigcup_{n\geq 1} \bar{k}((u^{1/n}))$. By Kummer theory, the absolute Galois group of $\bar{k}(t)_{b_i}$ is canonically isomorphic to $\hat{\mathbf{Z}}(1)_{\bar{k}} = \varprojlim_{n\geq 1} \boldsymbol{\mu}_n(\bar{k})$. The inclusion of fields $\bar{k}(t) \hookrightarrow \bar{k}(t)_{b_i}$ induces a continuous homomorphism in the reverse direction, well-defined up to conjugation, between their absolute Galois groups; hence it induces a continuous homomorphism $\hat{\mathbf{Z}}(1)_{\bar{k}} \to G$, well-defined up to conjugation by an element of G. This conjugacy class of homomorphisms $\hat{\mathbf{Z}}(1)_{\bar{k}} \to G$ is the analogue of the g_i from Corollary 2.3.2. We call it the *algebraic local monodromy* of π at b_i.

REMARKS 2.7.2. (i) When k is a subfield of $\mathbf{C}$, one can use the generator $\zeta_n = e^{2i\pi/n}$ of $\boldsymbol{\mu}_n(\bar{k})$ to identify $\hat{\mathbf{Z}}(1)_{\bar{k}}$ with $\hat{\mathbf{Z}}$ as topological groups (i.e. disregarding Galois actions). The algebraic local monodromy of π at b_i then becomes identified with a conjugacy class of G. One verifies that through this identification, the algebraic local monodromy of π at b_i coincides with the conjugacy class of g_i from Corollary 2.3.2.

(ii) The natural action of $\mathrm{Gal}(\bar{k}/k)$ on $\hat{\mathbf{Z}}(1)_{\bar{k}}$ induces an action of $\mathrm{Gal}(\bar{k}/k)$ on the set of conjugacy classes of homomorphisms $\hat{\mathbf{Z}}(1)_{\bar{k}} \to G$. It therefore makes sense to consider the conjugates, under this action of $\mathrm{Gal}(\bar{k}/k)$, of the algebraic local monodromy of π at b_i. (We recall that $b_i \in \mathbf{P}^1(k)$.) This feature of the algebraic point of view plays a crucial rôle in the rigidity method. It is not visible on the topological side of the identification of Remark 2.7.2 (i), except when $k = \mathbf{R}$. Indeed, in this case, complex conjugation acts by multiplication by -1 on $\hat{\mathbf{Z}}(1)_{\bar{k}}$, hence maps the algebraic local monodromy of π at b_i to its "inverse"; while on the topological side, it sends a local counterclockwise loop around b_i to a local clockwise loop around b_i, and hence again it maps the conjugacy class of g_i to the conjugacy class of g_i^{-1}.

(iii) For an arbitrary field k of characteristic 0, the choice of a topological generator of the procyclic group $\hat{\mathbf{Z}}(1)_{\bar{k}}$ should be thought of as an algebraic analogue of the choice of an orientation of the punctured unit disc (an insight of Grothendieck, see [**20**, (2.1)]), as is illustrated by Remark 2.7.2 (ii).

2.7.3. *Factoring ρ.* The natural continuous action of $\mathrm{Gal}(\bar{\mathbf{Q}}/\mathbf{Q})$ on the finite set $\mathrm{Hom}_{\mathrm{cont}}\left(\hat{\mathbf{Z}}(1)_{\bar{\mathbf{Q}}}, G\right)$ of continuous homomorphisms $\hat{\mathbf{Z}}(1)_{\bar{\mathbf{Q}}} \to G$ induces a continuous action of $\mathrm{Gal}(\bar{\mathbf{Q}}/\mathbf{Q})$ on the quotient of this finite set by the conjugation action of G. As the functor

$$\left(\text{reduced 0-dimensional varieties over } \mathbf{Q}\right) \to \left(\begin{array}{c}\text{finite sets endowed with a}\\ \text{continuous action of } \mathrm{Gal}(\bar{\mathbf{Q}}/\mathbf{Q})\end{array}\right)$$

that sends a variety Z to the set $Z(\bar{\mathbf{Q}})$ is an equivalence of categories, we can canonically associate with G a reduced 0-dimensional variety $\mathcal{C}_G$ over $\mathbf{Q}$ such that

$$\mathcal{C}_G(\bar{\mathbf{Q}}) = \mathrm{Hom}_{\mathrm{cont}}\left(\hat{\mathbf{Z}}(1)_{\bar{\mathbf{Q}}}, G\right)/(\text{conjugation by } G),$$

compatibly with the natural continuous actions of $\mathrm{Gal}(\bar{\mathbf{Q}}/\mathbf{Q})$ on both sides. One then has a canonical $\mathrm{Gal}(\bar{k}/k)$-equivariant identification

$$\mathcal{C}_G(\bar{k}) = \mathrm{Hom}_{\mathrm{cont}}\left(\hat{\mathbf{Z}}(1)_{\bar{k}}, G\right)/(\text{conjugation by } G)$$

for any field k of characteristic 0, with algebraic closure $\bar{k}$.

Given a G-cover $\pi : C \to \mathbf{P}^1_k$ over a field k of characteristic 0 and a rational branch point $b \in \mathbf{P}^1(k)$, the algebraic local monodromy of π at b defined in §2.7.1 is thus an element of $\mathcal{C}_G(k)$. More generally, if $b \in \mathbf{P}^1_k$ is an arbitrary branch point of π (i.e. a closed point, not necessarily rational), applying the definition of §2.7.1 to the G-cover over $k(b)$ obtained from π by scalar extension from k to $k(b)$ and to the rational branch point of this G-cover induced by b, we obtain an element of $\mathcal{C}_G(k(b))$, which we still call the *algebraic local monodromy of π at b*. Viewing this element as a morphism $\mathrm{Spec}(k(b)) \to \mathcal{C}_G$ and writing the branch locus $B \subset \mathbf{P}^1_k$ of π as the disjoint union, over the points $b \in B$, of the schemes $\mathrm{Spec}(k(b))$, we thus obtain a morphism of varieties $B \to \mathcal{C}_G$. We call it the *algebraic local monodromy morphism of π*.

For any $r \geq 1$, we denote by $\mathcal{V}^r \subset (\mathbf{P}^1_{\mathbf{Q}} \times \mathcal{C}_G)^r$ the inverse image of $\mathcal{U}^r$ by the projection $(\mathbf{P}^1_{\mathbf{Q}} \times \mathcal{C}_G)^r \to (\mathbf{P}^1_{\mathbf{Q}})^r$, and by $\mathcal{V}_r$ the quotient of $\mathcal{V}^r$ by the natural action of the symmetric group S_r. We have thus produced an étale covering $\nu : \mathcal{V}_r \to \mathcal{U}_r$ of smooth varieties over $\mathbf{Q}$. For any field k of characteristic 0, the set $\mathcal{V}_r(k)$ can be identified with the set of reduced 0-dimensional subvarieties of $\mathbf{P}^1_k \times \mathcal{C}_G$ of degree r that map isomorphically to their image in $\mathbf{P}^1_k$, or, what is the same, to the set of reduced 0-dimensional subvarieties B of $\mathbf{P}^1_k$ of degree r endowed with a morphism $B \to \mathcal{C}_G$.

For any field k of characteristic 0, associating with each G-cover its branch locus together with its algebraic local monodromy morphism provides us with a map $\mathcal{H}_{G,r}(k) \to \mathcal{V}_r(k)$. When $k = \mathbf{Q}(I)$ is the function field of a connected component I of $\mathcal{H}_{G,r}$, the image of the generic point of I by this map gives rise to a rational map $I \dashrightarrow \mathcal{V}_r$. Being a rational map between étale coverings of the normal variety $\mathcal{U}_r$, it is in fact a morphism. By letting I vary over all connected components of $\mathcal{H}_{G,r}$, we obtain, in this way, a morphism $\rho' : \mathcal{H}_{G,r} \to \mathcal{V}_r$ such that $\rho = \nu \circ \rho'$.

Let $\mathrm{Cl}(G)$ denote the set of conjugacy classes of G. Remark 2.7.2 (i) and (the proof of) Corollary 2.3.2 together imply the following explicit description of the complex fibres of ρ':

PROPOSITION 2.7.4. *Let $B \subset \mathbf{P}^1_{\mathbf{C}}$ be a reduced 0-dimensional subvariety of degree r. Write $B = \{b_1, \ldots, b_r\}$. Let $C = (C_1, \ldots, C_r)$ be an r-tuple of nontrivial conjugacy classes of G, viewed as a map $B(\mathbf{C}) \to \mathcal{C}_G(\mathbf{C})$ via the identification $\mathcal{C}_G(\mathbf{C}) = \mathrm{Cl}(G)$ of Remark 2.7.2 (i). Then the fibre of ρ' above the point of $\mathcal{V}_r(\mathbf{C})$ defined by B and C can be identified with the quotient $\mathrm{ni}_r^C(G)$ of the set of r-tuples $(g_1, \ldots, g_r) \in G^r$ satisfying the following three conditions by the action of G on this set by simultaneous conjugation:*

(1) $g_1 \cdots g_r = 1$;
(2) $g_1, \ldots, g_r$ *generate* G;
(3) $g_i \in C_i$ *for all* $i \in \{1, \ldots, r\}$.

PROOF. This would be a direct consequence of the quoted references if we knew that for a field k of characteristic 0 (here $k = \mathbf{C}$), the map $\mathcal{H}_{G,r}(k) \to \mathcal{V}_r(k)$ induced by ρ' sends the isomorphism class of any G-cover over k to its algebraic local monodromy morphism. By the definition of ρ', this is true for those G-covers whose branch locus is "generic", in the sense that it is a point of $\mathcal{U}_r(k)$ lying over the generic point of the variety $\mathcal{U}_r$ over $\mathbf{Q}$ (i.e. when viewed as a morphism $\mathrm{Spec}(k) \to \mathcal{U}_r$, its image is the generic point). Thus, Proposition 2.7.4 holds when B is "generic". As ρ' is a morphism between étale coverings of $\mathcal{U}_r$, the validity of Proposition 2.7.4 for arbitrary B follows. □

2.7.5. *Rational points of $\mathcal{C}_G$*. Viewing $\bar{\mathbf{Q}}$ as a subfield of $\mathbf{C}$, Remark 2.7.2 (i) also induces an identification $\mathcal{C}_G(\bar{\mathbf{Q}}) = \mathrm{Cl}(G)$. Via this identification, the natural action of $\mathrm{Gal}(\bar{\mathbf{Q}}/\mathbf{Q})$ on $\mathcal{C}_G(\bar{\mathbf{Q}})$ gives rise to the action of $\mathrm{Gal}(\bar{\mathbf{Q}}/\mathbf{Q})$ on $\mathrm{Cl}(G)$ given by the formula $\sigma(g) = g^{-\chi(\sigma)}$ for $\sigma \in \mathrm{Gal}(\bar{\mathbf{Q}}/\mathbf{Q})$ and $g \in G$, where $\chi : \mathrm{Gal}(\bar{\mathbf{Q}}/\mathbf{Q}) \twoheadrightarrow \hat{\mathbf{Z}}^*$ denotes the cyclotomic character. As a consequence, the set $\mathcal{C}_G(\mathbf{Q})$ of rational points of $\mathcal{C}_G$ gets identified with the set of *rational* conjugacy classes of G in the following sense:

DEFINITION 2.7.6. A conjugacy class C of a finite group G is *rational* if for every $g \in C$ and every integer $n \geq 1$ prime to the order of g, the element g^n belongs to C.

2.7.7. *Rational points of* $\mathcal{V}_r$. Here is a simple way to exhibit rational points of $\mathcal{V}_r$. Let $b_1, \dots, b_r \in \mathbf{P}^1(\mathbf{Q})$ be pairwise distinct. Let $B = \{b_1, \dots, b_r\}$. The rational points of $\mathcal{V}_r$ lying above the rational point of $\mathcal{U}_r$ defined by B are exactly the r-tuples of rational points of $\mathcal{C}_G$, i.e. they are the r-tuples of rational conjugacy classes of G.

2.7.8. *Rational points of* $\mathcal{H}_{G,r}$. The morphism $\rho' : \mathcal{H}_{G,r} \to \mathcal{V}_r$ is an étale covering. As such, it has a degree, which is a locally constant function on $\mathcal{V}_r$. This function is not constant on $\mathcal{V}_r$ in general—unlike the degree of ρ, which is constant, equal to the cardinality of $\mathrm{ni}_r(G)$, as we have seen in §2.6. A connected component of $\mathcal{V}_r$ over which ρ' has degree 1, i.e. over which ρ' restricts to an isomorphism, is said to be *rigid.*

It is a trivial but fruitful observation, which forms the basis of the rigidity method, that the existence of a rational point of a rigid connected component of $\mathcal{V}_r$ implies the existence of a rational point of $\mathcal{H}_{G,r}$.

By Proposition 2.7.4, the rigidity of a connected component can be verified by computing the set $\mathrm{ni}_r^C(G)$ associated with its complex points. This motivates the following definition:

DEFINITION 2.7.9. An r-tuple $C = (C_1, \dots, C_r)$ of nontrivial conjugacy classes of G is *rigid* if the set $\mathrm{ni}_r^C(G)$ defined in Proposition 2.7.4 has cardinality 1.

2.7.10. *Summing up.* We thus arrive at a down-to-earth condition that implies that the rational points of $\mathcal{V}_r$ constructed in §2.7.7 can be lifted to rational points of $\mathcal{H}_{G,r}$.

THEOREM 2.7.11. *Let G be a finite group with trivial centre. Let $r \geq 1$ be an integer. If there exists a rigid r-tuple $C = (C_1, \dots, C_r)$ of nontrivial rational conjugacy classes of G, then for any pairwise distinct $b_1, \dots, b_r \in \mathbf{P}^1(\mathbf{Q})$, the $\mathbf{Q}$-point of $\mathcal{U}_r$ defined by $\{b_1, \dots, b_r\}$ can be lifted to a $\mathbf{Q}$-point of $\mathcal{H}_{G,r}$. In particular, the regular inverse Galois problem admits a positive solution for G over $\mathbf{Q}$.*

Theorem 2.7.11 represents the base case of the rigidity method. It admits many variants; for instance, one can allow non-rational branch points. (Pro: this weakens the condition that the prescribed conjugacy classes be rational; con: these conjugacy classes cannot be chosen independently of one another any longer.) One could also work over a number field other than $\mathbf{Q}$, which would simultaneously weaken the conclusion of the theorem and the rationality assumption on the conjugacy classes C_i.

Even just the above base case is already unreasonably effective: the hypothesis of Theorem 2.7.11 has been shown to be satisfied, with $r = 3$, for at least 10 of the 26 sporadic simple groups, including the monster (by Thompson) and the baby monster (by Malle and Matzat); see [**58**, Chapter II, §9]. As another example, the variant of Theorem 2.7.11 in which $r = 3$ and b_1 is rational while b_2 and b_3 are conjugate quadratic points of the projective line can be applied to $\mathrm{PSL}_2(\mathbf{F}_p)$ for all primes p such that 2 or 3 is not a square modulo p (see [**87**, §8.3.3]), thus recovering the positive answer to the regular inverse Galois problem over $\mathbf{Q}$ for this infinite family of groups that had been obtained by Shih using modular curves rather than the rigidity method.

When the rigidity method is applicable, it is in principle possible to deduce from it an explicit polynomial that realises the desired regular Galois extension of $\mathbf{Q}(t)$ (see [**58**, Chapter II, §9]). This has some limits in practice (e.g. for the monster group, the degree of the polynomial cannot have less than 20 digits) but it leads to interesting computational challenges (see e.g. [**5**]).

3. Grunwald's problem and the Brauer–Manin obstruction

3.1. Looking for rational points. Despite its successes, the rigidity method, discussed in §2.7, often fails to be applicable. For instance, it fails, in general, for p-groups; indeed, the regular inverse Galois problem is still open for most p-groups over $\mathbf{Q}$, even though the inverse Galois problem itself is known to have a positive answer, over any number field, for all p-groups—and more generally, for all solvable groups, by a celebrated theorem of Shafarevich (see [**72**, Chapter IX, §6]). In other words, after learning, in §1, that the quotient variety $\mathbf{A}^n_k/G$ can fail to be rational or even retract rational, we now find ourselves unable, at least in practice, to salvage the Hilbert–Noether method by constructing rational curves in $\mathbf{A}^n_k/G$ over which to apply Hilbert's irreducibility theorem, as envisaged at the beginning of §2.1. This leads us to our next question: letting $Y \subseteq \mathbf{A}^n_k$ denote the locus where G acts freely, can we directly construct rational points of Y/G above which the fibre of the quotient map $Y \to Y/G$ is irreducible? An approach put forward by Colliot-Thélène consists in noting that Ekedahl's Theorem 1.3.5 reduces this question, in full generality, to the problem of finding rational points on Y/G subject to certain weak approximation conditions. In particular, if the variety Y/G satisfies weak approximation off a finite set of places of k, then the inverse Galois problem has a positive answer for G over k. Such a weak approximation property can be proved unconditionally in some cases; for instance, under the assumptions of the following remarkable theorem from [**71**], in which no place of k is excluded:

THEOREM 3.1.1 (Neukirch). *Let k be a number field. Let G be a finite solvable group, acting linearly on $\mathbf{A}^n_k$ for some $n \geq 1$. Let $Y \subseteq \mathbf{A}^n_k$ be the locus where G acts freely. Let $X = Y/G$. Assume that the order of G and the number of roots of unity contained in k are coprime. Then X satisfies weak approximation, i.e. the set $X(k)$ is dense in $X(k_\Omega)$. In particular, Grunwald's problem admits a positive answer for G over k, for any finite subset $S \subset \Omega$.*

We recall that Grunwald's problem is Problem 1.1.1. Without the assumption on the order of G, the conclusion of this theorem can fail, as we have seen in Proposition 1.6.2.

The validity of the weak approximation property is a problem of general interest that makes sense, and has been studied, for arbitrary smooth varieties. As we shall now explain, the tools that have been developed for its study on arbitrary smooth varieties turn out to be useful also in the special case of the quotient Y/G.

3.2. Brauer–Manin obstruction. A general mechanism, introduced by Manin [**59**] and now called the *Brauer–Manin obstruction*, explains, in some cases, why the weak approximation property fails for certain varieties over number fields. Let us recall it briefly. (For details, see [**18**, §13.3].) Let X be a smooth variety over a number field k. Let $\mathrm{Br}_{\mathrm{nr}}(X)$ denote its unramified Brauer group (see §1.6.5). We let $X(k_\Omega) = \prod_{v\in\Omega} X(k_v)$ and endow this set with the product of the v-adic topologies.

The *Brauer–Manin set* $X(k_\Omega)^{\mathrm{Br}_{\mathrm{nr}}(X)}$ is the set of all families $(x_v)_{v\in\Omega} \in X(k_\Omega)$ such that $\sum_{v\in\Omega} \mathrm{inv}_v\, \beta(x_v) = 0$ for all $\beta \in \mathrm{Br}_{\mathrm{nr}}(X)$. Here $\beta(x_v) \in \mathrm{Br}(k_v)$ denotes the evaluation of β at x_v, and $\mathrm{inv}_v : \mathrm{Br}(k_v) \hookrightarrow \mathbf{Q}/\mathbf{Z}$ is the invariant map from local class field theory. (To make sense of the sum, one checks that only finitely many of its terms are nonzero.) Manin's fundamental observation is that the image of the diagonal embedding $X(k) \hookrightarrow X(k_\Omega)$ is contained in the Brauer–Manin set, as a consequence of the reciprocity law of global class field theory. Thus, we have a sequence of inclusions

$$X(k) \subseteq X(k_\Omega)^{\mathrm{Br}_{\mathrm{nr}}(X)} \subseteq X(k_\Omega). \tag{3.2.1}$$

As $X(k_\Omega)^{\mathrm{Br}_{\mathrm{nr}}(X)}$ is a closed subset of $X(k_\Omega)$, the weak approximation property, i.e. the density of $X(k)$ in $X(k_\Omega)$, can hold only if $X(k_\Omega)^{\mathrm{Br}_{\mathrm{nr}}(X)} = X(k_\Omega)$. When this last equality fails, one says that there is a Brauer–Manin obstruction to weak approximation on X.

3.3. Reinterpreting the Grunwald–Wang theorem. Let us come back to the variety $X = Y/G$ considered in §3.1: G is a subgroup of S_n, which acts on $\mathbf{A}^n_k$ by permuting the coordinates, and $Y \subseteq \mathbf{A}^n_k$ is the locus where G acts freely. As Grunwald's problem has a negative answer for $G = \mathbf{Z}/8\mathbf{Z}$ and $k = \mathbf{Q}$ (see Proposition 1.6.2) and as $Y \to X$ is a versal G-torsor (see Example 1.5.2), the variety X cannot satisfy the weak approximation property in this case, according to Proposition 1.5.5. Hence a natural question: is there a Brauer–Manin obstruction to weak approximation on X when $G = \mathbf{Z}/8\mathbf{Z}$ and $k = \mathbf{Q}$? The answer is yes by the following theorem, which states, more precisely, that the weak approximation property on X is fully controlled by the Brauer–Manin obstruction as soon as G is abelian.

THEOREM 3.3.1 (Voskresenskiĭ, Sansuc). *Let k be a number field. Let G be a finite abelian group, acting linearly on $\mathbf{A}^n_k$ for some $n \geq 1$. Let $Y \subseteq \mathbf{A}^n_k$ be the locus where G acts freely. Let $X = Y/G$. The set $X(k)$ is dense in $X(k_\Omega)^{\mathrm{Br}_{\mathrm{nr}}(X)}$.*

Theorem 3.3.1 can be found in the literature by combining [**99**, §7.2, Theorem] with [**85**, Corollaire 8.13]. We shall explain a proof of it in §3.8 below.

Returning to an arbitrary finite group G and keeping k, Y and X as above, it is a general fact that the density of $X(k)$ in $X(k_\Omega)^{\mathrm{Br}_{\mathrm{nr}}(X)}$ implies the existence of a finite subset $S_0 \subset \Omega$ such that X satisfies weak approximation off S_0. When such an S_0 exists, a refinement of the arguments underlying the proof of Proposition 1.5.5 leads to the following conclusion (a point of view advocated in [**11**] and in [**38**, §1.2]): fully solving Grunwald's problem for G over k is in fact equivalent to describing the closure of $X(k)$ inside $X(k_\Omega)$. Thus, the Grunwald–Wang theorem, which indeed fully solves Grunwald's problem when G is abelian, can now be viewed, in retrospect, as the combination of Theorem 3.3.1 with an explicit computation of the Brauer–Manin set $X(k_\Omega)^{\mathrm{Br}_{\mathrm{nr}}(X)}$ when G is abelian.

3.4. Rationally connected varieties. Whether or not the abelianness hypothesis on G can be removed from Theorem 3.3.1 is a fundamental open question. When X is an arbitrary smooth variety possessing a rational point, the set $X(k)$ cannot be expected to be dense in $X(k_\Omega)^{\mathrm{Br}_{\mathrm{nr}}(X)}$ without strong assumptions on the geometry of X; for instance, Lang's conjectures predict that for d and N such that $d-2 \geq N \geq 4$, this density should fail for all smooth hypersurfaces of degree d in $\mathbf{P}^N$ that have a rational point (see [**77**, Appendix A]). The variety $X = Y/G$ that we have been considering in §3.1 and in §3.3, despite not being geometrically rational for an arbitrary finite group G (see §1.6.5), still has a reasonably tame geometry: it is unirational and therefore belongs to the class of *rationally connected varieties* according to the following definition.[6]

DEFINITION 3.4.1 (Campana, Kollár, Miyaoka, Mori). A smooth variety X over a field k is said to be *rationally connected* if for any algebraically closed field extension K/k and any two general K-points $x_0, x_1 \in X(K)$, there exists a rational map $f : \mathbf{A}^1_K \dashrightarrow X_K$ over K, defined in a neighbourhood of 0 and 1, such that $f(0) = x_0$ and $f(1) = x_1$. ("General" means that the set of pairs (x_0, x_1) satisfying the stated condition contains a dense Zariski open subset of $X(K) \times X(K)$.)

Theorem 3.3.1 conjecturally extends to all smooth rationally connected varieties:

CONJECTURE 3.4.2 (Colliot-Thélène). *Let X be a smooth, rationally connected variety, over a number field k. The set $X(k)$ is dense in $X(k_\Omega)^{\mathrm{Br}_{\mathrm{nr}}(X)}$.*

A number of known results towards this conjecture are listed in [**102**]. Conjecture 3.4.2 would imply that all smooth rationally connected varieties satisfy weak approximation off a finite set of places (see e.g. [**102**, Remarks 2.4 (i)–(ii)]). In particular, it would imply a positive answer to the inverse Galois problem in general, by Theorem 1.3.5 applied to the torsor of Example 1.5.2.

A list of groups G of small order for which Conjecture 3.4.2 is still open for the variety $X = Y/G$ appearing in Example 1.5.2 can be found in [**9**].

3.5. Determining the Brauer–Manin set. As discussed in §3.3, Conjecture 3.4.2 in the case $X = Y/G$ would, more precisely, reduce Grunwald's problem for G over k to the computation of the Brauer–Manin set of X. Even partial knowledge of the Brauer–Manin set can lead to concrete results, as the following theorem illustrates:

THEOREM 3.5.1 (Lucchini Arteche [**56**]). *Let k be a number field and G be a finite group acting linearly on $\mathbf{A}^n_k$. Let $Y \subseteq \mathbf{A}^n_k$ be the locus where G acts freely. Let $X = Y/G$. Let S_0 be the set of finite places of k that divide the order of G. If X satisfies Conjecture 3.4.2, then Grunwald's problem admits a positive answer for G over k, for any finite subset $S \subset \Omega$ disjoint from S_0.*

The proof of Theorem 3.5.1 consists in studying the evaluation of unramified Brauer classes at the local points of X, so as to deduce, from the density of $X(k)$ in $X(k_\Omega)^{\mathrm{Br}_{\mathrm{nr}}(X)}$, that X satisfies weak approximation off S_0; Proposition 1.5.5 then yields the desired statement.

The complete determination of $X(k_\Omega)^{\mathrm{Br}_{\mathrm{nr}}(X)}$, for X as in the statement of Theorem 3.5.1, is in general a difficult task. The case of a metabelian group G is investigated in [**23**]. In general, even the computation of $\mathrm{Br}_{\mathrm{nr}}(X)$ itself is a delicate problem. Over an algebraic closure $\bar{k}$ of k, one can apply Bogomolov's formula (Theorem 1.6.6). If the unramified Brauer group of $X_{\bar{k}}$ turns out to be nontrivial, one has to find out which classes of the finite group $\mathrm{Br}_{\mathrm{nr}}(X_{\bar{k}})$ are invariant under $\mathrm{Gal}(\bar{k}/k)$, and to determine the image of the natural map $\mathrm{Br}_{\mathrm{nr}}(X) \to \mathrm{Br}_{\mathrm{nr}}(X_{\bar{k}})^{\mathrm{Gal}(\bar{k}/k)}$; there is no general recipe for carrying this out. The kernel of the latter map, on the other hand, is now well understood: its quotient by the image of the natural map

[6] To be precise, Definition 3.4.1 coincides with the standard definition (found, e.g., in [**49**, Chapter IV]) only when X is proper.

$\mathrm{Br}(k) \to \mathrm{Br}_{\mathrm{nr}}(X)$ is finite and is described by a formula due to Harari [**38**, Proposition 4]. In the most favourable cases, the combination of these formulae can lead to the conclusion that the natural map $\mathrm{Br}(k) \to \mathrm{Br}_{\mathrm{nr}}(X)$ is onto, so that $X(k_\Omega)^{\mathrm{Br}_{\mathrm{nr}}(X)} = X(k_\Omega)$ and Grunwald's problem is then expected to have a positive solution for G over k with no restriction on the finite subset $S \subset \Omega$. See [**21**, Remarque 7] for a concrete example. In a different direction, by adapting the proof of Bogomolov's formula to non-algebraically closed ground fields, Colliot-Thélène [**13**, Corollaire 5.7] showed that the natural map $\mathrm{Br}(k) \to \mathrm{Br}_{\mathrm{nr}}(X)$ is onto, and hence that $X(k_\Omega)^{\mathrm{Br}_{\mathrm{nr}}(X)} = X(k_\Omega)$, whenever the order of G and the number of roots of unity contained in k are coprime, which explains, a posteriori, why the Brauer–Manin obstruction plays no rôle in Neukirch's Theorem 3.1.1.

3.6. Supersolvable groups. A finite group G is said to be *supersolvable* if there exists a filtration $1 = G_0 \subseteq G_1 \subseteq \cdots \subseteq G_n = G$ such that each G_i is a normal subgroup of G and each successive quotient G_{i+1}/G_i is cyclic. All nilpotent groups (in particular, all p-groups) are supersolvable. We proved, in [**41**], that Theorem 3.3.1 generalises to such groups:

THEOREM 3.6.1 (Harpaz, W.). *Let k be a number field. Let G be a finite supersolvable group, acting linearly on $\mathbf{A}^n_k$ for some $n \geq 1$. Let $Y \subseteq \mathbf{A}^n_k$ be the locus where G acts freely. Let $X = Y/G$. The set $X(k)$ is dense in $X(k_\Omega)^{\mathrm{Br}_{\mathrm{nr}}(X)}$.*

A positive answer to the inverse Galois problem for supersolvable groups results from this (via Theorem 1.3.5) but had already been established—more generally, for solvable groups—by Shafarevich, as mentioned in §3.1. As discussed in §§3.3–3.5, Theorem 3.6.1 refines this positive answer by bringing information about Grunwald's problem.

It may be that the strategy underlying the proof of Theorem 3.6.1 can be extended to all solvable groups. It will not, however, be of any help with non-abelian simple groups. In fact, to this day, no approach is known towards Grunwald's problem for non-abelian simple groups (with the exception of the two groups A_5 and $\mathrm{PSL}_2(\mathbf{F}_7)$, for which Noether's problem itself has a positive answer; see Example 1.4.4, [**66**] and Corollary 1.5.7).

3.7. Descent in a nutshell. Theorem 3.6.1 can be seen as a direct application of a general tool that is useful for proving cases of Conjecture 3.4.2, the so-called "descent" method. We now briefly discuss it. We shall illustrate it by proving Theorem 3.3.1 in §3.8, before indicating its applicability to other variants of the inverse Galois problem in §3.10.

To get started, we need to extend the notion of G-torsor from the case where G is a finite abstract group (Definition 1.2.1) to the case where G is an algebraic group (i.e. a group scheme over a field, possibly disconnected or of positive dimension).

DEFINITION 3.7.1. Let $\pi : Y \to X$ be a surjective morphism between smooth varieties over a field k of characteristic 0, with algebraic closure $\bar{k}$. Let G be an algebraic group over k, acting on Y in such a way that π is G-equivariant (for the trivial action of G on X). We say that π is a *G-torsor*, or that Y is a G-torsor over X, if π is smooth and $G(\bar{k})$ acts simply transitively on the fibres of the map $Y(\bar{k}) \to X(\bar{k})$ induced by π.

Unless otherwise specified, we now let k denote an arbitrary field of characteristic 0. (For the correct definition of a torsor without this assumption, see

[**89**, Definition 2.2.1].) As usual, by a *G-torsor over k* we shall mean a G-torsor over $\mathrm{Spec}(k)$. When $\pi : Y \to X$ is a G-torsor, the morphism π identifies X with the categorical quotient Y/G (see [**70**, Proposition 0.2, Proposition 0.1]).

EXAMPLE 3.7.2. Hilbert's Theorem 90, which we encountered in Example 1.5.2, is equivalent to the following statement: for any integer $n \geq 1$, any GL_n-torsor over k is isomorphic to GL_n. As an easy consequence, any SL_n-torsor over k is isomorphic to SL_n.

DEFINITION 3.7.3. Let X be a smooth variety over k. Let G be an algebraic group over k. Let $\pi : Y \to X$ be a G-torsor. The *twist* of Y by a G-torsor P over k is the quotient

$$_PY = (P \times Y)/G$$

of $P \times Y$ by the diagonal action of G, endowed with the natural morphism

$$_P\pi : {_PY} \to X$$

induced by the second projection $P \times Y \to Y$ and the identification $Y/G = X$. (We are not claiming that $_P\pi$ is a G-torsor. This is true when G is commutative, not in general. This point, which will be irrelevant for us, is discussed in [**89**, p. 21].)

The gist of the descent method is summarised in the following conjecture.

CONJECTURE 3.7.4. *Let X be a smooth variety over a number field k. Let G be a linear algebraic group over k. Let $\pi : Y \to X$ be a G-torsor, with Y rationally connected. Assume that for every twist Y' of Y by a G-torsor over k, the set $Y'(k)$ is dense in $Y'(k_\Omega)^{\mathrm{Br}_{\mathrm{nr}}(Y')}$. Then the set $X(k)$ is dense in $X(k_\Omega)^{\mathrm{Br}_{\mathrm{nr}}(X)}$.*

Conjecture 3.7.4, and the first significant cases in which it was established, appeared in a series of works by Colliot-Thélène and Sansuc. See [**89**] for an account. We content ourselves with mentioning the following positive result,[7] which in this form can be found in [**41**, Corollaire 2.2].

THEOREM 3.7.5 (Colliot-Thélène, Sansuc, Harpaz, W.). *Conjecture* 3.7.4 *holds true if G is a torus (i.e. an algebraic group such that $G_{\bar{k}} \simeq \mathbf{G}_{\mathrm{m},\bar{k}} \times \cdots \times \mathbf{G}_{\mathrm{m},\bar{k}}$).*

3.8. Sketch of proof of Theorem 3.3.1. We shall deduce Theorem 3.3.1 from Theorem 3.7.5. (More precisely, descent will be applied to a geometrically rational variety; Theorem 3.7.5 in the case of such varieties is due to Colliot-Thélène and Sansuc alone, see [**16**].)

Before starting the proof of Theorem 3.3.1, let us slightly change notation. We now fix an embedding $G \hookrightarrow \mathrm{SL}_n(k)$ for some $n \geq 1$, let Y be the algebraic group SL_n over k, and let G act on Y by right multiplication. As the resulting variety $X = Y/G$ is stably birationally equivalent to the variety X of Theorem 3.3.1 (see Remark 1.5.4), and as the density of $X(k)$ in $X(k_\Omega)^{\mathrm{Br}_{\mathrm{nr}}(X)}$ is a stable birational invariant among smooth, rationally connected varieties (see [**15**, Proposition 6.1 (iii)] and [**102**, Remark 2.4 (ii)]), this change of notation is harmless.

[7]More generally, after the writing of these notes, Conjecture 3.7.4 was proved by Nguyễn to be true as soon as G is connected; see [**73**], which builds on [**41**, §2] and on Borovoi's abelianisation technique.

Recall that G, by assumption, is a finite abelian group. Let us view it as a constant algebraic group over k. It is easy to see that G fits into a short exact sequence

$$1 \to G \to T \to Q \to 1 \tag{3.8.1}$$

of algebraic groups over k, where T and Q are tori and Q is quasi-trivial, i.e. the character group $\mathrm{Hom}(Q_{\bar{k}}, \mathbf{G}_{\mathrm{m},\bar{k}})$ of Q admits a basis over $\mathbf{Z}$ that is stable under the action of $\mathrm{Gal}(\bar{k}/k)$ (see [**87**, Proposition 4.2.1]). Letting G act on T by translation, we now consider the quotient $W = (\mathrm{SL}_n \times T)/G$ of $\mathrm{SL}_n \times T$ by the diagonal action, together with the morphism $\pi : W \to X = \mathrm{SL}_n/G$ induced by the first projection.

The action of T on $\mathrm{SL}_n \times T$ by multiplication on the second factor induces an action of T on W, with respect to which π is a T-torsor. According to Theorem 3.7.5 applied to π, it will suffice, in order to complete the proof of Theorem 3.3.1, to show that for every T-torsor P over k, the variety ${}_PW$ satisfies Conjecture 3.4.2.

We observe that ${}_PW = (\mathrm{SL}_n \times P)/G$, that the morphism $p : {}_PW \to P/G$ induced by the second projection is an SL_n-torsor (with respect to the action of SL_n on ${}_PW$ coming from its action on $\mathrm{SL}_n \times P$ by left multiplication on the first factor), and that P/G is a Q-torsor over k. By Hilbert's Theorem 90 (see Example 3.7.2), the generic fibre of p is isomorphic to SL_n, in particular it is rational. It also follows from Hilbert's Theorem 90 (case $n = 1$ of Example 3.7.2), combined with Shapiro's lemma, that any torsor under a quasi-trivial torus over k is rational; in particlar, the variety P/G is rational. These two remarks imply that ${}_PW$ is itself rational over k. Thus, it satisfies Conjecture 3.4.2 for trivial reasons, and Theorem 3.3.1 is proved.

3.9. Supersolvable descent. The ideas sketched in §3.8 are a starting point for the proof of the following theorem, established in [**42**, Corollary 3.3].

THEOREM 3.9.1 (Harpaz, W.). *Conjecture* 3.7.4 *holds true if G is finite and supersolvable.*

Here "supersolvable" means that $G(\bar{k})$ is supersolvable in the sense recalled at the beginning of §3.6, except that the filtration is now required, in addition, to be stable under the action of $\mathrm{Gal}(\bar{k}/k)$ on $G(\bar{k})$, in case this action is not trivial.

Theorem 3.9.1 implies Theorem 3.6.1: in the notation introduced at the beginning of §3.8, it suffices to apply Theorem 3.9.1 to the G-torsor $\mathrm{SL}_n \to \mathrm{SL}_n/G$ and to note that any twist of SL_n by a G-torsor over k is an SL_n-torsor (through left multiplication), hence is isomorphic to SL_n, by Hilbert's Theorem 90, hence is rational and satisfies Conjecture 3.4.2.

REMARK 3.9.2. Theorem 3.9.1 implies in particular that Conjecture 3.7.4 holds true for $G = \mathbf{Z}/2\mathbf{Z}$. We recall that Conjecture 3.7.4 assumes that the variety Y appearing in its statement is rationally connected. As was pointed out to the author by Mạnh Linh Nguyễn, this assumption cannot be dropped, even when $G = \mathbf{Z}/2\mathbf{Z}$. Indeed, there exist examples of double covers $Y \to X$, where Y is a $K3$ surface and X is an Enriques surface over $k = \mathbf{Q}$, such that $X(k)$ and the sets $Y'(k_\Omega)^{\mathrm{Br}_{\mathrm{nr}}(Y')}$ are all empty, while $X(k_\Omega)^{\mathrm{Br}_{\mathrm{nr}}(X)}$ is not (see [**4**, Theorem 1.2]).

3.10. Prescribed norms. Our last theorem is an application of supersolvable descent to a variant of the inverse Galois problem of a slightly different flavour, meant to demonstrate the flexibility of descent as a tool.

THEOREM 3.10.1 ([**42**, Theorem 4.16]). *Let G be a finite group. Let k be a number field. Let $\alpha_1, \dots, \alpha_m \in k^*$. If G is supersolvable, there exists a Galois field extension K of k such that* $\mathrm{Gal}(K/k) \simeq G$ *and* $\alpha_1, \dots, \alpha_m \in N_{K/k}(K)$.

The idea of the proof is to construct, in a formal and explicit way, a G-torsor $\pi : Y \to X$ together with invertible functions $\beta_1, \dots, \beta_m$ on Y whose norms along π are equal to the constant invertible functions $\alpha_1, \dots, \alpha_m$ on X. Namely, say $m = 1$ for simplicity, embed G into $\mathrm{SL}_n(k)$ and consider the subvariety Y of $\mathrm{SL}_n \times \prod_{g\in G} \mathbf{G}_{\mathrm{m}}$ consisting of all $(s, (t_g)_{g\in G})$ such that $\prod_{g\in G} t_g = \alpha_1$; the invertible function β_1 on Y given by projection onto the $\mathbf{G}_{\mathrm{m}}$ factor corresponding to $1 \in G$ has the required norm. One then checks that the twists of Y satisfy Conjecture 3.4.2 (despite not being rational in general, even when they have a rational point and G is assumed abelian), so that Theorem 3.9.1 implies the validity of Conjecture 3.4.2, and hence of weak approximation off a finite set of places, for X. As $X(k) \neq \varnothing$ (indeed even $Y(k) \neq \varnothing$), it follows, by Theorem 1.3.5, that there exists $x \in X(k)$ such that $\pi^{-1}(x)$ is irreducible, i.e. gives rise to a Galois field extension K/k with group G. Restricting $\beta_1, \dots, \beta_m$ to $\pi^{-1}(x)$ yields elements of K^* with norms $\alpha_1, \dots, \alpha_m$.

In the case where G is abelian, Theorem 3.10.1 was first shown to hold by Frei, Loughran and Newton [**30**], who established an asymptotic estimate for the number of such K.

References

[1] F. Amoroso and R. Dvornicich, *A lower bound for the height in abelian extensions*, J. Number Theory **80** (2000), no. 2, 260–272, DOI 10.1006/jnth.1999.2451. MR1740514 ↑298

[2] E. Artin and J. Tate, *Class field theory*, AMS Chelsea Publishing, Providence, RI, 2009. Reprinted with corrections from the 1967 original, DOI 10.1090/chel/366. MR2467155 ↑292

[3] M. Artin and D. Mumford, *Some elementary examples of unirational varieties which are not rational*, Proc. London Math. Soc. (3) **25** (1972), 75–95, DOI 10.1112/plms/s3-25.1.75. MR321934 ↑299

[4] F. Balestrieri, J. Berg, M. Manes, J. Park, and B. Viray, *Insufficiency of the Brauer-Manin obstruction for rational points on Enriques surfaces*, Directions in number theory. Proceedings of the 2014 Women in Numbers, WIN3 workshop. Banff, Alberta, Canada, April 21–25, 2014, 2016, pp. 1–31. ↑316

[5] D. Barth and A. Wenz, *Computation of Belyi maps with prescribed ramification and applications in Galois theory*, J. Algebra **569** (2021), 616–642, DOI 10.1016/j.jalgebra.2020.10.040. MR4187249 ↑311

[6] A. Beauville, J.-L. Colliot-Thélène, J.-J. Sansuc, and P. Swinnerton-Dyer, *Variétés stablement rationnelles non rationnelles* (French), Ann. of Math. (2) **121** (1985), no. 2, 283–318, DOI 10.2307/1971174. MR786350 ↑298

[7] G. Berhuy and G. Favi, *Essential dimension: a functorial point of view (after A. Merkurjev)*, Doc. Math. **8** (2003), 279–330. MR2029168 ↑296

[8] F. A. Bogomolov, *The Brauer group of quotient spaces of linear representations* (Russian), Izv. Akad. Nauk SSSR Ser. Mat. **51** (1987), no. 3, 485–516, 688, DOI 10.1070/IM1988v030n03ABEH001024; English transl., Math. USSR-Izv. **30** (1988), no. 3, 455–485. MR903621 ↑298

[9] E. Boughattas and D. Neftin, *The Grunwald problem and homogeneous spaces with non-solvable stabilisers*, Preprint, `arXiv:2404.09874`, 2024, to appear in Annales de l'Institut Fourier. ↑313

[10] J. Buhler and Z. Reichstein, *On the essential dimension of a finite group*, Compositio Math. **106** (1997), no. 2, 159–179, DOI 10.1023/A:1000144403695. MR1457337 ↑296

[11] V. I. Chernousov, *Galois cohomology and weak approximation for the quotient manifolds* A^n/G (Russian), Trudy Mat. Inst. Steklov. **208** (1995), no. Teor. Chisel, Algebra i Algebr.

Geom., 335–349. Dedicated to Academician Igor′Rostislavovich Shafarevich on the occasion of his seventieth birthday (Russian). MR1730273 ↑312

[12] J.-L. Colliot-Thélène, *Rational connectedness and Galois covers of the projective line*, Ann. of Math. (2) **151** (2000), no. 1, 359–373, DOI 10.2307/121121. MR1745009 ↑305

[13] J.-L. Colliot-Thélène, *Groupe de Brauer non ramifié de quotients par un groupe fini* (French, with English and French summaries), Proc. Amer. Math. Soc. **142** (2014), no. 5, 1457–1469, DOI 10.1090/S0002-9939-2014-11855-7. MR3168454 ↑314

[14] J.-L. Colliot-Thélène and M. Ojanguren, *Variétés unirationnelles non rationnelles: au-delà de l'exemple d'Artin et Mumford* (French), Invent. math. **97** (1989), no. 1, 141–158, DOI 10.1007/BF01850658. MR999316 ↑299

[15] J.-L. Colliot-Thélène, A. Pál, and A. N. Skorobogatov, *Pathologies of the Brauer-Manin obstruction*, Math. Z. **282** (2016), no. 3-4, 799–817, DOI 10.1007/s00209-015-1565-x. MR3473644 ↑315

[16] J.-L. Colliot-Thélène and J.-J. Sansuc, *La descente sur les variétés rationnelles. II* (French), Duke Math. J. **54** (1987), no. 2, 375–492, DOI 10.1215/S0012-7094-87-05420-2. MR899402 ↑315

[17] J.-L. Colliot-Thélène and J.-J. Sansuc, *The rationality problem for fields of invariants under linear algebraic groups (with special regards to the Brauer group)*, Algebraic groups and homogeneous spaces, Tata Inst. Fund. Res. Stud. Math., vol. 19, Tata Inst. Fund. Res., Mumbai, 2007, pp. 113–186. MR2348904 ↑296, 298, 299

[18] J.-L. Colliot-Thélène and A. N. Skorobogatov, *The Brauer-Grothendieck group*, Ergebnisse der Mathematik und ihrer Grenzgebiete. 3. Folge. A Series of Modern Surveys in Mathematics [Results in Mathematics and Related Areas. 3rd Series. A Series of Modern Surveys in Mathematics], vol. 71, Springer, Cham, [2021] ©2021, DOI 10.1007/978-3-030-74248-5. MR4304038 ↑299, 311

[19] P. Dèbes, *Arithmétique et espaces de modules de revêtements*, Number theory in progress. Proceedings of the international conference organized by the Stefan Banach International Mathematical Center in honor of the 60th birthday of Andrzej Schinzel, Zakopane, Poland, June 30–July 9, 1997. Volume 1: Diophantine problems and polynomials, 1999, pp. 75–102. ↑292

[20] P. Deligne, *La conjecture de Weil. I* (French), Publ. Math. I.H.É.S. **43** (1974), 273–307. MR340258 ↑308

[21] C. Demarche, *Groupe de Brauer non ramifié d'espaces homogènes à stabilisateurs finis* (French, with French summary), Math. Ann. **346** (2010), no. 4, 949–968, DOI 10.1007/s00208-009-0415-8. MR2587098 ↑314

[22] C. Demarche, G. Lucchini Arteche, and D. Neftin, *The Grunwald problem and approximation properties for homogeneous spaces* (English, with English and French summaries), Ann. Inst. Fourier (Grenoble) **67** (2017), no. 3, 1009–1033, DOI 10.5802/aif.3104. MR3668767 ↑293

[23] J. L. Demeio, *Ramified descent and transcendental Brauer-Manin obstruction*, Preprint, `arXiv:2112.00843`, 2021, to appear in Algebra & Number Theory. ↑313

[24] M. Dettweiler and S. Reiter, *An algorithm of Katz and its application to the inverse Galois problem*, J. Symbolic Comput. **30** (2000), no. 6, 761–798, DOI 10.1006/jsco.2000.0382. Algorithmic methods in Galois theory. MR1800678 ↑291

[25] L. Dieulefait, E. Florit, and N. Vila, *Seven small simple groups not previously known to be Galois over* $\mathbf{Q}$, Mathematics **10** (2022), no. 12, 2048, 10 pages. ↑292

[26] T. Ekedahl, *An effective version of Hilbert's irreducibility theorem*, Séminaire de Théorie des Nombres, Paris 1988–1989, Progr. Math., vol. 91, Birkhäuser Boston, Boston, MA, 1990, pp. 241–249. MR1104709 ↑295

[27] S. Endô and T. Miyata, *Invariants of finite abelian groups*, J. Math. Soc. Japan **25** (1973), 7–26, DOI 10.2969/jmsj/02510007. MR311754 ↑297

[28] E. Fischer, *Die Isomorphie der Invariantenkörper der endlichen Abelschen Gruppen linearer Transformationen*, Nachr. Ges. Wiss. Göttingen, Math.-Phys. Kl. **1915** (1915), 77–80. ↑295

[29] M. Florence, *On the essential dimension of cyclic p-groups*, Invent. math. **171** (2008), no. 1, 175–189, DOI 10.1007/s00222-007-0079-5. MR2358058 ↑296

[30] C. Frei, D. Loughran, and R. Newton, *Number fields with prescribed norms*, Comment. Math. Helv. **97** (2022), no. 1, 133–181, DOI 10.4171/cmh/528. With an appendix by Yonatan Harpaz and Olivier Wittenberg. MR4410726 ↑293, 317

[31] M. D. Fried and H. Völklein, *The inverse Galois problem and rational points on moduli spaces*, Math. Ann. **290** (1991), no. 4, 771–800, DOI 10.1007/BF01459271. MR1119950 ↑305, 306, 307

[32] W. Fulton, *Hurwitz schemes and irreducibility of moduli of algebraic curves*, Ann. of Math. (2) **90** (1969), 542–575, DOI 10.2307/1970748. MR260752 ↑306

[33] J.-P. Serre, *Cohomological invariants, Witt invariants, and trace forms*, Cohomological invariants in Galois cohomology, Univ. Lecture Ser., vol. 28, Amer. Math. Soc., Providence, RI, 2003, pp. 1–100. Notes by Skip Garibaldi. MR1999384 ↑296, 305

[34] P. Gille and T. Szamuely, *Central simple algebras and Galois cohomology*, Cambridge Studies in Advanced Mathematics, vol. 165, Cambridge University Press, Cambridge, 2017. Second edition of [MR2266528]. MR3727161 ↑298, 299

[35] A. Grothendieck (ed.), *Séminaire de géométrie algébrique du Bois Marie 1960-61 : revêtements étales et groupe fondamental (SGA 1)*, Documents Mathématiques, vol. 3, Société Mathématique de France, 2003. ↑301

[36] W. Grunwald, *A general existence theorem for algebraic number fields*, J. reine angew. Math. **169** (1933), 103–107. ↑292

[37] F. Häfner, *Braid orbits and the Mathieu group M_{23} as Galois group*, Preprint, arXiv:2202.08222, 2022. ↑307

[38] D. Harari, *Quelques propriétés d'approximation reliées à la cohomologie galoisienne d'un groupe algébrique fini* (French, with English and French summaries), Bull. Soc. Math. France **135** (2007), no. 4, 549–564, DOI 10.24033/bsmf.2545. MR2439198 ↑312, 314

[39] D. Harbater, *Galois coverings of the arithmetic line*, Number theory (New York, 1984), Lecture Notes in Math., vol. 1240, Springer, Berlin, 1987, pp. 165–195, DOI 10.1007/BFb0072980. MR894511 ↑304

[40] D. Harbater, *Fundamental groups of curves in characteristic p*, Proceedings of the International Congress of Mathematicians, August 3–11, 1994, Zürich, Switzerland. Vol. I, 1995, pp. 656–666. ↑304

[41] Y. Harpaz and O. Wittenberg, *Zéro-cycles sur les espaces homogènes et problème de Galois inverse* (French), J. Amer. Math. Soc. **33** (2020), no. 3, 775–805, DOI 10.1090/jams/943. MR4127903 ↑314, 315

[42] Y. Harpaz and O. Wittenberg, *Supersolvable descent for rational points*, Algebra Number Theory **18** (2024), no. 4, 787–814, DOI 10.2140/ant.2024.18.787. MR4708034 ↑316, 317

[43] D. Hilbert, *Ueber die Irreducibilität ganzer rationaler Functionen mit ganzzahligen Coefficienten* (German), J. reine angew. Math. **110** (1892), 104–129, DOI 10.1515/crll.1892.110.104. MR1580277 ↑292

[44] A. Hoshi, *On Noether's problem for cyclic groups of prime order*, Proc. Japan Acad. Ser. A Math. Sci. **91** (2015), no. 3, 39–44, DOI 10.3792/pjaa.91.39. MR3317750 ↑298

[45] A. Hoshi, *Noether's problem and rationality problem for multiplicative invariant fields: a survey*, Algebraic number theory and related topics 2016, RIMS Kôkyûroku Bessatsu, B77, Res. Inst. Math. Sci. (RIMS), Kyoto, 2020, pp. 29–53. MR4278977 ↑299

[46] A. Hurwitz, *Ueber Riemann'sche Flächen mit gegebenen Verzweigungspunkten* (German), Math. Ann. **39** (1891), no. 1, 1–60, DOI 10.1007/BF01199469. MR1510692 ↑306

[47] C. U. Jensen, A. Ledet, and N. Yui, *Generic polynomials. Constructive aspects of the inverse Galois problem*, Math. Sci. Res. Inst. Publ., vol. 45, Cambridge: Cambridge University Press, 2002. ↑292

[48] J.-P. Jouanolou, *Théorèmes de Bertini et applications* (French), Progress in Mathematics, vol. 42, Birkhäuser Boston, Inc., Boston, MA, 1983. MR725671 ↑300

[49] J. Kollár, *Rational curves on algebraic varieties*, Ergebnisse der Mathematik und ihrer Grenzgebiete. 3. Folge. A Series of Modern Surveys in Mathematics [Results in Mathematics and Related Areas. 3rd Series. A Series of Modern Surveys in Mathematics], vol. 32, Springer-Verlag, Berlin, 1996, DOI 10.1007/978-3-662-03276-3. MR1440180 ↑313

[50] J. Kollár, *Rationally connected varieties over local fields*, Ann. of Math. (2) **150** (1999), no. 1, 357–367, DOI 10.2307/121107. MR1715330 ↑305

[51] J. Kollár, *Fundamental groups of rationally connected varieties*, Michigan Math. J. **48** (2000), 359–368, DOI 10.1307/mmj/1030132724. Dedicated to William Fulton on the occasion of his 60th birthday. MR1786496 ↑305

[52] J. Kollár, *Rationally connected varieties and fundamental groups*, Higher dimensional varieties and rational points (Budapest, 2001), Bolyai Soc. Math. Stud., vol. 12, Springer, Berlin, 2003, pp. 69–92, DOI 10.1007/978-3-662-05123-8_4. MR2011744 ↑305

[53] W. Krull and J. Neukirch, *Die Struktur der absoluten Galoisgruppe über dem Körper* $\mathbf{R}(t)$ (German), Math. Ann. **193** (1971), 197–209, DOI 10.1007/BF02052391. MR289471 ↑304

[54] H. W. Lenstra Jr., *Rational functions invariant under a finite abelian group*, Invent. math. **25** (1974), 299–325, DOI 10.1007/BF01389732. MR347788 ↑297, 298

[55] H. W. Lenstra Jr., *Rational functions invariant under a cyclic group*, Proceedings of the Queen's Number Theory Conference, 1979 (Kingston, Ont., 1979), Queen's Papers in Pure and Appl. Math., vol. 54, Queen's Univ., Kingston, ON, 1980, pp. 91–99. MR634683 ↑298

[56] G. Lucchini Arteche, *The unramified Brauer group of homogeneous spaces with finite stabilizer*, Trans. Amer. Math. Soc. **372** (2019), no. 8, 5393–5408, DOI 10.1090/tran/7796. MR4014281 ↑313

[57] T. Maeda, *Noether's problem for* A_5, J. Algebra **125** (1989), no. 2, 418–430, DOI 10.1016/0021-8693(89)90174-9. MR1018955 ↑295

[58] G. Malle and B. H. Matzat, *Inverse Galois theory*, Springer Monographs in Mathematics, Springer, Berlin, 2018. Second edition [MR1711577]. MR3822366 ↑291, 292, 310

[59] Y. I. Manin, *Le groupe de Brauer-Grothendieck en géométrie diophantienne*, Actes du Congrès International des Mathématiciens (Nice, 1970), Gauthier-Villars Éditeur, Paris, 1971, pp. 401–411. MR427322 ↑311

[60] B. H. Matzat, *Konstruktive Galoistheorie* (German), Lecture Notes in Mathematics, vol. 1284, Springer-Verlag, Berlin, 1987, DOI 10.1007/BFb0098324. MR1004467 ↑292

[61] A. S. Merkurjev, *Essential dimension: a survey*, Transform. Groups **18** (2013), no. 2, 415–481, DOI 10.1007/s00031-013-9216-y. MR3055773 ↑296

[62] A. S. Merkurjev, *Essential dimension*, Bull. Amer. Math. Soc. (N.S.) **54** (2017), no. 4, 635–661, DOI 10.1090/bull/1564. MR3683628 ↑296

[63] J.-F. Mestre, *Extensions régulières de* $\mathbf{Q}(T)$ *de groupe de Galois* $\tilde{A}_n$ (French), J. Algebra **131** (1990), no. 2, 483–495, DOI 10.1016/0021-8693(90)90189-U. MR1058560 ↑305

[64] J.-F. Mestre, *Construction d'extensions régulières de* $\mathbf{Q}(t)$ *à groupes de Galois* $\mathrm{SL}_2(\mathbf{F}_7)$ *et* $\widetilde{M}_{12}$ (French, with English and French summaries), C. R. Acad. Sci. Paris Sér. I Math. **319** (1994), no. 8, 781–782. MR1300941 ↑305

[65] J.-F. Mestre, *Extensions* $\mathbf{Q}$*-régulières de* $\mathbf{Q}(t)$ *de groupe de Galois* $6.A_6$ *et* $6.A_7$ (French, with English summary), Israel J. Math. **107** (1998), 333–341, DOI 10.1007/BF02764017. MR1658587 ↑305

[66] J.-F. Mestre, *Correspondances compatibles avec une relation binaire, relèvement d'extensions de groupe de Galois* $\mathrm{L}_3(2)$ *et problème de Noether pour* $\mathrm{L}_3(2)$, 2005. ↑314

[67] J. S. Milne, *Class field theory*, 2020. Course notes, version 4.03, available from the author's webpage at `https://www.jmilne.org/math/CourseNotes/CFT.pdf`. ↑292

[68] L. Moret-Bailly, *Construction de revêtements de courbes pointées* (French, with English and French summaries), J. Algebra **240** (2001), no. 2, 505–534, DOI 10.1006/jabr.2001.8743. MR1841345 ↑304, 305

[69] D. Mumford, *Abelian varieties*, Tata Institute of Fundamental Research Studies in Mathematics, vol. 5, Published for the Tata Institute of Fundamental Research, Bombay; by Hindustan Book Agency, New Delhi, 2008. With appendices by C. P. Ramanujam and Yuri Manin; Corrected reprint of the second (1974) edition. MR2514037 ↑293

[70] D. Mumford, J. Fogarty, and F. Kirwan, *Geometric invariant theory*, 3rd ed., Ergebnisse der Mathematik und ihrer Grenzgebiete (2) [Results in Mathematics and Related Areas (2)], vol. 34, Springer-Verlag, Berlin, 1994. MR1304906 ↑315

[71] J. Neukirch, *On solvable number fields*, Invent. math. **53** (1979), no. 2, 135–164, DOI 10.1007/BF01390030. MR560411 ↑311

[72] J. Neukirch, A. Schmidt, and K. Wingberg, *Cohomology of number fields*, 2nd ed., Grundlehren der mathematischen Wissenschaften [Fundamental Principles of Mathematical Sciences], vol. 323, Springer-Verlag, Berlin, 2008, DOI 10.1007/978-3-540-37889-1. MR2392026 ↑292, 311

[73] M. L. Nguyễn, *On the descent conjecture for rational points and zero-cycles*, Preprint, `arXiv:2305.13228`, 2023, to appear in Journal de l'Institut de Mathématiques de Jussieu. ↑315

[74] E. Peyre, *Unramified cohomology of degree 3 and Noether's problem*, Invent. math. **171** (2008), no. 1, 191–225, DOI 10.1007/s00222-007-0080-z. MR2358059 ↑299

[75] B. Plans, *On Noether's rationality problem for cyclic groups over* $\mathbb{Q}$, Proc. Amer. Math. Soc. **145** (2017), no. 6, 2407–2409, DOI 10.1090/proc/13438. MR3626499 ↑298

[76] B. Poonen, *Rational points on varieties*, Graduate Studies in Mathematics, vol. 186, American Mathematical Society, Providence, RI, 2017, DOI 10.1090/gsm/186. MR3729254 ↑297

[77] B. Poonen and J. F. Voloch, *Random Diophantine equations*, Arithmetic of higher-dimensional algebraic varieties (Palo Alto, CA, 2002), Progr. Math., vol. 226, Birkhäuser Boston, Boston, MA, 2004, pp. 175–184, DOI 10.1007/978-0-8176-8170-8_11. With appendices by Jean-Louis Colliot-Thélène and Nicholas M. Katz. MR2029869 ↑312

[78] F. Pop, *Embedding problems over large fields*, Ann. of Math. (2) **144** (1996), no. 1, 1–34, DOI 10.2307/2118581. MR1405941 ↑304

[79] Z. Reichstein, *Essential dimension*, Proceedings of the International Congress of Mathematicians. Volume II, Hindustan Book Agency, New Delhi, 2010, pp. 162–188. MR2827790 ↑296

[80] Z. Reichstein, *From Hilbert's 13th problem to essential dimension and back*, Eur. Math. Soc. Mag. **122** (2021), 4–15, DOI 10.4171/mag-57. MR4417655 ↑296

[81] M. Romagny and S. Wewers, *Hurwitz spaces* (English, with English and French summaries), Groupes de Galois arithmétiques et différentiels, Sémin. Congr., vol. 13, Soc. Math. France, Paris, 2006, pp. 313–341. MR2316356 ↑302, 306

[82] D. J. Saltman, *Generic Galois extensions and problems in field theory*, Adv. in Math. **43** (1982), no. 3, 250–283, DOI 10.1016/0001-8708(82)90036-6. MR648801 ↑297, 299

[83] D. J. Saltman, *Noether's problem over an algebraically closed field*, Invent. math. **77** (1984), no. 1, 71–84, DOI 10.1007/BF01389135. MR751131 ↑298, 299

[84] D. J. Saltman, *The Brauer group and the center of generic matrices*, J. Algebra **97** (1985), no. 1, 53–67, DOI 10.1016/0021-8693(85)90073-0. MR812169 ↑299

[85] J.-J. Sansuc, *Groupe de Brauer et arithmétique des groupes algébriques linéaires sur un corps de nombres* (French), J. reine angew. Math. **327** (1981), 12–80, DOI 10.1515/crll.1981.327.12. MR631309 ↑312

[86] J.-P. Serre, *Lectures on the Mordell-Weil theorem*, 3rd ed., Aspects Math., vol. E15, Wiesbaden: Vieweg, 1997. ↑294

[87] J.-P. Serre, *Topics in Galois theory*, 2nd ed., Research Notes in Mathematics, vol. 1, A K Peters, Ltd., Wellesley, MA, 2008. With notes by Henri Darmon. MR2363329 ↑292, 304, 310, 316

[88] K.-y. Shih, *On the construction of Galois extensions of function fields and number fields*, Math. Ann. **207** (1974), 99–120, DOI 10.1007/BF01362150. MR332725 ↑291

[89] A. Skorobogatov, *Torsors and rational points*, Cambridge Tracts in Mathematics, vol. 144, Cambridge University Press, Cambridge, 2001, DOI 10.1017/CBO9780511549588. MR1845760 ↑315

[90] A. Speiser, *Zahlentheoretische Sätze aus der Gruppentheorie* (German), Math. Z. **5** (1919), no. 1-2, 1–6, DOI 10.1007/BF01203150. MR1544369 ↑296

[91] R. G. Swan, *Invariant rational functions and a problem of Steenrod*, Invent. math. **7** (1969), 148–158, DOI 10.1007/BF01389798. MR244215 ↑297

[92] R. G. Swan, *Noether's problem in Galois theory*, Emmy Noether in Bryn Mawr (Bryn Mawr, Pa., 1982), Springer, New York-Berlin, 1983, pp. 21–40. MR713790 ↑298

[93] T. Szamuely, *Galois groups and fundamental groups*, Cambridge Studies in Advanced Mathematics, vol. 117, Cambridge University Press, Cambridge, 2009, DOI 10.1017/CBO9780511627064. MR2548205 ↑292, 301, 305

[94] H. Völklein, *Groups as Galois groups: An introduction*, Cambridge Studies in Advanced Mathematics, vol. 53, Cambridge University Press, Cambridge, 1996, DOI 10.1017/CBO9780511471117. MR1405612 ↑292, 302

[95] H. Völklein, *The braid group and linear rigidity*, Geom. Dedicata **84** (2001), no. 1-3, 135–150, DOI 10.1023/A:1010341332396. MR1825350 ↑291

[96] V. E. Voskresenskiĭ, *Birational properties of linear algebraic groups* (Russian), Izv. Akad. Nauk SSSR Ser. Mat. **34** (1970), 3–19. MR262251 ↑297

[97] V. E. Voskresenskiĭ, *Rationality of certain algebraic tori* (Russian), Izv. Akad. Nauk SSSR Ser. Mat. **35** (1971), 1037–1046. MR286798 ↑298

[98] V. E. Voskresenskiĭ, *Fields of invariants of abelian groups* (Russian), Uspehi Mat. Nauk **28** (1973), no. 4(172), 77–102. MR392935 ↑297

[99] V. E. Voskresenskiĭ, *Algebraic groups and their birational invariants*, Translations of Mathematical Monographs, vol. 179, American Mathematical Society, Providence, RI, 1998. Translated from the Russian manuscript by Boris Kunyavski [Boris È. Kunyavskiĭ], DOI 10.1090/mmono/179. MR1634406 ↑312

[100] S. Wang, *On Grunwald's theorem*, Ann. of Math. (2) **51** (1950), 471–484, DOI 10.2307/1969335. MR33801 ↑292

[101] S. Wewers, *Construction of Hurwitz spaces*, Ph.D. Thesis, 1998. ↑306

[102] O. Wittenberg, *Rational points and zero-cycles on rationally connected varieties over number fields*, Algebraic geometry: Salt Lake City 2015, Proc. Sympos. Pure Math., vol. 97, Amer. Math. Soc., Providence, RI, 2018, pp. 597–635, DOI 10.1090/pspum/097.2/01717. MR3821185 ↑313, 315

[103] D. Zywina, *Inverse Galois problem for small simple groups*, 2013. Unpublished note, available from the author's webpage at `http://pi.math.cornell.edu/~zywina/papers/smallGalois.pdf`. ↑292

[104] D. Zywina, *The inverse Galois problem for* $\mathrm{PSL}_2(\mathbb{F}_p)$, Duke Math. J. **164** (2015), no. 12, 2253–2292, DOI 10.1215/00127094-3129271. MR3397386 ↑291, 292

Institut Galilée, Université Sorbonne Paris Nord, 99 avenue Jean-Baptiste Clément, 93430 Villetaneuse, France

Email address: `wittenberg@math.univ-paris13.fr`

Published Titles in This Series